Periodic Table of the Elements with the Gmelin System Numbers

1 H 2																	2 He 1
3 Li 20	4 Be 26											5 B 13	6 C 14	7 N 4	8 O 3	9 F 5	10 Ne 1
11 Na 21	12 Mg 27											13 Al 35	14 Si 15	15 P 16	16 S 9	17 Cl 6	18 Ar 1
19 * K 22	20 Ca 28	21 Sc 39	22 Ti 41	23 V 48	24 Cr 52	25 Mn 56	26 Fe 59	27 Co 58	28 Ni 57	29 Cu 60	30 Zn 32	31 Ga 36	32 Ge 45	33 As 17	34 Se 10	35 Br 7	36 Kr 1
37 Rb 24	38 Sr 29	39 Y 39	40 Zr 42	41 Nb 49	42 Mo 53	43 Tc 69	44 Ru 63	45 Rh 64	46 Pd 65	47 Ag 61	48 Cd 33	49 In 37	50 Sn 46	51 Sb 18	52 Te 11	53 I 8	54 Xe 1
55 Cs 25	56 Ba 30	57** La 39	72 Hf 43	73 Ta 50	74 W 54	75 Re 70	76 Os 66	77 Ir 67	78 Pt 68	79 Au 62	80 Hg 34	81 Tl 38	82 Pb 47	83 Bi 19	84 Po 12	85 At 8a	86 Rn 1
87 Fr 25a	88 Ra 31	89*** Ac 40	104 71	105 71													* NH_4 23

**Lanthanides 39	58 Ce	59 Pr	60 Nd	61 Pm	62 Sm	63 Eu	64 Gd	65 Tb	66 Dy	67 Ho	68 Er	69 Tm	70 Yb	71 Lu
***Actinides	90 Th 44	91 Pa 51	92 U 55	93 Np 71	94 Pu 71	95 Am 71	96 Cm 71	97 Bk 71	98 Cf 71	99 Es 71	100 Fm 71	101 Md 71	102 No 71	103 Lr 71

A Key to the Gmelin System is given on the Inside Back Cover

Gmelin Handbook of Inorganic and Organometallic Chemistry

8th Edition

Gmelin Handbook of Inorganic and Organometallic Chemistry

8th Edition

Gmelin Handbuch der Anorganischen Chemie

Achte, völlig neu bearbeitete Auflage

PREPARED AND ISSUED BY

Gmelin-Institut für Anorganische Chemie
der Max-Planck-Gesellschaft
zur Förderung der Wissenschaften
Director: Ekkehard Fluck

FOUNDED BY
Leopold Gmelin

8TH EDITION
8th Edition begun under the auspices of the Deutsche Chemische Gesellschaft by R. J. Meyer

CONTINUED BY
E. H. E. Pietsch and A. Kotowski, and by Margot Becke-Goehring

Springer-Verlag Berlin Heidelberg GmbH 1993

GMELIN HANDBOOK

Dr. J. von Jouanne

Dr. L. Berg, Dr. H. Bergmann, Dr. J. Faust, J. Füssel, Dr. H. Katscher, Dr. R. Keim, Dipl.-Phys. D. Koschel, Dr. A. Kubny, Dr. P. Merlet, Dr. M. Mirbach, Prof. Dr. W. Petz, Dr. H. Schäfer, Dr. F. A. Schröder, Dr. A. Slawisch, Dr. W. Töpper

Dr. R. Albrecht, Dr. G. Bär, D. Barthel, Dr. N. Baumann, Dr. K. Behrends, Dr. W. Behrendt, D. Benzaid, Dr. R. Bohrer, K. D. Bonn, Dipl.-Chem. U. Boßlet, Dr. U. Busch, A.-K. Castro, Dipl.-Ing. V. A. Chavizon, E. Cloos, A. Dittmar, Dipl.-Geol. R. Ditz, R. Dowideit, Dr. H.-J. Fachmann, B. Fischer, Dr. D. Fischer, Dipl.-Ing. N. Gagel, Dr. K. Greiner, Dipl.-Bibl. W. Grieser, Dr. R. Haubold, Dipl.-Min. H. Hein, H.-P. Hente, H. W. Herold, U. Hettwer, G. Hoell, Dr. G. Hönes, Dr. W. Hoffmann, G. Horndasch, Dr. W. Huisl, Dr. M. Irmler, B. Jaeger, Dr. R. Jotter, Dipl.-Chem. P. Kämpf, Dr. B. Kalbskopf, H.-G. Karrenberg, Dipl.-Phys. H. Keller-Rudek, Dipl.-Chem. C. Koeppel, Dr. M. Körfer, R. Kolb, Dr. M. Kotowski, E. Kranz, E. Krawczyk, Dipl.-Chem. I. Kreuzbichler, Dr. V. Kruppa, Dr. W. Kurtz, M. Langer, Dr. B. Ledüc, H. Mathis, E. Meinhard, M. Meßer, C. Metz, K. Meyer, E. Mlitzke, Dipl.-Chem. B. Mohsin, Dr. U. Neu-Becker, K. Nöring, Dipl.-Min. U. Nohl, Dr. U. Ohms-Bredemann, Dr. H. Pscheidl, Dipl.-Phys. H.-J. Richter-Ditten, E. Rudolph, G. Rudolph, Dr. B. Sarbas, Dr. R. Schemm, Dr. D. Schiöberg, V. Schlicht, Dipl.-Chem. D. Schneider, E. Schneider, A. Schwärzel, Dr. B. Schwager, R. Simeone, Dr. F. Stein, Dr. C. Strametz, Dr. G. Swoboda, Dr. D. Tille, A. Tuttas, Dipl.-Phys. J. Wagner, R. Wagner, M. Walter, Dr. E. Warkentin, Dr. C. Weber, Dr. A. Wietelmann, Dr. M. Winter, Dr. B. Wöbke, K. Wolff

GMELIN ONLINE

Dr. R. Deplanque

Dr. P. Kuhn, Dr. G. Olbrich

Dr. R. Baier, Dr. B. Becker, Dipl.-Chem. E. Best, Dr. H.-U. Böhmer, Dipl.-Phys. R. Bost, Dr. A. Brandl, Dr. R. Braun, Dr. T. Buck, Dipl.-Chem. R. Durban, R. Hanz, Dipl.-Phys. C. Heinrich-Sterzel, Dr. S. Kalwellis-Mohn, Dr. A. Kirchhoff, Dipl.-Chem. H. Köttelwesch, Dr. M. Kunz, Dipl.-Chem. R. Maass, Dr. K. Müller, Dr. A. Nebel, Dipl.-Chem. R. Nohl, Dr. M. Nohlen, H. Reinhardt, Dr. B. Rempfer, Dipl.-Ing. H. Vanecek

Volumes published on "Tungsten" (Syst. No. 54)

Tungsten
Main Volume — 1933

Tungsten Suppl. Vol. A 1
Metal, Technology — 1979

Tungsten Suppl. Vol. A 2
Physical Properties — 1987

Tungsten Suppl. Vol. A 3
Physical Properties — 1989

Tungsten Suppl. Vol. A 4
Surface Properties. Electron Emission — 1993 (present volume)

Tungsten Suppl. Vol. A 5a
Electrochemistry — 1990

Tungsten Suppl. Vol. A 5b
Metal, Chemical Reactions with Nonmetals Nitrogen to Arsenic — 1993

Tungsten Suppl. Vol. A 6a
Metal, Chemical Reactions with Metals Antimony to Barium — 1991

Tungsten Suppl. Vol. A 6b
Metal, Chemical Reactions with Metals Zinc to Lawrencium — 1988

Tungsten Suppl. Vol. A 7
Metal, Chemical Reactions with Inorganic and Organic Compounds — 1987

Tungsten Suppl. Vol. B 1
Systems with Noble Gases, Hydrogen, and Oxygen — 1978

Tungsten Suppl. Vol. B 2
Oxides — 1979

Tungsten Suppl. Vol. B 3
Compounds of Tungsten with Oxygen and Antimony, Bismuth; Alkali Tungstates — 1979

Tungsten Suppl. Vol. B 4
Tungstates of Group IIA and IIB Metals — 1980

Tungsten Suppl. Vol. B 5
Tungstates of Group IIIA and IIIB Metals — 1984

Tungsten Suppl. Vol. B 6
Anhydrous Tungstates of Group IVA and VIB Metals — 1984

Gmelin Handbook of Inorganic and Organometallic Chemistry

8th Edition

W

Tungsten

Supplement Volume A 4

Surface Properties. Electron Emission

With 53 illustrations

AUTHORS Gerhard Czack, Gerhard Kirschstein, Wolfgang Kurtz, Frank Stein

EDITORS Wolfgang Huisl, Wolfgang Kurtz, Frank Stein

CHIEF EDITOR Wolfgang Huisl

System Number 54

Springer-Verlag Berlin Heidelberg GmbH 1993

LITERATURE CLOSING DATE: 1992
IN SOME CASES MORE RECENT DATA HAVE BEEN CONSIDERED

Library of Congress Catalog Card Number: Agr 25-1383

ISBN 978-3-662-10156-8 ISBN 978-3-662-10154-4 (eBook)
DOI 10.1007/978-3-662-10154-4

Originally published by Springer-Verlag Berlin in 1993.
Softcover reprint of the hardcover 8th edition 1993

Preface

This volume continues the description of the physical properties of tungsten metal and covers specifically surface properties, electron emission, and field evaporation.

Tungsten surfaces are probably the most extensively studied metal surfaces. Recently, experimentalists and theorists have focused their interest on the atomic structure, lattice dynamics, and electronic properties of the W(100) surface. While the structure of the reconstructed low-temperature surface is well established, there are still unresolved problems concerning the structure at and above room temperature, the nature of the phase transition, and the driving force for the reconstruction. Other low-index tungsten surfaces have been studied in much less detail.

There are numerous and partly conflicting data on the surface energy, the self-diffusion parameters, and the work function of single-crystal and polycrystal tungsten surfaces. Electron emission, which is of fundamental importance for many applications, comprises thermionic, field, photofield, and photoelectric emission as well as emission induced by impinging electrons, atoms, or ions. Pioneering work in areas like thermionic or field emission is comprehensively discussed. Very recent studies of valence-band and core-level spectra moreover provide detailed information on intrinsic surface properties. Field evaporation phenomena are treated at the end of this volume.

Frankfurt am Main
November 1993

Wolfgang Huisl

Table of Contents

1 Atomic Structure, Lattice Dynamics, and Electronic Structure of Tungsten Surfaces

1.1 W(100)

1.1.1 Atomic Structure. Reconstruction and Relaxation

1.1.1.1 Survey

The W(100) surface is probably the most extensively studied metallic surface in the last 15 years. Numerous studies have shown that the clean surface is reconstructed at temperatures below about 220 K, and there is now general agreement concerning the structure of the low-temperature phase. On the other hand, there are still unresolved problems and controversies regarding the structure of the surface at higher temperatures.

Reconstruction and relaxation on metal surfaces are reviewed in [1], including a detailed discussion of the W(100) low-temperature (LT) structure and of the relationship between electronic structure and surface reconstruction [1]. There has been considerable confusion in the literature about W(100). In the literature on low-energy electron diffraction (LEED), often the necessary strict distinction between diffuse and spikelike reciprocal lattice rods has not been made (thus confusing short-range and long-range order). In addition, the disputable classification of the LT-to-high-temperature (HT) phase transition as an order-order transition (see e.g. [2]) was frequently taken to imply that the lateral structure of the HT surface phase of W(100) was simply a truncation of the bulk structure, the only modification being an inward relaxation of the first interlayer spacing d_{12}. Actually, the HT phase is more likely to be disordered, and the amplitude of the atomic displacements appears to be hardly affected by the LT-HT phase transition; see e.g. [3, 4]. Hence, probes sensitive to local structure or local order like Rutherford backscattering (RBS) do not distinguish between HT and LT W(100). In fact, deviations from simple bulk truncation were first indicated by the room-temperature RBS study [5] which had been clearly performed in the HT regime (see also the later RBS study [6]). A Monte Carlo (MC) and renormalization group study of the phase transition suggested that the interpretation of experimental results should take into account the range of the method applied (local probe vs. long-range probe), especially when comparing results of different methods obtained above the phase transition [7]. Melmed and Graham [8] concluded that LEED studies indicate at any temperature a reconstructed W(100) surface. Specifically, the studies [9] and [10] were cited for actually showing half-order spot intensities up to ~650 K and half-order spot widths up to ~550 K [8]. Broadened half-order beam profiles seen in LEED up to 750 K show that some reconstruction is preserved, though on the short-range scale only, up to temperatures much higher than the transition temperature T_c [11]. On the other hand, other workers take the view that the W(100) surface has an incommensurate structure at $T>T_c$ and transforms to the ordered, unreconstructed surface at about 450 K [12].

Instances where estimates of T_c exceed the best values (211 K by LEED [13], 226 ± 10 K by X-ray diffraction [14]) by 50% or more are common in the literature; thus, many observations ascribed to the LT phase actually were made in the HT regime. But to some extent, the transition temperature T_c is smeared out by surface steps; see e.g. [13].

Careful analysis of LEED intensities recorded below 190 K showed that the reconstructed domains have the plane group symmetry p2mg which unambiguously determines the possible surface structure type [15]. This implies that for surface atoms only shifts parallel to the surface have to be considered (no vertical shifts). The now generally accepted Debe-King model [15] consists of zigzag atom rows running along the ⟨11⟩ diagonals of the (1×1) unit mesh with the displacements confined to the surface plane [15]. LEED data taken at $\sim 120 \pm 8$ K [16] were in favor of this model and discounted the vertical-shift model by [17]. Later, vertical shifts were again advocated based on the results of field ion microscopy (FIM) studies [18]. Woodruff [19] rejected this view. He emphasized that the p2mg symmetry deduced by [15] and [10] from the LEED patterns confines the W atoms into a single plane parallel to the surface. Due consideration given to partial domain averaging, the maximum vertical shifts compatible with the Debe-King data ([8, 15]) are ≤ 0.006 Å, at any rate much less than 0.1 Å [19]. Support for the vertical-shift model by the electron reflectivity measured by [20] and the W 4f core-level shifts [21] seems to be spurious. LEED intensity vs. E_p spectra measured by [22] near 140 K were much better accounted for by lateral-shift models than by vertical-shift ones [22].

The zigzag chain model due to Debe and King (see **Fig. 1** [23]) is now generally accepted as a correct description of the W(100) surface structure at low temperatures.

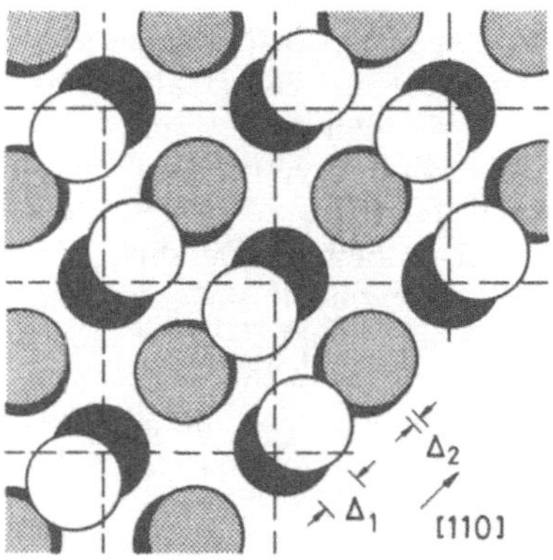

Fig. 1. Top view of the W(001) surface. Displacements in the first and second layers from ideal bcc bulk sites are denoted Δ_1 and Δ_2. Relative displacements are exaggerated three times allowing third- and fourth-layer atoms to be seen. The atoms are the more heavily shaded the deeper the layers lie to which they belong [23].

Some insight into the nature of the phase transition comes from molecular dynamics simulations, where the displacements of the atoms in ⟨011⟩ directions were assumed to decay with a penetration depth of the order of the bulk lattice spacing a = 3.16 Å [24]. The first-layer distortion amplitude Δ_1 varies weakly with temperature; see the $\Delta_1 = \langle \varrho \rangle$ vs. T curve in **Fig. 2** [24]. The order parameter ϱ^* is a two-dimensional vector, $\varrho^* = \varrho \cdot \exp(i\varphi)$; $\cos(\varphi) = (\sqrt{2}/2) \cdot ([\varrho_y + \varrho_z]/\varrho) \cdot \exp[i\vec{q}_0 \cdot \vec{R}_0]$, where $\vec{R}_0$ denotes the unperturbed position of an atom and $\vec{q}_0$ is the wave vector $(2\pi/a) \cdot (1/2\ 1/2)$; (i.e., ϱ and φ are the reconstruction staggered order parameter amplitude and phase relative to ⟨011⟩). Also shown in Fig. 2 are the susceptibilities for phase and amplitude, $\chi_\varphi = [\langle \cos^2(\varphi) \rangle - \langle \cos(\varphi) \rangle^2]/(k_B T)$ and $\chi_\varrho = [\langle \varrho^2 \rangle - \langle \varrho \rangle^2]/(a^2\, k_B T)$. The weak temperature dependence of $\langle \varrho \rangle$ and the distinct temperature dependence of $\langle \cos(\varphi) \rangle$ testify for an order-disorder type of the reconstruction phase transition [24].

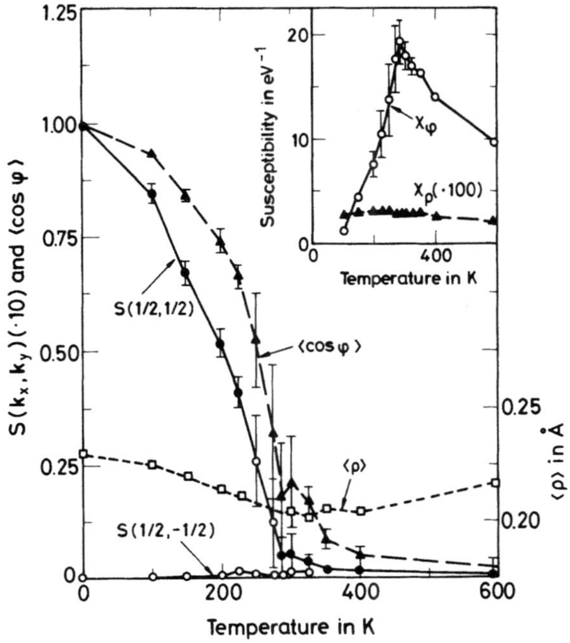

Fig. 2. Computed reconstruction structure factor (S) and order parameter of W(100). Inset: phase and amplitude susceptibilities χ_φ and χ_ϱ, characterizing fluctuations of the order parameter. The error bars represent the mean square deviation over 4000 steps in the molecular dynamics study [24].

This is consistent with conclusions from the above mentioned LEED studies [22] according to which both the zigzag and the dimer model (displacements along ⟨11⟩ and ⟨10⟩, respectively) gave acceptable fits [22]. At the phase transition, the phase coherence of the distortions is lost (e.g., to-and-fro switches of ⟨11⟩- and ⟨10⟩-displacements) while the amplitude is conserved [24, 25].

References for 1.1.1.1:

[1] Inglesfield, J. E. (Prog. Surf. Sci. **20** [1985] 105/64).
[2] King, D. A. (Phys. Scr. T **4** [1983] 34/43).
[3] Stensgaard, I.; Purcell, K. G.; King, D. A. (Phys. Rev. [3] B **39** [1989] 897/900).
[4] Robinson, I. K.; MacDowell, A. A.; Altman, M. S.; Estrup, P. J.; Evans-Lutterodt, K.; Brock, J. D.; Birgeneau, R. J. (Phys. Rev. Lett. **62** [1989] 1294/7).
[5] Bøgh, E. (Phys. Rev. Lett. **19** [1967] 61/3).
[6] Stensgaard, I.; Feldman, L. C.; Silverman, P. J. (Phys. Rev. Lett. **42** [1979] 247/50).
[7] Roelofs, L. D.; Hu, G. Y.; Ying, S. C. (Phys. Rev. [3] B **28** [1983] 6369/76).
[8] Melmed, A. J.; Graham, W. R. (Appl. Surf. Sci. **11/12** [1982] 470/8).
[9] Debe, M. K.; King, D. A. (J. Phys. C **10** [1977] L 303/L 308).
[10] Debe, M. K.; King, D. A. (Surf. Sci. **81** [1979] 193/237).

[11] Wendelken, J. F.; Wang, G.-C. (J. Vac. Sci. Technol. A **2** [1984] 888/92).
[12] Ernst, H.-J.; Hulpke, E.; Toennies, J. P. (Phys. Rev. [3] B **46** [1992] 16081/105).
[13] Wendelken, J. F.; Wang, G.-C. (Phys. Rev. [3] B **32** [1985] 7542/4).

[14] Evans-Lutterodt, K.; Birgeneau, R. J.; Specht, E. D.; Chung, J. W.; Brock, J. D.; Altman, M. S.; Estrup, P. J.; Robinson, I. K.; MacDowell, A. A. (J. Vac. Sci. Technol. A **7** [1989] 2209/16).
[15] Debe, M. K.; King, D. A. (Phys. Rev. Lett. **39** [1977] 708/11).
[16] Barker, R. A.; Estrup, P. J.; Jona, F.; Marcus, P. M. (Solid State Commun. **25** [1978] 375/9).
[17] Felter, T. E.; Barker, R. A.; Estrup, P. J. (Phys. Rev. Lett. **38** [1977] 1138/41).
[18] Tung, R. T.; Graham, W. R.; Melmed, A. J. (Surf. Sci. **115** [1982] 576/98).
[19] Woodruff, D. P. (Surf. Sci. **122** [1982] L 653/L 656).
[20] Stevens, M. A.; Russell, G. J. (Solid State Commun. **34** [1980] 785/8).

[21] Guillot, C.; Desjonquères, M. C.; Chauveau, D.; Tréglia, G.; Lecante, J.; Spanjaard, D.; Tran Minh Duc (Solid State Commun. **50** [1984] 393/6).
[22] Landskron, H.; Bickel, N.; Heinz, K.; Schmidtlein, G.; Müller, K. (J. Phys. Condens. Matter **1** [1989] 1/12).
[23] Altman, M. S.; Estrup, P. J.; Robinson, I. K. (Phys. Rev. [3] B **38** [1988] 5211/4).
[24] Wang, C. Z.; Parrinello, M.; Tosatti, E.; Fasolino, A. (Europhys. Lett. **6** [1988] 43/8).
[25] Fasolino, A.; Tosatti, E. (Phys. Rev. [3] B **35** [1987] 4264/83, 4276/7).

1.1.1.2 Phase Diagram

Low- and High-Temperature Phases. Yonehara and Schmidt [1] were the first to find superstructure reflections in W(100) LEED patterns corresponding to c(2 × 2) symmetry when cooling the sample from ~2500 K to below 300 K. While the integral-order beams had an almost constant intensity, half-order beams were observed only below 300 K, their intensity depending only on temperature [1]. These observations were confirmed in [2] where LEED was complemented by AES, low-energy AES fine structure, work function measurements, and characteristic energy loss spectroscopy. Above 400 K, there was a sharp (1 × 1) LEED pattern, while at lower temperatures (~200 K) a c(2 × 2) pattern developed [2]. The work function change $|\Delta\Phi| = 50 \pm 5$ meV developing on cooling down from high temperatures indicates that the low-temperature (LT) phase has a rougher surface [3]. Debe and King [2] established that the c(2 × 2) reconstruction is a *clean* surface effect and ruled out the possibility (proposed by [1]) that the c(2 × 2) structure formed by release of dissolved hydrogen from the bulk [2]. Paper [2] and the LEED study [4] performed between about 160 and 390 K [4] showed that the c(2 × 2) formation is a reversible, exclusively temperature-dependent process [2, 4]. The ease of the superstructure formation/destruction indicated that it is due to periodic displacements of the surface atoms [4].

At present, most authors take the view that there exists a single surface phase at temperatures above the transition temperature T_c and that the structure of this high-temperature (HT) phase is a disordered version of the LT phase. However, the existence of an incommensurate phase of W(100) at temperatures between T_c and about 450 K and of an unreconstructed surface phase at still higher temperatures has been claimed by [5] on the basis of He diffraction measurements. This suggestion is in conflict with results obtained by LEED, Ne, and X-ray diffraction; see p. 16.

Transition Temperature. LEED from a flat crystal face (nominal terrace widths larger than 1000 Å) showed reversible LT-HT transformations at $T_c \approx 210$ K [6]. For stepped and flat surfaces having average terrace widths of 30 ± 2 Å and >400 Å, the T_c values 217 and 211 K (±5 K) were obtained by LEED, respectively [7]. $T_c = 226 \pm 10$ K [8] ($T_c \approx 230$ K [9]) was derived from X-ray diffraction measurements [8, 9]. Below 200 K "flat" W(100) surfaces show a sharp c(2 × 2) LEED pattern which is fully developed at 180 K and disappears rapidly above 220 K [10]. At 220 K a new set of dispersion curves with a new zone boundary at $q = 0.70\ Å^{-1}$ appeared in the inelastic He atom scattering study [11]. $T_c \approx 200$ to 300 K was found by LEED [12]. The reconstruction is complete at 150 K according to the LEED intensities measured along with work function changes for $80\ K < T \leq 800\ K$ [3]. At 195 K the intensity of the half-order LEED spots (seen only below 300 K) was about 2/3 their value at 78 K [1]. $T_c \approx 200$ to 400 K was suggested by Debye-Waller plots of (00) and (1/2 1/2) He atom reflections [13].

High Electric Field Conditions. The field ion microscopy (FIM) study [14] suggests that the W(100) surface at strong applied fields is reconstructed over the whole temperature range studied, namely between 15 and 580 K. The W(100) surface under FIM conditions is characterized by vertical displacements. Gradual field evaporation produces first a p(2 × 2) structure, and eventually a c(2 × 2) pattern develops. The c(2 × 2) structure might consist of two sets of p(2 × 2) atom arrangements that differ in the amount of their vertical displacement [14]; see also [15]. On the other hand, though recognizing the unusual behavior of the W(100) plane under field evaporation, the FIM study [16], performed at 21 K and using tips with radii smaller than 150 Å, did not find any evidence for reconstruction (in particular there were no vertical displacements). Field evaporation of atoms from the interior of a plane (but close to its edge) is often observed to precede evaporation of kink site atoms; see Chapter 6, p. 271. Near the plane edge this results in what is called in [16] an "artifact c(2 × 2)" structure. Lateral shifts of the order of ~0.3 Å could be ruled out, while shifts in the 0.15 Å range would be at the resolution limit of the FIM study [16].

The FIM study [15] indicates an inherent instability of the W(100) surface. But there are considerable doubts [17] whether the structures seen by FIM are present also under conditions prevailing in LEED experiments, since the estimated energy difference between the HT and the LT structures seen by LEED is as small as ~43 meV per surface atom (see [18]) [17]. For an attempt to develop a unified picture of the W(100) reconstruction under both zero-field and high-field conditions, see [19].

Theoretical Studies. The surface reconstruction of W(100) was studied by [20], mainly for T = 0 K, by a lattice-dynamical approach using W-W pair potentials (force constants taken from [21]) that fit the bulk W phonon spectra. In order to have reconstruction, the force constants between the surface atoms are to be varied from their bulk values [20]. This model has also been used by [22] who, however, kept only the three uppermost atomic layers in the Hamiltonian. For the resultant phase diagram at T = 0 K as a function of the surface layer force constants α_s and β_s, see the paper [22]. Depending on the type of variation of the force constants, some phonon frequencies become imaginary at either the $\bar{M}$ point, the $\bar{L}$ point, or along the $\bar{\Lambda}$ line ($\bar{M} = (2\pi/a)\cdot(1/2\ 1/2)$, $\bar{L} = (2\pi/a)\cdot(1/2\ 0)$, $\bar{\Lambda} = (2\pi/a)\cdot(1/2\ \delta)$). The lattice instabilities signaled by the frequency going imaginary are indicated in a phase diagram at T = 0 K as a function of two surface force constants. With the twofold degenerate $\bar{M}_5$ zone-boundary phonon both the ⟨011⟩-type displacement pattern associated with clean W and the displacement pattern along ⟨010⟩ found at low hydrogen coverage are observed. A soft phonon of $\bar{M}_1$ symmetry with displacements along ⟨100⟩ (vertical displacements) is associated with the clean W c(2 × 2) structure seen by FIM in the presence of very strong fields (see [15]). Surface Brillouin zone boundary phonons at the $\bar{L}$ point lead to (2 × 1) or (2 × 2) reconstruction types that have not been observed. Phonons going soft on the $\bar{\Lambda}$ symmetry line might give rise to incommensurate reconstruction. The $\bar{M}_5$ distortion is said to result from attractive interactions between the surface-layer atoms (this is corrected in [23] where the converse statement is made that the c(2 × 2) reconstruction indicates a repulsion between the surface atoms) [20]. The stability chart (phase diagram at T = 0 K) given in [20] is discussed by [24] who enumerate and classify the phonons at high-symmetry

References for 1.1.1.2 on p. 6

points of the SBZ (i.e., they discuss the deformation modes important for surface reconstruction) [24].

The effective lattice Hamiltonian approach of [20] provides for the lattice entropy and thus enables one to calculate temperature effects. For the phase diagram as a function of temperature studied by the Monte Carlo (MC) method, see [25]; see also [26]. Finite-size effects result in a smearing out of the phase transition and agreement is found [25] with LEED results by [7].

References for 1.1.1.2:

[1] Yonehara, K.; Schmidt, L. D. (Surf. Sci. **25** [1971] 238/60, 246/55).
[2] Debe, M. K.; King, D. A. (J. Phys. C **10** [1977] L 303/L 308).
[3] Heilmann, P.; Heinz, K.; Müller, K. (Surf. Sci. **89** [1979] 84/94).
[4] Felter, T. E.; Barker, R. A.; Estrup, P. J. (Phys. Rev. Lett. **38** [1977] 1138/41).
[5] Ernst, H.-J.; Hulpke, E.; Toennies, J. P. (Phys. Rev. [3] B **46** [1992] 16081/105).
[6] Stensgaard, I.; Purcell, K. G.; King, D. A. (Phys. Rev. [3] B **39** [1989] 897/900).
[7] Wendelken, J. F.; Wang, G.-C. (Phys. Rev. [3] B **32** [1985] 7542/4).
[8] Evans-Lutterodt, K.; Birgeneau, R. J.; Specht, E. D.; Chung, J. W.; Brock, J. D.; Altman, M. S.; Estrup, P. J.; Robinson, I. K.; MacDowell, A. A. (J. Vac. Sci. Technol. A **7** [1989] 2209/16).
[9] Robinson, I. K.; MacDowell, A. A.; Altman, M. S.; Estrup, P. J.; Evans-Lutterodt, K.; Brock, J. D.; Birgeneau, R. J. (Phys. Rev. Lett. **62** [1989] 1294/7).
[10] Jupille, J.; Purcell, K. G.; King, D. A. (Phys. Rev. [3] B **39** [1989] 6871/9).

[11] Toennies, J. P. (Superlattices Microstructures **7** [1990] 193/200).
[12] Wendelken, J. F.; Wang, G.-C. (J. Vac. Sci. Technol. A **2** [1984] 888/92).
[13] Salanon, B.; Lapujoulade, J. (Surf. Sci. **173** [1986] L 613/L 618).
[14] Tung, R. T.; Graham, W. R.; Melmed, A. J. (Surf. Sci. **115** [1982] 576/98).
[15] Melmed, A. J.; Tung, R. T.; Graham, W. R.; Smith, G. D. W. (Phys. Rev. Lett. **43** [1979] 1521/4).
[16] Tsong, T. T.; Sweeney, J. (Solid State Commun. **30** [1979] 767/70).
[17] Walker, J. A.; Debe, M. K.; King, D. A. (Surf. Sci. **104** [1981] 405/18).
[18] Debe, M. K.; King, D. A. (Surf. Sci. **81** [1979] 193/237).
[19] Egorushkin, V. E.; Kul'ment'ev, A. I.; Savushkin, E. V. (Izv. Vyssh. Uchebn. Zaved. Fiz. **34** No. 8 [1991] 24/35; Sov. Phys. J. [Engl. Transl.] **34** [1992] 669/78).
[20] Fasolino, A.; Santoro, G.; Tosatti, E. (Phys. Rev. Lett. **44** [1980] 1684/7).

[21] Castiel, D.; Dobrzynski, L.; Spanjaard, D. (Surf. Sci. **59** [1976] 252/78).
[22] Hu, G. Y.; Ying, S. C. (Surf. Sci. **150** [1985] 47/54).
[23] Fasolino, A.; Tosatti, E. (Phys. Rev. [3] B **35** [1987] 4264/83).
[24] Pick, S.; Tomásek, M. (Surf. Sci. **130** [1983] L 307/L 310).
[25] Hu, G. Y.; Ying, S. C. (Physica A **140** [1987] 585/96).
[26] Ying, S. C.; Roelofs, L. D. (Surf. Sci. **125** [1983] 218/26).

1.1.1.3 Structure of the Low-Temperature (LT) Surface

The zigzag chain model due to Debe and King [1] is now generally accepted to describe the W(100) surface structure at low temperatures correctly; see Fig. 1, p. 2.

1.1.1.3.1 Lateral Atom Shifts (Reconstruction) and Interlayer Spacings (Relaxation)

The following tables compile selected data for the lateral atom shifts and surface relaxation values measured and calculated for LT W(100). The results of the 4-layer calculations

[2] and [3] are listed separately in a table. The calculations considered only Debe-King-type reconstruction. The index i in the atom shift parameter Δ_i refers to the layer number (i=1: topmost layer); similarly, the index ij appearing in the interlayer distances d_{ij} and relaxations $\delta d_{ij}/d = (d_{ij} - d)/d$ refers to the pair of adjacent layers (i, j); d is the bulk interlayer distance, 1.58 Å.

Experimental lateral atom shifts determined for the LT W(100) surface:

T in K	Δ_1 in Å	Δ_2 in Å	method	Ref.
175	0.24±0.025	0.046±0.016	X-ray	[4, 5]
~140	0.24±0.11	0.02±0.02	LEED	[6]
130	0.210±0.040	–	LEED	[7]
>80	0.14	–	RBS	[8]
70	0.27±0.03	–	He diffraction	[9]

Experimental interlayer distances d_{12} and d_{23} determined for the LT W(100) surface:

T in K	d_{12} in Å ($\delta d_{12}/d$ in %)	d_{23} in Å ($\delta d_{23}/d$ in %)	method	Ref.
~120±8	1.48 to 1.58 (−6.3 to 0)	–	LEED	[10, 11]
130	1.485±0.025 (−6±1.6)	–	LEED	[7]
~140	1.47±0.03 (−7±2)	1.60±0.03 (+1.2±2)	LEED	[6]
175	1.52±0.16 (−3.8±10)	–	X-ray	[4, 5]

Four-layer lateral shift (Δ_i, i=1 to 4) and four-layer relaxation values $\delta d_{ij}/d$ calculated by the tight-binding quenched molecular dynamics method [2] and by the linearized augmented plane wave method on a seven-layer slab [3] are given in the following table:

Δ_1 in Å ($\delta d_{12}/d$ in %)	Δ_2 in Å ($\delta d_{23}/d$ in %)	Δ_3 in Å ($\delta d_{34}/d$ in %)	Δ_4 in Å ($\delta d_{45}/d$ in %)	Ref.
0.20±0.02 (−1.9)	0.06 (−0.3)	0.03 (+0.7)	0.01 (+0.5)	[2]
0.268±0.005 (−6±0.5)	0.048±0.002 (0.5±0.5)	0.015±0.002 (–)	0.006±0.002 (–)	[3]

The following table lists the calculated lateral atom shifts Δ_1 and relaxation values $\delta d_{ij}/d$ for the LT W(100) surface (T=0 K):

Δ_1 in Å	$\delta d_{12}/d$ in %	$\delta d_{23}/d$ in %	$\delta d_{34}/d$ in %	method[a)]	Ref.
0.20	−2[b)]			LDF, LAPW	[12]
0.27	~−1			LDF, LAPW	[13]
0.234	−3[c)]			TB	[14]
0.18±0.01	~0			LDF, LAPW	[15]
0.225	~+2			ELH	[16]
–	−0.8	−0.7	+0.2		[17]

a) LDF: local density functional, LAPW: linearized augmented plane wave, TB: tight-binding, ELH: effective lattice Hamiltonian. – b) $\delta d_{12}/d = -4\%$ when sublayer reconstruction is included. – c) Assumed input value.

References for 1.1.1.3.1 on pp. 10/1

LEED Studies. Current-voltage curves from the c(2×2)R45° W(100) surface studied at T<190 K showed mm symmetry. The substrate has fourfold symmetry which implies preference for one rotational orientation over another of domains each having point group symmetry mm. Moreover, the plane group symmetry is p2mg since, apart from a common factor, the (±h/2±h/2) beam intensities equaled the (±h/2∓h/2) intensities. The plane group p2mg determines the possible surface structure type uniquely, and Debe and King [1] proposed a "parallel shift" model with the surface atom displaced along the ⟨011⟩ directions and confined to the surface plane (no vertical shifts). Their model consists of zigzag atom rows running along the diagonals of the (1×1) unit mesh [1]; see also [10, 18]. The lateral shifts Δ_1 measured at ~120±8 K ranged roughly between 0.15 and 0.3 Å, and the bulk (100) layer spacing of 1.58 Å was relaxed (contracted) to a first interlayer spacing of d_{12}=1.48 to 1.58 Å [10]; see also [11]. LEED at off-normal incidence ($\vartheta \approx 73.5°$) allowed one to discriminate between the zigzag chain and the dimer model, favoring the former. Using a Debye temperature of Θ_D=400 K, d_{12}=1.485 Å (6% contraction) was found at 130 K. The lateral shift Δ_1=0.210 Å was obtained from an R-factor-type comparison between 21 integer-order and 18 half-order beam profiles (50 to 400 eV) with full dynamical intensity calculations. This Δ_1=0.21 Å shift is consistent with the value of $\Delta_1 \approx 0.22$ Å derived from the ϑ-dependent (measured and calculated) ratio of the sum of all half-order intensities to the sum of all integer-order intensities [7]. Earlier, LEED spectra recorded near 140 K were fitted by lateral zigzag shifts and relaxation for the upper two layers. Using Θ_D=400 K, atom shifts along ⟨011⟩ of 0.24 Å in the first layer and of 0.02 Å in the second resulted along with interlayer spacings of d_{12}=1.47 Å and d_{23}=1.60 Å [6]. Melmed and Graham [19] have claimed that vertical displacements should be included in any model for the W(100) surface reconstruction [19]. Discussing this claim, Woodruff [20] assumed that the intensity of the reflections forbidden in p2mg would be due to vertical displacements breaking the p2mg symmetry rather than to the presence of an alternately oriented domain. In this case, the data of [18] would allow for vertical shifts by at most 0.006 Å [20]. Symmetry breaking, however, would additionally require that the "forbidden" LEED reflections are elongated like the allowed beams. But such an elongation was neither observed by [21] nor by [22]. Therefore, the "forbidden" beams are due to incomplete domain selection. Wendelken and Wang [21] achieved a degree as high as nearly 30:1 for the domain selection. Their data place an even lower limit on any vertical shifts than that calculated by [20] for the much more even domain distribution of the Debe-King data ([18]); see [21]. Comparing the accuracies achieved by LEED ([7]) and X-ray diffraction ([4, 5]) one finds LEED with Δd_{12}= ±0.025 Å to be more sensitive to d_{12} than X-ray diffraction (Δd_{12}= ±0.16 Å), whilst the reverse order was obtained for the lateral shifts (±0.025 Å with X-rays and ±0.040 Å with LEED) [7].

d_{12}=1.48 to 1.49 Å was determined by LEED for both (1×1) and c(2×2) surfaces [23] in agreement with the identical d_{12} values for LT and HT surfaces found by polarized LEED in [24]. Similarly, LEED between 80 and 800 K indicated d_{12} to be contracted by 10±2% [25].

X-Ray Diffraction. The surface crystallography of W(100) was studied at 175 K by grazing-incidence X-ray diffraction (Cu K_α radiation). The intensity I(hkl) along the reciprocal lattice rods (hkl) was measured for h=k=3/2 up to l=1.6, for h=k=1/2 up to l=1.5, and for (hk)=(3/2 5/2) up to l=1.2. The I(hkl) curves show that the zigzag chain (Debe-King type) reconstruction is not confined to the first atomic layer; rather, it involves at least also the second layer. From the I(hkl) curves the lateral shifts Δ_1=0.24 Å and Δ_2=0.046 Å [4, 5] were determined for the topmost and second layers, respectively; see Fig. 1, p. 2. The interlayer spacing d_{12} is 1.52 Å [4, 5]. The measured ratio $\Delta_2/\Delta_1 \approx 0.19$ is in good agreement with ~0.22 indicated by [16, fig. 5], where Δ_1=0.225 Å was calculated by an effective lattice

Hamiltonian approach [16]. There is also fair agreement with $\Delta_2/\Delta_1 = 0.06$ Å/0.20 Å = 0.3 calculated in the tight-binding quenched molecular-dynamics study [2].

σ_0, the "amplitude parameter" of the (super-) structure intensity (measured using synchrotron radiation), is shown in Fig. 5 a), p. 21 [26].

Rutherford Backscattering. The RBS study [8] yielded lateral shifts of 0.14 Å [8] (close to 0.16 Å evaluated by [23] from LEED data in [1, 18, 27]). The larger value of ~0.23 Å was derived in the earlier RBS study [28] from a crystal of unknown step density; however, only *half* of the surface atoms were found displaced from their ideal positions [28].

Calculated (Simulated) Relaxation and Reconstruction Parameters. The reconstruction of the surface layer strains the underlayer, as the surface atoms roll on top of the underlayer atoms. The strain energy is substantial because of the large bulk modulus and elastic constants of W. Much of this energy can be relieved through periodic displacements of the subsurface atoms from their bulk positions. Thus, the results of local-density-functional total-energy calculations are heavily modified by including sublayer reconstruction [12]: the contraction $\delta d_{12}/d$ is increased from −2% to −4% and the energy gain is doubled to 60 meV per pair of surface plus underlayer atoms despite a Debe-King-type shift of the underlayer atoms of only ~0.05 Å and only a slight increase of the surface atom displacements from 0.20 to 0.22 Å. The zigzag chain ⟨11⟩ single-layer reconstruction (subsurface layers kept fixed at their bulk truncation positions) reaches its energy minimum at 30 meV per surface atom for 0.20 Å shifts and $\delta d_{12}/d = -2\%$ (to be compared with −6% for the p(1×1) surface) [12]. LAPW ab initio calculations gave $\Delta_1 = 0.27$ Å and $\delta d_{12}/d \approx -1\%$ [13] for the Debe-King model. This corresponds to surface layer atomic distances of 2.80(2×) and 3.56(2×) Å and surface-to-subsurface atom distances of 2.52(1×), 2.75(2×), and 2.96(1×) Å [13]. A much smaller energy gain was found in the earlier calculation [15] because the 6p state contributions were underestimated [12]. But the reason for the high energy gain of 106 meV found by [29] remains unexplained [12]; this problem continues to exist [3]. A relaxation energy of 59 meV was found by tight-binding calculations for the unreconstructed surface and $\delta d_{12}/d = -5.9\%$ [30]. The W(100) surface (down to the fourth layer) was studied for T = 0 K by the tight-binding quenched molecular-dynamics method. As in [31] the unreconstructed surface was the most stable surface, if allowed rearrangements were confined to the first layer [2]. If the first four layers are free to rearrange themselves, lateral shifts with amplitudes of $\Delta_i = 0.20$, 0.06, 0.03, and 0.01 Å (i = 1 to 4) and relaxations of $\delta d_{ij}/d = -1.9$, −0.3, +0.7, and +0.5% for ij = 12, 23, 34, and 45, respectively, result [2]. First-principles total-energy and force calculations on a 7-layer slab using the LAPW method within the local-density approximation yielded $\Delta_1 = 0.268$ Å, $\Delta_2 = 0.048$ Å, $\Delta_3 = 0.015$ Å, and $\Delta_4 = 0.006$ Å; $\delta d_{12}/d = -6\%$ and $\delta d_{23}/d = 0.5\%$. Δ_i decays with the distance from the surface, the decay length being on the order of the interlayer distance [3]. Similarly, a decay rate of ~4/3 orders of magnitude per atomic layer was found for the square distortion amplitude [16].

Empirical N-body potentials were used by [17] for calculating surface energy and surface tension of W(100); see p. 74. Relaxation values of $\delta d_{ij}/d = -0.8\%$, −0.7%, and +0.2% were obtained for ij = 12, 23, and 34, respectively; but zigzag-type reconstruction was found to be less favored than an unreconstructed surface [17].

The surprisingly accurate early d-band tight-binding calculation [14] found atom shifts of 0.234 Å along ⟨11⟩ when taking the d-level shifts of the upper two atomic layers into account and assuming a 3% contraction of the first interlayer spacing (this $\delta d_{12}/d$ value was calculated for the unreconstructed surface in fair agreement with [32]) [14]. The instability against c(2×2) reconstruction found by [14] was contradicted by the tight-binding calculation [31]. But the disagreement was removed by [33] pointing out that the critique by [31]

 References for 1.1.1.3.1 on pp. 10/1

was based on a comparison between an unreconstructed surface and displacements Δ_1 as small as 0.08 Å. For such small shifts the local density functional 5-layer-slab calculation [15], employing the LAPW approach, also found a higher energy than for the (1 × 1) phase, while for larger shifts the relaxed, unreconstructed surface is unstable with respect to zigzag displacements. The calculated energy minimum occurred for $\Delta_1 = 0.18$ Å, essentially without relaxation [15]. For the combined effect of relaxation and ⟨11⟩-type reconstruction, see also an energy contour plot in the tight-binding study [34].

The twofold degenerate $\overline{M}_5$ zone-boundary phonon is associated with both ⟨011⟩- and ⟨010⟩-type displacement patterns [35] observed on clean W [1] and at low hydrogen coverage [40], respectively. The atomic-orbital approach [36] showed that both the Debe-King and dimer models result in a lowering of the surface energy [36]. The polarization vector of the $\overline{M}_5$ phonon is confined to the surface plane, either along or perpendicular to the wave vector, i.e., either parallel to [110] or parallel to [$\bar{1}$10]. In other words, the longitudinal and transverse $\overline{M}_5$ modes are degenerate (this degeneracy is lifted along $\overline{\Sigma}$ off $\overline{M}$) [37]. The ⟨11⟩ and the ⟨10⟩ distortion modes were studied on a 5-layer-slab model by a local-density-functional total-energy calculation [12]. Within the harmonic approximation, Debe-King and dimer models are energetically degenerate, and the rotational invariance is lifted only by anharmonic terms for the surface layer; see the effective lattice Hamiltonian study [16]. The difference of ~30 meV between zigzag chain- and dimer-type reconstruction is equivalent to the rotational anisotropic energy of the $\overline{M}$ modes [12] (this value is much larger than the value of ~6 meV indicated by the lattice dynamical study [16] (see [16, fig. 8]) and about half the energy gain for the zigzag reconstruction [12]). The calculated shifts along ⟨10⟩ (dimer model) are $\Delta_1 = 0.17$ Å and $\Delta_2 = 0.04$ Å [12]. Degeneracy was found in the 5-layer-slab LAPW calculation [13] for small displacements, but larger shifts favor the ⟨11⟩ mode, for which $\Delta_1 = 0.27$ Å was obtained (with $\delta d_{12}/d = -1\%$) [13]. The Monte Carlo (MC) simulation [38] finds that for $\Delta_1 = 0.20$ Å the ground state is lowered by 58.3 and 36.7 meV per atom by reconstruction for displacements along ⟨11⟩ and along ⟨10⟩, respectively. For larger shifts (0.24 Å), the difference is smaller [38].

The $\overline{M}_5$ distortion results from repulsive interactions between the surface layer atoms (see a correction in [16]) [35]. A fresh parametrization of the lattice potentials used in the effective-lattice-Hamiltonian study [35] resulted in $\Delta_1 = 0.225$ Å [16] (improving the earlier value of 0.33 to 0.38 Å of [35]) with an energy gain of 31.2 meV per surface atom [16]. The ~2% outward relaxation found by [16] may be compared with the essentially zero relaxation in the LAPW study [15], finding for $\Delta_1 = 0.18$ Å the energy minimum ~10 meV lower than for a relaxed, unreconstructed surface [15]. While the unreconstructed surface is stabilized by ~56 meV (for $\delta d_{12}/d \approx -6\%$), relaxation ($\delta d_{12}/d \approx -1\%$) only results in a negligible energy gain for lateral atom shifts of $\Delta_1 = 0.27$ Å [13]. Based on more recent multilayer relaxation and reconstruction calculations [3], the energy difference between a relaxed bulk-truncation surface ($\delta d_{12}/d = -10\%$, $\delta d_{23}/d = +5\%$, $\delta d_{34}/d \approx +1\%$) and a Debe-King-type reconstructed surface is ~110 meV [3] in good agreement with the values of 106 meV [29] and ~90 meV [13].

Two different sets of potential parameters were tried in [33], one producing potential wells for nonzero displacements and one with monotonous energy increase. The former choice corresponds to order-disorder behavior at the phase transition and achieves about twice the energy gain as the latter model (~74 meV per surface atom vs. ~35 meV) [39].

References for 1.1.1.3.1:

[1] Debe, M. K.; King, D. A. (Phys. Rev. Lett. **39** [1977] 708/11).
[2] Legrand, B.; Tréglia, G.; Desjonquères, M. C.; Spanjaard, D. (J. Phys. C: Solid State Phys. **19** [1986] 4463/72).

[3] Yu, R.; Krakauer, H.; Singh, D. (Phys. Rev. [3] B **45** [1992] 8671/8).
[4] Altman, M. S.; Estrup, P. J.; Robinson, I. K. (Phys. Rev. [3] B **38** [1988] 5211/4).
[5] Altman, M. S.; Estrup, P. J.; Robinson, I. K. (J. Vac. Sci. Technol. A **6** [1988] 630/1).
[6] Landskron, H.; Bickel, N.; Heinz, K.; Schmidtlein, G.; Müller, K. (J. Phys. Condensed Matter **1** [1989] 1/12).
[7] Schmidt, G.; Zagel, H.; Landskron, H.; Heinz, K.; Müller, K.; Pendry, J. B. (Surf. Sci. **271** [1992] 416/26).
[8] Stensgaard, I.; Purcell, K. G.; King, D. A. (Phys. Rev. [3] B **39** [1989] 897/900).
[9] Salanon, B.; Lapujoulade, J. (Surf. Sci. **173** [1986] L 613/L 618).
[10] Barker, R. A.; Estrup, P. J.; Jona, F.; Marcus, P. M. (Solid State Commun. **25** [1978] 375/9).

[11] Barker, R. A.; Estrup, P. J. (Phys. Rev. Lett. **41** [1978] 1307/10).
[12] Fu, C. L.; Freeman, A. J. (Phys. Rev. [3] B **37** [1988] 2685/8).
[13] Singh, D.; Krakauer, H. (Phys. Rev. [3] B **37** [1988] 3999/4006).
[14] Terakura, I.; Terakura, K.; Hamada, N. (Surf. Sci. **111** [1981] 479/91).
[15] Fu, C. L.; Freeman, A. J.; Wimmer, E.; Weinert, M. (Phys. Rev. Lett. **54** [1985] 2261/4).
[16] Fasolino, A.; Tosatti, E. (Phys. Rev. [3] B **35** [1987] 4264/83).
[17] Ackland, G. J.; Finnis, M. W. (Phil. Mag. [8] A **54** [1986] 301/15).
[18] Debe, M. K.; King, D. A. (Surf. Sci. **81** [1979] 193/237).
[19] Melmed, A. J.; Graham, W. R. (Appl. Surf. Sci. **11/12** [1982] 470/8).
[20] Woodruff, D. P. (Surf. Sci. **122** [1982] L 653/L 656).

[21] Wendelken, J. F.; Wang, G.-C. (Surf. Sci. **140** [1984] 425/36).
[22] Wang, G.-C.; Lu, T.-M. (Surf. Sci. **122** [1982] L 635/L 641).
[23] Walker, J. A.; Debe, M. K.; King, D. A. (Surf. Sci. **104** [1981] 405/18).
[24] Wang, G.-C.; Unguris, J.; Pierce, D. T.; Celotta, R. J. (Surf. Sci. **114** [1982] L 35/L 42).
[25] Heilmann, P.; Heinz, K.; Müller, K. (Surf. Sci. **89** [1979] 84/94).
[26] Evans-Lutterodt, K.; Birgeneau, R. J.; Specht, E. D.; Chung, J. W.; Brock, J. D.; Altman, M. S.; Estrup, P. J.; Robinson, I. K.; MacDowell, A. A. (J. Vac. Sci. Technol. A **7** [1989] 2209/16).
[27] Debe, M. K.; King, D. A. (J. Phys. C: Solid State Phys. **10** [1977] L 303/L 308).
[28] Stensgaard, I.; Feldman, L. C.; Silverman, P. J. (Phys. Rev. Lett. **42** [1979] 247/50).
[29] Singh, D.; Wei, S.-H.; Krakauer, H. (Phys. Rev. Lett. **57** [1986] 3292/5).
[30] Chan, C. T.; Louie, S. G. (Phys. Rev. [3] B **33** [1986] 2861/4).

[31] Tréglia, G.; Desjonquères, M.-C.; Spanjaard, D. (J. Phys. C: Solid State Phys. **16** [1983] 2407/19).
[32] Desjonquères, M. C. (J. Phys. Lett. **40** [1979] L 249/L 252).
[33] Roelofs, L. D.; Wendelken, J. F. (Phys. Rev. [3] B **34** [1986] 3319/36).
[34] Stephenson, P. C.; Bullett, D. W. (Surf. Sci. **139** [1984] 1/15).
[35] Fasolino, A.; Santoro, G.; Tosatti, E. (Phys. Rev. Lett. **44** [1980] 1684/7).
[36] Bullett, D. W.; Stephenson, P. C. (Solid State Commun. **45** [1983] 47/9).
[37] Terakura, I.; Terakura, K.; Hamada, N. (Surf. Sci. **103** [1981] 103/17).
[38] Han, W. K.; Ying, S. C. (Phys. Rev. [3] B **41** [1990] 9163/7).
[39] Roelofs, L. D. (Surf. Sci. **178** [1986] 396/405).
[40] King, D. A.; Thomas, G. (Surf. Sci. **92** [1980] 201/36, 229).

1.1.1.3.2 Reconstruction Domains

Domain Sizes, Shapes, and Orientations. The zigzag atom rows of the reconstructed W(100) surface may run either along [011] or [0$\bar{1}$1]. This choice gives rise to two different

References for 1.1.1.3.2 on p. 13

domain configurations. Size and shape of the reconstruction domains (regions of uniform orientation of the zigzag rows) depend on the step densities (terrace breadths) of the surface. They are deduced from the (anisotropic) broadening of the half-order Bragg spots. Terrace breadth and step orientation are derived from the amount and orientation of the integral-order spot splitting, respectively.

For $T<T_c$ (=226±10 K) the width of the half-order X-ray diffraction peaks indicates that the reconstruction domains are finite-sized [1]. At 175 K the FWHM domain size was found to be 185 Å [2]. This result may depend on the specimen and not be intrinsic to W(100), since in the companion study [3] the domain size remained near 200 Å (about the distance between steps due to the miscut of the specimen the orientation error of which was ±0.5°). The axial ratio $\xi_{\parallel}:\xi_{\perp}$ of the domain dimensions (∥ and ⊥ denote parallel and orthogonal orientation relative to the atomic displacements, respectively) ranged roughly at 1:2, the same value found above T_c [3]. Below T_c, the domains had a constant average size of $\xi_{\parallel}\approx210$ Å and $\xi_{\perp}\approx325$ Å (estimated) [1]; see Fig. 5 b), p. 21 (the values 4.72×10^{-4} and 3.08×10^{-4} $Å^{-1}$ actually given for $\xi_{\parallel}^{-1}$ and $\xi_{\perp}^{-1}$ in paper [1] obviously should be one order of magnitude larger).

The reconstructed domains on a surface with step edges running along [011] are oriented with the zigzag displacements perpendicular to the steps [4]. This orientation was also found in other LEED studies at 85 K by [5 to 7].

The seminal LEED study [8] found a terrace width of ~85±6 Å. The width of the (always) unsplit half-order beams, measured between ~190 and ~550 K, suggested that the c(2×2) phase starts to grow from the terrace centers, while the terrace edges inhibit growth within a margin of ~20 Å [8] (the same range was also indicated by the nitrogen adsorption study [9]). Similarly, the half-order LEED beams studied at 170 K were not split but uniformly broadened for all beam energies in the LEED study [4]. The split integral-order beams indicated an average terrace breadth of 28±2 Å for a surface cut 3.25±0.25° off the (100) planes with the step edges running parallel to [011]. The lack of splitting with the half-order beams is attributed to random antiphase domain effects [4] (i.e., the domains do not extend across step edges); see also [10]. On cooling, the domain growth on the terraces is limited by the steps [7]. Unsplit half-order beams from terraces running along [1$\bar{1}$0] also show that the reconstruction domains do not extend beyond the step edges. The 28±2 Å wide terrace limited the domain size on the terraces to a width of 28±5 Å [7] (23±3 Å in [4]). But parallel to the step edge the size was 128±20 Å, equal to the isometric domain size of 129±20 Å found for a flat surface at 90 to 100 K [7, 11] (nearly unchanged up to 150 K [7]). Hence, steps running along [011] have no influence on the domain dimension parallel to the step edge. The different result obtained by [4] who found a domain size limit of ~60 Å may be due to differences in the degree of domain selections achieved (5:1 in [4] vs. 30:1 in [7]) [7]; see also [6]. Steps do have an influence on the domain dimension parallel to the terrace edge, if the edge runs parallel to [100]. An average terrace width of 30±2 Å was prepared by cutting the crystal 3.25°±0.25° off W(001) slanting along [100]. At 85 K the domains had an elongated shape, the short side closely agreeing with the terrace width (~30 Å), but the long side was ~60 Å, i.e., about half the width of the isometric domains on a flat surface; see **Fig. 3** [11].

According to [4], the inhibiting effect of terrace edges on reconstruction, if present at all, is much smaller than claimed by Debe and King in [8, 12]. Inhibition ranges of 9 and 7 Å have been found by LEED at 85 K on two samples with terrace breadths of 29±13 and 36±13 Å, respectively, for step edges running along [011] [5]. Any prohibitive effect that lattice steps one atom high may have on reconstruction was questioned by [13] in

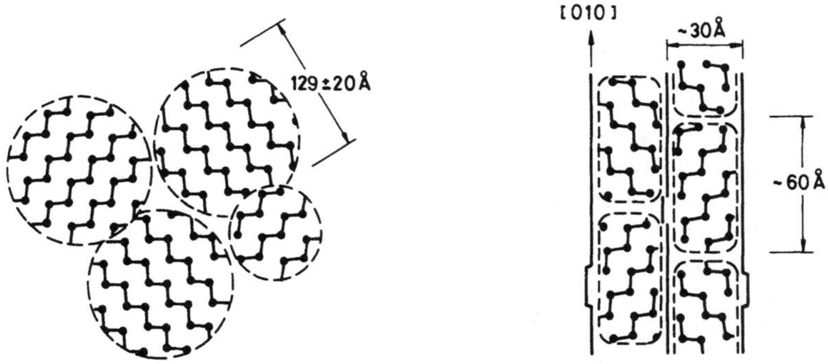

Fig. 3. Domain patterns derived from LEED data taken at 85 K. Left: flat W(001) surface; right: stepped 10(001) × [010] surface. Dashed lines indicate reconstruction domain boundaries; solid lines indicate step edges. The lateral shifts of the atoms (Debe-King zigzag chains) are highly exaggerated [11].

view of FIM evidence (though for an obviously different type of reconstruction), seen with planes as small as ~35 Å in diameter (see [14]).

References for 1.1.1.3.2:

[1] Evans-Lutterodt, K.; Birgeneau, R. J.; Specht, E. D.; Chung, J. W.; Brock, J. D.; Altman, M. S.; Estrup, P. J.; Robinson, I. K.; MacDowell, A. A. (J. Vac. Sci. Technol. A **7** [1989] 2209/16).

[2] Altman, M. S.; Estrup, P. J.; Robinson, I. K. (Phys. Rev. [3] B **38** [1988] 5211/4).

[3] Robinson, I. K.; MacDowell, A. A.; Altman, M. S.; Estrup, P. J.; Evans-Lutterodt, K.; Brock, J. D.; Birgeneau, R. J. (Phys. Rev. Lett. **62** [1989] 1294/7).

[4] Wang, G.-C.; Lu, T.-M. (Surf. Sci. **122** [1982] L 635/L 641).

[5] Zuo, J.-K.; Lu, T.-M.; Wang, G.-C.; Wendelken, J. F. (J. Vac. Sci. Technol. A **5** [1987] 777/81).

[6] Wendelken, J. F.; Wang, G.-C. (Surf. Sci. **140** [1984] 425/36).

[7] Wendelken, J. F.; Wang, G.-C. (J. Vac. Sci. Technol. A **2** [1984] 888/92).

[8] Debe, M. K.; King, D. A. (Surf. Sci. **81** [1979] 193/237).

[9] Griffiths, K.; King, D. A. (J. Phys. C: Solid State Phys. **12** [1979] L 755/L 759).

[10] Lu, T.-M.; Wang, G.-C. (Surf. Sci. **107** [1981] 139/47).

[11] Wendelken, J. F.; Wang, G.-C. (Phys. Rev. [3] B **32** [1985] 7542/4).

[12] Debe, M. K.; King, D. A. (Phys. Rev. Lett. **39** [1977] 708/11).

[13] Melmed, A. J.; Graham, W. R. (Appl. Surf. Sci. **11/12** [1982] 470/8).

[14] Melmed, A. J.; Tung, R. T.; Graham, W. R.; Smith, G. D. W. (Phys. Rev. Lett. **43** [1979] 1521/4).

1.1.1.4 Structure of the High-Temperature (HT) Surface

1.1.1.4.1 Lateral Structure of the HT Surface

Formerly, the HT surface was often conceived as a simple truncation of the bulk with the whole surface layer relaxed inward (see, e.g., the LEED [1 to 7], W 4f core level [8], and RBS [9] studies) [10]. But the RBS study [11] provided first evidence for a distorted room-temperature surface. Also, the agreement between measured and calculated LEED

References for 1.1.1.4.1 on pp. 17/8

spectra could be improved when disorder was assumed [12]. A disordered HT surface was suggested by the He atom diffraction study [13] which attributed the drop observed for the specular intensity around T_c to defect formation [13]. Also, analysis of the W $4f_{7/2}$ core level peak widths indicated a disorder-produced 12.4 meV shift for the LT-HT phase transition in very good agreement with the 15 ± 5 meV shift observed directly from the peak positions [43]. The RBS study [10] performed between 600 ± 50 K and ~80 K showed a smooth temperature dependence of the surface-peak yield consistent with (fairly temperature-independent) static lateral shifts of 0.14 Å superposed on a Gaussian distribution of thermal (dynamic) displacements. Stensgaard et al. [10] concluded that the surface structure is locally preserved on passing through the LT-HT phase transition [10]. The same conclusion (unchanged magnitude of the displacements) was drawn from the X-ray diffraction study [14, 15], where the integrated intensity of the superstructure beams (reciprocal lattice rods) stayed constant when T was changed through T_c; see the closed circles in Fig. 5 a), p. 21 [14]. Spin-polarized LEED rotation curves recorded upon cooling from 430 to 380 K showed the clean HT surface to be significantly more disordered than a hydrogen-saturated surface (obviously at room temperature) which is known to be ordered. The rotation technique, which emphasizes intraplanar scattering, rendered this experiment particularly sensitive to intralayer atom distances and disorder in these [16]. According to the molecular dynamics calculations [17], the distortion magnitude on heating above T_c remains nearly constant, ~0.2 Å, but the distortion direction becomes disordered [17]. For $T > T_c$ full C_{4v} symmetry is recovered on the time-average [18, 19]. About the loss of the phase coherence of the distortions, see also [20]. According to [14], the surface looses the LT long-range order above T_c ($= 226 \pm 10$ K) because of randomly placed domain walls (domain wall fluid model) [14].

Experimental studies on the displacement amplitudes on HT surfaces are scarce, evidently because of the short range of the ordering (for instance, the conclusions in the above cited X-ray studies are somewhat indirect and draw heavily on the LT structure determinations). Among the very few determinations of lateral shifts are the RBS study [10] cited above and the LEED study [12], where I-V profiles from (10), (11), (20), and (21) beams were recorded at 450 K and compared with calculated profiles. Allowing for relaxation and reconstruction, optimum agreement was achieved for diagonal shifts of the surface atoms by 0.16 Å and a relaxed first interlayer distance of $d_{12} = 1.45$ Å [12]. Also, early LEED I-V profiles recorded at 350 K showed that the peak appearing in the (10) profile for normal incidence at 85 eV was weaker than calculated for bulk termination. Since this peak is considerably attenuated in the Debe-King model (see [21]), there was some evidence for a persistence of the $c(2 \times 2)$ phase up to 350 K, despite the fact that no fractional-order beams were detected [22].

The LAPW [23] and Monte Carlo (MC) studies [24, 25] have been cited by Stensgaard et al. [10] to corroborate their view of the LT-HT transition to be an order-disorder-type transition [10]. This view is also consistent with the later LAPW study [26], but the most recent paper considered in the present review, the He atom scattering study [27], maintains that there is a sequence of order-order transitions: near 220 K from a Debe-King-type reconstructed LT surface to an incommensurate phase which transforms near 450 K to an (ordered) unreconstructed HT surface (even the data of [10], referred to above, are said to be compatible with purely dynamic rather than mainly static displacements) [27].

The distinction between an incommensurate and a commensurate phase is certainly not negligible. But it seems that any incommensurability will be restricted to the surface electrons (He atom scattering is sensitive to very small electron densities). Charge density waves (CDW) incommensurate with the lattice periodicity, if present at all, are too weak to cause any periodic lattice distortion (PLD) detectable by X-ray diffraction which is very sensitive to positional changes of the atom cores. The incommensurability of the W(100) surface as seen by He atom diffraction, therefore, will be treated in detail separately at the end of this section.

The significance of a distinction between disordered static displacements with $\langle u \rangle \neq 0$ ("disorder behavior") and dynamical displacements with $\langle u \rangle = 0$ and $\langle u^2 \rangle \neq 0$ ("displacive behavior") around ideal sites has been addressed by Roelofs and Wendelken [24] who emphasized that "displacive" and "disordered" represent only limiting cases and the actual behavior may share features of both of them [24]. Any short-range interaction model results in disorder around the transition temperature (T_c) and any reasonable choice of these interaction parameters results in a highly disordered HT surface. The magnitude of the displacements, on the other hand, depends strongly on the actual choice of the interaction parameters [24]. Using MC simulations, these authors have analyzed the measured T dependence of both integer and half-integer LEED beams (see **Fig. 4** [24]). The increase in integer beam intensity for increasing temperature around T_c seen in Fig. 4a) [24] (see also [28, fig. 2a]) is incompatible with a pure order-disorder transition and indicates *some* displacive behavior. With pure order-disorder behavior, the intensity vanishing from the half-order beams would be dispersed all over the reciprocal space (see the behavior indicated by Fig. 5a) [14], p. 21) [24].

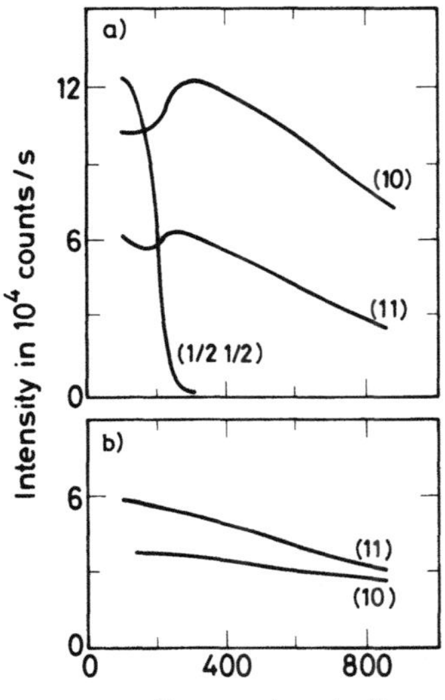

Fig. 4. Temperature dependence of (1/2 1/2), (10), and (11) LEED intensities from a flat W(001) surface (a) and (10) and (11) intensities from a stepped surface with an average terrace width of 30 Å (b) [24].

The amount of the LEED intensity increase around T_c strongly depends on the perfection of the surface; see Fig. 4b) [24]. Fig. 4 might explain some disagreement in published intensity vs. temperature curves as it shows that the integer beams from (even moderately) stepped surfaces do not show the maximum in the intensity vs. T curves seen with "flat" surfaces (terrace width at least 400 Å) [24]. This strong sensitivity for steps has been further studied by MC simulations in [29]. In agreement with LEED results by [28], these simulations show that the half-integer beam intensities are enhanced above T_c by steps with edges

References for 1.1.1.4.1 on pp. 17/8

running along ⟨100⟩. For ⟨110⟩-type steps and $T \geq 1.2\ T_c$, MC simulations indicate an exponential decay of the correlations with distance from the step edge. These MC results agree fairly well [29] with the experimental data of [28, 30, 31], the agreement being better with [30] than with [31].

UPS Evidence for Lateral Shifts in the HT Phase. The disagreement between the dispersion of the Swanson hump as observed in the angle-resolved UPS study [32] (see p. 48) and the dispersion curve calculated for this state in the local-density-functional study [33] may be attributed to lateral atom displacements of the surface atoms persisting above T_c (note that, on the other hand, there is fair agreement with several other resonances and states which are insensitive to lateral shifts) [32]. The self-consistent full-potential LAPW slab calculations in [34] describe very well measured energies and dispersions of the surface states and resonances throughout most of the first SBZ but failed near $\overline{M}$. This failure suggests that the bulk truncation model used is incorrect and that lateral and/or vertical rearrangements of the surface atoms along ⟨11⟩ rows exist [34]; see also the influence of lateral displacements as discussed in the tight-binding calculation [35].

"Incommensurate Structure". Controversial results have been reported on the periodicity of the W(100) surface layer at $T > T_c$, depending on whether it was studied by He diffraction or by other methods. Three groups using He diffraction, namely (i) [13], (ii) [27, 36 to 39], and (iii) [40, 41], found the superstructure reflection to shift towards the specular beam as the substrate was heated. This shift was observed in the temperature range between T_c and ~400 K and interpreted in terms of an incommensurate surface structure [27]. The group [40, 41] in addition to He diffraction also performed Ne diffraction experiments which did not show any indication for a shift of the Ne superstructure reflections and in contrast to their He data did not show any increase in peak width on heating through the phase transition [40, 41]. Since the He diffraction results were in sharp conflict with the reported LEED and X-ray data (and Ne diffraction), LEED and X-ray diffraction were repeated [42] confirming that any peak shift observed is due to W lattice expansion. According to Estrup et al. [42], the steeply decreasing structure factor of the surface accounts for the He diffraction peak shifts observed by the three above mentioned He diffraction groups [42]. As viewed by thermal He beams (the situation is different for Ne beams), the W(100) surface is only very weakly corrugated, implying a bell-shaped structure factor centered around the specular direction. The superstructure reflections, located on the wings of the bell, move towards the specular beam because of the increasing width of the reciprocal lattice rods [42]. In agreement with the results of [43, 44], such dramatic effects are not expected for Ne diffraction [42]. Ernst et al. [27] reject this attractive explanation, since the ratio of 10^{-3} found for the intensities of (00) to (11) beams in the He diffraction study [27] is some three orders of magnitude larger than would be required for the explanation by Estrup et al. [42] to hold [27]. Ernst et al., therefore, maintain the view that the peak shift seen in He diffraction reveals a periodic lattice distortion (PLD) related to a charge density wave (CDW); its propagation vector deviates the more from the commensurate (1/2 1/2) point the higher the temperature [27]. This behavior was discussed in terms of a lock-in phase transition between an incommensurate HT phase and a commensurate LT phase [38]. According to Ernst et al. [27], the broadened superstructure peaks seen by LEED and by He and Ne diffraction are superpositions of two satellite reflections. One component is associated to and scales in intensity with the specular beam, the other is associated to and scales in intensity with (11). The (11) beam being very weak for He diffraction, only the satellite to the specular beam would be visible in He diffraction. There remains, however, the discrepancy that any periodic lattice distortion (PLD) associated with the CDW would escape detection by X-ray diffraction only if its amplitude is ≤0.05 Å [27]. In fact, synchrotron X-ray diffraction had shown that in the whole T range studied (120 to 360 K [14]), the clean W(100) surface structure was commensurate with the bulk and any incommensurability (concluded from He diffraction) had been ruled out [14, 15]. These X-ray results need not be in conflict with the He diffraction data. As discussed by Ernst et al. [36], X-ray diffraction and LEED are sensitive to the atom core positions, while in He diffraction the He atoms are scattered by the electron charge density some 3 to 4 Å above the plane occupied by the surface atom cores [36]. Thus, the difference between He diffraction and the X-ray and LEED studies can be understood if there is a charge density wave incommensurate with the core positions of the surface atoms.

References for 1.1.1.4.1:

[1] van Hove, M. A.; Tong, S. Y. (Surf. Sci. **54** [1976] 91/100).
[2] Debe, M. K.; King, D. A.; Marsh, F. S. (Surf. Sci. **68** [1977] 437/47).
[3] Kirschner, J.; Feder, R. (Surf. Sci. **79** [1979] 176/88).
[4] Lee, B. W.; Ignatiev, A.; Tong, S. Y.; van Hove, M. (J. Vac. Sci. Technol. **14** [1977] 291/3).
[5] Read, M. N.; Russell, G. J. (Surf. Sci. **88** [1979] 95/108).
[6] Heilmann, P.; Heinz, K.; Müller, K. (Surf. Sci. **89** [1979] 84/94).
[7] Marsh, F. S.; Debe, M. K.; King, D. A. (J. Phys. C: Solid State Phys. **13** [1980] 2799/805).
[8] Guillot, C.; Thuault, C.; Jugnet, Y.; Chauveau, D.; Hoogewijs, R.; Lecante, J.; Tran Minh Duc; Tréglia, G.; Desjonquères, M. C.; Spanjaard, D. (J. Phys. C: Solid State Phys. **15** [1982] 4023/32).
[9] Feldman, L. C.; Kauffman, R. L.; Silverman, P. J.; Zuhr, R. A.; Barrett, J. H. (Phys. Rev. Lett. **39** [1977] 38/41).
[10] Stensgaard, I.; Purcell, K. G.; King, D. A. (Phys. Rev. [3] B **39** [1989] 897/900).
[11] Stensgaard, I.; Feldman, L. C.; Silverman, P. J. (Phys. Rev. Lett. **42** [1979] 247/50).
[12] Pendry, J. B.; Heinz, K.; Oed, W.; Landskron, H.; Müller, K.; Schmidtlein, G. (Surf. Sci. **193** [1988] L 1/L 6).
[13] Salanon, B.; Lapujoulade, J. (Surf. Sci. **173** [1986] L 613/L 618).
[14] Evans-Lutterodt, K.; Birgeneau, R. J.; Specht, E. D.; Chung, J. W.; Brock, J. D.; Altman, M. S.; Estrup, P. J.; Robinson, I. K.; MacDowell, A. A. (J. Vac. Sci. Technol. A **7** [1989] 2209/16).
[15] Robinson, I. K.; MacDowell, A. A.; Altman, M. S.; Estrup, P. J.; Evans-Lutterodt, K.; Brock, J. D.; Birgeneau, R. J. (Phys. Rev. Lett. **62** [1989] 1294/7).
[16] Venus, D. (Surf. Sci. **291** [1993] 418/28).
[17] Wang, C. Z.; Fasolino, A.; Tosatti, E. (Europhys. Lett. **7** [1988] 263/8).
[18] Wang, C. Z.; Parrinello, M.; Tosatti, E.; Fasolino, A. (Surf. Sci. **211/212** [1989] 323/8).
[19] Wang, C. Z.; Fasolino, A.; Tosatti, E. (Phys. Rev. [3] B **37** [1988] 2116/22).
[20] Fasolino, A.; Tosatti, E. (Phys. Rev. [3] B **35** [1987] 4264/83).
[21] Barker, R. A.; Estrup, P. J.; Jona, F.; Marcus, P. M. (Solid State Commun. **25** [1978] 375/9).
[22] Feder, R.; Kirschner, J. (Surf. Sci. **103** [1981] 75/102).
[23] Singh, D.; Wei, S.-H.; Krakauer, H. (Phys. Rev. Lett. **57** [1986] 3292/5).
[24] Roelofs, L. D.; Wendelken, J. F. (Phys. Rev. [3] B **34** [1986] 3319/36).
[25] Sugibayashi, T.; Hara, M.; Yoshimori, A. (J. Vac. Sci. Technol. A **5** [1987] 771/6).
[26] Yu, R.; Krakauer, H.; Singh, D. (Phys. Rev. [3] B **45** [1992] 8671/8).
[27] Ernst, H.-J.; Hulpke, E.; Toennies, J. P. (Phys. Rev. [3] B **46** [1992] 16081/105).
[28] Wendelken, J. F.; Wang, G.-C. (Phys. Rev. [3] B **32** [1985] 7542/4).
[29] Roelofs, L. D. (Phys. Rev. [3] B **34** [1986] 3337/44).
[30] Wang, G.-C.; Lu, T.-M. (Surf. Sci. **122** [1982] L 635/L 641).

[31] Debe, M. K.; King, D. A. (Surf. Sci. **81** [1979] 193/237).
[32] Holmes, M. W.; Gustafsson, T. (Phys. Rev. Lett. **47** [1981] 443/6).
[33] Posternak, M.; Krakauer, H.; Freeman, A. J.; Koelling, D. D. (Phys. Rev. [3] B **21** [1980] 5601/12; Erratum: Phys. Rev. [3] B **30** [1984] 4828).
[34] Ohnishi, S.; Freeman, A. J.; Wimmer, E. (Phys. Rev. [3] B **29** [1984] 5267/78).
[35] Bullett, D. W.; Stephenson, P. C. (Solid State Commun. **45** [1983] 47/9).
[36] Ernst, H.-J.; Hulpke, E.; Toennies, J. P. (Europhys. Lett. **10** [1989] 747/52).
[37] Ernst, H.-J.; Hulpke, E.; Toennies, J. P. (Phys. Rev. Lett. **58** [1987] 1941/4).
[38] Hulpke, E. (J. Electron Spectrosc. Relat. Phenom. **54/55** [1990] 299/308).
[39] Hulpke, E. (Nucl. Instrum. Methods Phys. Res. B **58** [1991] 347/53).
[40] Schweizer, E. K.; Rettner, C. T. (J. Vac. Sci. Technol. A **7** [1989] 1937/41).

[41] Schweizer, E. K.; Rettner, C. T. (Surf. Sci. **208** [1989] L 29/L 33).
[42] Estrup, P. J.; Robinson, I. K.; Tully, J. C. (Surf. Sci. **215** [1989] L 297/L 306).
[43] Jupille, J.; Purcell, K. G.; King, D. A. (Phys. Rev. [3] B **39** [1989] 6871/9).

1.1.1.4.2 Relaxation of the HT W(100) Interlayer Spacing

The reflectivity and polarization of the specular LEED beam studied between 300 and 1200 K showed that the top interlayer spacing d_{12} expands thermally by ~2% in this range, while the corresponding bulk expansion is only ~0.4% [1].

The following table compiles experimental relaxation data for the W(100) HT surface. The relaxation values $\delta d_{12}/d$ are referred to a bulk value of d = 1.58 Å. All of the reported results deal only with the first interlayer spacing, d_{12}. The only exception thereof is the LEED study [2] which allowed for relaxation also of the second interlayer spacing, d_{23}, which, however, did not differ from the bulk value.

$\delta d_{12}/d$ in %	T in K	method	remarks	Ref.
−6±6	~650±25	LEED	data from [3]	[4]
−10	550±100	LEED	polarized LEED; data from nonspecular beams [5] (similar results: [6])	[5]
−8±1.5	~470	LEED	revises the −4.4±3% found in [7]	[8]
−7.5±0.5	450	LEED	disordered HT surface	[2]
−5.5±1.5	~400	"LEED"	specular beam rotation diagrams for incidence under $\vartheta \approx 47.5°$	[9, 10]
−6.7±2	>370	LEED	few I-V profiles, but most data sets compare various beams for a single voltage	[11]
−7±1.5	350	LEED	I-V and polarization profiles; evidence for disorder	[12]
−6±3	<330	LEED	reevaluation of [13] data; revises the −11.4±2% found by [13]	[7]
−6	~300	RBS	RBS: He^+ ion Rutherford backscattering	[14]

The more recent $\delta d_{12}/d$ values given in the table tend to fall near −6% or −7%. This is remarkable insofar as according to [15] any $\delta d_{12}/d$ value between 0 and −15% would be warranted by the experimental (LEED) data available by 1979 (see in [15] a comparison of literature data from [3, 6, 7, 13] and fresh data measured at 650 K by [15]). These authors also compared their calculated LEED intensity profiles with profiles calculated by [4, 7, 13, 16]. There is a choice for the exchange-correlation parameter and also an uncertainty about electron scattering potentials [15]; see also paper [9] concerning a parameter correlation between the mean inner potential U_0 and $\delta d_{12}/d$. Calculated I-V profiles depend on the scattering potentials as much as on $\delta d_{12}/d$. This insufficient knowledge was said to impede any firm conclusion on the relaxation [15]. However, [8] and [11] consistently found U_0 to be at -11.5 ± 0.5 eV [8] and at -11 ± 1 eV [11] (see also a table in [11, p. 422] which compares the experimental LEED studies [4, 7, 9, 11, 13]; the same table compiles parameters used in the evaluation and gives results obtained from these LEED studies). Very similar I-V profiles were calculated in [9] when ion-core potentials from either [17] or [18] were used. The rotation LEED diagrams, on the other hand, were sufficiently different that a comparison with the experiment allowed for a discrimination [9] in favor of the potential given by [17].

Relaxation values $\delta d_{ij}/d$ calculated by tight-binding and LAPW methods are compiled for the unreconstructed W(100) surface in the following table. The first column, "m", indicates the number of interlayer spacings allowed to relax in the calculations. For the $\delta d_{12}/d \approx -6\%$ value of [19] an in-plane contraction of ~1% was determined by the LDF study [20].

m	$\delta d_{12}/d$	$\delta d_{23}/d$	$\delta d_{34}/d$	$\delta d_{45}/d$	method [a)]	Ref.
4	−3.8%	−0.1%	+0.7%	+0.5%	TB, quenched MD	[21]
1	−4%	–	–	–		[21]
3	−5.7%	+2.4%	+1.2%	–	LAPW, LDF, (5, 7)-LS	[22]
2	−5.8%	not given	–	–		[22]
1	−5.5% [b)]	–	–	–		[22]
3	−10%	+5%	~+1%	–	LAPW, LDF, 7-LS	[23]
2	−3.5%	−0.6%	–	–	TB, quenched MD	[24]
1	−3.75±0.25%	–	–		TB	[25]
1	−4% [b)]	–	–	–	LAPW, LDF, 5-LS	[26]
1	−5%	–	–	–	LAPW, LDF	[27]
1	−5.9%	–	–	–	TB, LDF	[28]
1	~−6%	–	–	–	LAPW	[19]
1	−10%	–	–	–	TB, 7-LS	[29]

[a)] TB: tight-binding, LDF: local-density functional, n-LS: n-layer slab, MD: molecular dynamics. — [b)] The difference between −5.5% [22] and −4% [26] stems from the treatment of the 5p electrons (as valence and core states in [26] and [22], respectively), see [26].

Relaxation values of $\delta d_{12}/d = -3.2\%$ and $\delta d_{23}/d = +0.9\%$ were calculated with the "equivalent crystal" theory [30]. Using Morse [31] and modified Morse [32] potentials, substantial expansion values $\delta d_{ij}/d$ were calculated; see also the study [33] where fitted values for the central interactions were used (any pair force model leads to an *expansion* of the topmost interlayer spacing; see [34]). An expansion of the first interlayer distance was also obtained by the embedded atom method [35]. These results are omitted here as they disagree with experimental evidence.

References for 1.1.1.4.2:

[1] Kirschner, J.; Feder, R. (Surf. Sci. **104** [1981] 448/62).
[2] Pendry, J. B.; Heinz, K.; Oed, W.; Landskron, H.; Müller, K.; Schmidtlein, G. (Surf. Sci. **193** [1988] L 1/L 6).
[3] Wei, P. S. P. (J. Chem. Phys. **53** [1970] 2939/45).
[4] van Hove, M. A.; Tong, S. Y. (Surf. Sci. **54** [1976] 91/100).

[5] Wang, G.-C.; Celotta, R. J.; Pierce, D. T. (Phys. Rev. [3] B **23** [1981] 1761/70).
[6] Kalisvaart, M.; O'Neill, M. R.; Riddle, T. W.; Dunning, F. B.; Walters, G. K. (Phys. Rev. [3] B **17** [1978] 1570/8).
[7] Debe, M. K.; King, D. A.; Marsh, F. S. (Surf. Sci. **68** [1977] 437/47).
[8] Marsh, F. S.; Debe, M. K.; King, D. A. (J. Phys. C: Solid State Phys. **13** [1980] 2799/805).
[9] Kirschner, J.; Feder, R. (Surf. Sci. **79** [1979] 176/88).
[10] Feder, R.; Kirschner, J. (Phys. Status Solidi A **45** [1978] K 117/K 119).

[11] Clarke, L. J.; Morales de la Garza, L. (Surf. Sci. **99** [1980] 419/39).
[12] Feder, R.; Kirschner, J. (Surf. Sci. **103** [1981] 75/102).
[13] Lee, B. W.; Ignatiev, A.; Tong, S. Y.; van Hove, M. (J. Vac. Sci. Technol. **14** [1977] 291/3).
[14] Feldman, L. C.; Kauffman, R. L.; Silverman, P. J.; Zuhr, R. A.; Barrett, J. H. (Phys. Rev. Lett. **39** [1977] 38/41).
[15] Read, M. N.; Russell, G. J. (Surf. Sci. **88** [1979] 95/108).
[16] Feder, R. (Phys. Rev. Lett. **36** [1976] 598/600).
[17] Mattheiss, L. F. (Phys. Rev. [2] **139** [1965] A 1893/A 1904).
[18] Christensen, N. E.; Feuerbacher, B. (Phys. Rev. [3] B **10** [1974] 2349/72).
[19] Singh, D.; Krakauer, H. (Phys. Rev. [3] B **37** [1988] 3999/4006).
[20] Chen, J.; Krakauer, H.; Singh, D. J. (Phys. Rev. [3] B **43** [1991] 2398/400).

[21] Legrand, B.; Tréglia, G.; Desjonquères, M. C.; Spanjaard, D. (J. Phys. C: Solid State Phys. **19** [1986] 4463/72).
[22] Fu, C. L.; Ohnishi, S.; Wimmer, E.; Freeman, A. J. (Phys. Rev. Lett. **53** [1984] 675/8).
[23] Yu, R.; Krakauer, H.; Singh, D. (Phys. Rev. [3] B **45** [1992] 8671/8).
[24] Luo, J. S.; Legrand, B. (Phys. Rev. [3] B **38** [1988] 1728/33).
[25] Desjonquères, M. C. (J. Phys. Lett. **40** [1979] L 249/L 252).
[26] Fu, C. L.; Freeman, A. J.; Wimmer, E.; Weinert, M. (Phys. Rev. Lett. **54** [1985] 2261/4).
[27] Singh, D.; Wei, S.-H.; Krakauer, H. (Phys. Rev. Lett. **57** [1986] 3292/5).
[28] Chan, C. T.; Louie, S. G. (Phys. Rev. [3] B **33** [1986] 2861/4).
[29] Stephenson, P. C.; Bullett, D. W. (Surf. Sci. **139** [1984] 1/15).
[30] Rodriguez, A. M.; Bozzolo, G.; Ferrante, J. (Surf. Sci. **289** [1993] 100/26).

[31] Jackson, D. P. (Can. J. Phys. **49** [1971] 2093/7).
[32] Wynblatt, P.; Gjostein, N. A. (Surf. Sci. **22** [1970] 125/36).
[33] Castiel, D.; Dobrzynski, L.; Spanjaard, D. (Surf. Sci. **59** [1976] 252/78).
[34] Inglesfield, J. E. (Rep. Prog. Phys. **45** [1982] 223/84).
[35] Guellil, A. M.; Adams, J. B. (J. Mater. Res. **7** [1992] 639/52).

1.1.1.5 Phase Transition

This section is focused on the thermodynamic aspects of the W(100) reconstruction transition which occurs around $T_c \approx 220$ K (see p. 4). For the driving mechanism, see p. 25.

Low-energy electron diffraction (LEED) showed that the high-temperature-low-temperature (HT-LT) phase transition does not depend on the cooling rate. The reversible and continuous behavior is consistent with a second-order phase transition [1]. The same conclusion was arrived at by X-ray diffraction performed with a temperature resolution of $\Delta T \approx 8$ K [2]; see also the Rutherford backscattering study [3].

Rather convincing experimental evidence has been presented recently that the W(100) surface is disordered at temperatures above T_c; see pp. 13/4. Three-dimensional molecular dynamics (MD) simulations also provide strong evidence for an order-disorder transition; see p. 2 and Fig. 2, p. 3 [5].

For lattice distortions conforming to $\overline{M}_5$ symmetry, Landau theory predicts a critical behavior described by the two-dimensional (2-D) XY model [6]. Regarding the application of the Landau theory to W(100) reconstruction, see the review [7, pp. 145/6]. The free energy for the 2-D XY model with cubic anisotropy and a discussion of Tosatti's study [6] was given by Bak [8]. He emphasized that the non-universality of this model prevents general predictions to be made about its critical behavior [8]. The critical properties of the model (studied for the isotropic system by renormalization group techniques in [9]) depend on the physical parameters of its actual realization, i.e., for W(100) they depend on the physical parameters relevant for this surface [8]. For this model, mean field theory adequately describes the temperature dependence of the order parameter below 0.95 T_c [21].

The temperature-dependent widths and peak intensities (σ_0) of the half-order X-ray reflections are accounted for by the two-dimensional XY model with cubic anisotropy [2]; see also [10]. The solid line drawn in **Fig. 5** a) [2] represents a power law fit of $\sigma_0(T)$, using the critical exponent 1.75 appropriate for the 2-D Ising model. Ising exponents were also consistent with the data shown in Fig. 5 b) [2] (for a discussion, see p. 24). There are deviations from a 2-D Ising behavior above 300 K; the deviations systematically increase with temperature [2].

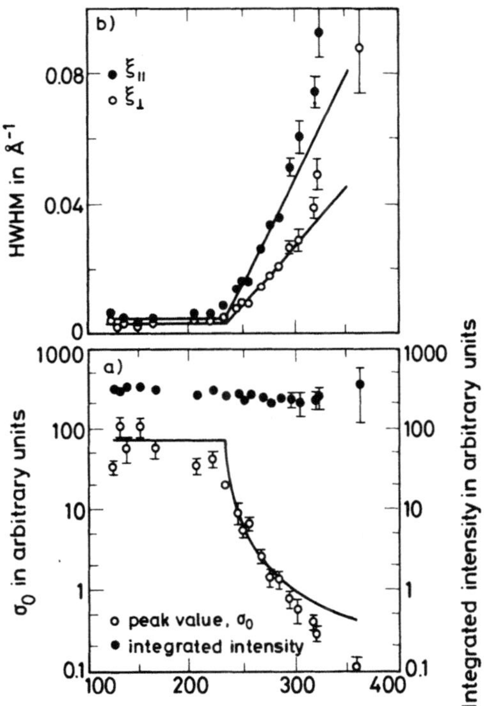

Fig. 5. Peak value, σ_0, and integrated intensity of the superstructure intensity plotted vs. T (a). Inverse correlation length given as half-width at half maximum of the superstructure peak parallel and perpendicular to the displacement vector (b) [2].

References for 1.1.1.5 on p. 25

A model similar to the 2-D XY model was applied in the MC study [11] where the temperature dependence of the thermal average of the Debe-King displacement vector amplitude (taken as order parameter) and the temperature dependence of the critical exponent for the coherence length were calculated [11].

The three-dimensional character of the LT-HT phase transition must not be neglected; coupling to surface relaxation and to surface phonon modes quantitatively modifies the acting forces and qualitatively the behavior of the fluctuations [5]. The MC study [12] showed that interactions with the subsurface layer may change the behavior from displacive to order-disorder. Han and Ying [4] confirmed this order-disorder behavior in their Monte Carlo (MC) simulation (for simulated snapshots of the displacement patterns at 110 and 240 K, see the paper [4]). Short-range interaction models were used in the MC study [12] of the surface reconstruction. Temperature-dependent LEED data suggest that any short-range interaction model will result in disorder around the transition temperature, and any reasonable choice of the interaction parameters results in a highly disordered high-temperature (HT) phase [12]. The model favored by [12] has potential wells at nonvanishing displacements. This model, however, yields a weaker decrease in the coherence length above T_c than was found by LEED [1]. The increase in integer beam intensity seen on perfect surfaces with increasing temperature around T_c (see Fig. 4 a), p. 15) is not compatible with the notion of a pure order-disorder transition, but indicates *some* displacive behavior [12]. The Hamiltonian used in the related statistical-mechanics calculation [13] is an oscillator potential for the vector displacements of the surface atoms and includes a term accounting for the interaction with the atoms in the subsurface layer. This latter potential consists of an isotropic anharmonic term and a cos(4 θ) term describing the cubic anisotropy. The anharmonic term has the form $A\ u^2+B\ u^4$ (u: displacements). When A and B are both positive, there is no potential well off $u=0$ (model II in [12]). For $A<0$, there are potential wells at $u>0$ resulting in order-disorder behavior at the phase transition (model I in [12]). Model I achieves about twice the energy gain for reconstruction as model II (~74 meV per surface atom vs. ~35 meV) [13].

According to the earlier lattice dynamics study [14], the W(100) reconstruction is a displacive phase transition, while paper [15] suggested that, depending on the interactions, the transition type may range from displacive to order-disorder. Support for the displacive picture has more recently come from surface-phonon dispersion studies. An anomalous, longitudinal surface mode along $\overline{\Gamma}$-$\overline{M}$ is seen by inelastic He scattering near 1200 K; it goes soft on cooling to ~500 K and even more so at ~300 K; see Fig. 8, p. 34. This soft mode behavior suggests a displacive transition. But in a broad range around T_c ($\Delta T>50$ K), disorder dominates. Ernst et al. [16] tend to interpret their findings with a sequence of two phase transitions, commensurate-incommensurate near 220 K and incommensurate-commensurate near 450 K. Above 450 K, they considered the HT phase to be a displacive phase [16]. Their conclusion is consistent with MD studies and also with the MC study [17] which showed that the peak in the probability distribution $P(u^2)$ of the squared displacements, occurring off-zero for 300 K, broadens and shifts toward $u^2=0$ on heating to 1150 K (where the maximum is still displaced off zero, while $P(u^2)$ for the high-symmetry phase in a displacive transition should be a Gaussian peaked at the origin) [17]. The molecular dynamics (MD) calculations [18] suggest a cross-over from phonon-mode softening (reminiscent of displacive behavior) away from T_c to mode broadening near T_c (consistent with order-disorder behavior) [18]. This would be consistent with the above mentioned results of He inelastic scattering presented by Ernst et al. [16].

The MD simulation by [19] is based on an effective lattice Hamiltonian and on the detailed W-W interactions used in the study of static, critical properties [5]. There is mainly displacive behavior for $T<0.7\ T_c$ and for $T>1.2\ T_c$, but for $0.7\leq T/T_c\leq 1.2$, the phase of the order parameter (the direction of the distortions) is gradually averaged out [19] with the distortion magnitude remaining essentially constant at ~0.2 Å [18]. The transverse mode softens, gets overdamped, and is totally ill-defined in the range 0.7 T_c to 1.1 T_c. This mode which is associated with the phase fluctuations is essentially destroyed in this order-disorder transition. The longitudinal wave, associated with the distortion amplitude, is strongly softened but is damped only in a small temperature range close to T_c [19]; see also [18].

The critical dynamics was studied for $T_c \le T \le 5\ T_c$ in [17] by combined analytic and MC calculations based on the two-dimensional model Hamiltonian already used by [4] to derive static properties that could be compared with LEED and X-ray diffraction data. On cooling down to T_c, the in-plane $\overline{M}$ point phonons soften and a central peak (around $\omega=0$) develops for the longitudinal modes which also soften more than the transverse shear modes do. The longitudinal mode becomes overdamped on approaching T_c from above; the associated central peak developing in the phonon spectrum is attributed to slow diffusive motion of large, ordered domains with LT structure [17]. This study was extended in [20] to the region below T_c and included cubic-order neighbor interactions to account for phonon band gapping below T_c. For the temperature range just below T_c, where the c(2×2) structure is already well developed, the quasi-harmonic approximation predicts sharp phonon peaks. The MC simulation, on the other hand, gives broad frequency widths in that range. Thus, anharmonicity and critical fluctuations render the lattice dynamics in the critical region below T_c very different from the behavior at low temperatures [20].

Finite-Size Effects. The LEED study [1] was looking for finite-size effects that are expected for the two-dimensional XY model with cubic anisotropy [1]. Properties of this model were studied as a function of the strength of the cubic anisotropy field h_4 in the MC study [22]. For the dependence $T_c=T_c(h_4)$ and the nonuniversal behavior of the critical exponent ν with respect to variation of h_4, see [22] (ν relates to the correlation length). The half-order LEED intensities calculated as a function of temperature for 4×4 and 12×12 blocks and for large cluster sizes (extrapolated to $\infty \times \infty$) agree for temperatures below 0.9 T_c, but differ considerably at and above T_c. These curves are compared in [22] with the (1/2 1/2) beam intensities measured by [23] on terraces ~100 Å wide and with the experimental data [1] of "flat" (terrace width: >400 Å) and stepped surfaces (terrace width: 30±2 Å). In agreement with MC [22] and MD [5] studies, the temperature dependence of the (1/2 1/2) LEED intensity (see **Fig. 6** a), p. 24) shows that the phase transition is quite sharp on flat surfaces but smeared-out significantly on surfaces with a high step density (at least for steps the edges of which run along [010]) [1]; the step density also affects the domain sizes (see below). The optimally fitted values for the critical temperature and critical exponent β of the order parameter for flat and stepped surface are T_c (±5 K)=211 and 217 K and $\beta=0.144\pm0.04$ and 0.050 ± 0.01, respectively. The big difference in the β values for flat and stepped surfaces emphasizes the nonuniversal behavior [1]. According to MC studies finite-size effects are pronounced for $T>T_c$, but below T_c they are significant only in a small temperature range a few % (of T_c) wide just below T_c [24].

Domain Sizes and Short-Range Order above T_c. A "domain wall fluid" [2] or domain wall disordering model [10] was used to interpret the HT X-ray data. Because of randomly placed domain walls, the HT phase possesses no long-range order of the still existing displacements [2].

In accordance with MC studies by [25], LEED studies by Wendelken and Wang [1, 26] show that short-range order exists even far above T_c. The domain size (129±20 Å near 100 K [26]) diminished to ~60 Å at 250 K [27], and about 15 Å were found for a flat surface

 References for 1.1.1.5 on p. 25

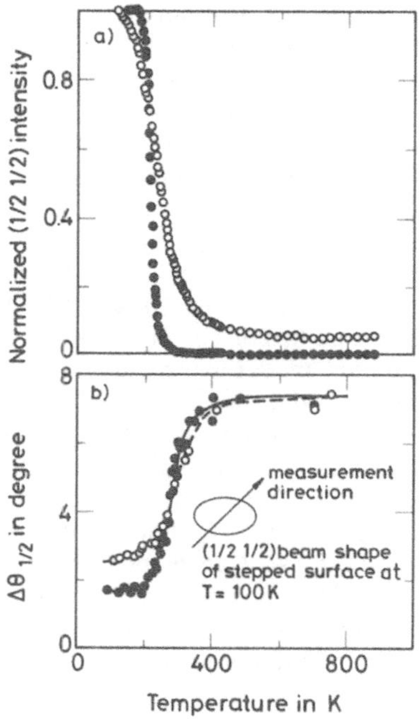

Fig. 6. Normalized intensity (a) and measured full width at half maximum (b) of (1/2 1/2) LEED beams from W(001); solid circles: flat surface, open circles: stepped surface 10(001) × [010]. Electron energy $E_p = 42$ eV [1].

at 350 K and for a stepped surface at 300 K [26]. A more detailed T dependence is given in [1], where a nearly constant domain size of ~12 Å was found between 400 and 750 K on both flat and stepped surfaces [1]; this value was derived from the broadening of the (1/2 1/2) LEED spots at $T>T_c$; see Fig. 6 b) [1]. For 300 K a correlation length ξ of 15 to 18 Å was estimated in the effective-lattice-Hamiltonian study [28]. For the dependence of the correlation length on type and density of steps, see [29].

Below T_c the domain size "saturates" at $\xi_{\parallel} = 210$ Å, a value slightly smaller than the resolution limit of the diffractometer used by [2]. Fig. 5 b) [2], p. 21, shows the T dependence of the inverse correlation lengths $\xi_{\parallel}^{-1}$ and $\xi_{\perp}^{-1}$ found by X-ray diffraction [2] on a surface the step density of which was about one step per 200 Å [10]. Here, $\xi_{\parallel}$ is parallel to the displacement vector $\langle 11 \rangle$ and $\xi_{\perp}$ is parallel to $\langle 1\bar{1} \rangle$. For $\xi_{\parallel}^{-1} \geq 4.72 \times 10^{-3}$ Å^{-1} and $\xi_{\perp}^{-1} \geq 3.08 \times 10^{-3}$ Å^{-1}, ξ^{-1} changes with temperature as follows: $\xi_{\parallel}^{-1} = (0.1474 \pm 0.006$ Å$^{-1}) \times ([T-T_c]/T_c)^{1.0}$ and $\xi_{\perp}^{-1} = (0.082 \pm 0.004$ Å$^{-1}) \times ([T-T_c]/T_c)^{1.0}$ [2]. This anisotropic behavior of ξ was duplicated fairly well by Han and Ying [4] within the temperature range ~230 to ~350 K in an MC study, where they used a two-dimensional effective Hamiltonian, the parameters of which referred to the p(1 × 1) bulk truncating surface [4]. Using parameters from the tight-binding study [30], Terakura and Hamada [31] analyzed the problem of ξ anisotropy considering interactions within a reconstructed environment. Their analysis accounts for the presence of reconstruction domains at a short-range-order level. Although this approach differs from that by Han and Ying [4], qualitatively the same result was obtained in [31].

References for 1.1.1.5:

[1] Wendelken, J. F.; Wang, G.-C. (Phys. Rev. [3] B **32** [1985] 7542/4).
[2] Evans-Lutterodt, K.; Birgeneau, R. J.; Specht, E. D.; Chung, J. W.; Brock, J. D.; Altman, M. S.; Estrup, P. J.; Robinson, I. K.; MacDowell, A. A. (J. Vac. Sci. Technol. A **7** [1989] 2209/16).
[3] Stensgaard, I.; Purcell, K. G.; King, D. A. (Phys. Rev. [3] B **39** [1989] 897/900).
[4] Han, W. K.; Ying, S. C. (Phys. Rev. [3] B **41** [1990] 9163/7).
[5] Wang, C. Z.; Parrinello, M.; Tosatti, E.; Fasolino, A. (Europhys. Lett. **6** [1988] 43/8).
[6] Tosatti, E. (Solid State Commun. **25** [1978] 637/40).
[7] Inglesfield, J. E. (Prog. Surf. Sci. **20** [1985] 105/64).
[8] Bak, P. (Solid State Commun. **32** [1979] 581/3).
[9] Kosterlitz, J. M. (J. Phys. C: Solid State Phys. **7** [1974] 1046/60).
[10] Robinson, I. K.; MacDowell, A. A.; Altman, M. S.; Estrup, P. J.; Evans-Lutterodt, K.; Brock, J. D.; Birgeneau, R. J. (Phys. Rev. Lett. **62** [1989] 1294/7).

[11] Sugibayashi, T.; Hara, M.; Yoshimori, A. (J. Vac. Sci. Technol. A **5** [1987] 771/6).
[12] Roelofs, L. D.; Wendelken, J. F. (Phys. Rev. [3] B **34** [1986] 3319/36).
[13] Roelofs, L. D. (Surf. Sci. **178** [1986] 396/405).
[14] Fasolino, A.; Santoro, G.; Tosatti, E. (Surf. Sci. **125** [1983] 317/25).
[15] Ying, S. C.; Roelofs, L. D. (Surf. Sci. **125** [1983] 218/26).
[16] Ernst, H.-J.; Hulpke, E.; Toennies, J. P. (Phys. Rev. [3] B **46** [1992] 16081/105).
[17] Han, W. K.; Ying, S. C.; Sahu, D. (Phys. Rev. [3] B **41** [1990] 4403/9).
[18] Wang, C. Z.; Fasolino, A.; Tosatti, E. (Europhys. Lett. **7** [1988] 263/8).
[19] Wang, C. Z.; Fasolino, A.; Tosatti, E. (Phys. Rev. [3] B **37** [1988] 2116/22).
[20] Han, W. K.; Ying, S.-C. (Phys. Rev. [3] B **46** [1992] 1849/52).

[21] Lau, K. H.; Ying, S. C. (Phys. Rev. Lett. **44** [1980] 1222/5).
[22] Hu, G. Y.; Ying, S. C. (Physica A **140** [1987] 585/96).
[23] Barker, R. A.; Estrup, P. J. (J. Chem. Phys. **74** [1981] 1442/52).
[24] Hu, G. Y.; Ying, S. C. (Surf. Sci. **150** [1985] 47/54).
[25] Roelofs, L. D.; Hu, G. Y.; Ying, S. C. (Phys. Rev. [3] B **28** [1983] 6369/76).
[26] Wendelken, J. F.; Wang, G.-C. (J. Vac. Sci. Technol. A **2** [1984] 888/92).
[27] Heilmann, P.; Heinz, K.; Müller, K. (Surf. Sci. **89** [1979] 84/94).
[28] Fasolino, A.; Santoro, G.; Tosatti, E. (Phys. Rev. Lett. **44** [1980] 1684/7).
[29] Roelofs, L. D. (Phys. Rev. [3] B **34** [1986] 3337/44).
[30] Terakura, I.; Terakura, K.; Hamada, N. (Surf. Sci. **111** [1981] 479/91).

[31] Terakura, K.; Hamada, N. (Springer Ser. Mater. Sci. **17** [1992] 159/67).

1.1.1.6 Driving Forces for Reconstruction and Relaxation

Up to now it is not clear whether the reconstruction on the W(100) surface is driven by a charge-density-wave instability related to nested segments of the surface Fermi contour lines (i.e., involving delocalized electronic band states) or by a Jahn-Teller-like mechanism based on local bonding interactions; see e.g. [1, 2].

The driving forces for the surface reconstruction were discussed in the review paper [3]. The band narrowing in the surface density of states (DOS) causes the experimentally found inward relaxation. This contraction enhances the interaction of the surface atoms with the substrate. Any atomic displacements which broaden or split the central peak near E_F in the surface DOS will lower the energy. The tendency of the surface atoms to increase their interactions with one another and the substrate makes the W(100) surface instable

References for 1.1.1.6 on pp. 29/30

[3, pp. 276/80]. The central peak in the density of states is made up of surface states and resonances over the whole of the SBZ. It seems that this peak destabilizes the (1 × 1) surface. The collective behavior, not that of *individual* surface states, drives the reconstruction transition according to [3, p. 257]. On the other hand, the surface state which was found by UPS in the HT phase crossing the Fermi level at 1.2 $Å^{-1}$ (see [4]) has a large component of d_{xy} orbitals that link atoms in the surface plane [5]. Upon reconstruction, that state splits into bonding and antibonding subbands. Whether or not the energetic contribution of this surface state band splitting is sufficient to drive the phase transition, its energetic contribution will be substantial in any case, since this state does not only exist on the $\bar{\Sigma}$ line but also over a considerable $\vec{k}$ space area around that line [5]. According to the tight-binding study [6], the reconstruction is driven by an attractive interaction between neighboring surface atom pairs, which is due to the high density of surface states near the center of the d band [6]. For the importance of the direct interaction of $d_{y^2-z^2}$ orbitals on the reconstruction of W(100), see also Pick et al. [7, 8] who treated the case of stepped W(100) surfaces in [9]. In contrast, repulsive interactions between surface atoms were found in the LAPW study [10], consistent with the weakly repulsive interactions assumed in the simulations in [11, 12]. Also, the main conclusion to be drawn from the anomalous, longitudinal phonon mode observed by [13] along $\bar{\Gamma}$-$\bar{M}$ is that the nearest surface atoms repel one another; see [14].

The LCAO studies by Tomásek and Pick [15 to 17] enumerated and discussed various distortion types. The above mentioned central peak (due to dangling bonds in the surface DOS; see [18]) remains confined to a narrow range around E_F for vertical atomic displacements, as these only moderately disturb the local symmetry. Horizontal displacements, on the other hand, broaden this peak, thus allowing for a decrease in the total electronic energy [15]. Within the tight-binding framework, the $\bar{X}_1$ and $\bar{X}_4$ phonon modes soften only for those parameter sets for which the $\bar{M}_1$- and $\bar{M}_5$-type reconstructions had already taken place [17]. The longitudinal in-plane $\bar{M}_5$ and $\bar{X}_3$ deformations are more probable than the buckled $\bar{M}_1$ and $\bar{X}_1$ modes (which result in an ionic superstructure) or the in-plane transversal $\bar{X}_4$ mode [16]. Thus, $\bar{M}_5$ modes are the most likely distortions. This is consistent with the molecular dynamics simulation at 500 K in [11] which showed that the most strongly softening phonon is an in-plane M_5 phonon [11].

While multilayer relaxation is dominated by electrostatic effects (see [19]), multilayer reconstruction is controlled by electronic properties for the surface layer and elastic properties for the subsurface layer. The subsurface-layer reconstruction involves a stiffening of the force constants above that of the bulk value (while soft mode behavior may be expected for the surface layer) [20]. Reconstruction implies that the force constants between the surface atoms differ from those between bulk atoms [21]. The lattice dynamics calculation [22] successfully modeled the W(100) reconstruction by an effective lattice Hamiltonian, distinguishing between atom pairs both partners of which are surface atoms and atom pairs at least one partner of which is a bulk atom [22]. The harmonic surface force constants determine either the stability of the unreconstructed surface or the type of surface reconstruction. The anharmonic surface force constants, on the other hand, determine the magnitude of the distortions at T = 0 K. Pure in-plane reconstruction lacking vertical displacements apparently requires that the harmonic second-order surface force constant is about −1/4 the corresponding bulk value [12]; (a soft phonon with vertical displacements and $\bar{M}_1$ symmetry is associated with the c(2 × 2) structure of clean W seen by FIM; see [23]).

The UPS studies [24] and [25] show a very curved Fermi line and indicate that the surface-state splitting can involve only a small number of electrons [24]. This observation is in conflict with the flat line of surface states at E_F calculated by [26] and the conclusion by [27] that surface states near E_F drive the phase transition. Hence, the authors of [24] conclude that the surface-state Fermi line plays no significant role in driving the reconstruction. Also in the review paper [28] it is concluded from the UPS studies [25] and [4] that the $\bar{\Gamma}$-$\bar{M}$ surface coupling is a result of rather than the driving force for the phase transition [28]. Thus, it appeared very unlikely that the charge density wave (CDW) mechanism suggested by [29] and [26] is operative [3]. Also in view of the UPS data of [4], there is no evidence for a single spanning wave vector that could support a CDW model for the W(100) reconstruction [30].

The tight-binding calculations of [31] suggest that the Fermi line is not particularly important for the W(100) reconstruction. Rather, there are substantial changes in the DOS in the whole d-band region (not only just around E_F) which favor the c(2×2) phase [31]. This is corroborated in the first-principles study [1] where an LAPW method was used within the local-density approximation. There is a great difference between the dispersion curves of a surface state pair for the relaxed ideal surface and the corresponding pair of curves for the reconstructed surface over a considerable wave vector range; see Fig. 15, p. 47 [1]. Because of enhanced interaction between pairs of surface atoms, the sharp surface DOS peak (close to E_F) separates into bonding and antibonding states. Thus, a large energy gain is expected for displacement patterns of frozen-in $\bar{M}_5$-zone-boundary phonons whose displacements are confined to the surface plane (see [31]). The main difference between the electronic driving mechanism in this Jahn-Teller-like picture to that of a CDW mechanism consists in the fact that now many electronic states (many $\vec{k}$ vectors) are involved and not only those near the Fermi level. The UPS data and the success of tight-binding calculations suggest that the "local" Jahn-Teller picture is more appropriate than the delocalized CDW model [30].

The tight-binding calculations by [32] of the surface DOS performed for both ⟨11⟩ (zigzag-chain) and ⟨10⟩ (dimer)-type displacements on a relaxed (10% contraction) surface [32] were extended in [31]. Surface states/resonances and electron-phonon interactions change the surface DOS near E_F such that the (1×1) structure is destabilized in favor of ⟨11⟩ zigzag-chain displacements [31]. The DOS peak of the unreconstructed surface around E_F is split into two peaks separated by 1.3 eV, considerably lowering the electronic energy. There is a strong coupling between $\bar{M}_5$ phonon modes and the surface states (but not for the $\bar{M}_1$ modes) [33]. The energy gain from reconstruction results from enhanced bonding between pairs of surface atoms, i.e., it is a strongly localized effect [31]. On the other hand, the energy gain from a 10% contraction of the first interlayer spacing is due to the broadening of the whole d band (mainly at the upper two layers) [31, 34]. Surface contractions by ~10% can be explained by d-band electron energy gains without surface state contributions [31]. These results were almost exactly reproduced by [35] using a self-consistent Hartree approximation, thus discounting objections by [36] criticizing that Terakura et al. did not strictly observe the charge neutrality condition in [31].

Also according to [22], half-filled surface states are responsible for the transition to occur. But a "strong-coupling CDW mechanism" was proposed (weak coupling, i.e., ordinary CDW is ruled out because of the negligible amount of surface nesting reported by [25] and [4]) being intermediate between weak coupling and outright chemical bond formation. Different interactions were assumed for atoms, at least one of which is in the bulk, and for atoms both of which are on the surface [22]. Short-range interaction models, which were applied in a Monte Carlo study of the surface reconstruction, led to the conclusion that interactions with atoms making up the second layer are significant for driving the transition [37].

References for 1.1.1.6 on pp. 29/30

The more recent UPS study [38] exploring the $\bar{\Sigma}$ line indicated large-scale nesting for W(100). The results were compatible with a CDW-driven reconstruction and an incommensurate surface structure. The results, being also consistent with a local-bonding model for the reconstruction, suggest that the reconstruction is driven by a combination of localized and delocalized electronic effects [38]; see also the review [39].

The importance of surface-state nesting is lessened by the fact that any reconstruction-related phonon that effects the p(1 × 1)-to-c(2 × 2) transition must be doubly degenerate, if reconstruction is to occur by a soft-phonon mechanism (the only doubly degenerate representation that changes the translational symmetry belongs to the $\bar{M}_5$ representation). Thus, the exact position in $\vec{k}$ space, where the $\bar{\Sigma}_2$ band of surface states crosses the Fermi level, is not relevant in determining the structure of the c(2 × 2) phase. Rather, charge densities calculated in [40] suggest that the W atoms are pulled off their p(1 × 1) positions towards the density maxima of the surface state, forming a pattern remarkably similar to the atom displacements of the Debe-King zigzag chain [40] (see Fig. 17, p. 54, taken from [1]). This charge density of the Σ_2 surface state band also plays an important role in the theory advanced by [41]. According to this study, the reconstruction transition can be rationalized by assuming a disordered HT phase, where the surface atoms are displaced both dynamically and statically (the latter by random occupation of potential wells existing off the ideal p(1 × 1) sites). The dynamic component of the displacements is due to oscillations of the atoms within their particular well. On the average, translational symmetry is preserved; hence, coupling of the oscillations results in phonon-like modes. Thus, in addition to the random static displacements, there exist dynamic displacements due to softening of phonon modes. As a result of interactions of the electrons with both phonons and static displacements, the random static displacements enhance the electron-phonon interaction (phonon softening), and the electron-phonon interaction increases the magnitude of the static displacements [41].

Very recently, Ernst et al. [13] also supported the CDW picture. Their elastic and inelastic He scattering experiments agreed with the surface phonon dispersion curves calculated for T = 0 K by [42]. Using a tight-binding scheme including long-range forces (taken from the energy-band calculations [43] and [33]), a whole region of soft phonon modes at the SBZ was calculated [42]. This softening stems from interactions with electronic surface states around E_F. Fermi surface nesting and d-orbital interactions soften in particular the frequency of the $\bar{M}_5$ phonon whose displacement pattern agrees with the structure of the c(2 × 2) phase [42]. The strong charge redistribution calculated by [42] increases the short-range force constants by 10 to 20% in agreement with the He scattering results of [13]. Local bonding effects of electrostatic origin (as suggested by [40]) appear to be unimportant. Rather, the driving mechanism seems to involve a Peierls-type (i.e., CDW-like) transition [42], and the soft phonon mode seen in He inelastic scattering is attributed to a giant Kohn anomaly [44]. The two-dimensional bands, found in [38] to be nondispersive close to E_F over a significant region of the SBZ, indicate a very high two-dimensional DOS close to E_F. These electrons being slow, strong nonadiabatic effects can be expected [2, 45]. (The tight-binding calculations [31] had already underlined the important role of the surface DOS near E_F and of electron-phonon interactions.)

The statistical-mechanics calculation [46] tried to bridge the gap between experimental and theoretical studies on the strength and nature of the driving force for the reconstruction. According to [46], the reconstruction energies calculated in the first-principles studies [31 to 33, 36, 43, 47, 48] do not suffice to maintain order below T_c. The simulation [46] showed that the energy substantially changes on heating from 0 K to just below T_c. Therefore, significant differences are to be expected between the theoretical ground-state (T = 0 K) studies and experimental investigations of the LT phase which were performed at and above ~120 K (≈T_c/2) [46].

References for 1.1.1.6:

[1] Yu, R.; Krakauer, H.; Singh, D. (Phys. Rev. [3] B **45** [1992] 8671/8).
[2] Elliott, G. S.; Smith, K. E.; Kevan, S. D. (Phys. Rev. [3] B **44** [1991] 10826/34).
[3] Inglesfield, J. E. (Rep. Prog. Phys. **45** [1982] 223/84).
[4] Holmes, M. W.; Gustafsson, T. (Phys. Rev. Lett. **47** [1981] 443/6).
[5] Stephenson, P. C.; Bullett, D. W. (Surf. Sci. **139** [1984] 1/15).
[6] Joubert, D. P. (J. Phys. C: Solid State Phys. **20** [1987] 1899/907).
[7] Pick, S.; Tomásek, M.; Luchini, M. U. (Int. J. Quantum Chem. **38** [1990] 329/38).
[8] Pick, S.; Tomásek, M.; Luchini, M. U. (Chem. Phys. Lett. **164** [1989] 345/7).
[9] Pick, S.; Tomásek, M.; Luchini, M. U. (Chem. Phys. Lett. **177** [1991] 45/8).
[10] Ohnishi, S.; Freeman, A. J.; Wimmer, E. (Phys. Rev. [3] B **29** [1984] 5267/78).

[11] Wang, C. Z.; Parrinello, M.; Tosatti, E.; Fasolino, A. (Surf. Sci. **211/212** [1989] 323/8).
[12] Fasolino, A.; Tosatti, E. (Phys. Rev. [3] B **35** [1987] 4264/83).
[13] Ernst, H.-J.; Hulpke, E.; Toennies, J. P. (Phys. Rev. [3] B **46** [1992] 16 081/105).
[14] Kulkarni, A. D.; de Wette, F. W.; Prade, J. (J. Electron Spectrosc. Relat. Phenom. **44** [1987] 397/402).
[15] Tomásek, M.; Pick, S. (J. Phys. Colloq. **45** [1984] C 5-125/C 5-129).
[16] Tomásek, M.; Pick, S. (Surf. Sci. **140** [1984] L 279/L 284).
[17] Pick, S.; Tomásek, M. (Czech. J. Phys. B **34** [1984] 1235/41).
[18] Friedel, J. (J. Phys. [Paris] **37** [1976] 883/94).
[19] Fu, C. L.; Ohnishi, S.; Jansen, H. J. F.; Freeman, A. J. (Phys. Rev. [3] B **31** [1985] 1168/71).
[20] Fu, C. L.; Freeman, A. J. (Phys. Rev. [3] B **37** [1988] 2685/8).

[21] Fasolino, A.; Santoro, G.; Tosatti, E. (Phys. Rev. Lett. **44** [1980] 1684/7).
[22] Fasolino, A.; Santoro, G.; Tosatti, E. (Surf. Sci. **125** [1983] 317/25).
[23] Melmed, A. J.; Tung, R. T.; Graham, W. R.; Smith, G. D. W. (Phys. Rev. Lett. **43** [1979] 1521/4).
[24] Campuzano, J. C.; Inglesfield, J. E.; King, D. A.; Somerton, C. (J. Phys. C: Solid State Phys. **14** [1981] 3099/113).
[25] Campuzano, J. C.; King, D. A.; Somerton, C.; Inglesfield, J. E. (Phys. Rev. Lett. **45** [1980] 1649/52).
[26] Krakauer, H.; Posternak, M.; Freeman, A. J. (Phys. Rev. Lett. **43** [1979] 1885/9).
[27] Posternak, M.; Krakauer, H.; Freeman, A. J.; Koelling, D. D. (Phys. Rev. [3] B **21** [1980] 5601/12; Erratum: Phys. Rev. [3] B **30** [1984] 4828).
[28] King, D. A. (Phys. Scr. T **4** [1983] 34/43).
[29] Tosatti, E. (Solid State Commun. **25** [1978] 637/40).
[30] Ying, S. C.; Roelofs, L. D. (Surf. Sci. **125** [1983] 218/26).

[31] Terakura, I.; Terakura, K.; Hamada, N. (Surf. Sci. **103** [1981] 103/17).
[32] Terakura, K.; Terakura, I.; Teraoka, Y. (Surf. Sci. **86** [1979] 535/41).
[33] Fu, C. L.; Freeman, A. J.; Wimmer, E.; Weinert, M. (Phys. Rev. Lett. **54** [1985] 2261/4).
[34] Desjonquères, M. C. (J. Phys. Lett. **40** [1979] L 249/L 252).
[35] Masuda-Jindo, K.; Hamada, N.; Terakura, K. (J. Phys. C: Solid State Phys. **17** [1984] L 271/L 274).
[36] Tréglia, G.; Desjonquères, M.-C.; Spanjaard, D. (J. Phys. C: Solid State Phys. **16** [1983] 2407/19).
[37] Roelofs, L. D.; Wendelken, J. F. (Phys. Rev. [3] B **34** [1986] 3319/36).

[38] Smith, K. E.; Elliott, G. S.; Kevan, S. D. (Phys. Rev. [3] B **42** [1990] 5385/7).
[39] Smith, K. E.; Kevan, S. D. (Prog. Solid State Chem. **21** [1991] 49/131, 81/8).
[40] Weinert, M.; Freeman, A. J.; Ohnishi, S. (Phys. Rev. Lett. **56** [1986] 2295/8).

[41] Egorushkin, V. E.; Kul'ment'ev, A. I.; Savushkin, E. V. (Izv. Vyssh. Uchebn. Zaved. Fiz. **34** No. 8 [1991] 24/35; Sov. Phys. J. [Engl. Transl.] **34** [1992] 669/78).
[42] Wang, X. W.; Weber, W. (Phys. Rev. Lett. **58** [1987] 1452/5).
[43] Mattheiss, L. F.; Hamann, D. R. (Phys. Rev. [3] B **29** [1984] 5372/81).
[44] Hulpke, E. (J. Electron Spectrosc. Relat. Phenom. **54/55** [1990] 299/308).
[45] Smith, G. S.; Kevan, S. D. (Phys. Rev. [3] B **43** [1991] 3986/93, 3991/2).
[46] Roelofs, L. D. (Surf. Sci. **178** [1986] 396/405).
[47] Terakura, I.; Terakura, K.; Hamada, N. (Surf. Sci. **111** [1981] 479/91).
[48] Stephenson, P. C.; Bullett, D. W. (Surf. Sci. **139** [1984] 1/15).

1.1.2 Lattice Dynamics

1.1.2.1 Surface Debye Temperature. Thermal Mean Square Displacements

Surface Debye Temperature. From the temperature dependence of the integral-order LEED spot intensities, the surface Debye temperatures $\Theta_D = 210 \pm 40$ K and 400 ± 100 K were derived for the high- (HT) and low-temperature (LT) phases, respectively. The change in the surface Debye temperature might be associated with a phonon becoming soft at $\bar{M}$ (see speculations in [1]) [2]. $\Theta_D = 183$ K was obtained from the (10) intensity between ~550 and ~800 K [3]. Using a simple oscillator model, $\Theta_D = 181$ and 191 K were derived for oscillations normal and parallel to the surface, respectively [4] in good agreement with $\Theta_D = 182 \pm 10$ K found for perpendicular oscillations by LEED between 400 and ~900 K [5]. The value $\Theta_D \approx 220$ K was arbitrarily chosen in the RBS study [6] to fit the data in the LT regime.

Thermal Mean Square Displacements. The slope of the integrated (3/2 3/2) X-ray intensity yielded a surface vibration amplitude of 0.19 ± 0.07 Å (mean square displacements: $\langle u^2 \rangle = 0.036$ Å^2) for HT W(100) at 300 K [7]. No reconstruction was considered in the calculation [8]. From Estrup's [3] Θ_D determination follows that the normal components of the mean-square displacements of the W(100) surface are 2.3 times those of the bulk (see [8]). The temperature dependence of the mean square displacements was calculated assuming Lennard-Jones potentials extended to fourth-nearest neighbors [8]. The model parameters were determined by fitting them to the bulk phonon dispersion curves measured by [9]. Theoretical ratios of $\langle u^2 \rangle_{surf} / \langle u^2 \rangle_{bulk}$ were plotted for both normal and parallel components between 0 and 700 K (unreconstructed surface) [8].

The Debye-Waller factor for LT W(100) was determined at 175 K by X-ray diffraction. It corresponds to a vibration amplitude of 0.05 Å ($\langle u^2 \rangle = 0.0025$ Å^2) [10]. Mean square displacements of surface atoms were calculated for normal vibrations (along [100]) and for in-plane vibrations along the inequivalent directions [011] and [0$\bar{1}$1]. Values for the ratio $R_{\langle u^2 \rangle} = \langle u^2 \rangle_{surf} / \langle u^2 \rangle_{bulk}$ are listed in the following table for T=0 and 200 K [11] (for $R_{\langle u^2 \rangle}$ values at 50, 100, and 150 K, see [11]):

temperature in K	$R_{\langle u^2 \rangle}$ along [100]	$R_{\langle u^2 \rangle}$ along [011]	$R_{\langle u^2 \rangle}$ along [0$\bar{1}$1]
0	1.20	1.35	2.43
200	1.75	1.82	5.20

References for 1.1.2.1:

[1] Tosatti, E. (Solid State Commun. **25** [1978] 637/40).
[2] Heilmann, P.; Heinz, K.; Müller, K. (Surf. Sci. **89** [1979] 84/94).
[3] Estrup, P. J. (in: Somorjai, G. A.; The Structure and Chemistry of Solid Surfaces, Wiley, New York 1969, pp. 19-1/19-9; C.A. **72** [1970] No. 47736).
[4] Jackson, D. P. (Surf. Sci. **43** [1974] 431/40).
[5] Boiko, B. A. (Ukr. Fiz. Zh. [Russ. Ed.] **20** [1975] 1707/14).
[6] Stensgaard, I.; Purcell, K. G.; King, D. A. (Phys. Rev. [3] B **39** [1989] 897/900).
[7] Robinson, I. K.; MacDowell, A. A.; Altman, M. S.; Estrup, P. J.; Evans-Lutterodt, K.; Brock, J. D.; Birgeneau, R. J. (Phys. Rev. Lett. **62** [1989] 1294/7).
[8] Wallis, R. F.; Cheng, D. J. (Solid State Commun. **34** [1980] 847/51).
[9] Chen, S. H.; Brockhouse, B. N. (Solid State Commun. **2** [1964] 73/7).
[10] Altman, M. S.; Estrup, P. J.; Robinson, I. K. (Phys. Rev. [3] B **38** [1988] 5211/4).

[11] Reinecke, T. L.; Ying, S. C. (Phys. Rev. [3] B **43** [1991] 12234/45).

1.1.2.2 Phonons

1.1.2.2.1 Survey

There is considerable interest in the lattice dynamics of W(100) from both the experimental and the theoretical standpoint. Elucidation of the dynamics should provide the key for understanding the LT-to-HT phase transition. According to Egorushkin et al. [1], the W atoms on HT W(100) occupy random potential wells around the ideal sites, and coupling of their oscillations results in phonon-like modes. Displacements caused by soft phonons as well as static displacements couple to the $\bar{\Sigma}_2$ surface state band, the electron density of which is shown in [2] and in Fig. 17, p. 54 [3]. The HT-to-LT phase transition appears as a synergistic effect of electron interaction with both soft phonons and random static displacements [1]. The importance of electron-phonon interactions for the phase transition was also demonstrated by an LCAO calculation [4]. A linearized augmented plane wave study [5] showed a strong coupling between $\bar{M}_5$ phonon modes and the surface states, while the $\bar{M}_1$ modes were not coupled [5].

With respect to the phase transition, the Rayleigh (R) mode may be considered irrelevant, since vertical displacements do not contribute to the two-dimensional Hamiltonian describing the dynamics of the surface [6]. In fact, molecular dynamics (MD) simulations show that the R waves remain harmonic and are hardly affected by the transition [7]; see also [8] and other simulations reported below which consistently attribute anomalous behavior only to in-plane modes rather than to R waves. Near T_c the dominant excitation is a central peak (around $\hbar\omega = 0$) associated with the longitudinal (L) modes and due to slow diffusion of large, ordered domains [6]. On the reconstructed surface, the two low-lying $\bar{M}$ modes L (longitudinal) and SH (shear horizontal) play the roles of "amplitude" and "phase" mode, respectively (referring to the amplitude and phase of the order parameter; SH is the "driving" mode and L the "driven" mode). MD simulations show that the amplitude mode softens substantially, while the phase mode mainly broadens [7].

The frequency distributions of low-lying surface modes calculated by MD for 100, 300, and 500 K suggest that there is a crossover from displacive mode softening away from T_c to mode broadening near T_c, i.e., from an order-order to an order-disorder behavior, respectively [9], corroborating the study of the phase transition by [10] (see Fig. 2, p. 3). The dispersion curves along $\bar{\Gamma}$-$\bar{X}$ hardly change with temperature for any type of mode. The behavior along the $\bar{\Gamma}$-$\bar{M}$ directions, however, is different. On approaching T_c, the SH

References for 1.1.2.2.1 on p. 33

and L waves soften in a wider region around $\overline{M}$. The MD simulations indicate that the L waves are anomalous, because they are lower in frequency than the R modes over the whole temperature range studied. The considerable broadening of the frequency distributions of R and L modes seen close to T_c has disappeared at 500 K. This softening/broadening and subsequent hardening/narrowing behavior between 100 and 500 K is consistent [9] with the results of [10]. Passing through T_c towards higher temperatures disorders the distortion direction (causing mode broadening), while the distortion magnitude remains essentially constant at ~0.2 Å [9]; see also [11].

With respect to the physical origin of reconstruction, Ernst et al. [12] find good agreement between their recent inelastic He scattering experiments (see pp. 33/5) and the calculations for unreconstructed W(100) by [13] who augmented their tight-binding scheme by including long-range forces using energy-band calculations from [5] and [14]. The long-range forces are the source of phonon anomalies or Peierls-type phonon instabilities [13]. The force constants derived in [13] from the frozen-phonon study [5] yielded unsatisfactory results; at least the nearest-neighbor force constants between atoms of the uppermost and second layers should be increased by ~10% or even ~20% [13]. According to these tight-binding

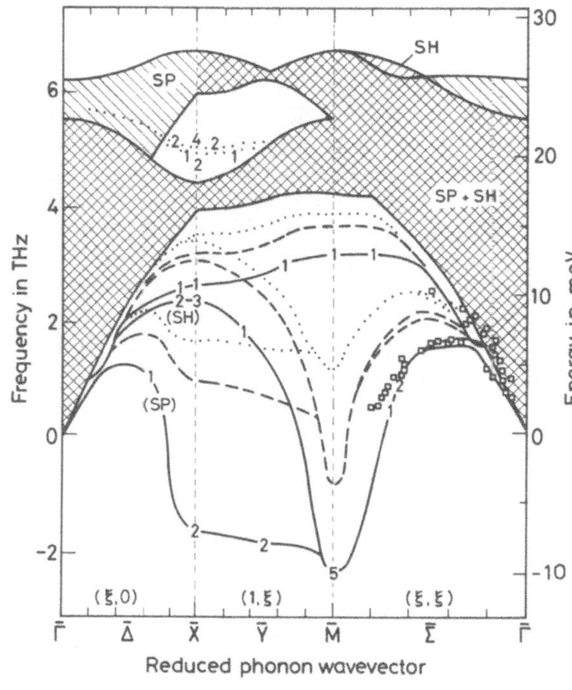

Fig. 7. Surface phonon dispersion of the (1 × 1)-W(100) surface [15]. Theoretical dispersion curves of Wang and Weber [13] are compared with the He inelastic scattering data (squares) of Ernst et al. [16]. The hatched areas indicate the bulk phonon continuum; the sagittal-plane (SP) and shear-horizontal (SH) modes are marked differently. The solid lines were calculated with bulk short-range forces, while the dashed and dotted lines show, respectively, the curves calculated using short-range forces increased by 10% and 20% for first-nearest-neighbor interlayer interactions. The Rayleigh modes are denoted by $\overline{\Delta}_1$-$\overline{X}_1$-$\overline{Y}_1$-$\overline{M}_1$-$\overline{\Sigma}_1$ throughout the surface Brillouin zone. The soft SP modes are labeled by $\overline{\Delta}_1$-$\overline{X}_2$-$\overline{Y}_2$-$\overline{M}_5$-$\overline{\Sigma}_1$ and the soft SH modes by $\overline{\Delta}_2$-$\overline{X}_3$-$\overline{Y}_1$-$\overline{M}_5$-$\overline{\Sigma}_2$ [13].

calculations performed for the unreconstructed surface, the SH and the (mainly) longitudinal L waves are expected to be softer than the R modes nearly throughout the whole surface Brillouin zone (SBZ) [13]; see **Fig. 7** [15, p. 269] including the earlier He inelastic scattering results of Ernst et al. [16]. According to [15] there is fair agreement between the He scattering data of [16] and the tight-binding study of [13] which attributes the predominant $\overline{M}_5$ phonon instability to coupling between $d_{x^2-y^2}$ and d_{xy} orbitals of the $\overline{\Sigma}_2$ surface states.

With respect to the temperature dependence of the dynamics, Ernst et al. [12] state good agreement between their experiments [12] and the MD studies of the Tosatti group ([7 to 11, 17, 18]; see also [19]) and the Monte Carlo (MC) study [6] which, however, because of the limited slab thickness used, did not reproduce the R mode. Yet, the softening and broadening of the L mode and its overdamping around T_c was excellently reproduced in the MC study [6]; see [12].

References for 1.1.2.2.1:

[1] Egorushkin, V. E.; Kul'ment'ev, A. I.; Savushkin, E. V. (Izv. Vyssh. Uchebn. Zaved. Fiz. **34** No. 8 [1991] 24/35; Sov. Phys. J. [Engl. Trans.] **34** [1992] 669/78).
[2] Weinert, M.; Freeman, A. J.; Ohnishi, S. (Phys. Rev. Lett. **56** [1986] 2295/8).
[3] Yu, R.; Krakauer, H.; Singh, D. (Phys. Rev. [3] B **45** [1992] 8671/8).
[4] Terakura, I.; Terakura, K.; Hamada, N. (Surf. Sci. **103** [1981] 103/17).
[5] Fu, C. L.; Freeman, A. J.; Wimmer, E.; Weinert, M. (Phys. Rev. Lett. **54** [1985] 2261/4).
[6] Han, W. K.; Ying, S. C.; Sahu, D. (Phys. Rev. [3] B **41** [1990] 4403/9).
[7] Wang, C. Z.; Fasolino, A.; Tosatti, E. (Phys. Rev. [3] B **37** [1988] 2116/22).
[8] Wang, C. Z.; Tosatti, E.; Fasolino, A.; Parrinello, M. (Surf. Sci. **189/190** [1987] 679/83).
[9] Wang, C. Z.; Fasolino, A.; Tosatti, E. (Europhys. Lett. **7** [1988] 263/8).
[10] Wang, C. Z.; Parrinello, M.; Tosatti, E.; Fasolino, A. (Europhys. Lett. **6** [1988] 43/8).
[11] Fasolino, A.; Tosatti, E. (Phys. Rev. [3] B **35** [1987] 4264/83).
[12] Ernst, H.-J.; Hulpke, E.; Toennies, J. P. (Phys. Rev. [3] B **46** [1992] 16081/105).
[13] Wang, X. W.; Weber, W. (Phys. Rev. Lett. **58** [1987] 1452/5).
[14] Mattheiss, L. F.; Hamann, D. R. (Phys. Rev. [3] B **29** [1984] 5372/81).
[15] Kern, K.; Comsa, G. (Adv. Chem. Phys. **76** [1989] 211/80, 265/9).
[16] Ernst, H.-J.; Hulpke, E.; Toennies, J. P. (Phys. Rev. Lett. **58** [1987] 1941/4).
[17] Fasolino, A.; Santoro, G.; Tosatti, E. (Phys. Rev. Lett. **44** [1980] 1684/7).
[18] Wang, C. Z.; Fasolino, A.; Tosatti, E. (Phys. Rev. Lett. **59** [1987] 1845/8).
[19] Wang, C. Z.; Fasolino, A.; Tosatti, E. (Surf. Sci. **211/212** [1989] 323/8).

1.1.2.2.2 Experimental Surface Phonon Dispersion Curves

Surface phonon dispersion curves have been measured by inelastic He atom scattering. The dispersion curves measured along $\overline{\Gamma}$-$\overline{X}$ are shown in **Fig. 8** a), p. 34 [1]. The upper curve is assigned to the Rayleigh (R) modes. Near 500 K the R mode has a zone boundary energy of $E(\overline{X}) \approx 11.5$ meV. The R mode is well separated from the lower-lying longitudinal (L) surface waves for $q > 0.4$ Å^{-1}. The L modes are expected to follow closely the surface-projected bulk band edge and, for small values of the wave vector q, to run slightly above the R branch. But near 0.4 Å^{-1}, the L and R branches cross, hybridize, and for larger q values exchange their roles [1].

Surface phonon dispersion curves along the $\overline{\Gamma}$-$\overline{M}$ (=⟨011⟩) azimuth of the HT W(100) surface were determined at 1200, 450, and 280 K [2, 3]; for dispersion curves at ~280

References for 1.1.2.2.2 on pp. 35/6

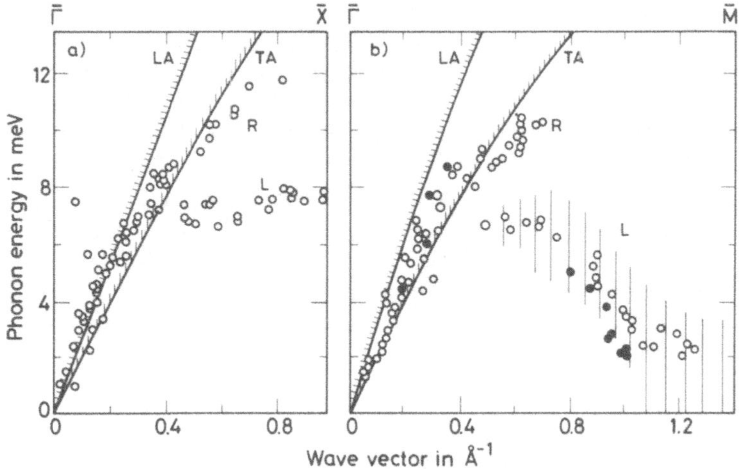

Fig. 8. Surface phonon dispersion curves of HT W(100) measured along the $\bar{\Gamma}$-$\bar{X}$ (a) and $\bar{\Gamma}$-$\bar{M}$ (b) directions at 500 K (open circles) and 300 K (closed circles). Hatched lines mark the band edges for transverse acoustic (TA) and longitudinal acoustic (LA) bulk modes [1].

and ~450 K, see also [4]. These and further studies are summarized in the comprehensive paper [1]. From the three types of surface waves, namely R, SH, and L, only the R and the L waves are detectable by the in-plane scattering used. The polarizations given are ideal ones, realized only at $\bar{\Gamma}$; away from $\bar{\Gamma}$, all these waves are elliptically polarized [2]; see also [3] and the molecular dynamics (MD) study [5]. The dispersion curves taken along $\bar{\Gamma}$-$\bar{M}$ at ~1200 K are very similar to those taken along $\bar{\Gamma}$-$\bar{X}$ at ~500 K. However, one branch along $\bar{\Gamma}$-$\bar{M}$, which showed only a shallow minimum near $q=1.1$ Å^{-1} at 1200 K, softened drastically upon cooling to ~500 K (and even more so at ~300 K); see Fig. 8 b) [1]. This anomalous behavior was attributed to a giant Kohn anomaly [2]. Based on a comparison with calculations by [6], the anomalous mode was identified with a mainly longitudinal mode [2]. The considerable energy spread (see the vertical bars in Fig. 8 b)) indicates an overdamped mode. The L branch approaches zero energy for $q>1.0$ Å^{-1} (i.e., before reaching $\bar{M}$). This position of the soft mode in the surface Brillouin zone (SBZ) is consistent with the observation of an incommensurate Bragg scattering seen by He diffraction [1]; see also [4].

Fig. 9 [1] shows surface phonon dispersion curves along $\bar{\Gamma}$-$\bar{M}$ at $T\approx220$ K, i.e., for the high-temperature limit of the LT surface. At this temperature, a new set of dispersion curves appears with a new zone boundary at $q=0.7$ Å^{-1} [1, 7]. The exact assignment of the polarization vectors was not possible, since the two modes seemed to merge in the intermediate region suggesting hybridization [1]. This situation is discussed in terms of an avoided crossing of Rayleigh and longitudinal modes in the calculation by [14]. The experimentally determined zone-center phonon energy of ~10 meV [1] is lower than the lowest optical modes calculated for 0 K by [8], namely ~16 meV for the shear vertical vibration and ~12 meV for the nearly degenerate in-plane longitudinal waves along ⟨011⟩ and ⟨0$\bar{1}$1⟩ [8]. These optical L modes at $E(\bar{\Gamma})\approx12$ meV [8] might correspond to the three weak signals shown as dashed circles in Fig. 9 [1].

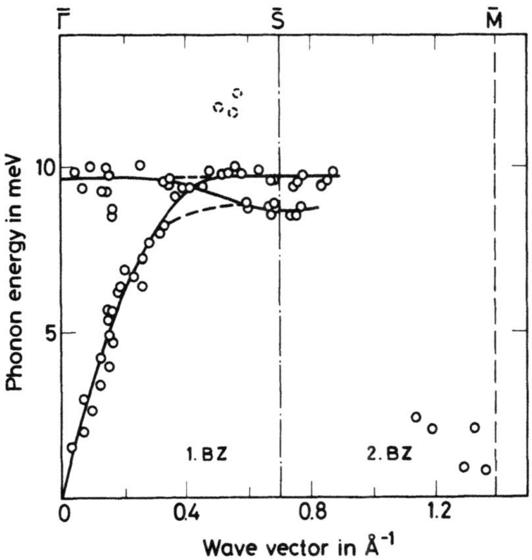

Fig. 9. Surface phonon dispersion curves of LT W(100) taken at 220 K along the ⟨011⟩ ($\bar{\Gamma}$-$\bar{M}$) direction of the HT surface. Dashed and solid curves indicate two possible groupings of the data points [1].

LT and HT W(100) surfaces have been studied by electron energy loss spectroscopy (EELS). Using primary electrons at $E_p = 4.5$ eV impinging along the ⟨011⟩ direction under $\vartheta_i = 60°$ on the reconstructed surface at ~100 K, an energy loss of 36±1 meV was found at $\bar{\Gamma}$ [9]; see also [10]. This loss peak corresponds to a surface vibration with an energy significantly above the highest bulk acoustic phonon frequencies which range near 28 meV for longitudinal acoustic phonons at the N point. The EELS peak at 36 meV shifted to 42 meV and weakened by about an order of magnitude on varying the exit angle from the specular position $\vartheta_e = \vartheta_i = 60°$ to $\vartheta_e = 45°$ [9]. A molecular dynamics (MD) simulation conducted by [11] suggested that EELS should find loss peaks in the 20 meV region in addition to a calculated, reconstruction-related mode near 31 meV. It exists only with the reconstructed surface [11] as does the EELS peak at ~36 meV found by [9, 10]; see also the MD study [12]. According to a study by [13], who used an effective lattice Hamiltonian, there are two higher longitudinal surface modes at 28.9 and 32.1 meV. The 32.1 meV mode moves the second layer by a factor of ~3.5 more than the uppermost layer. It was tentatively identified [13] with the EELS peak at 36 meV observed by [10]. However, the rather large discrepancy between calculated and measured mode energies renders this identification questionable [9].

References for 1.1.2.2.2:

[1] Ernst, H.-J.; Hulpke, E.; Toennies, J. P. (Phys. Rev. [3] B **46** [1992] 16081/105).
[2] Hulpke, E. (Nucl. Instrum. Methods Phys. Res. B **58** [1991] 347/53).
[3] Hulpke, E. (J. Electron Spectrosc. Relat. Phenom. **54/55** [1990] 299/308).
[4] Ernst, H.-J.; Hulpke, E.; Toennies, J. P. (Phys. Rev. Lett. **58** [1987] 1941/4).
[5] Wang, C. Z.; Parrinello, M.; Tosatti, E.; Fasolino, A. (Surf. Sci. **211/212** [1989] 323/8).
[6] Wang, X. W.; Weber, W. (Phys. Rev. Lett. **58** [1987] 1452/5).
[7] Toennies, J. P. (Superlattices Microstruct. **7** [1990] 193/200).
[8] Reinecke, T. L.; Ying, S. C. (Phys. Rev. Lett. **58** [1987] 242/5).

[9] Erskine, J. L.; Woods, J. P.; Kulkarni, A. D.; de Wette, F. W. (J. Electron Spectrosc. Relat. Phenom. **44** [1987] 27/36).
[10] Woods, J. P.; Erskine, J. L. (J. Vac. Sci. Technol. A **4** [1986] 1414/5).

[11] Wang, C. Z.; Fasolino, A.; Tosatti, E. (Phys. Rev. Lett. **59** [1987] 1845/8).
[12] Wang, C. Z.; Tosatti, E.; Fasolino, A.; Parrinello, M. (Surf. Sci. **189/190** [1987] 679/83).
[13] Fasolino, A.; Tosatti, E. (Phys. Rev. [3] B **35** [1987] 4264/83).
[14] Reinecke, T. L.; Ying, S. C. (Phys. Rev. [3] B **48** [1993] 5679/81).

1.1.2.2.3 Simulation of the Surface Dynamics

There are several models which fit the bulk phonon data. Yet, the calculated surface dynamical properties depend strongly on the particular model used [1]. The force constant model used in [2] for the unreconstructed surface incorporates radial and tangential pair interactions up to third-nearest neighbors and two angle-bending interactions. The force constants were determined by [2] by least-squares fits to the bulk phonon dispersion curves of [3], to the elastic constants of [4], and to restoring forces for surface atoms determined from the total energy calculations of [5] and [6]. The fit to experimental data distinguishes study [7] from earlier calculations by [1, 8 to 12]. But a similar approach was followed by [13] and [14]. Kulkarni and de Wette [2] demonstrated that an unmodified bulk interaction model does not account for the surface properties. The force constants in the surface layer differ from those in the bulk only for about half-filled 5d electron bands [15]. Fitting the parameters to the $\overline{\Gamma}$-$\overline{M}$ spectra measured by [16] allowed the identification of the upper, normally behaving and the lower, anomalously behaving branches in the measured data set as Rayleigh (R) and longitudinal (L) waves, respectively. On HT W(100), the calculated dispersion of the shear-horizontal (SH) wave nearly coincides with the dispersion of the anomalous L mode [7] (these conclusions are consistent with molecular dynamics (MD) studies; see e.g. [17]).

The anomalous behavior of the (acoustic) mainly longitudinally polarized L wave along $\overline{\Gamma}$-$\overline{M}$ implies also an anomalous behavior of the L wave along $\overline{\Gamma}$-$\overline{X}$, unless the model of the HT W(100) surface includes third-nearest (and possibly higher) neighbor interactions [18]. Since Ernst et al. [16] did not find anomalous behavior along $\overline{\Gamma}$-$\overline{X}$, realistic models for the W(100) surface have to include interactions at least up to third-nearest neighbors. The main conclusion to be drawn from the anomalous L mode along $\overline{\Gamma}$-$\overline{M}$ is that nearest neighbor surface atoms repel each other [18].

The phonon eigenvectors in the vicinity of $\overline{\Gamma}$ have been calculated for HT W(100) in [19] and surface Brillouin zone edge phonon frequencies were calculated using a 20-layer slab in [20]. For plots of the local density of surface phonon states at $\overline{\Gamma}$, see [19]; for dispersion curves along $\overline{X}$-$\overline{M}$, see [20].

A comparison with the three-parameter model of [12] suggested that the surface phonons along $\overline{\Gamma}$-$\overline{M}$ are affected mainly by short-range interactions [7]. This conclusion was also arrived at in the nonorthogonal tight-binding (NTB) study [14]. According to this study the strong $\vec{q}$ dependence of the R and SH dispersion curves found by [11] (see Fig. 7, p. 32) implies that long-range forces only modify the instability of the surface layer, but are not the cause for the instability [14]. At this point, the results by [14] contradict those of the quite similar NTB study [11].

The surface vibrations of reconstructed W(100) have been studied using total-energy results [21]. This paper includes information on only two surface distortions, namely (i) a surface layer distortion parallel to the surface and having the symmetry of the reconstruction and (ii) a uniform surface layer relaxation perpendicular to the surface [21]. Paper [22] is an extension of this earlier work. All qualitative features of [21] are reproduced in [22] which in addition to the two surface distortions (i, ii) considered in [21] includes information on (iii) a distortion of the second layer with the symmetry of the reconstructed surface, (iv) a distortion perpendicular to the surface with the two atoms of the reconstructed cell moving out of phase, and (v) a distortion in which surface atoms move parallel to the surface and perpendicular to the [011] directions [22]. All total-energy results available from the literature (especially [5, 6, 23 to 26]) were used to calculate the strain and force constants $\alpha_n = R_n^{-1}[dU/dR]$ and $\beta_n = [d^2U/dR^2]$, respectively [22].

Values for the strain and force constants α and β (all in eV/Å^2) are given in the following table for the reconstructed LT surface [22]. The meaning of the subscripts of α and β is evident from **Fig. 10** [22].

$\alpha_{1s}^{+}=4.42$	$\alpha_{1s}^{-}=1.10$	$\alpha_{1s}^{\perp}=2.76$	$\alpha_{2s}=-0.30$
$\beta_{1s}=5.18$	$\beta_{1s}'=0.20$	$\beta_{2s}=-1.19$	$\beta_{3s}=5.06$

Force constants derived from least-squares fits of models to the HT W(100) data of [16] are given in [18].

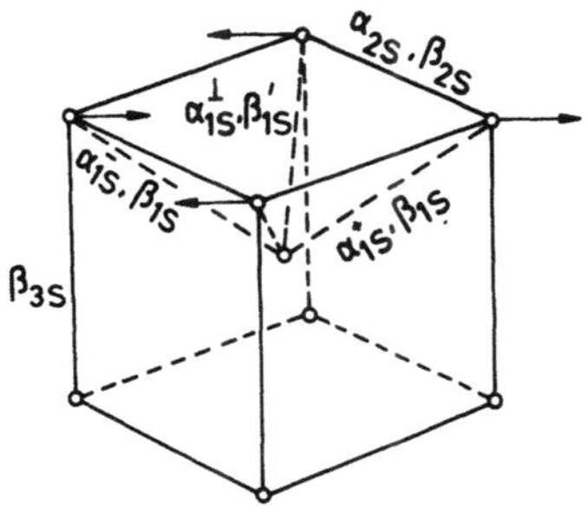

Fig. 10. This schematic indicates the surface force-constant parameters which were modified from their bulk values for the uppermost three W(100) layers in the calculations by [22]. The arrows indicate the directions of the displacements upon reconstruction for the topmost layer [22].

Qualitatively very similar results were obtained for the dispersion relations of the surface phonon modes regardless whether the surface strains α_n were included or omitted in the calculations for the reconstructed LT surface [22]; see also the very similar results obtained by [11] and shown in Fig. 7, p. 32. But inclusion of strains in the lattice-dynamical study modifies the earlier results obtained in [21] in so far as it suggests that the EELS peak at 36 meV (observed by [27]) is actually related to the subsurface mode at ~30 meV [22].

Phonon dispersion curves were calculated for reconstructed and unreconstructed surfaces in [13]; see the diagrams from 15-layer slab calculations given for $\hbar\omega < 35$ meV in [13, p. 4275]. For a 23-layer slab such curves were calculated for $\hbar\omega < 28$ meV as a function of d-electron density. While the surface modes behave normally for d-electron concentrations appropriate for Ta, they become increasingly softer if the d-electron density is increased; eventually they freeze in at $\bar{M}$ [14]. The SBZ boundary $\bar{M}$-$\bar{X}$ is a region of soft surface phonons [14]; see also Fig. 7, p. 32, summarizing the results of [11].

The following text concentrates on the low-lying modes which are most relevant with respect to the HT-LT phase transition. The dominant polarization character of the surface

References for 1.1.2.2.3 on p. 39

modes (in-plane or vertical) changes along the SBZ symmetry lines [13, p. 4275]. (Modes where the squared first-layer component of the eigenvector exceeds 30% were taken by [13] as surface modes.) The longitudinal $\bar{M}$ modes have a small uniform vertical component and the transversal vertical $\bar{M}$ modes have a small uniform longitudinal component; these components, which are important for He atom scattering, are estimated to be ~10% at 0 K and are expected to decrease with increasing temperature [28].

Using a two-dimensional Hamiltonian, the critical dynamics was studied by a combined analytical and Monte Carlo (MC) technique for $T_c < T \leq 5\,T_c$ [29]; the T range was extended to below T_c in the MC study [30]. The Hamiltonian used reproduced the main features of reconstruction and dynamics. Its potential energy terms were square- and fourth-order in the displacements. In addition to on-site terms, it contained harmonic terms for nearest and next-nearest neighbor interactions. Cubic-order interactions did not change the total energy, but enabled one to simulate phonon band gapping which reflects the inequivalence of the ⟨011⟩ and ⟨0$\bar{1}$1⟩ directions in the c(2×2) phase. At low temperatures, the behavior predicted by the quasi-harmonic approximation (QHA) agrees well with the MC simulations, but near T_c there are considerable deviations from the QHA [30]. The calculated temperature dependence for the dispersion of the soft L modes at $T \geq T_c$ is in fair agreement with He atom scattering data; see **Fig. 11** [29]. On approaching T_c from above, the soft L mode becomes overdamped near $\bar{M}$ and the above mentioned central peak develops for the L modes which soften more dramatically than the transverse SH modes [29]. According to the MD study [28], however, the L modes soften and broaden only in a small temperature range close to T_c. The transverse SH modes, on the other hand, being associated with the phase fluctuations, soften, get overdamped, and are totally ill-defined in the range 0.7 T_c to 1.1 T_c [28].

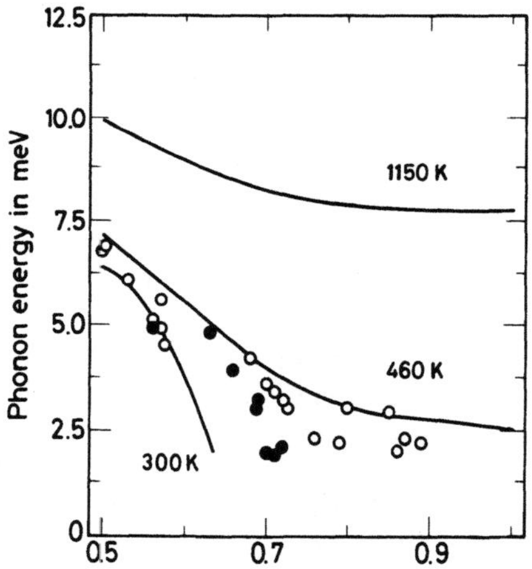

Fig. 11. Dispersion relation of the soft longitudinal phonon-mode branch for $0.5\,q_0 \leq q \leq q_0$ ($\vec{q}_0 = (\pi/a\ \pi/a)$) calculated for 300, 460, and 1150 K [29] and compared with He scattering data from [16] (see also Fig. 8b), p. 34) taken at 280 K (solid circles) and 450 K (open circles) [29].

The MD simulations of the temperature dependence of the lattice dynamics [17, 28] and the studies of the lattice dynamics at 0 K [8, 13] use the same type of effective lattice Hamiltonian (the interatomic forces assumed at the surface differ strongly from those in the bulk [17]; see the potential given in [13, p. 4266]). In [28] the same detailed W-W interactions are used as in the study [31] of static critical properties. At low T, the phonon spectra for the R, L, and the (transverse) SH modes peak at energies of 9, 10, and 5 meV, respectively [28] in agreement with the values for the lowest-lying optical $\overline{\Gamma}$ phonons calculated for 0 K: $\hbar\omega_R = 9.2$ meV, $\hbar\omega_L = 10.3$ meV, and $\hbar\omega_T = 4.7$ meV [13]. The two higher-lying longitudinal modes at 28.9 and 32.1 meV [13] have been mentioned above in conjunction with the EELS peak observed at 36 meV [27].

A frequency of ~4.5 THz (~18.6 meV) was derived for the vertical vibrations of the surface atoms by tight-binding and local density methods from the dependence of the relaxation energy on the magnitude of the relaxation of the first interlayer spacing [26].

References for 1.1.2.2.3:

[1] Black, J. E.; Campbell, D. A.; Wallis, R. F. (Surf. Sci. **115** [1982] 161/82).
[2] Kulkarni, A. D.; de Wette, F. W. (Surf. Sci. **186** [1987] 469/81).
[3] Chen, S. H.; Brockhouse, B. N. (Solid State Commun. **2** [1964] 73/7).
[4] Featherston, F. H.; Neighbours, J. R. (Phys. Rev. [2] **130** [1963] 1324/33).
[5] Fu, C. L.; Freeman, A. J.; Wimmer, E.; Weinert, M. (Phys. Rev. Lett. **54** [1985] 2261/4).
[6] Fu, C. L.; Ohnishi, S.; Wimmer, E.; Freeman, A. J. (Phys. Rev. Lett. **53** [1984] 675/8).
[7] Kulkarni, A. D.; de Wette, F. W.; Prade, J. (Phys. Rev. [3] B **36** [1987] 6190/3).
[8] Fasolino, A.; Santoro, G.; Tosatti, E. (Phys. Rev. Lett. **44** [1980] 1684/7).
[9] Fasolino, A.; Santoro, G.; Tosatti, E. (Surf. Sci. **125** [1983] 317/25).
[10] Roelofs, L. D.; Wendelken, J. F. (Phys. Rev. [3] B **34** [1986] 3319/36).

[11] Wang, X. W.; Weber, W. (Phys. Rev. Lett. **58** [1987] 1452/5).
[12] Castiel, D.; Dobrzynski, L.; Spanjaard, D. (Surf. Sci. **59** [1976] 252/78).
[13] Fasolino, A.; Tosatti, E. (Phys. Rev. [3] B **35** [1987] 4264/83).
[14] Joubert, D. P. (J. Phys. C: Solid State Phys. **21** [1988] 4233/8).
[15] Joubert, D. P. (J. Electron Spectrosc. Relat. Phenom. **39** [1986] 191/3).
[16] Ernst, H.-J.; Hulpke, E.; Toennies, J. P. (Phys. Rev. Lett. **58** [1987] 1941/4).
[17] Wang, C. Z.; Fasolino, A.; Tosatti, E. (Europhys. Lett. **7** [1988] 263/8).
[18] Kulkarni, A. D.; de Wette, F. W.; Prade, J. (J. Electron Spectrosc. Relat. Phenom. **44** [1987] 397/402).
[19] Bortolani, V.; Nizzoli, F.; Santoro, G.; Tosatti, E. (Solid State Commun. **26** [1978] 507/10).
[20] Black, J. E.; Campbell, D. A.; Wallis, R. F. (J. Phys. Colloq. **42** [1981] C 6-837/C 6-839).

[21] Reinecke, T. L.; Ying, S. C. (Phys. Rev. Lett. **58** [1987] 242/5).
[22] Reinecke, T. L.; Ying, S. C. (Phys. Rev. [3] B **43** [1991] 12234/45).
[23] Fu, C. L.; Freeman, A. J. (Phys. Rev. [3] B **37** [1988] 2685/8).
[24] Singh, D.; Wei, S.-H.; Krakauer, H. (Phys. Rev. Lett. **57** [1986] 3292/5).
[25] Singh, D.; Krakauer, H. (Phys. Rev. [3] B **37** [1988] 3999/4006).
[26] Chan, C. T.; Louie, S. G. (Phys. Rev. [3] B **33** [1986] 2861/4).
[27] Woods, J. P.; Erskine, J. L. (J. Vac. Sci. Technol. A **4** [1986] 1414/5).
[28] Wang, C. Z.; Fasolino, A.; Tosatti, E. (Phys. Rev. [3] B **37** [1988] 2116/22).
[29] Han, W. K.; Ying, S. C.; Sahu, D. (Phys. Rev. [3] B **41** [1990] 4403/9).
[30] Han, W. K.; Ying, S.-C. (Phys. Rev. [3] B **46** [1992] 1849/52).

[31] Wang, C. Z.; Parrinello, M.; Tosatti, E.; Fasolino, A. (Europhys. Lett. **6** [1988] 43/8).

1.1.3 Surface Electronic Structure

1.1.3.1 Introduction

The surface electronic structure of W(100) has been the subject of numerous theoretical papers. As far as experimental studies are concerned, mainly those using photoemission (see pp. 178/82) contributed to the present knowledge about the surface band structure below E_F. The groundbreaking work, however, were field emission studies reporting on what is now known as the "Swanson hump" (see pp. 144/5). Inverse photoemission and photofield emission (see pp. 159/61) gave information about the energy range between E_F and the vacuum level. LEED intensity profiles (see pp. 218/20) and secondary electron emission spectra helped to elucidate the electronic structure of W(100) above the vacuum level.

The considerable interest in the surface electronic structure of W(100) stems from the as-yet unsolved problem of reconstruction of this surface. There is broad agreement as to the electronic origin of the instability of the (1 × 1) surface (for a discussion, see [1]), but opinion is divided on whether localized (Jahn-Teller-like) or charge-density-wave (CDW)-type mechanisms drive the reconstruction transition; see e.g. [2]. The UPS study [3] showed that surface states crossing E_F near $\overline{M}$ are split by the phase transition (see [4]). Off $\overline{\Sigma}$ $(=\overline{\Gamma}\text{-}\overline{M})$, the Fermi line is very curved, implying poor Fermi-line nesting [3]. Surface states and resonances from a large part of the surface Brillouin zone (SBZ) contribute to the DOS peak at E_F involved in the reconstruction. But the UPS results mentioned did not allow to single out *individual* surface states as driving the transition [5].

The surface potential barrier, which is important for understanding the electron reflectivity data, is dealt with in Section 5.7.3, pp. 214/23. In the following paragraph, some values for the surface core-level shifts are given. Surface density of states (DOS) curves are treated on pp. 41/2. Dispersion relations of these states and their charge density contour maps will follow on pp. 43/52 and pp. 53/5, respectively, and finally the Fermi surface will be dealt with on pp. 55/6.

References for 1.3.1.1:

[1] Inglesfield, J. E. (Prog. Surf. Sci. **20** [1985] 105/64, 132/8).
[2] Smith, K. E.; Kevan, S. D. (Prog. Solid State Chem. **21** [1991] 49/131, 81/8).
[3] Campuzano, J. C.; Inglesfield, J. E.; King, D. A.; Somerton, C. (J. Phys. C **14** [1981] 3099/113).
[4] Inglesfield, J. E. (Vacuum **31** [1981] 663/9).
[5] Inglesfield, J. E. (Rep. Prog. Phys. **45** [1982] 223/84).

1.1.3.2 Surface Core-Level Shifts

The $4f_{7/2}$ core levels of the surface atoms are shifted towards lower binding energies in the case of W surfaces; see pp. 185/6. Values for this surface core-level shift (SCLS) of −0.35 to −0.40 eV were measured for the (1 × 1) unreconstructed high-temperature phase of W(100) [1 to 4]. For the c(2 × 2) reconstructed low-temperature phase, a similar value for the SCLS of −0.35 eV was found [1, 2, 5]. For more detailed information and a comparison with theoretical results, see Section 5.5.3, pp. 185/93.

References for 1.1.3.2:

[1] Jupille, J.; Purcell, K. G.; King, D. A. (Phys. Rev. [3] B **39** [1989] 6871/9).
[2] Guillot, C.; Desjonquères, M. C.; Chauveau, D.; Tréglia, G.; Lecante, J.; Spanjaard, D.; Tran Minh Duc (Solid State Commun. **50** [1984] 393/6).

[3] Wertheim, G. K.; Citrin, P. H.; van der Veen, J. F. (Phys. Rev. [3] B **30** [1984] 4343/7).
[4] van der Veen, J. F.; Himpsel, F. J.; Eastman, D. E. (Solid State Commun. **40** [1981] 57/60).
[5] Mullins, D. R.; Lyman, P. F. (Surf. Sci. Lett. **285** [1993] L 473/L 478).

1.1.3.3 Electronic Surface Density of States (SDOS)

The coordination number on the open W(100) surface is half that in the bulk; hence, the d bands of the surface atoms are narrowed. The resulting density-of-states (DOS) peak coincides with the Fermi energy E_F and occurs at a minimum of the bulk DOS. Relaxation enhances the interaction of the surface atoms with the substrate and increases the DOS near the d-band edges [1, 2]. UPS studies ([3, 4]) suggest that *individual* surface states play no part in driving the reconstruction transition, but surface states and resonances from a large part of the SBZ contribute to the DOS peak at E_F, which may be responsible for the instability of the (1×1) surface [1, p. 257]; see also [5]. Self-consistent calculations of the electronic structure at $\bar{\Gamma}$ indicated a large DOS peak which lies above E_F for Ta(100), while it is close to E_F for W(100). Ta(100) does not reconstruct, while W(100) does; therefore, Krakauer [6] concluded that in the case of W(100) surface states destabilize the (1×1) surface [6].

The surface DOS, calculated self-consistently in a seven-layer slab LAPW calculation, actually shows two DOS peaks for the outermost layer: one peak near 0.5 eV below E_F is partly related to a short line of surface states around $\bar{\Gamma}$ and partly to a double line of surface resonances for $k_{\parallel} > 0$. The second peak, located at 4.4 eV below E_F, stems from a surface resonance [7, 8].

An earlier calculation of the density of initial states confined to the surface [9] covered neither surface states nor surface resonances. It consisted of a mere scaleup of the bulk DOS, intended to facilitate comparison with photoemission from the surface [9]. For a recent attempt to correlate electron energy loss spectra from W(100) with those of bulk- and surface-confined DOS, see [10].

The central peak in the d-band surface DOS near E_F, predictable by tight-binding (LCAO) calculations, seems to be associated with the strong emission seen in the UPS studies [11 to 14]. It is not due to spin-orbit coupling which opens a gap within which a surface state might develop (as suggested by [15]). In fact, the peak would be less marked for larger spin-orbit couplings [16]. This conclusion was also drawn in the tight-binding study [17].

The DOS peak is due to dangling bonds and remains confined to a narrow range around E_F, even for vertical atomic displacements. These disturb the local symmetry at the surface to a minor degree only in contrast to horizontal displacements which broaden this peak [18] (similar arguments were used by [7] in the context of the Swanson hump state with respect to its sensitivity to Debe-King-type lateral displacements). The inward relaxation of the (1×1) surface increases the DOS near the band edges [2]. A surface contraction of $\delta d_{12}/d \approx -10\%$ can be explained by d-band electron energy gains without surface state contributions [19]. These tight-binding calculations underline the importance of electron-phonon interaction and of the surface DOS near E_F, destabilizing the (1×1) structure and favoring zigzag-chain formation. The energy gained by the reconstruction is attributed to enhanced bonding due to pairing of surface atoms [19]; see also [20, 21]. The two-dimensional bands found in [22] were nondispersive close to E_F over a significant region of the SBZ indicating a very high two-dimensional DOS close to E_F. Since the electrons of these flat bands are slow, strong nonadiabatic effects can be expected [23]; see also

References for 1.1.3.3 on pp. 42/3

[24]. Due to (local) electronic interactions (of dangling bonds) at the surface, the $\overline{M}_5$ phonon mode will soften more than do other phonon modes [18]. For the strong coupling between $\overline{M}_5$ phonon modes and the surface states, see also the LAPW study [25].

A local-density-functional five-layer-slab LAPW calculation showed a profound effect of the c(2×2) reconstruction on the surface DOS: the DOS peak around E_F of the (1×1) surface is split into two peaks separated by 1.3 eV, resulting in a considerably reduced electronic energy [25]. For a critical discussion of this paper, see [26, 27].

The surface layer DOS of unreconstructed W(100) shows a distinct peak due to surface states [26]. This DOS peak is eliminated as the surface reconstructs (~0.27 Å shifts); the associated removal of dangling bonds indicates a local (Jahn-Teller-like) driving mechanism for the reconstruction rather than the CDW mechanism invoked by [28]; [26].

Other, obviously obsolete DOS calculations are given in [29 to 31]. For the DOS of the outermost, second, third, fourth, and central layers of a 27-layer slab, see the tight-binding relativistic energy band calculation [32].

References for 1.1.3.3:

[1] Inglesfield, J. E. (Rep. Prog. Phys. **45** [1982] 223/84).
[2] Desjonquères, M. C. (J. Phys. Lett. **40** [1979] L 249/L 252).
[3] Campuzano, J. C.; Inglesfield, J. E.; King, D. A.; Somerton, C. (J. Phys. C **14** [1981] 3099/113).
[4] Campuzano, J. C.; King, D. A.; Somerton, C.; Inglesfield, J. E. (Phys. Rev. Lett. **45** [1980] 1649/52).
[5] Inglesfield, J. E. (Vacuum **31** [1981] 663/9).
[6] Krakauer, H. (Phys. Rev. [3] B **30** [1984] 6834/40).
[7] Posternak, M.; Krakauer, H.; Freeman, A. J.; Koelling, D. D. (Phys. Rev. [3] B **21** [1980] 5601/12).
[8] Wimmer, E.; Krakauer, H.; Freeman, A. J. (Adv. Electron. Electron Phys. **65** [1985] 357/434, 391/404).
[9] Christensen, N. E.; Feuerbacher, B. (Phys. Rev. [3] B **10** [1974] 2349/72).
[10] Czyzewski, J. J.; Krajniak, J.; Klein, S. (Acta Phys. Pol. A **81** [1992] 67/72).

[11] Plummer, E. W.; Gadzuk, J. W. (Phys. Rev. Lett. **25** [1970] 1493/5).
[12] Feuerbacher, B.; Fitton, B. (Phys. Rev. Lett. **29** [1972] 786/9).
[13] Feuerbacher, B.; Fitton, B. (Phys. Rev. Lett. **30** [1973] 923/6).
[14] Waclawski, B. J.; Plummer, E. W. (Phys. Rev. Lett. **29** [1972] 783/6).
[15] Sturm, K.; Feder, R. (Solid. State Commun. **14** [1974] 1317/20).
[16] Friedel, J. (J. Phys. [Paris] **37** [1976] 883/94).
[17] Desjonquères, M. C.; Cyrot-Lackmann, F. (J. Phys. F **6** [1976] 567/72).
[18] Tomásek, M.; Pick, S. (J. Phys. Colloq. [Paris] **45** [1984] C 5-125/C 5-129).
[19] Terakura, I.; Terakura, K.; Hamada, N. (Surf. Sci. **103** [1981] 103/17).
[20] Terakura, K.; Terakura, I.; Teraoka, Y. (Surf. Sci. **86** [1979] 535/41).

[21] Joubert, D. P. (J. Phys. C **20** [1987] 1899/907).
[22] Smith, K. E.; Elliott, G. S.; Kevan, S. D. (Phys. Rev. [3] B **42** [1990] 5385/7).
[23] Smith, K. E.; Kevan, S. D. (Phys. Rev. [3] B **43** [1991] 3986/93, 3991/2).
[24] Smith, K. E.; Kevan, S. D. (Prog. Solid State Chem. **21** [1991] 49/131, 81/8).
[25] Fu, C. L.; Freeman, A. J.; Wimmer, E.; Weinert, M. (Phys. Rev. Lett. **54** [1985] 2261/4).
[26] Singh, D.; Wei, S.-H.; Krakauer, H. (Phys. Rev. Lett. **57** [1986] 3292/5).
[27] Fu, C. L.; Freeman, A. J. (Phys. Rev. [3] B **37** [1988] 2685/8).
[28] Tosatti, E. (Solid State Commun. **25** [1978] 637/40).

[29] Laks, B.; Gonçalves da Silva, C. E. T. (Solid State Commun. **25** [1978] 401/3).
[30] Gurman, S. J.; Pendry, J. B. (Phys. Rev. Lett. **31** [1973] 637/9).
[31] Kawajiri, M.; Hermanson, J.; Schwalm, W. (Solid State Commun. **21** [1977] 327/9).
[32] Grise, W. R.; Dempsey, D. G.; Kleinman, L.; Mednick, K. (Phys. Rev. [3] B **20** [1979] 3045/50).

1.1.3.4 Surface States and Surface Resonances

1.1.3.4.1 Surface States and Resonances above E_F

Above the Vacuum Level. A band of surface resonances was identified by angle-resolved secondary electron emission (SEE) spectroscopy; it appeared for $\bar{\Gamma}$ near 10 eV above E_F (~5.5 eV above the vacuum level) and dispersed downward along $\bar{\Delta}$ (= $\bar{\Gamma}$-$\bar{X}$) to ~8.5 eV above E_F for $\bar{X}$ [1]. A structure found in the specular LEED beam near E_p = 4 eV at room temperature [2] was attributed to a surface resonance [3] and later resolved into a sequence of peaks [4]. For the dispersion of these resonances, see [5]. An apparent discrepancy was noted [6] between the LEED intensity profiles of [2 to 4] and the SEE results [7 to 9] for $k_{\parallel} < 0.6\pi/a$ ($k_{\parallel}$ values run along $\bar{\Delta}$) [6]. The LEED intensities for these $k_{\parallel}$ values are strongly damped, considerably broadened, and the resonance-like features appear to disperse to an upper bulk band gap edge [10]. The discrepancy between the LEED intensity profiles and SEE was due to an incorrect evaluation of the intensity profiles (the disagreement increased with decreasing $k_{\parallel}$: for $k_{\parallel} = 0.75\pi/a$ and amounted to ~0.25 eV, while it was ~0.75 eV for $k_{\parallel} = 0.5\pi/a$) [10].

Resonance scattering on the average involves a large number of scattering events (multiple scattering between substrate and surface potential barrier) and affects the spin polarization of the reflected electrons [5]. Hence, spin polarization profiles may be used to identify resonances.

Between Vacuum Level and E_F. An inverse photoemission (IPE) study found a manifold of unoccupied states between E_F and the vacuum level; see **Fig. 12**, p. 44. Because of their strong adsorption sensitivity, the features B to G are attributed to transitions to surface states or resonances [11].

The rather broad feature D is a resonance. State G is identified with an image state (a state in which electrons are bound by their image potential in the direction normal to the surface and behave like free electrons in the direction parallel to the surface) [11]. Its parabolic dispersion (see Fig. 12) corresponds to an effective mass of $m^* = (1.8 \pm 0.4)$ m_e. For $\bar{\Gamma}$, this image-potential surface state occurred near 4.2 eV above E_F (at a binding energy of $E_B = 0.4 \pm 0.2$ eV) [11]. A similar image state at 3.9 eV above E_F (E_B = 0.7 eV) was found in the IPE study [12].

Feature E in Fig. 12 disperses downward and approaches E_F near the middle of $\bar{\Sigma}$ where the UPS studies [13 to 15] had found a state dispersing from below [11]. Therefore, one may expect that feature E is strongly affected by the reconstruction. Instead, new states appeared, and feature E remained unchanged after (hydrogen-induced) reconstruction [11].

There is disagreement between the surface state band F in Fig. 12 and the (x^2-y^2)-type $\bar{\Sigma}_2$ band of surface states (see below), calculated to disperse upward and to cross E_F near (0.95 ± 0.25) $Å^{-1}$ in [16 to 22]; [11].

In a more recent IPE study, a surface state was found just above the Fermi level, approaching halfway along $\bar{\Sigma}$ and dispersing upwards to ~2 eV above the Fermi level at $\bar{M}$ [23].

References for 1.1.3.4.1 on pp. 44/5

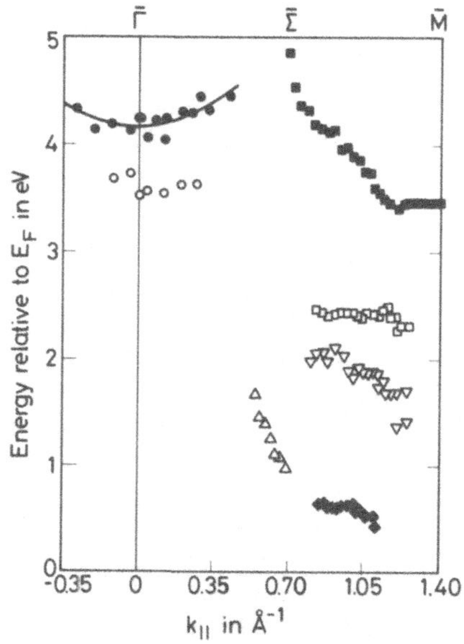

Fig. 12. Experimental band structure of the unreconstructed W(100) surface above the Fermi level found by inverse photoemission (isochromat mode). ■: state A (bulk feature), □, ▽, ○, △, ◆, ●: states B to G [11].

A nondispersing band of surface features seen in IPE around $\bar{\Gamma}$ at 0.25 eV above E_F may be due to mixing of the Swanson hump state (see below) with bulk bands [12].

References for 1.1.3.4.1:

[1] Willis, R. F.; Feuerbacher, B.; Christensen, N. E. (Phys. Rev. Lett. **38** [1977] 1087/91).
[2] Edwards, D. E., Jr.; Probst, F. M. (J. Chem. Phys. **56** [1972] 3184/5).
[3] McRae, E. G.; Wheatley, G. H. (Surf. Sci. **29** [1972] 342/54).
[4] Adnot, A.; Carette, J. D. (Phys. Rev. Lett. **38** [1977] 1084/7).
[5] McRae, E. G. (Rev. Mod. Phys. **51** [1979] 541/68).
[6] McRae, E. G. (Phys. Rev. [3] B **17** [1978] 907/8).
[7] Willis, R. F.; Feuerbacher, B.; Fitton, B. (Solid State Commun. **18** [1976] 1315/9).
[8] Willis, R. F.; Christensen, N. E. (Phys. Rev. [3] B **18** [1978] 5140/61).
[9] Christensen, N. E.; Willis, R. F. (J. Phys. C **12** [1979] 167/207, 185/90).
[10] Willis, R. F. (Phys. Rev. [3] B **17** [1978] 909/11).

[11] Krainsky, I. L. (J. Vac. Sci. Technol. [2] A **5** [1987] 735/8).
[12] Drube, W.; Straub, D.; Himpsel, F. J.; Soukiassian, P.; Fu, C. L.; Freeman, A. J. (Phys. Rev. [3] B **34** [1986] 8989/92).
[13] Holmes, M. W.; Gustafsson, T. (Phys. Rev. Lett. **47** [1981] 443/6).
[14] Campuzano, J. C.; King, D. A.; Somerton, C.; Inglesfield, J. E. (Phys. Rev. Lett. **45** [1980] 1649/52).
[15] Campuzano, J. C.; Inglesfield, J. E.; King, D. A.; Somerton, C. (J. Phys. C **14** [1981] 3099/113).
[16] Bullett, D. W.; Stephenson, P. C. (Solid State Commun. **45** [1983] 47/9).
[17] Stephenson, P. C.; Bullett, D. W. (Surf. Sci. **139** [1984] 1/15).

[18] Grise, W. R.; Dempsey, D. G.; Kleinman, L.; Mednick, K. (Phys. Rev. [3] B **20** [1979] 3045/50).
[19] Ohnishi, S.; Freeman, A. J.; Wimmer, E. (Phys. Rev. [3] B **29** [1984] 5267/78).
[20] Posternak, M.; Krakauer, H.; Freeman, A. J.; Koelling, D. D. (Phys. Rev. [3] B **21** [1980] 5601/12).
[21] Posternak, M.; Krakauer, H.; Freeman, A. J. (Phys. Rev. [3] B **25** [1982] 755/61).
[22] Mattheiss, L. F.; Hamann, D. R. (Phys. Rev. [3] B **29** [1984] 5372/81).
[23] Collins, I. R.; Laine, A. D.; Andrews, P. T. (J. Phys. Cond. Matter **4** [1992] 2891/901).

1.1.3.4.2 Surface States and Resonances below E_F or Dispersing through E_F

1.1.3.4.2.1 Synoptic Characterization of States and Resonances; Dispersion Curves

High-Temperature Phase. Numerous surface states and resonances were found in experimental and theoretical studies. However, the picture is considerably simplified when attention is confined to those states and resonances, where the charge density is concentrated in the outermost atomic layer [1]. The calculations of [2] indicate just two such states along $\bar{\Sigma}$ ($=\bar{\Gamma}$-$\bar{M}$), identifiable with features observed by UPS [1]: near $\bar{\Gamma}$ at 0.32 eV below E_F there is a surface state of d_{z^2} character and a $\bar{\Sigma}_2$ state band starting near $\bar{\Gamma}$ at about 0.5 eV below E_F. All other states found extend over several layers [1].

Elliott et al. [3] studied the surface electronic structure along the $\bar{\Sigma}$ ($=\bar{\Gamma}$-$\bar{M}$), $\bar{Y}$ ($=\bar{M}$-$\bar{X}$), and $\bar{\Delta}$ ($=\bar{\Gamma}$-$\bar{X}$) lines using UPS at room temperature [3]. A similarly complete map of the W(100) surface electronic structure had been reported earlier by Campuzano et al. [4, 5] who found significantly different results, however. But there is broad agreement between [3] and the more limited UPS study [6] which was confined to the $\bar{\Sigma}$ line [3].

The surface states or resonances found by UPS are given in **Fig. 13** (taken from [3]) in addition to the bulk band structure projected onto the (100) SBZ. These bulk bands were calculated using a Slater-Koster three-center nonorthogonal tight-binding scheme fitted

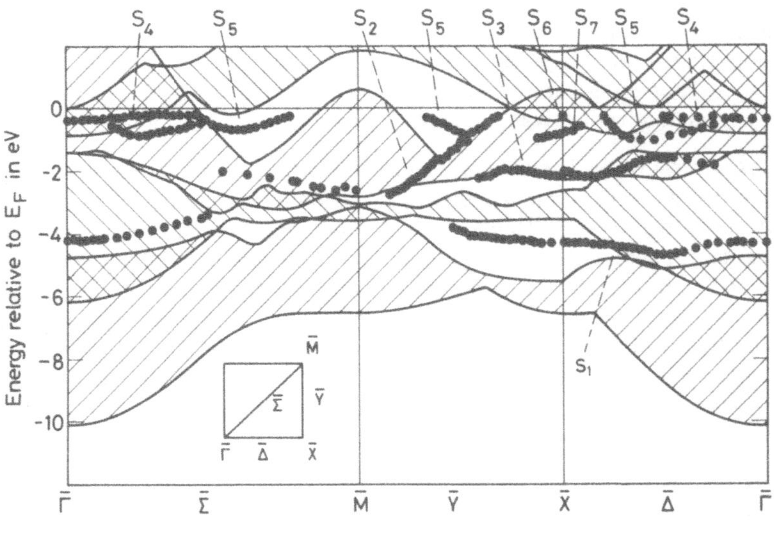

Fig. 13. Bulk band structure projected onto the (100) surface Brillouin zone. The experimental points (from UPS) show the dispersion of the surface localized states [3].

References for 1.1.3.4.2.1 on p. 48

to an APW band structure modified by inclusion of the spin-orbit interaction [3]. Spin-orbit interaction plays a key role in understanding the behavior of the surface state bands labeled S_4 (the Swanson hump state) and S_5 in Fig. 13 [3]. The qualitative agreement with the fully relativistic calculations [7] shows that the gap extending near E_F along $\bar{\Sigma}$ in Fig. 13 results from a superposition of a broad and a narrow gap having odd and even symmetry, respectively. Each of these gaps supports a surface state band. According to scalar relativistic calculations, the dispersions of these bands nearly coincide. Therefore, these bands hybridize and, because of the large spin-orbit interaction, may be split such that they no longer resemble the bands calculated in scalar relativistic theory. Two bands of mixed symmetry result, one in the middle of the gap and the other close to E_F, such as the bands labeled S_4 and S_5 in Fig. 13 which are the focus of discussions about the W(100) reconstruction. Since these bands are sensitive to spin-orbit interaction, the authors conclude that spin-orbit induced hybridization of surface state bands has a profound effect on the c(2×2) reconstruction [3].

The dispersion of state S_5 around $\bar{M}$ is nearly isotropic. Parabolic extrapolation suggests a band origin at 0.25 eV above E_F; the effective mass is $m^*=0.8\ m_e$. Around $\bar{X}$, on the other hand, the band is distinctly anisotropic with effective masses of $m^*=0.2\ m_e$ and $0.5\ m_e$ along $\bar{\Delta}$ and $\bar{Y}$, respectively. Here, the band origin is predicted at 0.45 eV above E_F [3].

The somewhat ambiguous spectra along $\bar{\Delta}$ suggest that the bands S_3 and S_5 either exchange intensity or hybridize, leaving an avoided crossing some 0.3 eV wide (this latter possibility is plotted in Fig. 13) [3].

The dispersion of the surface state band labeled S_1 in Fig. 13 had been studied earlier by UPS [4 to 6, 8]. For $\bar{\Gamma}$ this state occurs near 4.3 eV below E_F [6] and is composed of d_{z^2} (and/or s) orbitals [4], [8]; s-, p_z-, or d_{z^2}-orbital character was deduced for this state by [9].

The surface state bands labeled S_6 and S_7 in Fig. 13 were found only near X [3].

The bands labeled S_2 and S_3 in Fig. 13 are best defined along $\bar{Y}$, but they extend also into the other directions studied. The observation of band S_2 is in fair agreement with the UPS results reported in [4, 5] and can be identified with a band calculated to occur in this region by [10]. Band S_3, existing in a narrow gap that in the absence of spin-orbit interaction would close at $\bar{X}$ and $\bar{M}$, is expected to decay slowly into the bulk. This may explain why this resonance was not found in the UPS studies [4, 5] using only a limited photon energy range and why it was not predicted in any slab calculation [3].

Low-Temperature Phase. Theoretical electronic band dispersions for the reconstructed W(100) surface are shown in **Fig. 14**, where the symmetry lines $\bar{\Delta}$ and $\bar{\Delta}'$ correspond to the Σ line of unreconstructed W(100), $\bar{\Delta}$ and $\bar{\Delta}'$ denoting directions along and normal to the zigzag chains, respectively. The solid and dashed curves indicate surface features that are even and odd with respect to *z* reversal, respectively. The curves drawn indicate states that have ≥65% of their electron density located in the two outermost layers. Among these is the Swanson hump state occurring for $\bar{\Gamma}$ near 0.3 eV below E_F as an even/odd doublet, the even and odd components of which have ~75% and ~90% of their charge concentrated in the surface layers [11]. Prior to paper [11] the theoretical surface band structure had been studied from first principles for a bulk-truncated surface only (see [10] and [7]). There seems to be only one UPS study on the low-temperature phase, namely that by Campuzano et al. [4, 5]; [11]. The attenuation of the surface feature close to E_F at the midpoint of the $\bar{\Gamma}$-$\bar{M}$ line in the high-temperature phase (the $\bar{X}$ and $\bar{X}'$ points of the low-temperature phase) observed by Campuzano et al. upon cooling [4, 5] is in conflict with the seven-layer-slab first-principles total-energy calculation [11] of the low-temperature phase, since there

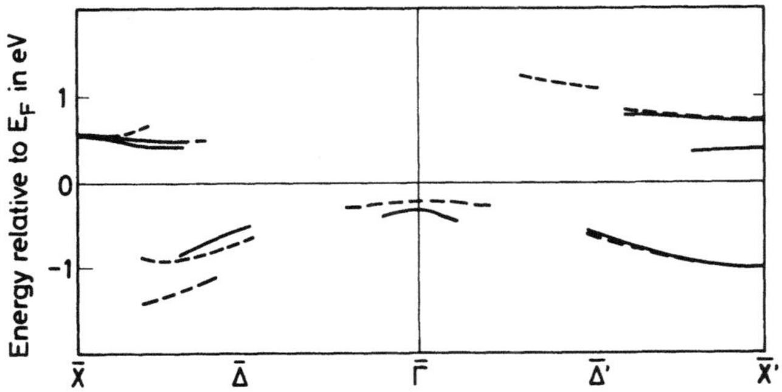

Fig. 14. Dispersion of odd-symmetry (dashed lines) and even-symmetry (solid lines) surface states for the reconstructed W(100) surface along the high-symmetry directions. $\bar{\Delta}$ and $\bar{\Delta}'$ are directions along and perpendicular to the zigzag chains, respectively. They correspond to the $\bar{\Sigma}$ line of the (1×1) surface. Only the states highly localized on the surface are plotted [11].

is a strong bonding state near 1 eV below E_F around $\bar{X}'$; see Fig. 14. This state should be observable by UPS [11].

Fig. 15 compares the calculated odd-symmetry surface-state dispersion of the relaxed ideal surface with that of the reconstructed surface. It is apparent that the band structure is affected over a large portion of the SBZ. This is characteristic for a bonding-type mechanism of reconstruction; a CDW-related mechanism would be expected to be confined to a small SBZ region close to $\bar{X}'$. The conclusion, therefore, is that the reconstruction is driven mainly by localized interactions [11].

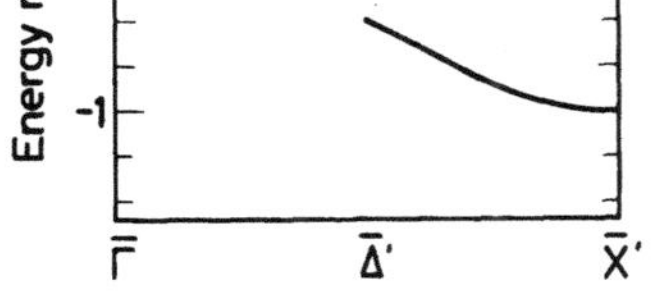

Fig. 15. Comparison of the odd-symmetry surface state dispersion of the relaxed, ideal surface (dashed lines) and the reconstructed surface (solid lines) [11].

While a bonding state was found for $\bar{X}'$ near 1 eV below E_F in the LAPW study [11] which was not seen by UPS, this first-principles study, on the other hand, found no theoretical counterpart to the features near 2.5 and 4.5 eV below E_F, observed by UPS on the low-temperature phase (see [5]) and known from the high-temperature phase (for which they were also predicted by [10] and [7]). Perhaps, these two states were not found in [11], because the seven-layer slab may not have been thick enough [11].

References for 1.1.3.4.2.1 on p. 48

$\bar{\Delta}'$ line (normal to the zigzag chains): In addition to the Swanson hump state doublet around $\bar{\Gamma}$, there are nearly degenerate, even- and odd-symmetry surface state bands at 1.0 eV below E_F close to $\bar{X}'$; see Fig. 14. Both of these nearly degenerate bands are odd with respect to (1$\bar{1}$0) reflection ($\bar{\Delta}'_2$ symmetry). They consist mainly of $d_{x^2-y^2}$ and d_{xz-yz} orbitals, the former providing the bonds between neighboring atoms in the zigzag chains (see Fig. 17, p. 54). This pair of surface states evolved from a surface resonance on the (1×1) surface. The reconstruction splits this resonance into bonding and antibonding states with predominantly $d_{x^2-y^2}$ and d_{xz-yz} character, respectively [11]. According to the tight-binding studies of Bullett and Stephenson [12 to 14], these subbands are separated by ~1 eV at the new zone boundary; see also [1]. For $\bar{X}'$, the antibonding state is located at 0.7 eV above E_F. For a charge density contour plot, see the paper [11]. The lowest-lying antibonding state near $\bar{X}'$ occurs at ~0.4 eV above E_F with ~70% of the charge localized in the surface layers [11].

$\bar{\Delta}$ line (parallel to the zigzag chains): There is a surface state with $\bar{\Delta}_2$ symmetry for binding energies <1 eV (with respect to E_F); about 75% of its charge is localized on the surface layers (for 0.7 $k_{\bar{\Gamma}\bar{X}}$) [11]. At $\bar{\Gamma}$, the bonding and antibonding subbands, discussed above in conjunction with the $\bar{\Delta}'$ line, are pinned together by symmetry constraints [12, 13]; see also [14].

References for 1.1.3.4.2.1:

[1] Drube, W.; Straub, D.; Himpsel, F. J.; Soukiassian, P.; Fu, C. L.; Freeman, A. J. (Phys. Rev. [3] B **34** [1986] 8989/92).
[2] Ohnishi, S.; Freeman, A. J.; Wimmer, E. (Phys. Rev. [3] B **29** [1984] 5267/78).
[3] Elliott, G. S.; Smith, K. E.; Kevan, S. D. (Phys. Rev. [3] B **44** [1991] 10826/34).
[4] Campuzano, J. C.; King, D. A.; Somerton, C.; Inglesfield, J. E. (Phys. Rev. Lett. **45** [1980] 1649/52).
[5] Campuzano, J. C.; Inglesfield, J. E.; King, D. A.; Somerton, C. (J. Phys. C **14** [1981] 3099/113).
[6] Holmes, M. W.; Gustafsson, T. (Phys. Rev. Lett. **47** [1981] 443/6).
[7] Mattheiss, L. F.; Hamann, D. R. (Phys. Rev. [3] B **29** [1984] 5372/81).
[8] Weng, S.-L.; Plummer, E. W.; Gustafsson, T. (Phys. Rev. [3] B **18** [1978] 1718/40).
[9] Lapeyre, G. J.; Anderson, J.; Smith, R. J. (Surf. Sci. **89** [1979] 304/18).
[10] Posternak, M.; Krakauer, H.; Freeman, A. J.; Koelling, D. D. (Phys. Rev. [3] B **21** [1980] 5601/12).
[11] Yu, R.; Krakauer, H.; Singh, D. (Phys. Rev. [3] B **45** [1992] 8671/8).
[12] Bullett, D. W.; Stephenson, P. C. (Solid State Commun. **45** [1983] 47/9).
[13] Stephenson, P. C.; Bullett, D. W. (Surf. Sci. **139** [1984] 1/15).
[14] Bullett, D. W. (Philos. Mag. [8] B **51** [1985] 223/41, 234/5).

1.1.3.4.2.2 Characterization of Prominent Surface Bands

Swanson Hump State. Field emission (FE) energy distributions at 78 K showed an adsorption-sensitive feature ("Swanson hump") at 0.37 eV below E_F [1]; see also [2] and [3, 4]. The hump seen in dark FE near 0.36 eV below E_F appeared in photofield emission (PFE) spectra at 0.34 eV below E_F [5]; see also [6 to 8].

Along $\bar{\Sigma}$ the Swanson hump state comes close to E_F (see S_4 in Fig. 13, p. 45) [9]. In their UPS study confined to $\bar{\Sigma}$ and performed at or slightly above room temperature, Holmes and Gustafsson [10] observed this state for $k_{\parallel}$ as high as 0.69 Å^{-1}, for higher $k_{\parallel}$ values the state no longer existed [10]. The intensity of the corresponding UPS peak strongly decreased with increasing $k_{\parallel}$ along $\bar{\Sigma}$ and $\bar{\Delta}$ [10, 11]. Discussing the polarization effects observed with UPS [12], Hermanson [13] supported Kasowski's conclusion [14] that the Swanson hump is a $z^2(\Delta_1)$ state and discounted Feder's and Sturm's [15] suggestion of a surface resonance in a spin-orbit induced gap [13]. The mainly d_{z^2} character [16, 17] or s, p_z, or d_{z^2} character [18] was confirmed for this state by UPS; see also the review [19]. The UPS studies of the orbital character superseded the calculations by [20] and [21]; [17]. According to an LAPW study [22], as much as 93% of this state is localized in the surface layer; it has $3z^2-r^2$ character and is a true surface state only at $\bar{\Gamma}$ [22]. Moving away from $\bar{\Gamma}$, it becomes a surface resonance as the symmetry restriction (which prevented interactions with the Δ_2, Δ_2', and Δ_5 bands) no longer applies [19].

$\bar{\Sigma}$ Band of $d_{x^2-y^2}$ Surface States. The IR reflectance spectrum of clean W(100) was measured by Reutt et al. [23] (superseding the earlier paper [24]) at 300 K between 900 and ~3500 cm^{-1} using p-polarized radiation incident under $\alpha \approx 85°$. A dip occurred with $\Delta R/R \approx -0.7\%$ centered at 0.36 eV. This ~0.25 eV-wide feature is ascribed to a surface state. The rather narrow absorption profile suggests that fairly localized electrons, d electrons, are involved. The IR features involve excitations of intraband transitions from states with xy or x^2-y^2 symmetry and are attributed to electron-hole pair generation [23] in the $\bar{\Sigma}$ band of $d_{x^2-y^2}$ surface states calculated to disperse through E_F by [25, p. 5379]; [23]. When irradiated with light below ~1.5 eV, the surface reflectance showed a distinct absorbance which was quenched by rather low adsorbate coverages [26]; see also [27], the surface electromagnetic wave spectroscopy studies [24, 28], and the review [29]. These absorption bands are seen near 2900 cm^{-1} (0.36 eV) in IR reflection-adsorption spectra also in the low-temperature phase (at 100 K) [30].

The band of surface resonances occurring near $\bar{\Gamma}$ at about 0.8 eV below E_F exists only for $k>0$ and is mainly composed of $d_{x^2-y^2}$ and $d_{zx,zy}$ orbitals according to polarization-dependent UPS [16]; see also [18]. It is not clear whether these resonances are the weakly adsorption-sensitive features which were detected around 0.78 eV below E_F in field emission studies at 78 K [1]. The LAPW calculations [22, 31] showed that the band found by [16] actually is a doublet of states, even and odd with respect to the (110) mirror plane normal to (1 × 1)-W(001) [22, 31]. The components differ in the degree of localization: the antisymmetric band ($\bar{\Sigma}_2$) is about 20 to 25% more localized in the two outermost layers than is the symmetric band ($\bar{\Sigma}_1$). Similarly, along $\bar{\Delta}$ ($=\bar{\Gamma}$-$\bar{X}$) this band extends as an even/odd doublet of resonances, symmetric and antisymmetric with respect to the (100) mirror plane. The $\bar{\Sigma}_1$ resonance has largely xy and (xz+yz) character, while the $\bar{\Sigma}_2$ resonance has (x^2-y^2) and (xz−yz) character [22, 31]. The only surface states around E_F are the odd parity $\bar{\Sigma}_2$ bands; the even $\bar{\Sigma}_1$ bands are resonances. Away from the $\bar{\Sigma}$ line, even and odd states mix and all states are more or less resonances [32].

The $\bar{\Sigma}_2$ state band first dispersed moderately upward, crossing E_F near 0.6($\bar{\Gamma}$-$\bar{M}$) [33, 34]. The unoccupied branch of the $\bar{\Sigma}_2$ band was found to disperse upwards to 2 eV above E_F at $\bar{M}$ [33 to 35]. This strong upward shift occurring near the SBZ boundary was attributed to a bulk-like atom arrangement assumed in the calculations [34]. The IPE study [36] indicated a surface state near $\bar{M}$ at 1.4 eV above E_F. The results of [36] and [35] are in very poor agreement with those of [37]; see Fig. 12, p. 44.

According to UPS, it is uncertain whether the $\bar{\Sigma}_2$ band crosses E_F. But it clearly exists close to E_F for ~$0.31\ Å^{-1} < k_{\parallel} < 0.64\ Å^{-1}$. This weak dispersion was seen not only along $\bar{\Sigma}$ but also over disks centered on $\bar{\Sigma}$ around $k_{\parallel}=0.5\ Å^{-1}$ [38, 39]; see also the review [40]. The minimum and maximum spanning vectors for Fermi surface nesting associated with this band were 1.0 and 1.28 $Å^{-1}$, falling short of the 1.41 $Å^{-1}$ required by the classic CDW model [38, 39]. For a transition between commensurate phases, symmetry arguments

References for 1.1.3.4.2.2 on pp. 50/1

showed that the exact position of the Fermi level crossing is immaterial. The fact that the twofold degenerate $\bar{\Sigma}$ symmetry splits into two nonequivalent lines, $\bar{\Delta}$ and $\bar{\Delta}'$, implies that the soft phonon is doubly degenerate. This confines the choice of wave vectors for the soft mode to $\vec{k}=0$ and $\bar{\Gamma}\bar{M}$, with only the latter providing for a doubling of the unit cell [41]. Strong nonadiabatic effects are expected, and strong electron-phonon coupling with the nesting regions of nondispersing surface states close to E_F may drive the W(100) reconstruction [38, 39].

For the contribution of the $\bar{\Sigma}_2$ band to the bonding of the surface atoms, see the local-density-functional total-energy calculations by [41, 42].

Surface State near 4.3 eV below E_F at $\bar{\Gamma}$. For this band of surface states, see the feature labeled S_1 in Fig. 13, p. 45. UPS suggests d_{z^2} (and/or s) [16] or s, p_z, or d_{z^2} orbital character [18]. According to a self-consistent LAPW study [22], the wave function of this resonance is symmetric, antisymmetric, and symmetric (with respect to the corresponding mirror planes) along $\bar{\Delta}$, $\bar{Y}$, and $\bar{\Sigma}$, respectively [22].

References for 1.1.3.4.2.2:

[1] Plummer, E. W.; Gadzuk, J. W. (Phys. Rev. Lett. **25** [1970] 1493/5).
[2] Plummer, E. W.; Young, R. D. (Phys. Rev. [3] B **1** [1970] 2088/109, 2101).
[3] Swanson, L. W.; Crouser, L. C. (Phys. Rev. Lett. **16** [1966] 389/92).
[4] Swanson, L. W.; Crouser, L. C. (Phys. Rev. [2] **163** [1967] 622/41).
[5] Lopinski, G. P.; Donders, P. J.; Lee, M. J. G. (Surf. Sci. **201** [1988] 294/310).
[6] Venus, D.; Lee, M. J. G. (Phys. Rev. [3] B **34** [1986] 4449/63).
[7] Venus, D.; Lee, M. J. G. (Surf. Sci. **172** [1986] 477/95).
[8] Radon, T.; Jaskólka, S. (Surf. Sci. **231** [1990] 160/4).
[9] Elliott, G. S.; Smith, K. E.; Kevan, S. D. (Phys. Rev. [3] B **44** [1991] 10826/34).
[10] Holmes, M. W.; Gustafsson, T. (Phys. Rev. Lett. **47** [1981] 443/6).

[11] Feuerbacher, B.; Willis, R. F. (J. Phys. C **9** [1976] 169/216, 204/8).
[12] Feuerbacher, B.; Fitton, B. (Solid State Commun. **15** [1974] 295/9).
[13] Hermanson, J. (Solid State Commun. **22** [1977] 9/11).
[14] Kasowski, R. V. (Solid State Commun. **17** [1975] 179/83).
[15] Feder, R.; Sturm, K. (Phys. Rev. [3] B **12** [1975] 537/48).
[16] Weng, S.-L.; Plummer, E. W.; Gustafsson, T. (Phys. Rev. [3] B **18** [1978] 1718/40).
[17] Smith, R. J.; Lapeyre, G. J.; Anderson, J.; Williams, G. P. (Conf. Ser.-Inst. Phys. No. 39 [1978] 220/32).
[18] Lapeyre, G. J.; Anderson, J.; Smith, R. J. (Surf. Sci. **89** [1979] 304/18).
[19] Inglesfield, J. E. (Rep. Prog. Phys. **45** [1982] 223/84).
[20] Smith, N. V.; Mattheis, L. F. (Phys. Rev. Lett. **37** [1976] 1494/7).

[21] Weng, S.-L. (Phys. Rev. Lett. **38** [1977] 434/7).
[22] Posternak, M.; Krakauer, H.; Freeman, A. J.; Koelling, D. D. (Phys. Rev. [3] B **21** [1980] 5601/12).
[23] Reutt, J. E.; Chabal, Y. J.; Christman, S. B. (Phys. Rev. [3] B **38** [1988] 3112/32, 3121, 3128).
[24] Riffe, D. M.; Hanssen, L. M.; Sievers, A. J. (Phys. Rev. [3] B **34** [1986] 692/703).
[25] Mattheiss, L. F.; Hamann, D. R. (Phys. Rev. [3] B **29** [1984] 5372/81).
[26] Anderson, J.; Rubloff, G. W.; Passler, M. A.; Stiles, P. J. (Phys. Rev. [3] B **10** [1974] 2401/15).
[27] Restorff, J. B.; Drew, H. D. (Surf. Sci. **88** [1979] 399/426).
[28] Riffe, D. M.; Hanssen, L. M.; Sievers, A. J. (Surf. Sci. **176** [1986] 679/90).

[29] Chabal, Y. J. (Surf. Sci. Rep. **8** [1988] 211/357, 314/9).
[30] Riffe, D. M.; Sievers, A. J. (Surf. Sci. **210** [1989] L 215/L 222).

[31] Wimmer, E.; Krakauer, H.; Freeman, A. J. (Adv. Electron. Electron Phys. **65** [1985] 357/434, 391/404).
[32] Wang, X. W.; Weber, W. (Phys. Rev. Lett. **58** [1987] 1452/5).
[33] Krakauer, H.; Posternak, M.; Freeman, A. J. (Phys. Rev. Lett. **43** [1979] 1885/9).
[34] Ohnishi, S.; Freeman, A. J.; Wimmer, E. (Phys. Rev. [3] B **29** [1984] 5267/78).
[35] Collins, I. R.; Laine, A. D.; Andrews, P. T. (J. Phys. Condens. Matter **4** [1992] 2891/901).
[36] Drube, W.; Straub, D.; Himpsel, F. J.; Soukiassian, P.; Fu, C. L.; Freeman, A. J. (Phys. Rev. [3] B **34** [1986] 8989/92).
[37] Krainsky, I. L. (J. Vac. Sci. Technol. [2] A **5** [1987] 735/8).
[38] Smith, K. E.; Elliott, G. S.; Kevan, S. D. (Phys. Rev. [3] B **42** [1990] 5385/7).
[39] Smith, K. E.; Elliott, G. S.; Kevan, S. D. (Appl. Surf. Sci. **48/49** [1991] 135/8).
[40] Smith, K. E.; Kevan, S. D. (Prog. Solid State Chem. **21** [1991] 49/131, 81/8).

[41] Weinert, M.; Freeman, A. J.; Ohnishi, S. (Phys. Rev. Lett. **56** [1986] 2295/8).
[42] Fu, C. L.; Freeman, A. J. (Phys. Rev. [3] B **37** [1988] 2685/8).

1.1.3.4.3 Theoretical Studies

Two points emerge from the following paragraphs: First, realistic theories of the W(100) surface states should account for spin-orbit interaction. Second, formulating a theory may be easier for the low-temperature surface than for the high-temperature surface, since the presumably disordered nature of the latter seriously flaws any theory based on structural models that assume (1 × 1) symmetry on a microscopic level.

Surface band structures were calculated for an unrelaxed seven-layer (1 × 1)-W(100) slab with a self-consistent LAPW method [1]. This study included all relativistic corrections except for the spin-orbit interaction. It is extensively reviewed in [2]; see also the study of slab thickness effects [3] and the review [4]. Qualitative agreement with [1] was found for similar scalar LAPW slab calculations by Mattheiss and Hamann [5]. These authors also performed fully relativistic LAPW calculations. The Fermi level crossings along $\bar{\Sigma}$ they calculated fully relativistically were in slightly better agreement with the UPS data measured by [6] and [7] than were their scalar relativistic calculations [5]. This suggests that the discrepancies between the dispersion curves for the surface resonances calculated by [1] and the measured data from [6] and [7] are due to neglecting the spin-orbit interaction in scalar relativistic calculations [5]. Elliott et al. [8] suspected that calculations which neglect the spin-orbit interaction are seriously flawed, at least with respect to surface bands near E_F. This conclusion was drawn from the fact that the UPS results were best approximated by the fully relativistic calculations of Mattheiss and Hamann [5]. See also the qualitative discussion in [8] based on tight-binding calculations which included to some extent the spin-orbit interaction.

When Posternak et al. [9] tried to improve their earlier calculation of bulk-terminating W(100) [1] by including a 6% relaxation of the first interlayer spacing, it had a surprisingly small effect on the bands [9] (conflicting with the outdated slab calculation [10] but in agreement with the tight-binding study [11]). Motivated by this result, Ohnishi et al. [3] investigated the effect of slab thickness. They found that the seven-layer slabs used earlier were sufficiently thick to describe surface phenomena and excellently reproduced the UPS data of surface states having predominantly d_{z^2} character. By contrast, surface states with important components of $d_{xz,yz}$ character were not reproduced. In particular, the calculations did not

References for 1.1.3.4.3 on pp. 52/3

account for the band of states found by UPS near E_F along $\bar{\Sigma}$. They attributed this failure to the bulk truncation model used. Lateral and/or vertical rearrangements of the surface atoms along the ⟨110⟩ directions would prevent the strong upward shift of the surface state energy that occurs with the (1 × 1) structure on going from $\bar{\Gamma}$ to $\bar{M}$ [3]. This conclusion is consistent with the findings of a tight-binding calculation [11] that lateral displacements of the surface atoms strongly influence the $\bar{\Sigma}_2$ surface state band [3].

An LAPW study [12] of the low-temperature phase performed within the local-density approximation treated core states fully relativistically and the valence states semirelativistically (neglecting the spin-orbit interaction). On account of the highly asymmetric potential near the surface, the 5p semicore states were treated variationally in a second energy window [12]. A comparison with the earlier paper [13], where the 5p semicore states were treated as core states, showed a significant difference in the results of these procedures: an ~1% relaxation of the surface layer in the low-temperature phase was found in the *ab initio* LAPW calculation [13] in contrast to the 6% contraction calculated in [12]. (Fu and Freeman [14] found a much smaller effect, however, when they compared the results obtained by treating the 5p electrons either as core states or as valence states in a separate energy window; for a separate treatment of 5p electrons, see also [15].)

Tight-binding calculations indicate that surface states play an important role in bringing about the c(2 × 2) structure. The pairing of surface atoms is favored by the enhanced bonding involving these states. The tight-binding calculations underline the importance of the surface DOS near E_F as well as that of electron-phonon interaction [16]. The latter, involving surface states around E_F, results in a whole range of soft SBZ boundary phonon modes on (1 × 1)-W(100) [17].

The thin-film tight-binding relativistic energy bands calculated in [18], including spin-orbit interaction for $\bar{\Delta}$, $\bar{Y}$, and $\bar{\Sigma}$, compare only poorly with the results of the UPS study [19]. The tight-binding calculations [20, 21] are obviously outdated.

The transferability of surface states (between different surfaces of a crystal) invoked in [22] is further discussed in [23, 24]. Matching Green function calculations (see [25]) predict a doublet of surface states on W(100) centered around 0.4 eV above E_F at $\bar{M}$. The lower state consists of d_{z^2} orbitals. On W(100) this is an empty state (above E_F) which on the W(110) surface exists as an occupied state giving rise to a strong UPS peak [23].

References for 1.1.3.4.3:

[1] Posternak, M.; Krakauer, H.; Freeman, A. J.; Koelling, D. D. (Phys. Rev. [3] B **21** [1980] 5601/12).
[2] Wimmer, E.; Krakauer, H.; Freeman, A. J. (Adv. Electron. Electron Phys. **65** [1985] 357/434, 391/404).
[3] Ohnishi, S.; Freeman, A. J.; Wimmer, E. (Phys. Rev. [3] B **29** [1984] 5267/78).
[4] Inglesfield, J. E. (Rep. Prog. Phys. **45** [1982] 223/84).
[5] Mattheiss, L. F.; Hamann, D. R. (Phys. Rev. [3] B **29** [1984] 5372/81).
[6] Campuzano, J. C.; King, D. A.; Somerton, C.; Inglesfield, J. E. (Phys. Rev. Lett. **45** [1980] 1649/52).
[7] Holmes, M. W.; Gustafsson, T. (Phys. Rev. Lett. **47** [1981] 443/6).
[8] Elliott, G. S.; Smith, K. E.; Kevan, S. D. (Phys. Rev. [3] B **44** [1991] 10826/34).
[9] Posternak, M.; Krakauer, H.; Freeman, A. J. (Phys. Rev. [3] B **25** [1982] 755/61).
[10] Kasowski, R. V. (Solid State Commun. **17** [1975] 179/83).
[11] Bullett, D. W.; Stephenson, P. C. (Solid State Commun. **45** [1983] 47/9).
[12] Yu, R.; Krakauer, H.; Singh, D. (Phys. Rev. [3] B **45** [1992] 8671/8).

[13] Singh, D.; Krakauer, H. (Phys. Rev. [3] B **37** [1988] 3999/4006).
[14] Fu, C. L.; Freeman, A. J. (Phys. Rev. [3] B **37** [1988] 2685/8).
[15] Weinert, M.; Freeman, A. J.; Ohnishi, S. (Phys. Rev. Lett. **56** [1986] 2295/8).
[16] Terakura, I.; Terakura, K.; Hamada, N. (Surf. Sci. **103** [1981] 103/17).
[17] Wang, X. W.; Weber, W. (Phys. Rev. Lett. **58** [1987] 1452/5).
[18] Grise, W. R.; Dempsey, D. G.; Kleinman, L.; Mednick, K. (Phys. Rev. [3] B **20** [1979] 3045/50).
[19] Weng, S.-L.; Plummer, E. W.; Gustafsson, T. (Phys. Rev. [3] B **18** [1978] 1718/40).
[20] Bisi, O.; Calandra, C.; Flaviani, P.; Manghi, F. (Solid State Commun. **21** [1977] 121/3).
[21] Sturm, K.; Feder, R. (Solid State Commun. **14** [1974] 1317/20).
[22] Holmes, M. W.; King, D. A.; Inglesfield, J. E. (Phys. Rev. Lett. **42** [1979] 394/7).
[23] Holmes, M. W.; Inglesfield, J. E. (Surf. Sci. **89** [1979] 133/41).
[24] Holmes, M. W.; King, D. A. (Proc. R. Soc. [London] A **376** [1981] 565/83, 580/2).
[25] Inglesfield, J. E. (J. Phys. C **12** [1979] L 349/L 354).

1.1.3.4.4 Charge Density Contour Plots

The total valence charge density at a bulk-terminating W(100) surface is given as a contour plot in a self-consistent seven-layer LAPW slab calculation [1, p. 5605] (see also the review [2]). The t_{2g} orbitals (i.e. the bonding orbitals d_{xy}, d_{xz}, and d_{yz}) of atoms in the second and deeper layers form fairly localized lobes pointing towards the nearest neighbors. For the top-layer atoms, these localized lobes are also present at the vacuum side, where the neighbors are missing, but substantial weight is removed from the lobes (although the maximum value of the d-bond charge density is still about the same in these directions). The rather uniform metallic-bond charge density in the interstitial regions of the bulk is drastically changed in the topmost layer [1, 2]. **Fig. 16** (taken from [3]) compares the charge densities of W(100) surfaces, both unreconstructed (relaxed inward by 5%) and reconstructed ($\sim$0.27 to 0.28 Å Debe-King-type atom shifts; relaxation $<1\%$), obtained by a self-consistent LAPW method applied on supercells consisting of five-layer slabs separated by empty

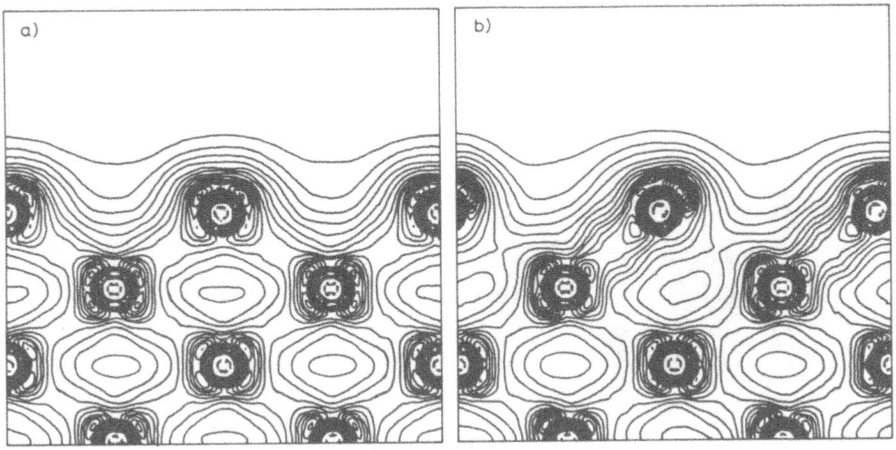

Fig. 16. Charge-density contour plots for a (011) plane perpendicular to the W(100) surface: a) unreconstructed surface, relaxation -5%; b) reconstructed surface (lateral atomic shifts: 0.27 to 0.28 Å), relaxation less than -1%. Adjacent contour lines are separated by 0.008 e/a.u.^{3} (1 a.u. = 0.529 Å) [3].

References for 1.1.3.4.4 on p. 55

spaces five atomic layers thick [3]. These valence charge density contour plots show that the dangling bonds present on the bulk-terminating surface are replaced on the (Debe-King) reconstructed surface by largely nondirectional, bond-like features between the surface atoms that approach each other during reconstruction [4].

The so-called Swanson hump surface state (near 0.3 eV below E_F) with $\bar{\Delta}_1$ and $\bar{\Sigma}_1$ symmetry is localized (to ~93%) in the surface layer and projects quite far into the vacuum region, as shown in a charge-density contour plot in [1, p. 5608]. A similar charge-density contour plot given in [5, p. 4918] presents the results of an all-electron local-density-functional full-potential LAPW five-layer slab calculation [5]. Somewhat different results were obtained for a nine-layer slab [6, p. 5275]. In the low-temperature phase, this state is split into an even/odd doublet (also near 0.3 eV below E_F). About 75% of the even-symmetry state is localized on the surface layers. The odd-symmetry component at $\bar{\Gamma}$ has ~90% of its charge concentrated in the surface layers and is similar to the Swanson hump state found for the bulk-terminating surface by [1] (except for the distortion arising from the reconstruction) [7].

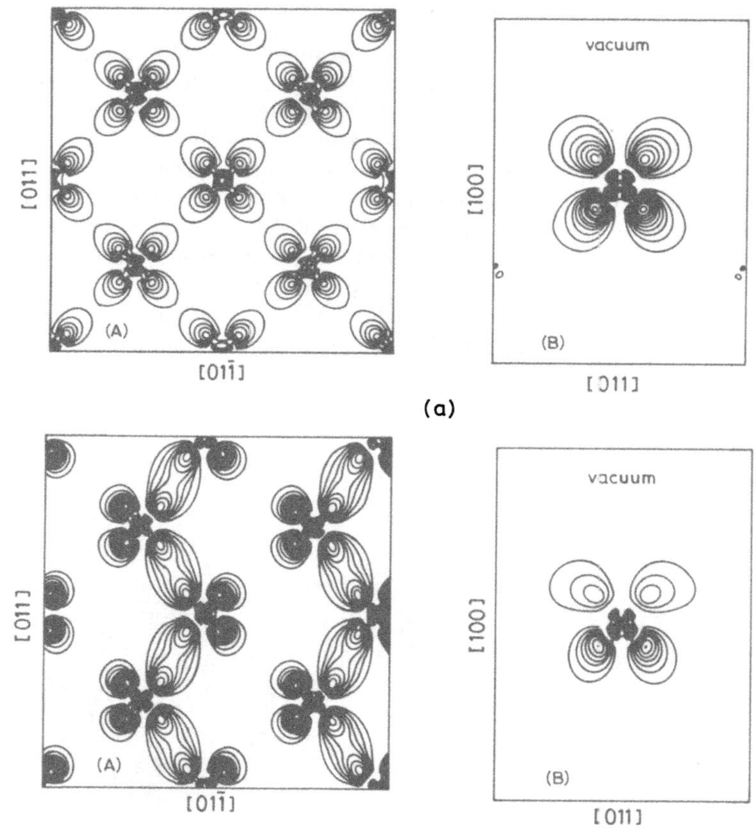

Fig. 17. Comparison of the charge densities of (a) the $\bar{\Sigma}_2$ surface state of a relaxed, ideal (1 × 1) surface at $k_{\parallel} = 0.5(\bar{\Gamma}-\bar{M})$ and (b) the corresponding $\bar{X}_2'$ bonding state of the reconstructed surface. The contours in the left panels (A) are in the plane of the surface and in the right panels (B) are in a vertical (011) plane passing through a surface atom. The contour lines are separated by 0.0015 electrons per c(2 × 2) cell [7].

On the (1 × 1) surface there are two high-lying resonances (some 0.5 eV below the Swanson state) which are nearly degenerate. Their strong localization is shown in [1] by charge-density contour plots which also indicate that the degree of localization for the two resonances is different: the antisymmetric resonance ($\overline{\Sigma}_2$) is about 20 to 25% more localized in the two outermost layers than is the symmetric one ($\overline{\Sigma}_1$) [1, 2]. Upon reconstruction, this band splits into a bonding and an antibonding subband. Charge density contour plots are given in [7] for $\vec{k}_{\parallel}$ perpendicular to the zigzag chains for the bonding component (which is a true surface state at the new zone boundary $\overline{X}'$). **Fig. 17** compares the charge density of the $\overline{\Sigma}_2$ surface state of the relaxed (1 × 1) surface at $k_{\parallel} = 0.5(\overline{\Gamma}-\overline{M})$ with the corresponding contour plot of the reconstructed surface at $\overline{X}'$ ($\vec{k}_{\parallel}$ perpendicular to the zigzag chains) [7].

References for 1.1.3.4.4:

[1] Posternak, M.; Krakauer, H.; Freeman, A. J.; Koelling, D. D. (Phys. Rev. [3] B **21** [1980] 5601/12).

[2] Wimmer, E.; Krakauer, H.; Freeman, A. J. (Adv. Electron. Electron Phys. **65** [1985] 357/434, 391/404).

[3] Singh, D.; Wei, S.-H.; Krakauer, H. (Phys. Rev. Lett. **57** [1986] 3292/5).

[4] Singh, D.; Krakauer, H. (Phys. Rev. [3] B **37** [1988] 3999/4006).

[5] Soukiassian, P.; Riwan, R.; Lecante, J.; Wimmer, E.; Chubb, S. R.; Freeman, A. J. (Phys. Rev. [3] B **31** [1985] 4911/23, 4918).

[6] Ohnishi, S.; Freeman, A. J.; Wimmer, E. (Phys. Rev. [3] B **29** [1984] 5267/78).

[7] Yu, R.; Krakauer, H.; Singh, D. (Phys. Rev. [3] B **45** [1992] 8671/8).

1.1.3.5 Fermi Surface

The structure of the two-dimensional Fermi surface of W(100) was investigated by photoemission [1, 2]; see also the review [3]. **Fig. 18** (taken from [2]) presents the measured

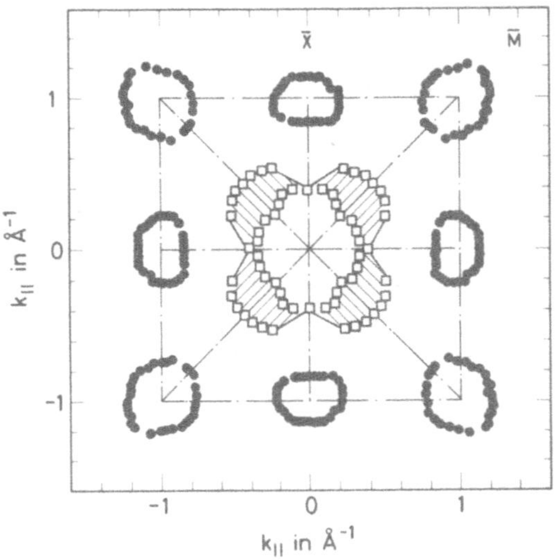

Fig. 18. Measured two-dimensional Fermi surface of W(100). The hatched areas mark the regions where the Swanson-hump related surface state is nondispersive and at or close to the Fermi level [2].

References for 1.1.3.5 on p. 56

Fermi surface. The actual measurements were restricted to one quarter of the SBZ and mapped over the full zone by symmetry. The Fermi surface consists of two classes of structures. On the one hand, there are well-defined hole pockets around $\bar{X}$ and $\bar{M}$ mimicking projections of the bulk Fermi surface. On the other hand, there are the shaded areas in Fig. 18 where the energies of the photoemission peaks are very close to or at E_F. Due to the limitations of the photoemission measurements, these regions cannot rigorously be included as part of the Fermi surface, since no crossing was observed. However, the high density of states close to E_F allows for strong electron-phonon interactions to occur. This point is of great interest for a discussion of the driving mechanism for the reconstruction of W(100); see Section 1.1.1.6, pp. 25/9. The observed large-scale Fermi surface nesting supports a collective electronic origin for reconstruction [1 to 3]. Earlier photoemission studies [4, 5] had revealed no significant nesting.

References for 1.1.3.5:

[1] Smith, K. E.; Elliott, G. S.; Kevan, S. D. (Phys. Rev. [3] B **42** [1990] 5385/7).
[2] Smith, K. E.; Elliott, G. S.; Kevan, S. D. (Appl. Surf. Sci. **48/49** [1991] 135/8).
[3] Smith, K. E.; Kevan, S. D. (Prog. Solid State Chem. **21** [1991] 49/131, 81/8).
[4] Campuzano, J. C.; Inglesfield, J. E.; King, D. A.; Somerton, C. (J. Phys. C **14** [1981] 3099/113).
[5] King, D. A. (Phys. Scr. T **4** [1983] 34/43).

1.2 W(110)

1.2.1 Atomic Structure. Reconstruction and Relaxation

The bulk-truncated W(110) surface has a centered rectangular c2mm symmetry. The possibility of a lateral reconstruction was critically discussed for a long time (see e.g. [1]), but more recent experimental and theoretical studies led to the conclusion that the clean W(110) surface is unreconstructed (see e.g. [2, 3]).

Van Hove and Tong [4] pointed out that a rigid displacement of the surface along the [$\bar{1}$10] direction shifts the surface atoms from the twofold coordinated bridge position of the unreconstructed surface to a threefold coordinated position (see **Fig. 19**). This position may energetically be more favorable, because the most stable site is generally the site with maximum coordination. (Although such a shift of the surface as a whole does not alter the symmetry and the dimensions of the two-dimensional surface unit cell, it is termed reconstruction (in contrast to relaxation), because one of the two mirror planes of the bulk-truncated surface vanishes if the subsurface layers are included (see Fig. 19). Similar lateral displacements of the surface layer as a whole, observed on W surfaces with lower symmetries as, for example, W(211) (cf. Section 1.4.1), are termed relaxation, because in these cases no symmetry element is lost.) Although the threefold coordinated position seemed to be energetically favored, a comparison of experimental LEED data with theoretically calculated intensities for both positions (i.e., the twofold and threefold coordinated) led to the conclusion that the surface maintains the bulk structure [4]. On the other hand, an analysis of the angular anisotropy of the $4f_{7/2}$ core-level photoemission spectra from W(110) suggested that the surface atoms occupy threefold coordinated positions [5]. However, in both cases the agreement between experimental and theoretical curves is not fully convincing. An LEED study using constant-momentum-transfer averaging (CMTA) of LEED intensities (a method to extract single-scattering intensities) provided no conclusive evidence for a lateral reconstruction of W(110). It was only pointed out that a lateral reconstruction is hypothetically possible, but highly unlikely [6, 7]. W(110) in ultrahigh vacuum (about

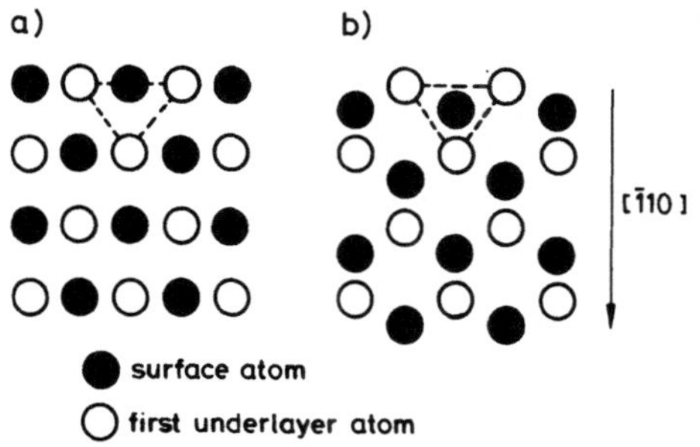

Fig. 19. Model of reconstruction of the W(110) surface: a) unreconstructed centered rectangular c2mm surface structure with surface atoms in bridge positions, b) reconstructed surface with surface atoms displaced in the [$\bar{1}$10] direction and going to ternary positions [1].

2×10^{-10} Torr) and in an O atmosphere was studied with LEED in the temperature range 293 to 2500 K. Reconstruction was observed on the O-covered surface, but not on the clean one [8].

In an early tight-binding calculation of the binding energy of W adatoms on a W surface an equilibrium position intermediate between two- and threefold coordinated sites was obtained [9, 10], whereas another theoretical study using a tight-binding approximation found the unreconstructed surface structure to be the stable structure at T = 0 K [1]. In the latter work it was noted that the occurrence of a lateral reconstruction may depend on the d-band filling. Keeping all model parameters fixed and changing the d-band filling from 4.6 to 7 electrons causes the unreconstructed surface to become unstable [1].

The stability of the unreconstructed state of the clean W(110) surface was confirmed in some more recent experimental [2, 11 to 13] and theoretical [3] studies: LEED experiments measuring simultaneously the intensity of the (01), (0$\bar{1}$), (1$\bar{1}$), and ($\bar{1}$1) beams (providing the possibility to detect intensity variations arising from symmetry changes) indicate that the clean W(110) surface is unreconstructed [11, 12]. Moreover, the results of high-energy He^{+} ion backscattering experiments (0.5 MeV $\leq E \leq$ 2.0 MeV) are in excellent agreement with structural models suggesting no lateral reconstruction [13]. A comparison of photoelectron diffraction patterns [2] with theoretical calculations within a single-scattering plane-wave theory [14] allows the conclusion that the W(110) surface is unreconstructed when clean [2].

Using a quenched molecular-dynamics approach to minimize the total energy in a tight-binding framework, the multilayer relaxation of different low-index surfaces was calculated. It was found that, even when starting from a reconstructed geometry, the W(110) surface spontaneously rearranges itself to the unreconstructed position [3]. A calculation of static electronic generalized susceptibilities supported the conclusion that a reconstruction on W(110) is very unlikely [31].

A two-dimensional effective-Hamiltonian method was used in [15] to study the surface structure of W(110) at T = 0 K as a function of the next-neighbor force constants. The author pointed out that the (110) surface might reconstruct if the surface force constants were reduced. He proposed that the clean W(110) surface is unreconstructed, but a reduction of the surface force constants by the adsorption of H should lead to a p(2 × 1) reconstruction involving in-plane distortion at a certain coverage [15].

The theoretical prediction of an adsorption-induced reconstruction was confirmed by LEED studies on hydrogen-covered W(110) surfaces at temperatures ranging from 85 to

References for 1.2.1 on pp. 58/9

300 K. Comparing the integrated intensities of the (01), (0$\bar{1}$), (1$\bar{1}$), and ($\bar{1}$1) beams, an asymmetry was observed for H coverages θ higher than about 0.5 monolayer indicating that one of the mirror planes had disappeared. This observation is consistent with a lateral shift of the whole top layer of W atoms parallel to the [$\bar{1}$10] direction [11, 12]. Photoemission results on clean and hydrogen-covered W(110) surfaces support this picture [16 to 18]. The variation of the core-level shifts as a function of coverage clearly exhibits two different regimes for θ below and above 0.5. In addition, a valence-band peak 4 eV below E_F is observed at a coverage of 0.5 monolayer [16]. The same conclusion was drawn from photoelectron diffraction measurements on a hydrogen-saturated W(110) surface [2], whereas in a VLEED (very low energy electron diffraction, $0<E\leq20$ eV) study at 300 K, no reconstruction was observed for H coverages up to saturation [19].

Recently, a strong surface phonon anomaly, characterized by a sharply defined dip in the surface phonon dispersion at $q\approx0.9$ Å^{-1} along the [001] direction ($=\bar{\Sigma}$), was found during high-resolution inelastic helium-atom scattering measurements on a hydrogen-saturated W(110) surface. This anomaly was not observed on the clean surface. The relationship between the pronounced phonon anomaly at an incommensurate wave vector and the possible reconstruction is not yet clear [20]. Angle-resolved photoemission measurements predict Kohn anomalies along $\bar{\Sigma}$ for 0.42, 0.91, 1.47, and 2.52 Å^{-1} [21].

Surface relaxation, i.e., a change of the interlayer spacing at the surface, is significantly smaller on the (110) surface than on the (100) and (111) surfaces of W; see e.g. [3]. He^+ ion backscattering studies at 300 K led to the conclusion that there is less than 2% reduction in the first interlayer spacing [13] in good agreement with results of previous LEED studies [4, 6]. A tight-binding study based on the quenched molecular-dynamics approach gave a reduction of 1.4% for the first interlayer distance and 0.4% for the second one [3]. The value for the first interlayer relaxation agrees well with a value of 1.2% from a semiempirical approach (with an empirical N-body potential) [22]. In a more recent calculation [23] based on the short-range analytic EAM model ("Embedded Atom Method") of [24], a reduction of 0.49% and a small expansion of 0.04% were obtained for the first and second interlayer separation, respectively [23]. Contractions of 2.11% [32], 2% [25], and <1% [26] were calculated in further theoretical studies, whereas a tight-binding approach taking into account correlation effects showed no modification of the first interlayer spacing [1]. The same result, i.e., no relaxation at all, was found in an earlier calculation, where empirical interatomic forces were employed [27], and even older calculations showed small expansions [28, 29].

The atomic-scale morphology of the W(110) surface in air was investigated using grazing incident X-ray diffraction. The conclusion from a series of well-defined extra reflections was that the surface contains steps with step heights of 1.4, 2, 4, and 7 atomic layers along the [001] direction [30].

References for 1.2.1:

[1] Bourdin, J. P.; Tréglia, G.; Desjonquères, M. C.; Ganachaud, J. P.; Spanjaard, D. (Solid State Commun. **47** [1983] 279/82).

[2] Chauveau, D.; Guillot, C.; Villette, B.; Lecante, J.; Desjonquères, M. C.; Sébilleau, D.; Spanjaard, D.; Tréglia, G. (Solid State Commun. **69** [1989] 1015/8).

[3] Luo, J. S.; Legrand, B. (Phys. Rev. [3] B **38** [1988] 1728/33).

[4] Van Hove, M. A.; Tong, S. Y. (Surf. Sci. **54** [1976] 91/100).

[5] Tran Minh Duc; Spanjaard, D.; Lecante, J.; Lassailly, Y. (Vide Couches Minces No. 201 Suppl. [1980] 1204/6).

[6] Buchholz, J. C.; Wang, G.-C.; Lagally, M. G. (Surf. Sci. **49** [1975] 508/28).

[7] Lagally, M. G.; Buchholz, J. C.; Wang, G.-C. (J. Vac. Sci. Technol. **12** [1975] 213/21).
[8] Zhurtov, Z. M.; Shebzukhov, A. A. (Fiz. Khim. Poverkhn. **1982** 48/56).
[9] Allan, G. (Phys. Lett. A **47** [1974] 167/8).
[10] Desjonquères, M. C.; Spanjaard, D. (J. Phys. C **15** [1982] 4007/21).

[11] Chung, J. W.; Ying, S. C.; Estrup, P. J. (Phys. Rev. Lett. **56** [1986] 749/52).
[12] Altman, M.; Chung, J. W.; Estrup, P. J.; Kosterlitz, J. M.; Prybyla, J.; Sahu, D.; Ying, S. C. (J. Vac. Sci. Technol. [2] A **5** [1987] 1045/8).
[13] Smith, R. J.; Hennessy, C.; Kim, M. W.; Whang, C. N.; Worthington, M.; Mingde, Xu (Phys. Rev. Lett. **58** [1987] 702/5).
[14] Sébilleau, D.; Tréglia, G.; Desjonquères, M. C.; Spanjaard, D.; Guillot, C.; Chauveau, D.; Lecante, J. (J. Phys. [Paris] **49** [1988] 227/36).
[15] Hu, G. (Chin. Phys. Lett. **2** [1985] 361/4).
[16] Riffe, D. M.; Wertheim, G. K.; Citrin, P. H. (Phys. Rev. Lett. **65** [1990] 219/22).
[17] Gaylord, R. H.; Kevan, S. D. (Phys. Rev. [3] B **37** [1988] 8491/4).
[18] Aiura, Y.; Yanashima, H.; Fukutani, H.; Kato, H.; Sakisaka, Y.; Maruyama, T.; Edamoto, K. (Surf. Sci. **283** [1993] 344/8).
[19] Herlt, H.-J.; Bauer, E. (Surf. Sci. **175** [1986] 336/68).
[20] Hulpke, E.; Lüdecke, J. (Phys. Rev. Lett. **68** [1992] 2846/9; Surf. Sci. **272** [1992] 289/93; Surf. Sci. **287/288** [1993] 837/41).

[21] Gaylord, R. H.; Jeong, K. H.; Dhar, S.; Kevan, S. D. (J. Vac. Sci. Technol. [2] A **7** [1989] 2203/6).
[22] Ackland, G. J.; Finnis, M. W. (Philos. Mag. [8] A **54** [1986] 301/15).
[23] Guellil, A. M.; Adams, J. B. (J. Mater. Res. **7** [1992] 639/52).
[24] Johnson, R. A.; Oh, D. J. (J. Mater. Res. **4** [1989] 1195/201).
[25] Stephenson, P. C.; Bullett, D. W. (Surf. Sci. **139** [1984] 1/15).
[26] Cheng, D. J.; Wallis, R. F.; Dobrzynski, L. (Surf. Sci. **43** [1974] 400/16).
[27] Johnson, R. A.; White, P. J. (Phys. Rev. [3] B **13** [1976] 5293/302).
[28] Jackson, D. P. (Can. J. Phys. **49** [1971] 2093/7).
[29] Wynblatt, P.; Gjostein, N. A. (Surf. Sci. **22** [1970] 125/36).
[30] Jeong, S. M.; Kim, H. W.; Kim, C. Y.; Chung, S. M.; Lee, K. B.; Chung, J. W. (Surf. Sci. **276** [1992] 353/9).

[31] Man'kovskii, S. V.; Ostroukhov, A. A.; Tomilenko, V. N.; Cherepin, V. T. (Metallofizika [Kiev] **13** No. 11 [1991] 54/9).
[32] Rodríguez, A. M.; Bozzolo, G.; Ferrante, J. (Surf. Sci. **289** [1993] 100/26).

1.2.2 Surface Debye Temperature. Surface Atom Vibrations

The effective surface Debye temperatures (denoted $\Theta_\perp$ normal to the surface and $\Theta_\parallel^{(i)}$ in the two principle directions [001] and [1$\bar{1}$0] parallel to the surface) were calculated with a simple decoupled oscillator model with Morse potentials to be $\Theta_\perp = 186$ K, $\Theta_\parallel^{(1)} = 308$ K, and $\Theta_\parallel^{(2)} = 263$ K. The bulk Debye temperature was found to be $\Theta_B = 307$ K [1] (for an overview of bulk Debye temperatures, see "Tungsten" Suppl. Vol. A 2, 1987, pp. 148/50). Values from LEED experiments, $\Theta_\perp = 200 \pm 20$ K [2] and 193 ± 20 K [3], are in good agreement with the calculation above, whereas an He atom scattering study gave a value of $\Theta_\perp = 350$ K [4]. Thus, $\Theta_\perp$ of W(110) is somewhat larger than the corresponding value for W(100); see p. 30. This is to be expected, since the surface atom concentration on the (110) surface is about 1.4 times larger than that on W(100) [3].

References for 1.2.2 on p. 61

An electron-spectroscopic method based on the temperature dependence of the amplitudes of the diffraction peaks in the disappearance potential spectrum is proposed and applied for determining the Debye temperature of the W(110) surface. Θ_D determined in this way depends on the primary electron beam energy E_p. It differs by no more than 20 K from that obtained by LEED [5].

The frequencies of surface modes at high symmetry points ($\bar{H}$, $\bar{P}$, $\bar{N}$) were presented in a theoretical study of the dynamics of atom motion on bcc metal (110) surfaces. The interactions between atoms were modeled with central and angle-bending forces calculated by fitting the model frequencies to bulk frequencies obtained in neutron scattering experiments [6]. The results were found to be qualitatively very similar to the (100) results which had been calculated in a preceding study [7], [6].

Earlier, a similar force model was used to calculate the ratios of the atomic mean-square displacements on the W(110) surface to those in the bulk as a function of temperature. The ratio $\langle u^2\rangle_{Surf}/\langle u^2\rangle_{Bulk}$ is largest for displacements along [110]. At T > 300 K it is about 1.9 for [110], 1.4 for [1$\bar{1}$0], and 1.3 for [001]. This strong anisotropy, which was not found on the W(100) surface, is due to the special site symmetry of an atom on the (110) surface [8]. From the calculations of [1], the corresponding values $\langle u_\perp^2\rangle_{Surf}/\langle u_\perp^2\rangle_{Bulk}=2.7$, $\langle u_{\parallel[1\bar{1}0]}^2\rangle_{Surf}/\langle u_{\parallel[1\bar{1}0]}^2\rangle_{Bulk}=1.4$, and $\langle u_{\parallel[001]}^2\rangle_{Surf}/\langle u_{\parallel[001]}^2\rangle_{Bulk}=1.0$ were derived. Elastic continuum calculations for mean-square displacements of W surface atoms gave ratios of 2.5 [9] and 2.36 [10] for the normal component and 2.0 [9, 10] for the parallel component.

From an LEED study, the enhancement of $\langle u^2\rangle$ normal to the (110) surface was calculated to be between 2 and 3 for the first layer atoms [11]. In further LEED investigations, ratios of the mean square displacements normal to the (110) surface of 2.0 ± 0.2 [3] and 2.4 ± 0.3 [2] were determined.

According to He^+ ion backscattering studies at 300 K, there is a large anisotropy in the thermal vibration amplitudes for the surface atoms with the component normal to the surface, 2.6 times larger than the one-dimensional bulk amplitude (0.05 Å), while the parallel component in the surface plane shows very little enhancement [12].

The mean effective displacement $\Delta_\perp=\langle u_\perp^2\rangle^{1/2}$ normal to the surface was estimated from LEED studies on W(110) to be 0.08 ± 0.008 Å with an oscillation frequency of $(4.0\pm0.4)\times10^{12}\ s^{-1}$ [3].

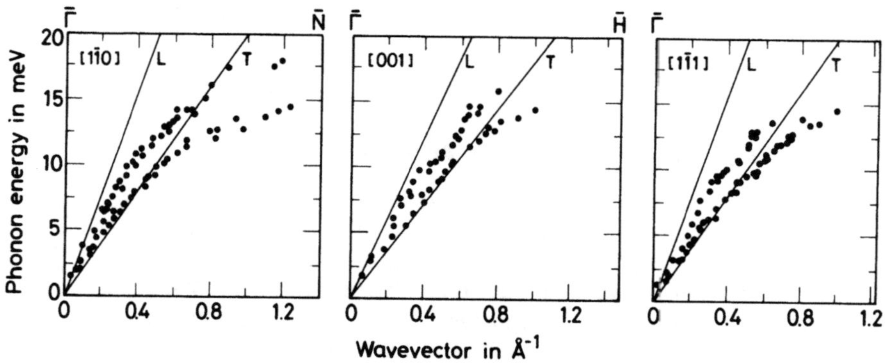

Fig. 20. Surface phonon dispersion curves of the clean W(110) surface along the three high-symmetry directions indicated in each panel. The solid lines represent the slope of the lower edges of the surface-projected longitudinal (L) and transverse (T) bulk phonon bands at $\bar{\Gamma}$ [13].

Collective vibrations of the W(110) surface atoms were investigated by inelastic He atom scattering at temperatures in the range 200 to 1000 K. The data reveal the existence of two phonon branches in each of the high-symmetry directions [001], [$1\bar{1}0$], and [$1\bar{1}1$]; see **Fig. 20**. The mode at lower energies was identified as the transversally polarized Rayleigh mode and the other one as a longitudinally polarized surface resonance. No anomalies or pronounced temperature dependence were found for the clean surface, whereas the hydrogen-saturated surface showed a sharply defined surface phonon anomaly at $q \approx 0.9$ Å^{-1} along the [001] direction which could not be explained [13].

References for 1.2.2:

[1] Jackson, D. P. (Surf. Sci. **43** [1974] 431/40).
[2] Stern, R. M. (Trans. Am. Crystallogr. Assoc. **4** [1968] 14/44).
[3] Babanskaya, L. N.; Boyko, B. A.; Vasilýev, M. A.; Gorodetskii, S. D.; Cherepin, V. T. (Poverkhnost **1983** No. 5, pp. 76/83; Phys. Chem. Mech. Surf. **2** [1984] 1385/99).
[4] Weinberg, W. H. (J. Chem. Phys. **57** [1972] 5463/6).
[5] Kobilanskii, A. V.; Martinyuk, A. V.; Sklar, V. K. (Poverkhnost **1984** No. 3, pp. 47/50; Phys. Chem. Mech. Surf. **3** [1985] 720/7).
[6] Black, J. E.; Huynh, V. T.; Cheng, D. J.; Wallis, R. F. (Surf. Sci. **192** [1987] 541/54).
[7] Black, J. E.; Campbell, D. A.; Wallis, R. F. (Surf. Sci. **115** [1982] 161/82).
[8] Wallis, R. F.; Cheng, D. J. (Solid State Commun. **34** [1980] 847/51).
[9] Dennis, R. L.; Huber, D. L. (Phys. Rev. [3] B **5** [1972] 4717/20).
[10] Wallis, R. F.; Maradudin, A. A.; Dobrzynski, L. (Phys. Rev. [3] B **15** [1977] 5681/5).
[11] Lagally, M. G.; Buchholz, J. C.; Wang, G.-C. (Bull. Am. Phys. Soc. **20** [1975] 490); Buchholz, J. C.; Wang, G.-C.; Lagally, M. G. (Surf. Sci. **49** [1975] 508/28).
[12] Smith, R. J.; Hennessy, C.; Kim, M. W.; Whang, C. N.; Worthington, M.; Mingde, Xu (Phys. Rev. Lett. **58** [1987] 702/5).
[13] Hulpke, E.; Lüdecke, J. (Phys. Rev. Lett. **68** [1992] 2846/9; Surf. Sci. **272** [1992] 289/93).

1.2.3 Surface Electronic Structure

The core-level shifts of the surface atoms, the surface density of states, electronic states and resonances at the surface, and the Fermi surface are dealt with on the following pages. The surface potential barrier, i.e. the form of the potential in the neighborhood of the surface, will be discussed in Section 5.7.3, pp. 214/7.

1.2.3.1 Surface Core-Level Shifts

Photoemission experiments showed that the $4f_{7/2}$ core electron energy levels of the surface atoms are shifted by -0.30 eV [1], -0.29 ± 0.01 eV [2], and -0.321 ± 0.001 eV [3]. The $4f_{7/2}$ core-level spectra of W(110) are well fitted by two peaks (one from the surface and one from the bulk) in all cases. For more detailed information, see Section 5.5.3, pp. 185/93.

References for 1.2.3.1:

[1] Tran Minh Duc; Guillot, C.; Lassailly, Y.; Lecante, J.; Jugnet, Y.; Vedrine, J. C. (Phys. Rev. Lett. **43** [1979] 789/92).
[2] Purcell, K. G.; Jupille, J.; King, D. A. (Surf. Sci. **208** [1989] 245/66).
[3] Riffe, D. M.; Wertheim, G. K.; Citrin, P. H. (Phys. Rev. Lett. **63** [1989] 1976/9).

1.2.3.2 Surface States. Electronic Band Structure

Experimental determinations of surface states and resonances were done predominantly by ultraviolet photoemission spectroscopy [1 to 10]. A detailed description of the measured spectra will be given in Section 5.5.2.3, pp. 173/7. Here, the main results of these studies are summarized.

The normal photoemission results of [1] gave evidence for a narrowing of the d bands at the surface. A simple model introduced in the companion theoretical paper [11] accounted for the d-band narrowing at the surface as a consequence of the reduced d-d overlap resulting from the lower number of neighbor atoms as compared to the bulk. Fairly good agreement was achieved between the surface DOS (density of states) derived from this model and the experimental energy distribution of the electrons emitted by the surface effect [1]; cf. also the review [12]. No surface state or resonance were detected in these early studies [1, 11, 12].

A surface state on W(110) was observed for the first time in the angle-resolved photoemission measurements of Holmes et al. [2 to 4]. This surface state was identified and characterized throughout the surface Brillouin zone as having predominantly m = 1 d-orbital character. At the zone center the surface state does not exist, it could only be observed at emission angles well away from the surface normal [2 to 4]. In [4] the dispersion of this state was found to correlate with the band gap between the third and fourth bands (about 1.2 and 0.6 eV below E_F at Γ) of the bulk band structure from the RAPW (relativistic augmented plane wave) band structure calculation of [11]; see "Tungsten" Suppl. Vol. A 2, 1987, Fig. 6, p. 121.

In a more recent, angle-resolved photoemission study of Gaylord and Kevan [5], a surface resonance was observed (i.e., a state that is not entirely located at the surface as a true surface state but couples to bulk states of the same energy and symmetry for several parallel momenta; see e.g. [12, pp. 171/3]). This resonance is located near the center $\bar{\Gamma}$ of the surface Brillouin zone at 1.18 ± 0.04 eV below E_F in a pseudogap between Γ_8^+ and Γ_7^+ (that corresponds to the abovementioned gap between the third and fourth bands at the zone center of the bulk band structure) opened by the spin-orbit interaction. It is degenerate with a bulk band which in absence of the spin-orbit interaction would have a different symmetry than a state in the gap. The resultant weak coupling leads to the formation of a well-defined resonance. This surface feature was observed to vanish rapidly for emission in both symmetry azimuths away from the normal, which presumably explains why it had not been found before [5]. In subsequent papers of the same group, a variety of surface-localized states along $\bar{\Sigma}$ [6 to 8] and $\bar{\Delta}$ [9, 10] were detected. A considerable shift to higher binding energies with hydrogen adsorption was observed and interpreted to be related to the hydrogen-induced reconstruction of the W(110) surface (cf. Section 1.2.1, pp. 56/8) [6 to 10].

From measurements with other experimental methods no conclusive results concerning surface states or resonances on W(110) were obtained. The total energy distribution of field-emitted electrons from W(110) exhibits some structure around 0.5 eV below E_F, but this feature could not be unequivocally related to a surface state [13]. Photofield emission data revealed a weak structure at 0.37 eV below E_F in the energy distribution curve [14]. In another photofield emission study some features at 0.47, 0.58, 0.79, and 0.89 eV below E_F were observed [15]. None of these features was identified as a surface state or a surface resonance [14, 15].

A peak at 1.8 eV above E_F in the inverse photoemission spectrum was interpreted as a surface state. It is very sensitive to contamination [16]. An unoccupied surface state at

7.3 eV above E_F, lying in the large band gap between 5.5 and 10 eV above E_F of bulk W along Σ, was observed in another inverse photoemission study [17]; cf. "Tungsten" Suppl. Vol. A 3, 1989, p. 267.

Surface reflectance spectroscopy with s-polarized photons in the energy range 0.6 to 6 eV reveals a negative feature at 3.9 eV and a positive one at 5.5 eV in the optical conductance curves that had been derived from an inversion of the changes measured in reflectance. The results seem to be ambiguous and do no correlate with any other data [18].

Electronic structure calculations of [19] in a self-consistent atomic-orbital approach (all interatomic overlaps and hopping matrix elements were calculated by direct integration over the appropriate atomic spheres) give a number of surface states which are all confined to small regions of the surface Brillouin zone. One of these states has the correct energy range and symmetry to be identified with the experimentally observed state of Holmes et al. [2 to 4], but while the "Holmes state" spreads over a large part of the surface Brillouin zone including $\bar{\Delta}$ as well as $\bar{\Sigma}$, the calculated state only is found over a small region along $\bar{\Delta}$. Because they are confined to small regions of the surface Brillouin zone, none of the highly localized resonances and states make significant contributions to the peaks in the DOS. These peaks, which occur at 0.8, 1.5, 2.5, and 3.5 eV below E_F, are made up of weakly localized states distributed throughout the zone. The close-packed nature of the W(110) surface and the corresponding small change in the number of neighbors with respect to the bulk make the difference between the surface DOS and the bulk DOS much smaller than for the open (100) surface [19].

A further tight-binding calculation [20] with a nonorthogonal basis set yields a strong surface feature located at 4.62 eV below E_F at $\bar{\Gamma}$, also appearing along the $\bar{\Delta}$ and $\bar{\Sigma}$ direction. This peak is mainly made up of orbitals s and $3z^2r^2$. It is interpreted as a surface resonance in spite of the lack of experimental evidence. Other peaks in the spectral DOS correspond to the dominant surface resonance found by [2 to 4]. In addition to the total DOS for the surface, the DOS for the next three interior planes was calculated. The spectral analysis of the W(110) surface shows that no surface resonance crosses the Fermi level in contrast to the case of the (100) surface. The Fermi level lies in a DOS minimum, i.e., while the DOS is large at E_F for W(100), it is small for W(110) [20]. A semiempirical, topological method (essentially a classical molecular orbital approach) was used to calculate the local DOS for the W(110) surface. A comparison of the results with the experimental photoemission data of [1] is given, but the agreement is not very good [21].

Self-consistent band structure calculations were performed for a five-layer film of W(110) using the linear-augmented-plane-wave method without spin-orbit coupling and assuming a local-density approximation for exchange and correlation [22]. The layer-projected DOS curves for the subsurface and second layers below the surface show pronounced similarities and are also remarkably similar to the results for W(100) surfaces. The Fermi energy lies in a minimum in the pronounced d structure (in agreement with the results of [20]) and the occupied d states are separated by another deep minimum around 4 to 5 eV below E_F. The surface-projected DOS also shows some similarity to that of the W(100) surface. The lower d states are removed and some states are pushed up into a valley at about 5 eV below E_F, forming surface states or resonances. The valley near the Fermi energy has disappeared at the surface due to the appearance of surface states. This is also similar to the (100) surface, even though the structure is broader and lower. Two distinct surface-state peaks may be seen at 0.7 and 1.6 eV below E_F [22]. The projected surface DOS published by [19] and [20], resulting from non-self-consistent model calculations, bear only superficial resemblance to these results [22].

References for 1.2.3.2 on p. 64

The corrugation of the electron density, i.e., its periodic dependence on x and y (z is the axis normal to the surface) is least on close-packed planes, as e.g. W(110), and larger on more open planes, as e.g. W(100). This was proven by a density functional calculation using a pseudopotential [23].

References for 1.2.3.2:

[1] Feuerbacher, B.; Christensen, N. E. (Phys. Rev. [3] B **10** [1974] 2373/90).
[2] Holmes, M. W.; King, D. A.; Inglesfield, J. E. (Phys. Rev. Lett. **42** [1979] 394/7).
[3] Holmes, M. W.; Inglesfield, J. E. (Surf. Sci. **89** [1979] 133/41).
[4] Holmes, M. W.; King, D. A. (Proc. R. Soc. London A **376** [1981] 565/83).
[5] Gaylord, R. H.; Kevan, S. D. (Phys. Rev. [3] B **36** [1987] 9337/40).
[6] Gaylord, R. H.; Kevan, S. D. (Phys. Rev. [3] B **37** [1988] 8491/4).
[7] Kevan, S. D.; Gaylord, R. H. (J. Vac. Sci. Technol. A **6** [1988] 636/7).
[8] Jeong, K. H.; Gaylord, R. H.; Kevan, S. D. (J. Vac. Sci. Technol. A **7** [1989] 2199/202).
[9] Gaylord, R. H.; Jeong, K. H.; Kevan, S. D. (Phys. Rev. Lett. **62** [1989] 2036/9).
[10] Gaylord, R. H.; Jeong, K. H.; Dhar, S.; Kevan, S. D. (J. Vac. Sci. Technol. A **7** [1989] 2203/6).

[11] Christensen, N. E.; Feuerbacher, B. (Phys. Rev. [3] B **10** [1974] 2349/72).
[12] Feuerbacher, B.; Willis, R. F. (J. Phys. C **9** [1976] 169/216, 209).
[13] Plummer, E. W.; Gadzuk, J. W. (Phys. Rev. Lett. **25** [1970] 1493/5).
[14] Lopinski, G. P.; Donders, P. J. (Surf. Sci. **201** [1988] 294/310).
[15] Venus, D.; Lee, M. J. G. (Surf. Sci. **172** [1986] 477/95).
[16] Funnemann, D.; Merz, H. (J. Phys. Colloq. [Paris] **48** [1987] C9-1037/C9-1040).
[17] Artamanov, O. M.; Belkina, G. M.; Samarin, S. N.; Yakovlev, I. I. (Fiz. Tverd. Tela [Leningrad] **29** [1987] 753/5; Sov. Phys.-Solid State [Engl. Transl.] **29** [1987] 430/1).
[18] Restorff, J. B.; Drew, H. D. (Surf. Sci. **88** [1979] 399/426).
[19] Stephenson, P. C.; Bullett, D. W. (Surf. Sci. **139** [1984] 1/15).
[20] López Sancho, M. P.; López Sancho, J. M.; Rubio, J. (J. Phys. C **18** [1985] 1803/15).

[21] Pancíř, J.; Haslingerova, I. (Chem. Phys. Lett. **107** [1984] 598/602).
[22] Jepsen, O.; Jones, R. O. (Phys. Rev. [3] B **34** [1986] 6695/8).
[23] Gubanov, A. I.; Dunayevskiy, S. M. (Fiz. Met. Metalloved. **47** No. 6 [1979] 1141/6; Phys. Met. Metallogr. [Engl. Transl.] **47** No. 6 [1979] 12/8).

1.2.3.3 Fermi Surface

Angle-resolved photoemission measurements of the Fermi surfaces of several localized states on clean and hydrogen-covered W(110) are reported by [1 to 3]. With respect to crossings of the Fermi level, the W(110) surface differs from W(100) in that the two-dimensional surface states or resonances cross E_F at single points; thus, with W(110) there are no extended regions close to E_F [1 to 3]. **Fig. 21** (from [1]) shows a two-dimensional map of the Fermi-level crossings. The shaded areas are the projection of the measured bulk Fermi surface (from [4]) onto the (110) plane. There is a nesting wave vector along $\bar{\Delta}$ of magnitude 0.3 $Å^{-1}$ that couples the distorted cross-shaped structure around $\bar{\Gamma}$ in Fig. 21 to the structure around $\bar{N}$ [1, 3]. From the Fermi surface data it is concluded that there may exist several Kohn anomalies, the more pronounced of which are listed in [2].

Four closed Fermi surface orbits are found, three elliptical hole orbits (enclosing unoccupied levels) and one electron orbit of irregular shape (see Fig. 21). The electron orbit, centered around $\bar{\Gamma}$, extends across the first surface Brillouin zone boundary line $\bar{P}$-$\bar{H}$ and,

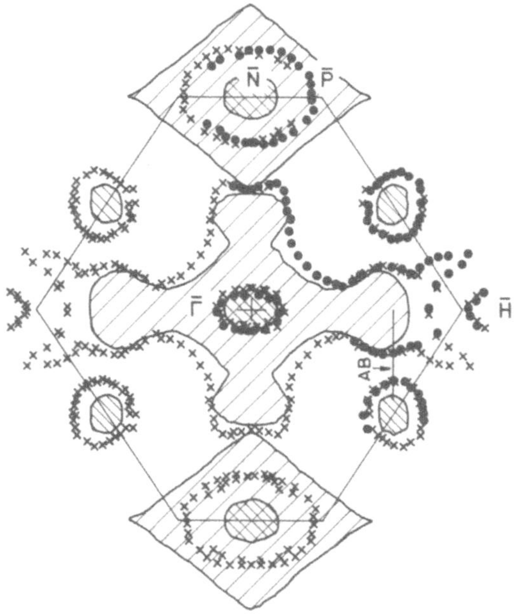

Fig. 21. Fermi surface of clean W(110). Solid circles represent real data, the crosses were produced by mirror plane symmetry. The shaded areas result from a projection of the bulk Fermi surface (taken from [4]). Along the line AB angle-resolved UPS line widths were studied; see p. 177. Scale: $\bar{\Gamma}$-$\bar{N}$ = 1.407 Å^{-1} [1].

for the most part, surrounds the projection of a bulk electron pocket (commonly referred to as a "jack") taken from the de Haas-van Alphen results of Girvan et al. [4]; see "Tungsten" Suppl. Vol. A 2, 1987, Fig. 7, p. 126. Of the three hole orbits one is centered around $\bar{\Gamma}$, the second is located symmetrically on the line $\bar{P}$-$\bar{N}$-$\bar{P}$, and the third is located on the line $\bar{P}$-$\bar{H}$. All three hole orbits surround the projections of bulk hole ellipsoids which are centered at the N points of the bulk Brillouin zone; cf. "Molybdenum" Suppl. Vol. A 2a, 1985, Fig. 72, p. 341. These bulk hole ellipsoids project onto three separate regions of the surface Brillouin zone. The first is located along the edge of the surface Brillouin zone between $\bar{H}$ and $\bar{P}$. The surface hole orbit around this bulk projection is located in a true gap and is thus a surface state. At the $\bar{N}$ and $\bar{\Gamma}$ points, projections of the ellipsoids are embedded in projections of the bulk hole octahedra and the bulk electron jack, respectively. The surface hole orbits are also embedded in these projections and are thus resonances [1]. The observation of these well-defined resonances is, in some cases, related to the spin-orbit interaction [5]. A comparison of the Fermi surface of W(110) from Fig. 21 with that of Mo(110), derived from analogous measurements, reveals strong similarities [6].

The hole-orbit states are rapidly attenuated by hydrogen, while the electron-orbit states are shifted to higher binding energies resulting, initially, in an expansion of the Fermi surface and, ultimately, in its conversion to two hole orbits [1, 2, 7].

References for 1.2.3.3:

[1] Gaylord, R. H.; Jeong, K. H.; Kevan, S. D. (Phys. Rev. Lett. **62** [1989] 2036/9).

[2] Gaylord, R. H.; Jeong, K. H.; Dhar, S.; Kevan, S. D. (J. Vac. Sci. Technol. A **7** [1989] 2203/6).

[3] Smith, K. E.; Kevan, S. D. (Prog. Solid State Chem. **21** [1991] 49/131, 88/92).
[4] Girvan, R. F.; Gold, A. V.; Phillips, R. A. (J. Phys. Chem. Solids **29** [1968] 1485/502).
[5] Gaylord, R. H.; Kevan, S. D. (Phys. Rev. [3] B **36** [1987] 9337/40).
[6] Jeong, K. H.; Gaylord, R. H.; Kevan, S. D. (Phys. Rev. [3] B **39** [1989] 2973/81).
[7] Jeong, K. H.; Gaylord, R. H.; Kevan, S. D. (J. Vac. Sci. Technol. A **7** [1989] 2199/202).

1.3 W(111)

1.3.1 Atomic Structure. Relaxation

In contrast to W(100) and W(110), the (111) surface of W does not reconstruct with or without hydrogen [1].

The relaxation of the relatively open W(111) surface structure (compared to W(100) and W(110)) is the largest among the three low-index surfaces [2 to 5]. Theoretical considerations imply that the relaxations normal to the surface are related to the openness or roughness of the surfaces (the roughness of a surface is defined as the inverse of the fraction of the area in the surface plane occupied by atoms with a diameter equal to the bulk nearest-neighbor distance) [2]. The increase of top-layer relaxation with increasing roughness had been experimentally observed before on six different bcc-Fe surfaces [6]. Relaxations of 19%, 10%, and 2% were predicted for W(111), W(100), and W(110), respectively, from a total energy calculation using a Born-Mayer-type repulsive interaction [3]. A more recent calculation of multilayer relaxation within the tight-binding scheme, using the quenched molecular dynamics method, gives the following changes δd_{ij} in the distance between the i-th and j-th layer (relative to the bulk interlayer distance d): $\delta d_{12}/d = -16.3\%$, $\delta d_{23}/d = -7.8\%$, $\delta d_{34}/d = +7.3\%$, $\delta d_{45}/d = -1.1\%$, $\delta d_{56}/d = -1.9\%$, and $\delta d_{67}/d = 0$ [4]. The "contraction-contraction-expansion-contraction" sequence for d_{12}, d_{23}, d_{34}, and d_{45} of the (111) face appeared in all bcc transition metals investigated (Nb, Ta, Mo, W) [4] and was experimentally demonstrated for Fe with LEED [7]. A different sequence, (slight) expansion-contraction-expansion, was reported in an older calculation where empirical interatomic forces were used [8]. This sequence, except for a much larger expansion of the first interlayer spacing d_{12} (17.78%), was also found with another theoretical approach by summing pairwise interacting Morse potentials [5]. Very small expansions of $\delta d_{12}/d = -2.98\%$ and $\delta d_{23}/d = -4.15\%$ were found in a more recent theoretical calculation [9].

The surface structure of W(111) was analyzed with high-resolution proton energy loss spectroscopy using glancing scattering geometry for the incident 100 keV protons. Computer simulations of the spectra for various spacings of the first interlayer were compared with the measured loss spectra. The best results were obtained, when a contraction of 0.05 Å corresponding to about 5% of the bulk interlayer spacing was assumed [10].

References for 1.3.1:

[1] Riffe, D. M.; Wertheim, G. K.; Citrin, P. H. (Phys. Rev. Lett. **65** [1990] 219/22).
[2] Jiang, P.; Marcus, P. M.; Jona, F. (Solid State Commun. **59** [1986] 275/80).
[3] Stephenson, P. C.; Bullett, D. W. (Surf. Sci. **139** [1984] 1/15).
[4] Luo, J. S.; Legrand, B. (Phys. Rev. [3] B **38** [1988] 1728/33).
[5] Jackson, D. P. (Can. J. Phys. **49** [1971] 2093/7).
[6] Sokolov, H. D.; Jona, F.; Marcus, P. M. (Solid State Commun. **49** [1984] 307/12).
[7] Sokolov, J.; Jona, F.; Marcus, P. M. (Phys. Rev. [3] B **33** [1986] 1397/400).
[8] Johnson, R. A.; White, P. J. (Phys. Rev. [3] B **13** [1976] 5293/302).
[9] Rodríguez, A. M.; Bozzolo, G.; Ferrante, J. (Surf. Sci. **289** [1993] 100/26).
[10] Matsunami, N. (Toyoda Kenkyu Hokoku No. 43 [1990] 69/75; C.A. **114** [1991] No. 150956).

1.3.2 Surface Debye Temperature

Effective surface Debye temperatures of $\Theta_\perp = 184$ K normal to the surface and $\Theta_\parallel = 191$ K for both principle directions parallel to the surface were calculated for W(111) with a simple decoupled oscillator model [1]. By comparing the computer-simulated and experimental proton energy loss spectra from W(111), values for the surface Debye temperature $\Theta_\perp$ of 100, 200, 300, and 400 K were used as fit parameters. The best result was obtained by assuming $\Theta_\perp = 200$ K [2].

References for 1.3.2:

[1] Jackson, D. P. (Surf. Sci. **43** [1974] 431/40).
[2] Matsunami, N. (Toyoda Kenkyu Hokoku No. 43 [1990] 69/75; C.A. **114** [1991] No. 150956).

1.3.3 Surface Electronic Structure

1.3.3.1 Surface Core-Level Shifts

Shifts of the $4f_{7/2}$ core electron energy levels of the surface atoms were determined by fitting either three or four lines to the photoemission spectra. By representing the measured data with three lines, surface core level shifts of -0.43 ± 0.01 eV [1] and -0.439 ± 0.009 eV or -0.436 ± 0.003 eV (depending on the used fit parameters for the line shapes) [2] were obtained, whereas a four-line fit resulted in shifts of -0.46 ± 0.03 eV [3] and -0.484 ± 0.007 eV [2]. For more detailed information, see Section 5.5.3, pp. 185/93.

References for 1.3.3.1:

[1] Van der Veen, J. F.; Heimann, P.; Himpsel, F. J.; Eastman, D. E. (Solid State Commun. **37** [1981] 555/9).
[2] Wertheim, G. K.; Citrin, P. H. (Phys. Rev. [3] B **38** [1988] 7820/3).
[3] Purcell, K. G.; Jupille, J.; Derby, G. P.; King, D. A. (Phys. Rev. [3] B **36** [1987] 1288/91).

1.3.3.2 Surface States. Electronic Band Structure

Electronic states and the band structure of the W(111) surface were experimentally investigated by different methods such as photoemission [1 to 4], field emission [5], photofield emission [6 to 8], and surface reflectance spectroscopy [9].

Especially angle-resolved ultraviolet photoemission spectroscopy (ARUPS) provides detailed information about the electronic surface structure; see also Section 5.5.2.5, p. 184. Polarized photons were used by Cerrina et al. [1 to 3] to discriminate between the symmetry of the initial states. According to the symmetry rules formulated by [10], normal-emission, s(p)-polarized radiation gives rise to excitations from $\bar{\Gamma}_3$ ($\bar{\Gamma}_1$) initial states only, and $\bar{\Sigma}_1$ ($\bar{\Sigma}_2$) states are found with s-polarized light and a polarization vector parallel (perpendicular) to the mirror plane. At the zone center four surface-sensitive structures were detected, two states of $\bar{\Gamma}_1$ symmetry at 0.5 and 2.0 eV below E_F and two states of $\bar{\Gamma}_3$ symmetry at about 0.4 and 1 eV below E_F. Measurements away from the zone center revealed several flat bands, suggesting strongly localized wave functions in real space with well-defined d character. It appears that the open structure of the W(111) surface, which gives a very small overlapping between the d orbitals, causes the surface atoms to interact more with underlayer atoms than with other surface atoms [1 to 3].

In angular-integrated energy distribution curves, a strong surface feature at about 0.7 eV below E_F was observed [3]. A corresponding peak had been detected before in photoemis-

References for 1.3.3.2 on pp. 69/70

sion experiments of Feuerbacher and Christensen [4]. They had measured energy distribution spectra of photoelectrons emitted normal to the W(111) surface. Several peaks were found, but only the two highest lying features at 0.7 and 1.2 eV below E_F were attributed to surface emission, because their position was independent of the photon energy and their peak size was fairly constant [4].

In a field emission study two features at 0.70 and 1.33 eV below E_F were observed in total energy distribution curves in good agreement with the photoemission results mentioned above. Whereas the lower lying feature is believed to be a surface state, the higher one was found to be not as sensitive to surface contamination [5].

Photofield emission measurements with p-polarized light revealed two anomalies in the current-voltage characteristics at about 0.35 and 0.85 eV below E_F which were attributed to surface states [6]. The energy distribution of photofield electron emission from W(111) was measured in [7, 8]. In good agreement with the photoemission [3, 4] and field emission [5] studies mentioned above, some structure in the distribution curves was observed at energies about 0.7 eV below the Fermi level (0.688 eV [7], 0.69 eV [8]). In addition, features at 0.925, 1.30, and 1.56 eV below E_F were detected [7].

Surface reflectance spectroscopy with s-polarized light and photon energies ranging from 0.6 to 6 eV was applied to several W surfaces including W(111) [9]. The changes $\Delta R/R$ measured in reflectance were converted into surface optical conductance curves using a technique developed by [11] in order to relate the spectra to changes in the electronic structure. The converted data show two negative features at 1.4 and 2.8 eV and two sharp positive features at 3.5 and 4.0 eV, but the analysis of the curve is not cogent. The correlation with the photoemission, field emission, and photofield emission data cited above is very poor [9].

The projected bulk DOS calculated with the relativistic-augmented-plane-wave method [12] shows an absolute band gap between 0.6 and 1.2 eV below E_F. A crude estimate for the surface DOS was derived from the bulk band structure by narrowing the d bands based on the effective number of neighbors per surface atom [12]. The difference between the surface and bulk DOS taken from tight-binding calculations shows large positive and negative peaks (in contrast to W(110)) due to the rather open nature of the W(111) surface [13].

The data from photoemission experiments were compared with the results of a tight-binding calculation by Cerrina et al. [1 to 3]. In this theoretical approach the W(111) surface is simulated by a slab of sixteen regularly stacked layers. A semiempirical LCAO Hamiltonian was obtained by fitting this model to the relativistic bulk band structure calculated by [14]. Surface relaxation effects were not included. Six surface bands departing from $\bar{\Gamma}$ (four along $\bar{\Sigma}_1$ and two along $\bar{\Sigma}_2$) were found. Although several discrepancies remain between experiment and theory, good agreement was found concerning the very small dispersion of all surface state and resonance bands in the surface Brillouin zone [1 to 3].

By simulating the surface by slabs of four, seven, or nine atomic layers, the electronic structure of the W(111) surface was calculated along the edges of the irreducible zone employing a self-consistent atomic orbital approach where the matrix elements are derived by direct numerical integration over an atomic sphere [13]. The results were compared with the experimental and theoretical data of [1 to 3]. Several surface states were found, of which the most prominent one is 90% localized at the surface at $\bar{\Gamma}$ where it is located 0.45 eV below E_F. It disperses upwards along $\bar{\Gamma}$-$\bar{M}$(=$\bar{\Sigma}$) and $\bar{\Gamma}$-$\bar{K}$ (note that the line $\bar{\Gamma}$-$\bar{K}$ is labeled $\bar{T}$ in [1 to 3] and $\bar{T}'$ in [13], and the line $\bar{M}$-$\bar{K}$ is labeled $\bar{T}'$ in [1 to 3] and $\bar{T}$ in [13]). At the zone center two other states were found at 1.3 and 1.15 eV below E_F. The state lying 1.3 eV below E_F is made up of d_{z^2} and $d_{x^2-y^2}$ orbitals which are even with respect to reflection in the $\bar{\Sigma}$ mirror plane. The photoemission experiments of [1 to 3] indicate the existence of a state of the same symmetry with an energy of 1.0 eV below E_F at $\bar{\Gamma}$ and for a considerable distance towards $\bar{M}$. The third state at 1.15 eV below E_F at $\bar{\Gamma}$ has $\bar{\Gamma}_3$ symmetry and gives rise to a $\bar{\Sigma}_2$ band dispersing upwards [13]. This compares well with a theoretical state found in the LCAO approach of [1 to 3] with an energy of 1.6 eV below E_F at $\bar{\Gamma}$, but the corresponding experimental state at 2.0 eV below the Fermi level occurs only at $\bar{\Gamma}$ [1 to 3]. Along the other two edges of the surface Brillouin zone $\bar{T}$ and $\bar{T}'$, many states were found by [13] and [1 to 3], but it is difficult to identify the symmetry of these states, and there are few experimental data available besides [1 to 3] for comparison [13].

A tight-binding calculation of the one-dimensional DOS for the two bulk band symmetries Λ_1 and Λ_3 was performed for the first, second, fourth, and eighth layer and for the bulk [15]. The symmetry line appropriate to normal emission extends from Γ to P along Λ and from P to H along F in the bulk Brillouin zone; see "Molybdenum" Suppl. Vol. A 2a, 1985, Fig. 80, p. 366, Fig. 72, p. 341. See also the energy band diagram of bulk W in "Tungsten" Suppl. Vol. A 2, 1987, Fig. 1, p. 107. The atomic orbitals s, p_z, and d_{z^2} form a basis for $\Lambda_1(F_1)$ and p_x, d_{xz}, and d_{xy} for $\Lambda_3(F_3)$, if the z axis is in (111) direction. The results clearly show the removal of bulk singularities near the surface and the emergence of a surface resonance in the spd gap. This peak in the Λ_1 DOS 1.36 eV below E_F is degenerate with band states of F_1 symmetry into which it can decay; thus, it is not a true surface state. A state in the Λ_3 DOS at 0.75 eV below E_F on the other hand is due to a surface state in the absolute gap formed by the Λ_3 bands. These two peaks at 1.36 and 0.75 eV below E_F were assigned to the normal emission structures found in [4] at 1.2 and 0.7 eV which are stationary over a wide range of photon energies [15].

A semiempirical topological method (essentially a classical molecular orbital approach) was used to calculate the local DOS for several W surfaces including the (111) face. The results were compared with the experimental data of [4] giving no satisfactory agreement, perhaps because of the simplicity of the model [16].

References for 1.3.3.2:

[1] Cerrina, F.; Anderson, J. R.; Lapeyre, G. J.; Bisi, O.; Calandra, C. (Phys. Rev. [3] B **25** [1982] 4949/61).

[2] Cerrina, F.; Williams, G. P.; Anderson, J. R.; Lapeyre, G. J.; Bisi, O.; Calandra, C. (Recent Dev. Condens. Matter Phys. Pap. 1st Gen. Conf., Antwerp 1980 [1981], Vol. 2, pp. 391/7; C.A. **96** [1982] No. 11953).

[3] Cerrina, F.; Williams, G. P.; Anderson, J. R.; Lapeyre, G. J.; Bisi, O.; Calandra, C. (Vide Couches Minces No. 201 [1980] 1162/5; C.A. **93** [1980] No. 214196).

[4] Feuerbacher, B.; Christensen, N. E. (Phys. Rev. [3] B **10** [1974] 2373/90).

[5] Plummer, E. W.; Gadzuk, J. W. (Phys. Rev. Lett. **25** [1970] 1493/5).

[6] Radoń, T.; Jaskółka, S. (Surf. Sci. **200** [1988] 199/208).

[7] Venus, D.; Lee, M. J. G. (Surf. Sci. **172** [1986] 477/95).

[8] Lopinski, G. P.; Donders, P. J.; Lee, M. J. G. (Surf. Sci. **201** [1988] 294/310).

[9] Restorff, J. B.; Drew, H. D. (Surf. Sci. **88** [1979] 399/426).

[10] Hermanson, J. (Solid State Commun. **22** [1977] 9/11).

[11] Prange, R. E.; Drew, H. D.; Restorff, J. B. (J. Phys. C **10** [1977] 5083/8).

[12] Christensen, N. E.; Feuerbacher, B. (Phys. Rev. [3] B **10** [1974] 2349/72).

[13] Stephenson, P. C.; Bullett, D. W. (Surf. Sci. **139** [1984] 1/15).

[14] Mattheiss, L. F. (Phys. Rev. [2] A **139** [1965] 1893/904).

[15] Kawajiri, M.; Hermanson, J.; Schwalm, W. (Solid State Commun. **25** [1978] 303/5).
[16] Pancíř, J.; Haslingerová, I. (Chem. Phys. Lett. **107** [1984] 598/602).

1.4 W(211) and Other Surfaces

1.4.1 Atomic Structure. Relaxation

If the structure of a crystal surface has a symmetry with less than two mirror planes, as for example all bcc surfaces except (100), (110), and (111), the relaxation may have a component parallel to the surface in addition to the normal component. Experimental results are only available for W(211) [1 to 3] and W(310) [4]. The W(211) surface has an open, corrugated structure. It consists of parallel, close-packed rows of atoms along $[1\bar{1}\bar{1}]$ (i.e., parallel to the only existing mirror plane) separated by wide channels. According to an LEED-I-V profile analysis, a perpendicular relaxation (contraction) of the top layer of 0.16 Å ($\delta d_{12}/d = -12.4\%$) and a parallel relaxation of 0.09 Å along the close-packed rows in $[1\bar{1}\bar{1}]$ direction occurs [1]. Recently, by using a pulsed 4 keV Ar^+ ion beam, time-of-flight scattering and recoiling spectrometry (TOF-SARS) with detection of both neutrals and ions was applied for the structural analysis of a clean W(211) surface [2]. The results show, in good agreement with LEED [1], that the first- to second-layer spacing is contracted by 0.12±0.07 Å ($\delta d_{12}/d = -9.3\%$) and that there is a lateral shift from the bulk-truncated surface along the $[1\bar{1}\bar{1}]$ direction such that the layer registry is shifted by 0.10±0.07 Å. This shift brings the center of the first-layer atoms nearer to the symmetrical bridge position between two second-layer atoms. Scattering structural contour maps and scattering structural plots representing the W(211) surface symmetry in the (α,δ) space (α = polar beam incident angle, δ = surface azimuthal angle) are given by [2]; see also [3].

A calculation of multilayer relaxation for the (211), (210), and (310) surfaces of different bcc metals including W, based on the quenched molecular-dynamics approach in which the total energy is minimized in a tight-binding framework, is given by [5]. Results are reported for perpendicular as well as for parallel relaxations. The values for the relative perpendicular relaxation between the i-th and the j-th layer $\delta d_{ij}/d$ (in %) are given in the following table and compared with the abovementioned experimental values (from [1], [2], and [4]) and with further theoretical results [6 to 9]. In the calculations of [6, 7], the energy was minimized according to a modified point-ion model. In addition, results for (410) and (510) are given in [7]. A second-moment approach of the tight-binding scheme with hopping integrals between nearest neighbors and repulsive Born-Mayer interactions was used in [8], whereas the data for the (310) relaxation from [9] were obtained by a semiempirical calculation using empirical N-body potentials.

ij	(211) $\delta d_{ij}/d$ in %					(210) $\delta d_{ij}/d$ in %		(310) $\delta d_{ij}/d$ in %			
12	−10.8	−10.6	−8.3	−12.4	−9.3	−19.9	−22.0	−16.9	−16.8	−6.0	−18.3
23	+3.3	+4.1	−1.5			−5.0	−9.8	+5.2	+4.2	+2.3	−0.7
34	−1.2	−1.6	+0.6			+7.7	+10.5	−1.2	−1.5	−2.6	−0.6
45	−0.5	+0.6				−1.3	−4.7	−1.0	+1.1	+0.7	−1.7
56		−0.2				−5.8	+0.8	+0.1	−0.6		
67		+0.1				+2.4					
Ref.	[5]	[6]	[8]	[1]	[2]	[5]	[7]	[5]	[7]	[9]	[4]

Despite the fact that the roughness of the (310) surface (defined as the inverse of the fraction of the area in the surface plane occupied by atoms of diameter equal to the bulk nearest-neighbor distance) is smaller than that of the (111) surface, the contraction of the first layer was calculated to be slightly larger for the unsymmetrical (310) surface (−16.9%) than for (111) (−16.3%). The same result was obtained for Nb and Mo, but not for Ta [5]. This theoretically predicted behavior was experimentally observed on Fe [10] and was supported by the calculations mentioned before based on a modified point-ion model [6]. A more recent theoretical study using the equivalent crystal theory also confirmed this result giving contractions of 4.44% for W(310) in comparison to 2.98% for W(111). In the same study contractions of 4.49%, 2.54%, 1.67%, and 2.84% were calculated for the (211), (210), (311), and (331) surfaces, respectively [16]. By comparing LEED intensity-energy spectra with spectra calculated with the multiple-scattering theory, a large contraction of the first interlayer spacing of 18.3% was found for the W(310) surface [4].

In some older studies calculations were performed for W(211) [11 to 13] and for W(321) [13]. In [11] empirical interatomic forces were used, whereas in the other studies pairwise interaction models with a Mie potential [12] or with a modified (Girifalco and Weizer) Morse potential [13] were applied. For the top layer of the (211) surface, outward normal relaxations of about 0.13 Å ($\delta d_{12}/d = 10\%$) [12] and 0.18 Å ($\delta d_{12}/d = 14.0\%$) [13] were obtained, whereas in [11] normal relaxation was found to be negligible. The calculation for the (321) surface resulted in an outward relaxation of 0.16 Å ($\delta d_{12}/d = 19.3\%$) [13].

The values given for the surface-parallel relaxation are more difficult to compare, because different conventions were used for the base directions and the sign of the shift. The direction of the displacement depends on the crystallographic orientation of the surface. For the (211) surface, the parallel shift was found to be towards a higher symmetry of the first layer with respect to the second layer [5]. This is in agreement with experimental results on W(211) [1, 2] and Fe(211) [14], and with theoretical calculations for W(211) [6, 11 to 13] and Na(211) [15], but it is contrary to the results given in [8], where a horizontal shift of 0.12 Å of the topmost layer towards higher symmetry with respect to the third-layer atoms located directly below (and hence towards lower symmetry with respect to second-layer atoms located aside) was calculated. In the older theoretical studies [11 to 13], surface layer displacements for W(211) of 0.14 Å [11], 0.41 Å [12], and 0.49 Å [13] parallel to the [$1\bar{1}\bar{1}$] direction were obtained. In the more recent, abovementioned studies [5] and [6], parallel relaxations of 0.16 Å and 0.35 Å, respectively, in the same direction were found. It is remarkable that all the theoretically calculated shifts are larger than the experimentally determined values 0.09 Å [1] and 0.10 Å [2].

Besides W(211) other W surfaces were theoretically investigated in regard to surface-parallel relaxations. For the W(321) surface a lateral shift of 0.54 Å parallel to [$4\bar{5}\bar{2}$] was obtained [13]. For W(210) and W(310) displacements towards the less symmetrical position of the first layer relative to the second layer (in contrast to the (211) surface) were found [5, 7]. Values of 0.19 Å [5] and 0.48 Å [7] for W(210) and 0.17 Å [5] and 0.18 Å [7] for W(310) were calculated. From LEED a shift of 0.09 Å was determined for the surface-parallel relaxation on W(310) [4]. Looking at the step profile of bcc (210), (310), (410), and (510) surfaces, all displacements during relaxation were concluded to be perpendicular to the step edges in the direction of reducing the step height and smoothing the sharp edge profile [7].

References for 1.4.1:

[1] Davis, H. L.; Wang, G.-C. (Bull. Am. Phys. Soc. **29** [1984] 221).

[2] Grizzi, O.; Shi, M.; Bu, H.; Rabalais, J. W.; Hochmann, P. (Phys. Rev. [3] B **40** [1989] 10127/46).

[3] Rabalais, J. W.; Grizzi, O.; Shi, M.; Bu, H. (Phys. Rev. Lett. **63** [1989] 51/4).

[4] Adams, D. L.; Andersen, S. P. (Springer Ser. Surf. Sci. **24** [1991] 395/400).

[5] Luo, J. S.; Legrand, B. (Phys. Rev. [3] B **38** [1988] 1728/33).
[6] Jiang, P.; Marcus, P. M.; Jona, F. (Solid State Commun. **59** [1986] 275/80).
[7] Jiang, P.; Jona, F.; Marcus, P. M. (Phys. Rev. [3] B **35** [1987] 7952/8).
[8] Tománek, D.; Bennemann, K. H. (Surf. Sci. **163** [1985] 503/15).
[9] Ackland, G. J.; Finnis, M. W. (Philos. Mag. [8] A **54** [1986] 301/15).
[10] Sokolov, H. D.; Jona, F.; Marcus, P. M. (Solid State Commun. **49** [1984] 307/12; Phys. Rev. [3] B **29** [1984] 5402/9).
[11] Johnson, R. A.; White, P. J. (Phys. Rev. [3] B **13** [1976] 5293/302).
[12] Bonneton, F.; Drechsler, M. (Surf. Sci. **22** [1970] 426/32).
[13] Wynblatt, P.; Gjostein, N. A. (Surf. Sci. **22** [1970] 125/36).
[14] Sokolov, H. D.; Shih, H. D.; Bardi, U.; Jona, F.; Marcus, P. M. (J. Phys. C **17** [1984] 371/83).
[15] Barnett, R. N.; Landman, U.; Cleveland, C. L. (Phys. Rev. Lett. **51** [1983] 1359/61).
[16] Rodríguez, A. M.; Bozzolo, G.; Ferrante, J. (Surf. Sci. **289** [1993] 100/26).

1.4.2 Surface Debye Temperature

Debye temperatures for the (211) surface of W were determined by He atom scattering assuming a depth of the attractive helium-surface potential of 300 cal/mol. For vibrations along the $[\bar{1}11]$ direction, i.e., along the close-packed rows of W atoms parallel to the surface, $\Theta_{\parallel}$ = 350 K was found, whereas in the $[0\bar{1}1]$ direction a lower value of $\Theta_{\parallel}$ = 255 K was derived.

Reference for 1.4.2:

Weinberg, W. H. (J. Chem. Phys. **57** [1972] 5463/6).

1.4.3 Surface Electronic Structure

1.4.3.1 Surface Core-Level Shifts

Surface core-level shifts for the W $4f_{7/2}$ level were determined by photoemission for three stepped planes along the ⟨100⟩ zone, W(320), (310), and (610), and were related to results for the low-index surfaces W(110), (100), and (111) [1]; cf. Section 5.5.3.2.1, pp. 187/90. The spectra were decomposed into four components which were assigned on the basis of the bulk truncation structures of each surface. The (320) surface (stepped W $[6(110)\times(1\bar{1}0)]$) had already been experimentally and theoretically investigated by [2]. There is very good agreement between the core-level shifts of atoms with a given coordination on any of the surfaces. The importance of coordination is emphasized by the strong correlation between core-level shift and effective coordination [1]; cf. Fig. 42, p. 190.

The intensities of component peaks were shown to be consistent with a distinct contribution from underlayer atoms on the (110) terraces of stepped surfaces. This demonstrates the importance of the third coordination shell of W atoms in determining the core-level shift and explains the relatively large shift observed for the underlayer on W(100); cf. p. 188. Good agreement between experiment and theory was found with a bond-breaking model which considers the influence of the first three shells on the core-level binding energy [1]; cf. Section 5.5.3.2.2, pp. 190/3.

References for 1.4.3.1:

[1] Purcell, K. G.; Jupille, J.; King, D. A. (Surf. Sci. **208** [1989] 245/66).
[2] Chauveau, D.; Roubin, P.; Guillot, C.; Lecante, J.; Tréglia, G.; Desjonquères, M. C.; Spanjaard, D. (Solid State Commun. **52** [1984] 635/9).

1.4.3.2 Surface Electronic States

The total energy distribution of field-emitted electrons revealed features at 0.69 and 1.50 eV below E_F for a W(211) surface and at 0.39 and 1.45 eV below E_F for a (310) surface [1]. These results (and those of a subsequent field emission study [2]) for a (211) surface are in good agreement with photofield emission data obtained with p-polarized light at grazing incidence [3, p. 491]. The (211) surface was also treated in another photofield emission study, where an anomalous feature in the total energy distribution was observed at 0.66 eV below E_F [4]. Six surface-density-of-states peaks for (211) lying between 0.31 and 0.69 eV below E_F were observed with p-polarized light [5].

Since the [510] and [100] directions are separated by only 11°, the broad peak in the enhancement factor of photofield emission for the (510) surface observed by [3] at 0.35 eV below E_F is likely due to the same surface resonance as the well-known Swanson hump at 0.35 eV below E_F in the emission from the (100) plane. The width of the peak suggests that the surface resonance is not as well localized on the lower symmetry face [3].

References for 1.4.3.2:

[1] Plummer, E. W.; Gadzuk, J. W. (Phys. Rev. Lett. **25** [1970] 1493/5).
[2] Plummer, E. W.; Bell, A. E. (J. Vac. Sci. Technol. **9** [1972] 583/90).
[3] Venus, D.; Lee, M. J. G. (Surf. Sci. **172** [1986] 477/95).
[4] Lopinski, G. P.; Donders, P. J. (Surf. Sci. **201** [1988] 294/310).
[5] Radoń, T.; Jaskółka, S. (Surf. Sci. **247** [1991] 106/10).

2 Surface Free Energy. Surface Tension

The surface tension γ and the (specific) surface free energy σ of a crystal face with area A are related by $\gamma=\sigma+A\cdot(d\sigma/dA)$ [1]. There is some confusion in the literature in properly using γ and σ; for a more recent discussion of the difference between σ and γ, see [2 to 4]. According to Kumikov and Khokonov [5], cracking, zero-creep, and field emission microscopy (FEM) studies yield values for the surface tension equal to the surface free energy, $\gamma=\sigma$. Inert gas bubble precipitation, grain boundary grooving, and scratch relaxation rate studies can be used to determine σ. Contact angle and multiphase equilibrium studies, on the other hand, yield γ values [5].

There is a large spread in the reported σ values, both calculated and experimental. For instance, values obtained in the period from 1959 to 1977 range between ~2 and ~8 J/m^2; see a table in [6]. The methods for measuring surface energies of solids are reviewed in [7]. Paper [8] critically reviews the experimental work on liquid W.

2.1 Single-Crystal Studies Involving Several Surfaces

Theoretical Studies and Empirical Relationships. σ and γ have been calculated for (110), (100), and (310) surfaces at T=0 K by [4] using the empirical N-body potentials derived in [9]. Because of the large effect relaxation has on γ, many surface layers have to be considered in calculating γ, while the influence of relaxation on σ is much weaker and only amounts to less than 0.5%. σ and γ values (in J/m^2) of relaxed surfaces compare as follows [4]:

$$\sigma(110)=2.575,\ \sigma(100)=2.924,\ \text{and}\ \sigma(310)=3.036;$$

$$\gamma(110)=1.328,\ \gamma(100)=3.032,\ \text{and}\ \gamma(310)=2.142\ [4].$$

This confirms the values of $\sigma(100)=2.923$ and 2.920 J/m^2 for unrelaxed and relaxed surfaces in [10] who also used the N-body potentials by [9]. For values calculated with unrelaxed surfaces, see [4]. The surface stress on W(110) and W(310) is anisotropic (in contrast to the (100) surface), and the γ values quoted are half the traces of the respective surface stress tensors [4].

The "embedded atom" method (see [11, 12]) and the "equivalent crystal" theory (see [13]) may be used for calculating the surface energy. A large range of σ(hkl) values have been calculated by the equivalent crystal method for both relaxed and unrelaxed W surfaces [14] assuming unreconstructed surfaces. The following table gives the first eleven relaxed σ(hkl) values (in J/m^2) rounded to four digits from a table covering 36 surfaces [14]:

hkl:	110	211	100	111	310	321	411	210	332	431	510
σ(hkl):	3.258	3.932	5.808	4.290	5.122	3.809	4.996	4.665	4.151	3.702	5.454

With the embedded atom method, $\sigma(110)=2.599$ and $\sigma(100)=2.809$ J/m^2 were obtained [15]. A modified version of this method including directional binding yielded the values $\sigma(110)=2.232$, $\sigma(100)=2.646$, and $\sigma(111)=2.247$ J/m^2 at T=0 K [16].

σ values were calculated for relaxed surfaces assuming an unrelaxed vacancy formation energy of 4 eV. σ varied smoothly in the spherical triangle (110)-(111)-(100)-(110). Empirical short-range potentials were used for the calculation augmented by a volume-dependent term that provided for the lattice stability [17]. A recalculation using a value of 4.35 eV for the unrelaxed vacancy formation energy gave $\sigma=2.86$, 3.2, 2.82, and 3.43 J/m^2 for (110), (211), (100), and (111), respectively [18].

σ was calculated from Morse potentials (obviously for T=0 K) to be σ(hkl)=2.04, 2.26, and 2.499 J/m^2 for (110), (100), and (111), respectively [19]. Also using Morse potentials, a contour plot of σ normalized to the high σ(hkl=110) value 7.734 J/m^2 was calculated for more than 50 orientations [20]; see also the quotations in [17, p. 5299] and [21]. Using modified Morse potentials, σ(hkl)=7.7, 8.3, 7.8, and 8.4 J/m^2 were calculated for (110), (211), (100), and (111), respectively [22]; see also [23]. In close agreement with [24], facet size ratios observed by field ion microscopy (FIM) could be well fitted by Mie potentials (for these, see [25] and especially [26]) if a lattice expansion of 1.5% was assumed [27]. Use of pairwise interaction potentials (for these, see [28] and [29]) yielded σ=3.48 and 3.72 J/m^2 for (110) and (100), respectively [21].

From a relationship involving Young's modulus, interplanar distances, and (assumed) ranges for the attractive forces, σ(hkl)=3.32, 4.68, and 8.13 J/m^2 were calculated for (110), (100), and (111), respectively [30].

σ (T=0 K ?) was calculated from elastic and optical data to be 1.71, 1.76, and 1.82 J/m^2 for W(110), (100), and (111), respectively. The relative order of these systematically low values is considered reliable [31].

The interplanar force constants derived from the acoustic phonon dispersion curves measured by [32] were used to calculate σ [33 to 35]. Values of 3.69, 4.05, and 3.34 J/m^2 were obtained for W(110), (100), and (111), respectively, while for the average surface an experimental σ value of 3.07 J/m^2 was given [33]. Taking into account contributions from second- and third-nearest-neighbor interplanar force constants, σ(hkl)=2.41, 2.19, and 1.54 J/m^2 were calculated for W(110), (100), and (111), respectively [35].

σ(hkl) has been calculated for the $\langle 100 \rangle$ zone by [36] applying a theory developed in [3]. The results for T=0 K are presented graphically. Values of $\sigma(100) \approx 3.44$ J/m^2 and $\sigma(110) < 3.2$ J/m^2 may be read from the plot given. Between (100) and (110), σ(hkl) exhibits several maxima reaching nearly 4 J/m^2 [36].

Rigid-fracture surface energies have been calculated using the vacancy formation energy as input to give σ=3.33 and 5.88 J/m^2 for (110) and (100) [37].

Using the empirical relationship $\sigma(hkl) \approx \Phi_{hkl} \cdot 1150 \cdot (z/V^*)^{5/6}$ between the number of valence electrons z, the molar volume V* (in cm^3/mol), and literature data for the work function Φ_{hkl}, σ(hkl)=4.39, 3.59, 3.59, 3.22, 3.3, 3.15, and 3.19 J/m^2 were derived for (110), (211), (100), (111), (311), (611), and (310), respectively [38].

σ(T=0 K)=2.73 and 3.06 J/m^2 and ($10^3 \cdot d\sigma/dT$)= −0.114 and −0.128 $J \cdot m^{-2} \cdot K^{-1}$ were calculated for W(110) and W(100), respectively, using a value of 844 kJ/mol for the heat of sublimation at T=0 K [39].

The ratio $\sigma(110):\sigma(111):\sigma(100)=0.68:0.89:1$ was estimated in [40].

Experimental Studies. The surface tension γ was determined from the steady-state shape of a field emitter tip resulting from the balance between the compressing forces of the surface tension and the expanding forces of the applied field. A thermofield treatment at T=1780 K resulted in γ=5.317 and 4.868 J/m^2 for the regions between (100) and (310) and between (100) and (411), respectively [41].

The anisotropy of σ was studied by FIM at 1780 K. The results could be satisfactorily fitted by Mie potentials (see above). From the distribution of observed facet sizes, the ratio σ(hkl)/σ(110) was found to range between 1 and 1.043; ratios of 1.03, 1.028, and 1.032 were given for (211), (100), and (310), respectively [27]. This is in close agreement with the FIM results found near 2600 K by [24] who studied the shape of W tips with long shanks

References for 2 on pp. 79/81

and cone angles < 1.6°. Relative values for the surface energy were deduced from the distance the various faces have with respect to the center of the hemisphere forming the tip apex (application of Wulff's theorem for equilibrium shapes of crystals). The ratios σ(hkl)/σ(110) plotted in **Fig. 22** [24] agree with calculations where a Mie potential was used. σ varies only slightly (≤3%) with (hkl), but the shallow minima observed near (310), (211), and (111) are considered real [24]; see also [42]. Also see the FEM and FIM studies mentioned under "Polycrystals. Average Field Emitter Tips" below.

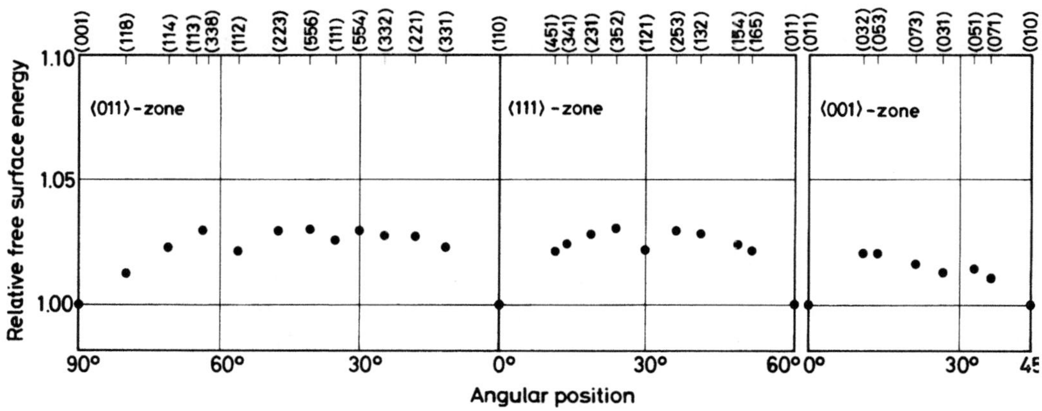

Fig. 22. Relative free surface energy σ(hkl)/σ(110) determined from the W tip profile in the FIM study [24]. The orientation of the surface regions is indicated by Miller indices at the top of the panels; at the bottom the angular distance is shown from the (011) pole for the ⟨001⟩ zone and from the (110) pole for the ⟨011⟩ and ⟨111⟩ zones.

Polyhedral grain boundary gas bubbles observed in chemical-vapor-deposited W after heating in vacuum above ~1600 °C are bounded by (110)-, (211)-, (100)-, and (210)-type faces. Polyhedral grain boundary holes formed in powder-metallurgical W during high-temperature creep (2200 °C in vacuum under a load of ~13.8 MPa) were also bounded by these faces. The (210)-type faces may be composite faces formed from two (211)-type faces. The ratio of σ values determined from the shape of the polyhedra by Wulff plots was σ(110):σ(211):σ(100) = 1.080:1.052:1.000 [43]. This ordering of σ values differs from that found by [27] (see above).

2.2 Single-Crystal Studies on Particular Surfaces

W(110). Cracks were introduced into ⟨110⟩ axis tensile specimens by spark-cutting. In one instance, a specimen, spark-cracked at 295 K and cleft at 77 K, showed a nearly pure (110) cleavage plane (while the cracks usually consisted of a combination of (100), (110), and (010) cleavage planes). This one test gave σ(110) ≈ 4.3 J/m^2 [44], lower than σ(100) = 6.0 J/m^2 found in [45] under the same experimental conditions (which tended to yield high σ values obviously due to plastic flow at room temperature) [44]. The "surface energy" was calculated from microhardness data for W(110) [46] (apparently, γ(110) was actually calculated).

W(100). σ values around 3 J/m^2 best fit the results from semi-brittle fracture tests of precracked, ⟨100⟩-oriented W crystals in liquid N_2. No large differences in the fracture stress along [010] and [110] directions were found for large cracks [6].

Tensile specimens of ⟨100⟩-oriented crystal sheets were spark-precracked in liquid N_2 and immediately tested at 77 K. Specimens thicker than t=0.4 mm cracked along the (100) plane, while thinner ones tended to crack along (110) steps. For $t \gtrsim 0.5$ mm, $\sigma \approx 1.7$ J/m^2 was obtained [47]; see also the value of $\sigma = 1.7 \pm 0.3$ J/m^2 reported by these same authors in [45]. Cleavage stresses were measured up to 478 K. Above ~150 K, they sharply increased, obviously due to plastic flow. Large effective σ values were determined in [48] where cracks were introduced at room temperature and their propagation studied at 77 K and higher temperatures [48]. At 77 K, $\sigma(100) = 6.3$ J/m^2 was determined from the cleavage crack propagation (brittle fracture), while 4.68 J/m^2 was expected from the theoretical relationship $\sigma = (E/a) \cdot (r/\pi)^2$ used by [30], where E stands for Young's modulus, a for the interplanar spacing, and r for the atomic radius [48] (a more recent E value than that used by [48] in the evaluation of their experiments would lower their result to 6 J/m^2 [45]). Crack initiation at 295 K in paraffin oil and crack propagation in liquid N_2 or liquid He resulted in effective σ values of 6 ± 0.9 J/m^2 which were high presumably because of crack blunting that occurred during spark initiation [45].

A value of $\sigma = 3.09$ eV per surface atom (~4.94 J/m^2) was obtained for W(100) in a self-consistent tight-binding calculation using pseudopotentials. The calculations involved sheets of W(100) 3, 5, and 7 atomic layers thick. Relaxing the unreconstructed surface resulted in a contraction of 5.9% for the top layer and gave rise to a relaxation energy of 59 meV per atom (<2% of the total surface energy). The ratio of the calculated surface energy to the cohesive energy, ~3.1 eV per atom to 8.9 eV per atom, is close to the ratio 5:14 of the number of missing nearest neighbors to the total number of nearest neighbors (counting both first and second nearest neighbors) [49].

The linearized augmented plane-wave (LAPW) method was applied on a periodic stack of thin W(100) slabs five atomic layers thick alternating with an empty space five layers thick. $\sigma = 5.54$ J/m^2 was obtained for the unreconstructed surface which relaxed inward by 5% with a relaxation energy of 3.1 mRy (~42.2 meV) per surface atom [50]. $\sigma(100) = 5.1$ J/m^2 was calculated for T=0 K with the local-density-functional theory and using the all-electron full-potential LAPW method applied to thin films five and seven layers thick. The relaxation energy involved amounted to less than 2% of σ [51].

Using Born-Mayer-type repulsive potentials and taking the band term in the tight-binding approximation into account, $\sigma(T = 0\ K) \approx 3$ J/m^2 was obtained for the unreconstructed top-layer relaxed inward by ~3% of the interlayer spacing [52].

2.3 Polycrystals. Average Field Emitter Tips

Experimental Studies. This section treats cracking, creep, FEM/FIM, and multiphase equilibration studies. Using fine-grained, polycrystalline W wires annealed near 2000 °C, the minimum σ measured on pre-cracked specimens at 77 K was $\sigma = 4.5$ J/m^2 [53]. $\sigma(T = 3460 \pm 80\ K) = 2.69 \pm 0.02$ J/m^2 determined by the zero-creep method [54] is recommended in the review [5]. Measurement of the blunting rate of a ⟨110⟩-oriented FEM emitter tip by the ring-stopping method (see p. 98) yielded $\sigma(T = 2100\ K) = 2.8$ J/m^2 ±10% [55]; similarly, $\sigma(T = 2000\ K) = 2.9 \pm 0.3$ J/m^2 [56]. FIM showed that σ decreases linearly from ~5.38 to 3.88 J/m^2 when T is increasing from 1200 to ~1880 K [57]. These rather high values are likely due to an inappropriate field enhancement factor β determined from the threshold of field emission (by [57]) rather than from F-N plots. When [58] repeated this experiment,

References for 2 on pp. 79/81

values of 3.9 J/m^2 were obtained for the (111) region on determining β from the emission threshold. Using F-N plots, however, a much lower value, 2.25 ± 0.05 J/m^2, resulted for the same region [58].

An average σ was determined at 2000 °C from grain boundary grooving studies performed in 1.05 atm Ar, in Ar plus Sn vapor, or in liquid Sn. From these multiphase equilibration studies (which according to [5] yield γ rather than σ), the rather low value $\sigma = 1.81 \pm 0.63$ J/m^2 was obtained [59], significantly lower than $\sigma(T = 1723\ K) = 2.83 \pm 0.47$ J/m^2, found in a similar study using Ar and Cu [60]. $\sigma = 1.81$ J/m^2 found in an Ar-Sn study [59] may be low because of surface contamination by sulfur. A *positive* temperature coefficient, $d\sigma/dT \approx +0.55 \times 10^{-3}\ J \cdot m^{-2} \cdot K^{-1}$ (suggesting the presence of surface-active impurities), was estimated in these studies from the temperature dependence of the grain boundary groove angles [59]. The study of an Ag droplet on a W wire at 1233 K in vacuum indicated $\sigma = 3$ J/m^2 [61].

Theoretical Studies and Empirical Relationships. According to a table in [62, p. 494], σ decreases from 4.435 J/m^2 at 0 K to 3.938 J/m^2 at 2100 K; the entries were evaluated from elastic stiffness data as reported in [63, 64].

$\sigma(T = 0\ K) = 3.3$ J/m^2 and $\partial\sigma/\partial T = -0.15 \times 10^{-3}\ J \cdot m^{-2} \cdot K^{-1}$ [65] were obtained from an empirical relationship between the electronic density at the cell boundary and the surface tension of liquid metals developed by [66].

σ for rough surfaces ("average" high-index surfaces) at the melting temperature T_m has been estimated by multiplying the σ value for the liquid at T_m by 1.18. By considering the surface entropy, $\sigma(T)$ was calculated between T_m and 0 K; see a plot in [67] where calculated [67] and experimental data from the compilation [68] are compared.

$\sigma(15\ °C) = 5.96$ J/m^2 was calculated for polycrystalline W using Buckingham potentials. Using Morse potentials, $\sigma(15\ °C) = 8.85$ J/m^2 was obtained [69]. Also using Morse potentials, $\sigma(T = 0\ K) = 2.266$ J/m^2 was derived from $1/3[\sigma(110) + \sigma(100) + \sigma(111)]$; see [19].

$\sigma = 3.24$ J/m^2 was calculated for $T = 3460$ K [70]. $\sigma(T_m) = 2.68$ J/m^2 was estimated using a correlation between σ and the heat of sublimation (or also other physical properties) [71]. Similarly, $\sigma(T = 1723\ K) = 3.25$ J/m^2 was predicted in [72].

An empirical relationship between the excess surface free enthalpy (η) and the bulk heat of atomization ($\Delta H'$), $\eta = \alpha \cdot \Delta H'$, is discussed in [73] citing a value of $\alpha = 0.179$ at $T = 0$ K derived in [74]; see also [75]. α depends on the temperature according to $\partial\alpha/\partial T = -(R \cdot \ln m)/\Delta H'$, where m is the number of surface layers between vacuum and bulk. $\partial\alpha/\partial T = -1.5 \times 10^{-5}\ K^{-1}$ was calculated [76]. The calculated standard value (25 °C), $\sigma^0 = 3.468$ J/m^2 [75], revises the value 2.918 J/m^2 of [77] who in addition had calculated $\sigma(T = 2000\ K) = 2.727$ J/m^2 [77]. $\sigma(T = 3683\ K) = 2.487$ J/m^2 was calculated in [75] (the theoretical methods used in [74, 75] were later improved in [78]).

Interfacial Energy. The interfacial free energy between solid W and liquid Sn was determined at 2273 K by grain boundary grooving studies to be 0.916 ± 0.47 J/m^2 [79, p. 1624] which may be compared with 0.86 J/m^2 calculated for this system at 2300 K in [80]. An interfacial energy between solid and liquid W (at T_m) of 0.59 J/m^2 was estimated whereby the contributions of solid and liquid surface layers are 0.192 and 0.398 J/m^2, respectively [80]. An empirical relationship had suggested only 0.145 J/m^2 [81].

2.4 Melt

The surface tension (which for the liquid equals the specific surface free energy σ) is quoted in [82] to be 2.343 ± 0.143 J/m^2 (at T_m) as the mean value from three (cited) experimental determinations. With ultrapure W, σ may be as high as 2.5 J/m^2 (cited from [83] ?) [82]. An experimental value of $\sigma=2.22$ J/m^2 was quoted by [54] and $\sigma=2.5$ J/m^2 was recommended by [5].

$\sigma=2.316$ J/m^2 ($\pm1.3\%$) was measured by the pendant drop technique [8]. This paper critically reviews the earlier experimental work by [83 to 88]. The results by [83 to 86] are criticized in [8] for use of high densities ($D\approx17.5$ g/cm^3 rather than 16.74 g/cm^3) [8]. (Densities of 16.65 and 16.9 g/cm^3 have been used by [87, 88], respectively. For a compilation of density values, see "Tungsten" Suppl. Vol. A 3, 1989, pp. 63/4.) The values $\sigma=1.913$, 2.099, 2.11, 2.193 to 2.215, 2.278, 2.363 to 2.434, and 2.542 J/m^2 were recalculated with corrected D values in [8] from the data in [84], [83], [85], [86], [87] (weight-of-drop method), [88] (drawing-of-crucible method), and [87] (suspended-drop method), respectively [8].

The pendant drop technique gave $\sigma(T_m)=2$ J/m^2 [86]. Similarly, $\sigma=2.3$ J/m^2 ($\pm5\%$ estimated error) assuming $D=17.65$ g/cm^3 [84]. The pendant drop and drop weight methods yielded $\sigma(T_m)=2.545$ and 2.471 J/m^2, respectively, (assuming $D=17.5$ g/cm^3) [83]; these values reasonably agree with the theoretical estimate 2.68 J/m^2 cited from [71].

$d\sigma/dT$ for the liquid was estimated from the bulk expansion coefficient of the solid (extrapolated to T_m). The calculated value $d\sigma/dT=-0.00047$ $J\cdot m^{-2}\cdot K^{-1}$ [89] has the same order of magnitude as the estimate -0.00029 $J\cdot m^{-2}\cdot K^{-1}$ from [83, 90]. A value of -0.000189 $J\cdot m^{-2}\cdot K^{-1}$ was calculated for T_m in [91].

References for 2:

[1] Shuttleworth, R. (Proc. Phys. Soc. [London] A **63** [1950] 444/57).
[2] Zadumkin, S. N.; Temrokov, A. I. (Izv. Vyssh. Uchebn. Zaved. Fiz. **11** No. 9 [1968] 40/51; Sov. Phys. J. [Engl. Transl.] **11** No. 9 [1968] 35/45).
[3] Zadumkin, S. N.; Shebzukhova, I. G. (Fiz. Met. Metalloved. **28** [1969] 434/9; Phys. Met. Metallogr. [Engl. Transl.] **28** No. 3 [1969] 50/6).
[4] Ackland, G. J.; Finnis, M. W. (Philos. Mag. [8] A **54** [1986] 301/15).
[5] Kumikov, V. K.; Khokonov, Kh. B. (J. Appl. Phys. **54** [1983] 1346/50).
[6] Liu, J.; Bilello, J. C. (Philos. Mag. [8] **35** [1977] 1453/72).
[7] Linford, R. G. (Chem. Soc. Rev. **1** [1972] 445/64).
[8] Martsenyuk, P. S.; Ivashchenko, Yu. N.; Eremenko, V. N. (Teplofiz. Vys. Temp. **12** [1974] 1310/2; High Temp. [Engl. Transl.] **12** [1974] 1161/3).
[9] Finnis, M. W.; Sinclair, J. E. (Philos. Mag. [8] A **50** [1984] 45/55).
[10] Matthai, C. C.; Bacon, D. J. (Philos. Mag. [8] A **52** [1985] 1/3).
[11] Daw, M. S.; Baskes, M. I. (Phys. Rev. Lett. **50** [1983] 1285/8).
[12] Daw, M. S.; Baskes, M. I. (Phys. Rev. [3] B **29** [1984] 6443/53).
[13] Smith, J. R.; Perry, T.; Banerjea, A.; Ferrante, J.; Bozzolo, G. (Phys. Rev. [3] B **44** [1991] 6444/65).
[14] Rodríguez, A. M.; Bozzolo, G.; Ferrante, J. (Surf. Sci. **289** [1993] 100/26).
[15] Guellil, A. M.; Adams, J. B. (J. Mater. Res. **7** [1992] 639/52).
[16] Baskes, M. I. (Phys. Rev. [3] B **46** [1992] 2727/42).
[17] Johnson, R. A.; White, P. J. (Phys. Rev. [3] B **13** [1976] 5293/302).
[18] Johnson, R. A.; White, P. J. (Phys. Rev. [3] B **18** [1978] 2939/40).
[19] Taskaev, I. P. (Izv. Akad. Nauk SSSR Met. **1982** No. 4, pp. 153/7; Russ. Metall. [Engl. Transl.] **1982** No. 4, pp. 139/44).
[20] Nicholas, J. F. (Austral. J. Phys. **21** [1968] 21/34).
[21] Tyson, W. R.; Ayres, R. A.; Stein, D. F. (Acta Metall. **21** [1973] 621/7).
[22] Vingsbo, O. (Mater. Sci. Eng. **8** [1971] 32/40).

[23] Schweitz, J. Å.; Vingsbo, O. (Mater. Sci. Eng. **8** [1971] 275/83).
[24] Müller, A.; Drechsler, M. (Surf. Sci. **13** [1969] 471/90).
[25] Drechsler, M.; Nicholas, J. F. (J. Phys. Chem. Solids **28** [1967] 2597/607).
[26] Drechsler, M.; Nicholas, J. F. (J. Phys. Chem. Solids **28** [1967] 2609/27).
[27] Kumar, R.; Grenga, H. E. (Surf. Sci. **59** [1976] 612/8).
[28] Johnson, R. A. (Phys. Rev. [2] **145** [1966] 423/33).
[29] Johnson, R. A. (Phys. Rev. [2] A **134** [1964] 1329/36).
[30] Gilman, J. J. (Fract. Proc. Int. Conf. At. Mech. Fract., Swampscott, Mass., 1959, pp. 193/222, 204/5, discuss. pp. 222/4).

[31] Zaremba, E. (Solid State Commun. **23** [1977] 347/50).
[32] Chen, S. H.; Brockhouse, B. N. (Solid State Commun. **2** [1964] 73/7).
[33] Tyson, W. R. (J. Appl. Phys. **47** [1976] 459/65).
[34] Caglioti, G.; Rizzi, G.; Bilello, J. C. (J. Appl. Phys. **42** [1971] 4271/6).
[35] Caglioti, G.; Rizzi, G.; Bilello, J. C. (J. Appl. Phys. **43** [1972] 3600/1).
[36] Shebzukhova, I. G. (Fiz. Khim. Poverkhn. **1982** 27/31; C.A. **100** [1984] No. 180533).
[37] Smith, J. R.; Banerjea, A. (Phys. Rev. Lett. **59** [1987] 2451/4).
[38] Richman, M. H. (ASM Trans. Q. **60** [1967] 719/21).
[39] Avraamov, Yu. S.; Gvozdev, A. G. (Fiz. Met. Metalloved. **23** [1967] 405/8; Phys. Met. Metallogr. [Engl. Transl.] **23** No. 3 [1967] 19/22).
[40] Desjonquères, M. C.; Cyrot-Lackmann, F. (Surf. Sci. **50** [1975] 257/61).

[41] Vlasov, J. A. [Yu. A.]; Golubev, O. L.; Shrednik, V. N. (J. Phys. Colloq. [Paris] **49** [1988] C 6-131/C 6-136).
[42] Drechsler, M.; Müller, A. (J. Cryst. Growth **3/4** [1968] 518/22).
[43] Farrell, K.; Loh, B. T. M.; Stiegler, J. O. (ASM Trans. Q. **60** [1967] 485/93).
[44] Cordwell, J. E.; Hull, D. (Philos. Mag. [8] **26** [1972] 215/24).
[45] Cordwell, J. E.; Hull, D. (Philos. Mag. [8] **19** [1969] 951/66).
[46] Dzherakhov, N. I. (Izv. Vyssh. Uchebn. Zaved. Tsvetn. Metall. **1987** No. 1, pp. 22/8; C.A. **106** [1987] No. 180246).
[47] Cordwell, J. E.; Hull, D. (Philos. Mag. [8] **27** [1973] 1183/92).
[48] Hull, D.; Beardmore, P.; Valintine, A. P. (Philos. Mag. [8] **12** [1965] 1021/41).
[49] Chan, C. T.; Louie, S. G. (Phys. Rev. [3] B **33** [1986] 2861/4).
[50] Singh, D.; Wei, S.-H.; Krakauer, H. (Phys. Rev. Lett. **57** [1986] 3292/5).

[51] Fu, C. L.; Ohnishi, S.; Jansen, H. J. F.; Freeman, A. J. (Phys. Rev. [3] B **31** [1985] 1168/71).
[52] Tréglia, G.; Desjonquères, M.-C.; Spanjaard, D. (J. Phys. C **16** [1983] 2407/19).
[53] Valintine, A. P.; Hull, D. (J. Less-Common Met. **17** [1969] 353/61).
[54] Digilov, R. M.; Zadumkin, S. N.; Kumykov, V. K.; Khokonov, Kh. B. (Fiz. Met. Metalloved. **41** [1976] 979/82; Phys. Met. Metallogr. [Engl. Transl.] **41** No. 5 [1976] 68/71).
[55] Bettler, P. C.; Barnes, G. (Surf. Sci. **10** [1968] 165/76).
[56] Barbour, J. P.; Charbonnier, F. M.; Dolan, W. W.; Dyke, W. P.; Martin, E. E.; Trolan, J. K. (Phys. Rev. [2] **117** [1960] 1452/9).
[57] Dranova, Zh. I.; D'yachenko, A. M.; Mikhailovskii, I. M. (Fiz. Met. Metalloved. **31** [1971] 1108/11; Phys. Met. Metallogr. [Engl. Transl.] **31** No. 5 [1971] 223/6).
[58] Butenko, V. G.; Golubev, O. L.; Shrednik, V. N. (Pis'ma Zh. Tekh. Fiz. **18** No. 19 [1992] 80/5; Sov. Tech. Phys. Lett. [Engl. Transl.] **18** [1992] 642/4).
[59] Allen, B. C. (J. Less-Common Met. **29** [1972] 263/82).
[60] Hodkin, E. N.; Nicholas, M. G.; Poole, D. M. (J. Less-Common Met. **20** [1970] 93/103).

[61] Skorik, V. I.; Solyanyk, Z. V.; Stetsura, M. M. (Fiz. Elektron. [L'vov] No. 21 [1980] 113/5).
[62] Wawra, H. (Z. Metallkd. **66** [1975] 492/8).

[63] Wawra, H. (Z. Metallkd. **66** [1975] 395/401).
[64] Wawra, H. H. (Materialprüfung **14** [1972] 413/9).
[65] Miedema, A. R. (Z. Metallkd. **69** [1978] 287/92).
[66] Miedema, A. R.; Boom, R. (Z. Metallkd. **69** [1978] 183/90).
[67] Tyson, W. R.; Miller, W. A. (Surf. Sci. **62** [1977] 267/76).
[68] Tyson, W. R. (Can. Metall. Q. **14** [1975] 307/14).
[69] Linford, R. G.; Mitchell, L. A. (Surf. Sci. **27** [1971] 142/56).
[70] Ivanov, O. A.; Karyakin, V. S. (Izv. Akad. Nauk SSSR Met. **1989** No. 3, pp. 176/80; Russ. Metall. [Engl. Transl.] **1989** No. 3, pp. 169/73).

[71] Taylor, J. W. (Metallurgia **50** [1954] 161/5).
[72] Chatterjee, B. (Indian J. Technol. **15** [1977] 221/2).
[73] Kristyan, S.; Olson, J. A. (J. Phys. Chem. **95** [1991] 921/32).
[74] Mezey, L. Z.; Giber, J. (Acta Phys. Pol. A **66** [1984] 525/38).
[75] Mezey, L. Z.; Giber, J. (Jpn. J. Appl. Phys. **21** [1982] 1569/71).
[76] Kristyan, S.; Giber, J. (Surf. Sci. **201** [1988] L 532/L 538).
[77] Mezei [Mezey], L. Z.; Marton, D.; Giber, Ya. [J.] ["Hiber, H."] (Zh. Fiz. Khim. **52** [1978] 850/3; Russ. J. Phys. Chem. [Engl. Transl.] **52** [1978] 489/91).
[78] Mezey, L. Z.; Giber, J. (Surf. Sci. **234** [1990] 210/9).
[79] Allen, B. C. (Trans. Metall. Soc. AIME **245** [1969] 1621/32).
[80] Miedema, A. R.; den Broeder, F. J. A. (Z. Metallkd. **70** [1979] 14/20).

[81] Zadumkin, S. N.; Dokhov, M. P. (Izv. Akad. Nauk SSSR Met. **1968** No. 1, pp. 91/5; Russ. Metall. [Engl. Transl.] **1968** No. 1, pp. 52/4).
[82] Lang, G. (Z. Metallkd. **67** [1976] 549/58).
[83] Allen, B. C. (Trans. Metall. Soc. AIME **227** [1963] 1175/83).
[84] Calverley, A. (Proc. Phys. Soc. [London] B **70** [1957] 1040/4).
[85] Pekarev, A. I. (Izv. Vyssh. Uchebn. Zaved. Tsvetn. Metall. **1963** No. 6, pp. 111/3; C.A. **60** [1964] 12973).
[86] Flint, O. (J. Nucl. Mater. **16** [1965] 260/70).
[87] Elyutin, V. P.; Kostikov, V. I.; Maurakh, M. A.; Mitin, B. S.; Pen'kov, I. A. (Poverkhn. Yavleniya Rasplavakh **1968** 155/9 from C.A. **71** [1969] No. 116775 and cited in [8]).
[88] Agaev, A. D. (Author's abstract of candidate's Diss., Moscow 1973; cited in [8]).
[89] Papazian, H. A. (Scr. Metall. **18** [1984] 1401/3).
[90] Allen, B. C. (in: Beer, S. Z.; Liquid Metals Chemistry and Physics, Dekker, New York 1972, pp. 161/212, 188).

[91] Digilov, R. M. (Adgez. Rasplavov Paika Mater. No. 23 [1990] 8/9; C.A. **113** [1990] No. 198626).

3 Surface Self-Diffusion

3.1 Atomic Displacement Studies

3.1.1 Survey

The surface self-diffusion of tungsten was studied mainly by field ion microscopy (FIM). The atomic displacements observed have been evaluated in terms of diffusion coefficients. With more than one adatom on a plane at a time, interactions between adatoms occur, and binding energies may be deduced. Diffusion of single adatoms as well as of dimers and larger clusters was observed. In addition to FIM studies with atomic resolution, some effort was made to elucidate the dynamics of various tungsten surfaces also by field electron microscopy (FEM). In the following, a summary of reviews on FIM studies of tungsten self-diffusion on various crystallographic surfaces is given. Then, measurements of correlations between fluctuations in the field-electron current (flicker noise) are described. The bulk of the following chapter involves the diffusion coefficient, $D = D_0 \cdot \exp(-Q/k_BT)$, which was determined on the various surfaces W(hkl) listed in order of increasing roughness.

FIM Studies. FIM studies of W displacements on W surfaces have been reviewed by [1 to 7]. The surface self-diffusion of single atoms (monomers), dimers, and trimers is reviewed in [5]; see also [7]. For the monomer diffusion on (110), (211), (111), and (321) surfaces, see [6]. For dimers on W(110), see [6] and for dimers on W(211), see [6, 7]. Trimer diffusion on W(211) is reviewed by [7].

An elucidation of the diffusion mechanism requires knowledge of the direction and length of the elementary jumps performed by the diffusing particles. These jump lengths can be deduced from the displacement distribution observed. This is rather difficult for two-dimensional diffusion, but for one-dimensional diffusion of a W adatom on channeled surfaces, fairly reliable results may be obtained. W atoms diffusing on W(211), for instance, perform about nine out of ten jumps to adjacent sites and only one to the next-nearest sites [3, 8].

Early FIM studies were falsified by the fact that dimers rather than monomers were actually studied [5]; see field ion micrographs from [9], reexamined in [5]. The more recent results by [10] and [11] are compared in [5] with recalculated diffusion parameters from [21]; see also [4]. Anomalous behavior with respect to diffusion was noted on W(110): no diffusion was observed at low temperatures [5]; also, there was an anomaly with (110) in the sequence of adatom mobilities: $(211) > (321) \approx (110) > (111) \approx (310)$, as deduced by directly observing the atomic displacements in [21]. Odd results associated with the (110) surface are also discussed at length in [12].

According to [13], the values of the preexponential factor D_0 found for W(110), W(211), W(111), and W(321) show that the entropy of activation is small on these surfaces.

Diffusing W adatoms arriving at a plane edge were never observed to descend down the next step. Rather, they always bounced back [5]. However, adatom losses from planes were observed, as the FIM study [4] showed. The height of the extra barrier causing adatom reflection at lattice steps has been deduced from the isothermal rate of adatoms lost from the (topmost) atom plane, and attempts were made to determine the extra barrier height directly from the temperature dependence of the loss rate. The reflection efficiency for W adatoms diffusing on W(110) is high, as deduced from the mean lifetime τ of the adatoms determined from the fraction of heating cycles for which no adatom loss from the surface occurred [4].

The random walk problem is treated in [7] with attention paid to boundary effects which are important in FIM studies. Atomic interactions determined either by equilibrium (radial distribution function) studies or kinetic experiments (dissociation experiments, diffusion coefficient measurements, and transition rate determinations) are reviewed in [7]. The random walk of adatoms was studied by FIM with the diffusion step at elevated temperatures, followed by quenching to and imaging at cryogenic temperatures. Biased diffusion, referred to as directional walk, occurs under the experimental condition of turned-on imaging field. From such diffusion studies on W(110), the polarizability of adatoms has been deduced [6, 7].

Flicker Noise Studies. At least with the more recent FIM studies there is little doubt about the nature of the diffusing entity. In the FEM flicker noise studies described below, on the other hand, the corresponding entities have remained obscur. At this writing, the flicker noise studies reported appear to be only partially understood.

Kinks and atoms on terraces act like defects. The FEM emission properties of defects differ from those of the unperturbed environment. The dynamics of the defects is reflected in the flicker noise of the field emission [14]. The emission current fluctuations reflect the fluctuations in the number of "loose" atoms (or other emission centers). "Loose" W atoms are thermally generated by detachment from steps, kinks, or other surface irregularities [15]. Probe-hole flicker noise studies allow one to conduct measurements at the diffusion temperature, making quenching unnecessary (in contrast to FIM studies) [14]. The frequency, current, and temperature dependence of the emission current fluctuations from the various planes support a mechanism involving the generation of mobile atoms on a terrace. Current fluctuations set in at threshold temperatures specific for the faces studied. The increase of the threshold temperature in the sequence (310) < (211) < (100) suggests that the energetics of formation of mobile terrace atoms from ledge sites is the rate-controlling factor [16].

The flicker noise study [15] confirms that single-atom W diffusion occurs along and across (011) terraces and steps approximately equally (see [14]), but the adatoms do not climb or descend across [001]-oriented steps (see [17]).

3.1.2 Diffusion on W(110)

This closest-packed W surface usually is considered smooth. According to Wang and Tsong [18], this requires some qualification, however. They studied the directional walk of W adatoms by FIM at 280 K in fields ranging from 2.0 to 2.8 V/Å (field dependence study) and at 2.4 V/Å with T between 270 and 290 K (temperature dependence study). Within a heating period, only negligible random walk displacements occurred at 280 K. 444 adatoms were observed to drift from near the (110) plane center to the plane edge. Of these, 273 adatoms moved along the surface channels on W(110) running along $[1\bar{1}1]$ or $[1\bar{1}\bar{1}]$ directions. This shows that these channels act as constraints for the walk (i.e., the diffusion on W(110) has mainly one-dimensional character). These results have been evaluated in terms of polarizability and dipole moment of the adatom [18] (using the fact that at room temperature the walk of adatoms on W(110) turns into directional walk at fields of 2.4 to 2.8 V/Å [19]).

The adatom diffusion became appreciable ($D \approx 10^{-17}$ cm²/s) at 304 ± 2 K on W(110) [12]. From Arrhenius plots of the D values taken between ~304 and ~354 K, the values Q = 19.9 ± 2 kcal/mol and $D_0 = 0.0021$ cm²/s have been deduced [12], confirming $D_0 = 0.03$ cm²/s, $\Delta S = 4 \pm 4$ e.u., and Q = 22 kcal/mol obtained from both the jump frequency $\nu = \nu_0 \cdot \exp(\Delta S/R) \cdot \exp(-Q/RT)$ with $\nu_0 \approx k_B T/h$ and the atomic displacements measured from 288 to 337 K [21]. The values of D_0 found imply a small entropy of activation ΔS [13]. The parameters $D_0 = 0.0026$ cm²/s, Q = 21.2 ± 1.1 kcal/mol, and ΔS = 2 ± 4 e.u. have been recalculated by [20] from the data in [21]; see also [5, 22]. An estimate of Q ≈ 20 kcal/mol is given in [5].

References for 3.1 on pp. 92/3

Arrhenius plots for single W adatom and W_2 cluster diffusion coefficients D have been derived from FIM observations on a W(110) plane in the low 10^{-11} Torr range. The data for W_1 have been collected in the range 298 to 338 K and for W_2 between ~308 and ~348 K [6]. The activation energies for monomer and dimer diffusion virtually agree [7]. After correcting a temperature calibration error in [23], the D parameters (see also [7, 24]) are as follows [6]:

	W_1	W_2
Q in kcal/mol	20.8 ± 1.6	21.2 ± 3.2
D_0 in cm^2/s	$0.0062 \times 13^{\pm 1}$	$0.0014 \times 160^{\pm 1}$

Q = 17 kcal/mol was derived from the calculated binding energy $U \approx -157$ kcal/mol [25], while Q = 11.1 kcal/mol had been calculated with the pair interaction theory [26] using the Morse potential functions given in [27]. Using modified Morse potentials (yielding $U = -141$ kcal/mol), Q = 12.5 kcal/mol was calculated for jumps between adjacent equilibrium sites separated by true saddle points [28]. For single adatom displacements, Q = 11.9 kcal/mol was calculated using Morse potentials [22]. Also with Morse potentials, Q = 10.2 and 10.8 kcal/mol had been calculated for the relaxed and unrelaxed surface in [29, p. 1477]. Similarly, Q = 10.6 and 11.5 kcal/mol were calculated in [30]. Using the Lennard-Jones potential $V(r) = 4\,\varepsilon[(\sigma/r)^{12} - (\sigma/r)^6]$ with the parameters $\varepsilon/k_B = 12391$ K and $\sigma = 2.562$ Å (from [31]), Monte Carlo simulations gave $D_0 = 0.00364$ cm^2/s and Q = 22.8 kcal/mol [32], closely reproducing the diffusion coefficients reported by [33] and [20]. This simulation was extended, and relaxation of either no substrate layer or of the topmost layer was included. The activation energy $Q = E_{saddle} - E_{min}$ and the frequency factor ν_0 were calculated for randomly oriented hops between adjacent binding sites by minimizing the total energy for the relaxing substrate-adatom system. Transition-state-theoretical results from a Monte Carlo simulation and from a simplified version of this theory (MCTST and STST, respectively) [34] are given in the following table; MD denotes the molecular dynamics results cited from [32]:

method	STST	MCTST	MD
Q in kcal/mol	16.9	16.2 ± 1.7	22.7 ± 3.0
D_0 in cm^2/s	–	$0.0011 \times 1.31^{\pm 1}$	$0.0036 \times 1.77^{\pm 1}$

Surface stresses related with steps on the W(110) surface were modeled in [35]. Q = 25 kcal/mol was obtained for the activation of surface adatom migration under the assumption that the crystal is isotropic with respect to its mechanical responses [35].

The value Q = 19.8 kcal/mol was deduced for the activation of adatom displacements on terraces by FIM. $Q'_{exp} = 20.6 \pm 2.4$ kcal/mol is the energy barrier to adatom displacement over steps down on a W(110) surface and includes the extra barrier causing adatom reflection at the terrace edges. The (obviously conflicting) value of 12 kcal/mol for the extra barrier was calculated from the frequency of visits at edge sites and the frequency of adatom losses [4, 13]. Adatoms approaching step edges were never observed to descend down the steps. Rather, they always bounced back [5]. The edges of W(110) were not absorbing [21].

A probe-hole version of the ring-rate method (see p. 98) indicates a Q value of 27.4 ± 2.5 kcal/mol which is attributed to the separation of edge atoms from the terraces [36]. This Q value is ~7 kcal/mol higher than the activation energy for single adatom diffusion determined by FIM in [5, 12, 20, 21]. It is much lower, however, than Q = 41.5 kcal/mol determined by FIM for W atom displacements from (111)-type kinks on (110) terraces. Displacements from (100)-type kinks required even Q=48.4 kcal/mol [37] (obviously revising the earlier paper [38]).

Atom clusters were prepared on W(110) by field evaporation above 140 K. The chain-like clusters had FIM images consisting of oval spots lengthened in the [001] direction and were arranged in zigzag chains along [1$\bar{1}$0]. When heated to 270 K in zero field, the zigzag chain reformed from the ends. The activation energies for the rearrangements observed at 290 and 330 K were 19.8 and 21.4 kcal/mol, respectively. Stable rhombic clusters formed frequently from the zigzag clusters on heating at 270 to 300 K. The rhombic clusters were imaged as four oval spots at the ends of diagonals 12 Å and 6 Å wide along [001] and [1$\bar{1}$0], respectively [39].

More random structures evolved from rhombic clusters on heating at 330 to 360 K. Rarely, at still higher temperatures, did the random clusters move by surface self-diffusion without change of shape. These clusters were stable up to at least 500 K [39]. Encounters of two or more adatoms usually resulted in cluster formation. A W_6 cluster has a "phase transition" at 390 K between a symmetric cluster stable below 390 K and a less symmetric one above 390 K. The W_2 clusters on W(110) became mobile around 300 K when occasional changes of their orientations occurred [19].

On condensing W vapor on the large (110) face of a built-up FIM tip near 80 K, the adatoms diffused radially outward in the presence of a radial field of the order of 1.5 V/Å, but did not reach the edge. Rather, a ring-like arrangement resulted with the adatoms kept some 10 Å off the plane edge. This behavior is explained by repulsive forces between the induced dipoles of the edge atoms and the induced dipoles on the adatoms. If the field is switched off and the tip warmed to 300 K in order to allow for thermal diffusion, clusters form which remain stable when after cooling the high field is restored [40].

3.1.3 Diffusion on W(211)

Single Adatoms. Adatom diffusion becomes appreciable ($D \approx 10^{-17}$ cm²/s) at 268±2 K [12]. The edges of W(211) are nonabsorbing [21], and the adatom diffusion is faster along the close-packed atom rows than across [5]. The adsorption sites for single adatoms appear to be lattice sites [41]. On the channeled W(211) surface, the measured distances between the adsorption sites of the diffusing adatoms corresponded closely to integer multiples of a common factor identified as the jump length of 2.74 Å [12]. The directional walk of W adatoms was studied near 270 K [18].

The effect of a (211) lattice step on the behavior of an adatom nearby was studied by FIM, emphasizing those configurations where the walls of a [$\bar{1}$11] channel formed by two ridges (atom rows) have square ends. The probability for finding an adatom was observed to be much higher at the channel ends than at the center of the channel. Moreover, the two ends of the channel were different: the end facing towards (111) binds the adatom more strongly than does the end facing (100). The lattice step exerts an effect on sites a considerable distance away from the plane edge [42]. The FIM data from single adatoms diffusing between 264 and 319 K along the channels on W(211) support the simplest models for surface diffusion, namely jumps to nearest neighbors [43] (which comprise about 90% of the jumps performed [8]).

The following table lists the more recent diffusion parameters D_0 and Q determined by FIM of single adatoms.

References for 3.1 on pp. 92/3

D_0 in cm^2/s	Q in kcal/mol	Ref.
0.001	17.7	[44][a)]
0.0003	17.5 ± 1.7	[5]
$0.00012 \times 20^{\pm 1}$	16.8 ± 2.1	[6]
$0.015 \times 10^{\pm 1.5}$	19.7 ± 2	[41][c)]
0.02	19.6	[42]
$0.0077 \times 1.9^{\pm 1}$	19.0 ± 0.4	[43][b)]

[a)] ~260 to 295 K. – [b)] 264 to 319 K. – [c)] 266 to 303 K.

$\Delta S = -5.2 \pm 5.3$ e.u. is given for the entropy of activation in [5]. $D_0 = 0.015 \times 10^{\pm 1.5}$ cm^2/s (see the table) implies $\Delta S \approx 0$ [41]. These results [41] were later slightly modified to $D_0 = 0.019 \times 7.3^{\pm 1}$ cm^2/s and $Q = 19.7 \pm 1.1$ kcal/mol in a reanalysis [43] of the reported data. The activation energy Q is related to the heat of vaporization at the boiling point, ΔH_v; the ratio $Q:\Delta H_v = 0.197 \pm 0.003$ was found for W(211), in agreement with similar ratios found on other metals [43]; see also [45].

The results of early studies were falsified by the fact that dimers rather than monomers had actually been investigated, as a reexamination of FIM micrographs from [9] showed [5]. Earlier FIM studies [12, 21] did not take sufficient precaution to assure that only one adatom was present on a plane at one time, thus causing errors in the resulting data [13]. FIM observations between ~268 and ~318 K yielded the values $D_0 = 3.8 \times 10^{-7}$ cm^2/s and $Q = 13.0 \pm 2$ kcal/mol [12]. The diffusion parameters $D_0 = 2 \times 10^{-7}$ cm^2/s, $Q = 13$ kcal/mol, and $\Delta S = -20 \pm 10$ e.u. were derived from the atomic displacements measured between 224 and 288 K [21]. A recalculation of the data from [21] still yielded low values: $D_0 = 3.0 \times 10^{-8}$ cm^2/s, $\Delta S = -22 \pm 4$ e.u., and $Q = 12.3 \pm 0.9$ kcal/mol [5, 20, 22].

FIM investigations (at 78 K) on the surface diffusion at 600 to 1000 K gave an activation energy of 55 kcal/mol for the removal of W atoms from W(211) edges adjacent to (411) [46], much higher than 28.1 kcal/mol, the amount of energy required for diffusion along steps in this region [47]. FIM between 286 and 323 K yielded for ledge atoms $D_0 = (0.17$ to $42.9) \times 10^{-7}$ cm^2/s and $Q = 13.4 \pm 1.6$ kcal/mol [48].

Using modified Morse potentials, $Q = 9.96$ kcal/mol was obtained assuming that the jumps are between adjacent equilibrium sites separated by true saddle points [28]. Similarly, a value of 9.56 kcal/mol was calculated with a central-force model [13].

Lennard-Jones 6-12 potentials were employed in a transition-state-theoretical (TST) study of the surface self-diffusion. The topmost 0, 1, or 2 substrate layers were allowed to relax [34]. Relaxation of both adatoms and substrate was considered using a Morse potential [30]. The activation energy $Q = E_{saddle} - E_{min}$ and the frequency factor ν_0 were calculated for randomly oriented hops between adjacent binding sites. The results obtained by minimizing the total energy for the relaxing substrate-adatom system using TST both in its simple version (STST) and by Monte Carlo simulations (MCTST) are given in the following table [34]. The molecular dynamics (MD) results have been calculated in a Monte Carlo study for temperatures between 2000 and 3000 K [49]. These TST, STST, and MD studies used the Lennard-Jones parameters $\varepsilon/k_B = 12391$ K and $\sigma = 2.562$ Å (from [31]) and the results given in the table are those cited in [34]:

method	STST	MCTST	MD[a)]
Q in kcal/mol	16.9	22.0 ± 4.2	20.5 ± 5.2
D_0 in cm^2/s	0.0016	$0.0028 \times 1.99^{\pm 1}$	$0.00218 \times 2.2^{\pm 1}$

[a)] cited from [49]

Dimers and Larger Clusters. On the channeled W(211) surface one has to distinguish in-channel and cross-channel dimers. When the adatoms are separated by at least two empty channels (distance $R \geq 8.95$ Å), no interactions are detectable by statistical analysis of the distance distribution data [50]. W adatoms in adjacent channels ($R = 4.48$ Å) are subjected to strong attractive interactions resulting in a correlated motion of adatom pairs [50]. Such cross-channel-dimer motion was studied above 255 K in [5, 11]. At 295 K cross-channel dimers diffused along the surface channels [50]. In-channel dimers became mobile only above 420 K and dissociated above 430 K [19]. The FIM study of the cooperative walk of the (cross-channel) chain-like clusters W_2, W_3, and W_4 on W(211) indicated similar or identical mobilities for these clusters [19]. Cross-channel dimers have been observed to have bond lengths of 4.47 and 5.24 Å (straight and staggered configuration, respectively) [51]. The analysis of the frequency, with which straight and staggered configurations occur, showed that the straight configuration is 0.959 ± 0.148 kcal/mol more stable than the staggered one [6]. The frequency ratio of ~2.5 for straight to staggered dimers found in [6] was confirmed in [3, p. 633] reevaluating data from [11].

The stability of dimers and tetramers indicates a bond strength at 5.24 Å of at least 12.9 kcal/mol which is a lower limit for the activation energy of surface diffusion for W atoms [51]. Pair motion consists of jumps performed by one atom at a time. The jumps performed by both atoms during the heating interval at higher temperatures are well explained by two single, independent jumps. Also, trimer diffusion at 298 K showed that the motion of triplets is a sequence of single atom jumps as that of pairs [5].

After depositing ten W atoms on an FIM tip and heating it for 1 min at 310 K, a chainlike tetramer W_4 and a cluster W_6 were observed. The chain structure migrated as a unit, occasionally changing its bond length from 4.47 Å to 5.24 Å [51].

The parameters $Q = 18.9 \pm 1.8$ kcal/mol and $D_0 = 0.0007 \times 20^{\pm 1}$ cm²/s for dimers are given *in [6]*, where the evaluation of an FIM study of W_2 and W_3 diffusion between 255 and 298 K by [11] was criticized. In [11] nearly the same values for the diffusion parameters were reported for dimer and trimer (W_2: $Q = 8.5 \pm 0.8$ kcal/mol, $D_0 = 2 \times 10^{-11}$ cm²/s; W_3: $Q = 8.3 \pm 1.3$ kcal/mol, $D_0 \approx 10^{-11}$ cm²/s), while much larger values, $Q = 17.5 \pm 1.5$ kcal/mol and $D_0 \approx 10^{-3}$ cm²/s, were obtained for the monomer [11]; see also [5, 44]. According to [6, p. 517], a difference of 1 to 8 orders of magnitude between the diffusivities of single adatoms and their dimers, and a factor of 2 between the activation energies quoted in [5, 11, 44] are artifacts due to improper data analysis [6]. A recalculation of the data of [11] for center-of-mass motion yielded $Q = 15.7 \pm 0.83$ kcal/mol and $D_0 = 6.7 \times 10^{-6}$ cm²/s [3, p. 633], still in considerable disagreement with the values 18.9 kcal/mol and 0.0007 cm²/s found by [6].

Flicker Noise Studies. These FEM studies are treated separately from the FIM papers above, since the cause for the noise signal often is uncertain.

The noise spectrum (in the frequency range 50 Hz to 1 MHz) of the field-emitted current from W(211) was measured up to ~1500 K. Current fluctuations set in at a threshold temperature of 650 K. The noise amplitude vs. T curve showed maxima near 1100 K. From Arrhenius plots of the spectral density, $Q = 9.7 \pm 1.2$ kcal/mol was obtained. This activation energy may be related to surface diffusion of adatoms or to diffusion-mediated flip-flops between two states with different dipole moments [16]. The frequency dependence of the noise at 905 K observed in [16] was satisfactorily reproduced by the calculation [52].

The in-channel and cross-channel diffusion of W adatoms was studied at 3×10^{-11} Torr on a [110]-oriented tip flashed at 2500 K before the runs. The diffusion *across* the channels was measured with the probe-hole slit parallel to the channels on W(211), while the diffusion

References for 3.1 on pp. 92/3

along the channels was measured with the probe-hole slit normal to the channels [53] (this study extended and improved the earlier study [54] where a [211]-oriented tip was used). The long-term ($t \geq 10$ ms) autocorrelation function of the flicker noise indicated that self-diffusion at 700 K is two-dimensional [55]. Gomer and co-workers [53, 55] determined the following Q and D_0 values for in-channel and cross-channel diffusion:

Q in kcal/mol	D_0 in cm^2/s	T in K	Ref.
in-channel diffusion			
16.8 ± 0.5	$(3 \pm 1) \times 10^{-5}$	500 to 800	[53]
13.2	7.1×10^{-7}	568 to 694	[55]
22 ± 0.5	$\sim 5 \times 10^{-5}$	826 to $\sim$950	[53]
cross-channel diffusion			
6.6	4.2×10^{-9}	526 to 752	[53]
6.6	1.8×10^{-9}	568 to 694	[55]
24	5×10^{-4}	>752	[53]

The in-channel parameters determined by [55] closely agree with the parameters determined by FIM in the study [21] where no attention was paid to have just one diffusing atom on the surface at a time [55].

3.1.4 Diffusion on W(100)

Emitter tips were annealed at up to 2300 K and studied by FIM. Surface self-diffusion seemed to occur by migration of the W atoms from one kink site to the next. Displacements along the edges of (100) terraces at 600 to 700 K require Q = 53 kcal/mol [37]. Arrhenius plots of the diffusion coefficients determined by a probe-hole version of the (FEM) ring-rate method yielded $Q = 31.1 \pm 1.8$ kcal/mol [36] in excellent agreement with the flicker-noise-derived value 32 ± 14 kcal/mol from [16]. This activation energy may be related to flip-flop processes between atoms located at net plane terrace and ledge sites [36]. The spectrum of the flicker noise in the field-emitted current was measured below $\sim$1400 K at frequencies ranging from 50 Hz to 1 MHz. Current fluctuations started at a threshold temperature of 1000 K. The noise amplitude vs. T curve showed a maximum near 1200 K [16].

An attempt was made to calculate the activation energy Q for single adatom diffusion with a Morse potential. Relaxation of both adatoms and substrate was considered [30]. Q = 62 kcal/mol was determined from the binding energy $U \approx -194$ kcal/mol calculated using a tight-binding approximation and Born-Mayer-type repulsive potentials [25] ($U = -148$ kcal/mol resulted with modified Morse potentials [28]). For the temperature region around the (1×1) to $c(2 \times 2)$ phase transition temperature T_c, an attenuation for the diffusion coefficient by several orders of magnitude was predicted. The D values are damped by $|1 - T_c/T|^x$ with $x \approx 0.9$ to 1.8 when compared with D values expected in the absence of critical fluctuations [56]; see also [57].

3.1.5 Diffusion on W(111)

W adatom migration on (111) occurs only at much higher temperatures than are required on (110), (211), and (321) [21]. According to FIM, 71% of the adatoms were adsorbed into lattice sites, while 29% of the adatoms were adsorbed into sites of lower symmetry, namely deep [211] edge sites and bridge [110] edge sites. The adatoms adsorbed into the normal lattice sites were not field-desorbed even at higher than best (He) image fields (BIF), and no diffusion was observed even at 450 K. The adatoms at the low-symmetry sites, on the other hand, were desorbed at ~88% of the He BIF. The [211]-edge-site adatoms readily diffused at 280 K (between 263 and 296 K the frequency of displacements increased by an order of magnitude). The bridge atoms on [110] edge sites already diffused at 263 K. With both types of edge-site adatoms, the diffusion was always off the (111) plane and into a lattice site of the next layer [41]. Earlier, the adatoms had been found located on lattice sites and seen to hop between adjacent lattice sites [5, 10].

Surface self-diffusion seems to occur by adatom migration from one kink site to the next [37]. Displacements along the edges of (111) terraces occurred at 600 to 700 K and required Q = 50.7 kcal/mol [37]. The diffusion of adatoms from [211] edge sites is activated by ~18 kcal/mol, assuming a preexponential factor of $D_0 \approx 10^{-3}$ cm^2/s. The energy required for the diffusion of bridge atoms on [110] edge sites is ~3 ± 1 kcal/mol lower (i.e., amounts to ~15 kcal/mol) [41].

The value Q ≈ 41 kcal/mol for a single adatom was derived from results over a limited temperature range, assuming a normal preexponential factor (*not* derived from an Arrhenius plot) [5, 10]. It amounted to about 2/3 of the theoretical value of 65.25 kcal/mol, deduced from calculations based on pair-wise additive potentials and taking into account the relaxation of the lattice as described in [28]; see [4].

Flicker noise studies indicated a single regime of two-dimensional diffusion and the low values Q = 10.6 kcal/mol and $D_0 = 5.8 \times 10^{-8}$ cm^2/s were found. At 750 K, D decreased and at higher temperatures increased again [14]; see also [58].

3.1.6 Diffusion on W(321)

The diffusion of monomers on W(321) was reviewed in [6]. Diffusion along close-packed atom rows is faster than across [5, 21]; see also [12, 39, 50]. The adatom diffusion became appreciable ($D \approx 10^{-17}$ cm^2/s) in FIM at 300 ± 2 K. The distances measured for the one-dimensional diffusion corresponded closely to integer multiples of the jump length between nearest-neighbor adsorption sites (2.74 Å) [12]. Published Q values from FIM studies cluster around 20 kcal/mol. Near W(321) the flicker noise study [15] found activation energies of 17 and 37 kcal/mol for "fast" and "slow" diffusion processes, respectively, along the step edges and 9.3 and 17 kcal/mol for the corresponding diffusion across the step edges [15]. These results refer to a region *around* W(321) and the nature of the process studied appears to be unclear. Therefore, the results from the flicker noise studies [14, 15, 58] will not further be discussed here, but will be treated together with related studies on pp. 90/1 in the context of the (011)-to-(112) zone.

The distance distribution was measured by FIM at 20 K for W adatoms diffusing at 295 K in adjacent channels (R = 7.08 Å). No interactions were detected by statistically analyzing the data [50]. But observations of cross-channel dimers (at the closest possible distance of ~7 Å) suggest that at least a weak interaction exists between these adatoms. Tungsten atoms within a channel quickly combine forming clusters. No dissociation of dimers occurred below 375 K. Trimers were even more stable than dimers and did not dissociate below 420 K [39].

Similar to the case of W(111) described above, there are two different surface sites for W adatoms on W(321). Some 30% adsorb into a "weakly" bound surface site (and field-desorb at ~85% BIF), the remainder (70%) adsorbs into a substantially more stable surface site (and field-desorbs at ~110% BIF). The (~70%) fraction of W adatoms adsorbed

 References for 3.1 on pp. 92/3

into the deep sites does not depend on temperature between 10 and 200 K. Also, these adatoms do not diffuse below ~300 K. The displacements of the weakly adsorbed adatoms (at the "shallow sites") are restricted to jumps normal to the walls of the surface channels and end at the stable adsorption sites. Shallow-to-shallow jumps are extremely rare if they occur at all and deep-to-shallow jumps were never observed [41]. The edges of W(321) are not absorbing [21].

The one-dimensional random walk of single adatoms was studied at 330 K. During a heating cycle lasting for $\tau=45$ s, an adatom executed a total of N=9 jumps over 2.74 Å. Assuming $N \approx \tau(k_BT/h) \cdot \exp(-Q/RT)$, an activation energy of Q=20.5 kcal/mol was obtained [39]. This is close to the ~20 kcal/mol found from the onset temperature of 300 K for motion along the deep sites (while shallow-to-stable jumps were activated by ~15 kcal/mol) [41]. This also agrees with the FIM results $D_0=0.0012$ cm^2/s and $Q=19.4\pm2$ kcal/mol obtained between ~300 and ~350 K [12] and with measurements between 290 and 310 K which yielded Q=20.5 kcal/mol and $D_0=0.026\times10^{\pm2.4}$ cm^2/s [59]. $D_0=0.0001$ cm^2/s, $Q=18.8\pm1.8$ kcal/mol, and $\Delta S=-7.5\pm4.4$ e.u. are given in [5]. The values $D_0=0.00037$ cm^2/s, $Q=20.1\pm1.8$ kcal/mol, and $\Delta S=-4\pm6$ e.u. [20, 22] were recalculated from 0.001 cm^2/s, 20 kcal/mol, and -3 ± 3 e.u. derived from FIM at 288 to 337 K by [21].

Between 270 and 350 K, the surface diffusion on terraces and ledges of (321) planes was studied by FIM using 0.01 Pa of He as an imaging gas [48].

An attempt was made to calculate the activation energy for single adatom diffusion on (321) surfaces with a Morse potential. Relaxation of both adatoms and substrate was considered [30]. 11.7 kcal/mol resulted from pairwise additive potentials and taking into account the relaxation of the lattice (as described in [28]) [4, p. 79]; see also [13].

3.1.7 Diffusion on Other Surfaces

This section first reports data on (311), (431), and (433) surfaces, then in the context of the (011)-to-(112) zone data on (257), (123), and (235), and last, in the context of the (011)-to-(001) zone, data on (023), (012), (013), and (017). Only the studies on (311), (431), and (433) employed FIM techniques. The other single-crystal surfaces were studied by analyzing the flicker noise of the field emitter current. This section is concluded with an account of a tracer study of diffusion along a W polycrystal.

The FIM study [21] found W adatom migration on W(311) only at much higher temperatures than were required for migration on W(110), W(211), and W(321) [21]. Using 0.01 Pa of He as an imaging gas and base pressures of 10^{-7} Pa, the surface self-diffusion parameters for terrace adatoms on W(431) were determined by FIM between 278 and 339 K to be $D_0=(0.45$ to $6.65)\times10^{-6}$ cm^2/s and $Q=60.9\pm3.3$ kcal/mol [48]. FIM studies of W(433) showed that at 450 K entire rows of edge atoms jump simultaneously [19].

Flicker noise from the interior region of the **(011)-to-(112) zone** was studied at about 3×10^{-11} Torr [14, 15, 58, 60], the diameter of the probed circular regions being ~60 Å in [14, 58] and ~100 Å in [60]. The tip region consisted of (011) terraces with (101)-type step edges, one atom layer high and running along $[\bar{1}\bar{1}1]$ [14]; see also [60]. The regions around (257), (123), (235) were studied by [14, 58] and those around (123) and (257) between ~500 and ~1000 K by [15]. The latter work used a rotatable, rectangular probe hole (24 Å × 160 Å) rather than a circular one. This enabled the study of anisotropies in the surface diffusion. The diffusion direction sampled was perpendicular to the alignment of the slit [15] (the theoretical background about this method for determining the anisotropy of D is given in [61] and [62]).

The correlation functions observed at 80 to 300 K are attributed to to-and-fro movements ("flip-flop") on the terraces. Above 450 K the behavior was diffusive and the correlation functions f indicated coexistence of one- and two-dimensional diffusion (when a circular probe hole was used for two-dimensional diffusion, f was proportional to t^{-1}; for one-dimensional diffusion: $f \propto t^{-1/2}$; t being time). Up to ~700 K, two-dimensional diffusion with correlation times in the 10 ms range dominated the spectra ("fast" process), while above 750 K a one-dimensional diffusion with correlation times of ~1s prevailed ("slow" process) [14].

The slow process is attributed to diffusion along the close-packed steps of the (011)-to-(112) zone and perhaps corresponds to diffusion of negative kinks or positive-negative kink combinations. At 875 K the diffusion coefficient associated with the slow process decreased sharply and increased again at 925 K, where the diffusion was still one-dimensional. The lattice steps running along ⟨111⟩ disorder sufficiently above 875 K to impede one-dimensional diffusion; in fact, diffusion turned two-dimensional at 950 K. Between 925 and 950 K the disorder was so great that diffusion no longer proceeded along step edges but instead over the terraces [58, 60].

For fast diffusion (ms range), the diffusion parameters below 600 K were Q=8 to 9.5 kcal/mol and $D_0=(2 \text{ to } 5)\times10^{-8}$ cm²/s, depending on the position of the region probed, and Q=16 kcal/mol and $D_0=(1 \text{ to } 2.8)\times10^{-5}$ cm²/s resulted above 600 K [14]. For the slow, one-dimensional diffusion dominating above 750 K, Q=13 kcal/mol [14] and $D_0\approx5\times10^{-9}$ cm²/s [15] were obtained.

Correlation functions for the flicker noise sampled around (123) were determined near 615 and at 871 K for diffusion along and perpendicular to the step edges. Near 615 K the "fast" process dominated, at 871 K the "slow" one [15]. The very complex results obtained for the region around W(257) in [15] are reflected by the various activation energies between 6.8 and 33 kcal/mol given in [15].

The surface diffusion on stepped surfaces from the interior of the **(011)-to-(001) zone** was studied by analyzing the flicker noise in the field emission currents from (023), (012), (013), and (017). The (023) and (012) surfaces consist of (011) terraces, and (017) is made up of (001) terraces. The (013) surface is intermediate [17]. On W(310) the noise spectrum was measured in the range 50 Hz to 1 MHz for temperatures up to ~1500 K. Current fluctuations set in at a threshold temperature of 300 K, and the noise amplitude peaked near 1000 K. The low activation energy of 3.2±1.2 kcal/mol determined was tentatively attributed to flip-flops between atoms located at net plane terrace and ledge sites [16]. The region around (023) was studied by [15] and most experiments were performed with the tip annealed at 1000 K in a field of 20 MV/cm, since the tip history affected the results at high temperatures. The diffusion was studied on (011) terraces of the (023) surfaces. These terraces are separated by non-close-packed steps running along [001]. Diffusion parallel to the steps is characterized by Q=8 kcal/mol and $D_0=6\times10^{-8}$ cm²/s below 520 K (in agreement with [17]), but Q=17 kcal/mol and $D_0=4\times10^{-4}$ cm²/s above 520 K. Near 600 K, D decreased sharply (as already seen by [17]). Below 600 K the results appeared to be independent of tip history. However, between 740 and 870 K, D varied with the emitter treatment, and the deepest dip in the Arrhenius plots of D occurred in runs that started around 800 K. Above 900 K, the D values again appeared to be independent of tip history. It was assumed that the results below 600 K involve single-atom or kink diffusion, while the value of Q=25 kcal/mol determined above ~800 K may correspond to single-atom diffusion on a very disordered surface. The behavior in the intermediate regime (740 to 870 K) presumably was effected by surface roughening, the details of the hysteretic behavior being not understood [15].

References for 3.1 on pp. 92/3

Polycrystals. D values (in 10^{-6} cm^2/s) of 0.362 and 1.97 were measured under $p=10^{-7}$ Torr at 2002 and 2526 K, respectively, for the diffusion of ^{185}W deposited on a W ribbon. An Arrhenius plot yielded the parameters $D_0=0.0011$ cm^2/s and $Q=31.6$ kcal/mol for the ^{185}W diffusion along the surface [63]. Apparently, the diffusion had more the character of random-walk atomic displacements ("tracer diffusion" in the proper sense of the word) rather than the character of spreading adlayers discussed in Section 3.2.1.

References for 3.1:

[1] Ehrlich, G. (Surf. Sci. **246** [1991] 1/12).
[2] Ehrlich, G. (Scanning Microsc. **4** [1990] 829/41 discussion on pp. 841/2).
[3] Ehrlich, G.; Stolt, K. (Annu. Rev. Phys. Chem. **31** [1980] 603/37).
[4] Bassett, D. W. (Surf. Sci. **53** [1975] 74/86).
[5] Graham, W. R.; Ehrlich, G. (Thin Solid Films **25** [1975] 85/96).
[6] Kellogg, G. L.; Tsong, T. T.; Cowan, P. (Surf. Sci. **70** [1978] 485/519).
[7] Tsong, T. T.; Cowan, P. (Crit. Rev. Solid State Mater. Sci. **7** [1978] 289/316).
[8] Ehrlich, G. (J. Vac. Sci. Technol. **17** [1980] 9/14).
[9] Ehrlich, G. (Br. J. Appl. Phys. **15** [1964] 349/64 and plates I to VI).
[10] Graham, W. R.; Ehrlich, G. (Surf. Sci. **45** [1974] 530/52, 545/9).

[11] Graham, W. R.; Ehrlich, G. (J. Phys. F **4** [1974] L 212/L 214).
[12] Bassett, D. W.; Parsley, M. J. (J. Phys. D **3** [1970] 707/16).
[13] Bassett, D. W.; Chung, C. K.; Tice, D. (Vide No. 176 [1975] 39/43).
[14] Gong, Y. M.; Gomer, R. (J. Chem. Phys. **88** [1988] 1359/69).
[15] Choi, D. S.; Kim, S. K.; Gomer, R. (Surf. Sci. **234** [1990] 262/72).
[16] Swanson, L. W. (Surf. Sci. **70** [1978] 165/80).
[17] Gong, Y. M.; Gomer, R. (J. Chem. Phys. **88** [1988] 1370/2).
[18] Wang, S. C.; Tsong, T. T. (Phys. Rev. [3] B **26** [1982] 6470/5).
[19] Tsong, T. T.; Cowan, P.; Kellogg, G. (Thin Solid Films **25** [1975] 97/106).
[20] Ayrault, G.; Ehrlich, G. (J. Chem. Phys. **60** [1974] 281/94, 281).

[21] Ehrlich, G.; Hudda, F. G. (J. Chem. Phys. **44** [1966] 1039/49).
[22] Ayrault, G. (Diss. Univ. Illinois Urbana-Champaign 1974, pp. 1/231; Diss. Abstr. B **35** [1974/75] 447).
[23] Cowan, P.; Tsong, T. T. (Phys. Lett. A **53** [1975] 383/4).
[24] Cowan, P. L.; Tsong, T. T. (Surf. Sci. **67** [1977] 158/79).
[25] Desjonquères, M. C.; Spanjaard, D. (J. Phys. C **15** [1982] 4007/21, 4019/20).
[26] Gjostein, N. A. (13th Sagamore Army Mater. Res. Conf. Proc. [1966/67] 271/304; C.A. **70** [1969] No. 14631).
[27] Girifalco, L. A.; Weizer, V. G. (Phys. Rev. [2] **114** [1959] 687/8).
[28] Wynblatt, P.; Gjostein, N. A. (Surf. Sci. **22** [1970] 125/36).
[29] Ehrlich, G.; Kirk, C. F. (J. Chem. Phys. **48** [1968] 1465/80).
[30] Flahive, P. G.; Graham, W. R. (Surf. Sci. **91** [1980] 449/62).

[31] Halicioglu, T.; Pound, G. M. (Phys. Status Solidi A **30** [1975] 619/23).
[32] McDowell, H. K.; Doll, J. D. (Surf. Sci. **121** [1982] L 537/L 540).
[33] Bassett, D. W.; Parsley, M. J. (J. Phys. D **2** [1969] 13/6).
[34] Voter, A. F.; Doll, J. D. (J. Chem. Phys. **80** [1984] 5832/8).
[35] van der Merwe, J. H.; Kunert, H. (Phys. Rev. [3] B **39** [1989] 5017/24).
[36] Stępień, Z. M.; Kukułka, J. (Surf. Sci. **216** [1989] 505/8).
[37] Bassett, D. W. (Proc. R. Soc. [London] A **286** [1965] 191/203).
[38] Bassett, D. W. (Nature **198** [1963] 468/9).
[39] Nishigaki, S.; Nakamura, S. (Jpn. J. Appl. Phys. **14** [1975] 769/77).
[40] Audiffren, M.; Drechsler, M. (Surf. Sci. **34** [1973] 471/6).

[41] Flahive, P. G.; Graham, W. R. (Surf. Sci. **91** [1980] 463/88).
[42] Fink, H.-W.; Ehrlich, G. (Surf. Sci. **143** [1984] 125/44).
[43] Wang, S. C.; Ehrlich, G. (Surf. Sci. **206** [1988] 451/74, 465/7).
[44] Graham, W. R.; Ehrlich, G. (Phys. Rev. Lett. **31** [1973] 1407/8).
[45] Wang, S. C.; Ehrlich, G. (J. Phys. Colloq. [Paris] **49** [1988] C 6-263).
[46] Nishigaki, S.; Nakamura, S.; Kuroda, T. (Jpn. J. Appl. Phys. **17** [1978] 79/83).
[47] Nishigaki, S.; Nakamura, S. (Jpn. J. Appl. Phys. **15** [1976] 1647/54).
[48] Sadakane, Y.; Abe, T.; Hirano, K. (Nippon Kinzoku Gakkaishi **45** [1981] 20/8; C.A. **94** [1981] No. 143855).
[49] Doll, J. D.; McDowell, H. K. (Surf. Sci. **123** [1982] 99/105).
[50] Graham, W. R.; Ehrlich, G. (Phys. Rev. Lett. **32** [1974] 1309/11).

[51] Tsong, T. T. (J. Chem. Phys. **55** [1971] 4658/9).
[52] Bakhtizin, R. Z.; Valeyev [Valeev], V. G.; Shikhovtseva, E. S. (J. Phys. Colloq. [Paris] **50** [1989] C 8-117/C 8-122).
[53] Choi, D.-S.; Gomer, R. (Surf. Sci. **230** [1990] 277/82).
[54] Tringides, M.; Gomer, R. (J. Chem. Phys. **84** [1986] 4049/61).
[55] Gong, Y. M. (Surf. Sci. **266** [1992] 30/4).
[56] Ala-Nissila, T.; Han, W. K.; Ying, S. C. (Phys. Rev. Lett. **68** [1992] 1866/8).
[57] Ala-Nissila, T.; Han, W. K.; Ying, S. C. (J. Electron Spectrosc. Relat. Phenom. **54/55** [1990] 245/53).
[58] Gong, Y. M.; Gomer, R. (J. Phys. Colloq. [Paris] **48** [1987] C 6-15/C 6-20).
[59] Casanova, R.; Tsong, T. T. (Surf. Sci. **94** [1980] L 179/L 183).
[60] Gomer, R.; Gong, Y. M. (Scanning Microsc. **1** [1987] 1571/3 discussion on pp. 1573/4).

[61] Bowman, D. R.; Gomer, R.; Muttalib, K.; Tringides, M. (Surf. Sci. **138** [1984] 581/98).
[62] Tringides, M.; Gomer, R. (Surf. Sci. **155** [1985] 254/78).
[63] Neumann, G. M.; Hirschwald, W.; Stranski, I. N. (Z. Naturforsch. **21a** [1966] 807/11).

3.2 Surface Self-Diffusion with Appreciable Mass Transport

3.2.1 Diffusion Studies on W Adlayers

Bright regions, rings, or collars formed when 0.1 to 100 atomic W layers were deposited onto FEM tips. Between 650 and 1200 K, collars formed mainly around (110), (211), and (100) [1]. Near 1070 K the adatoms diffused from the interior of (110), (211), and (100) planes to the plane edges where they formed asymmetric collars [2]. This confirms the earlier observations by [3] even though [3] gave a different interpretation. Between ~10 and 80 monolayers of W were deposited with less than 10^{-11} Torr active gases being present on smooth, round, thermal endform FEM tips above 1200 K. The adatoms deposited at high temperatures accumulated at the edges of the planar faces, and the remolded tips closely resembled tips built-up in strong fields [4]; see also [5].

When exposed to a W atom beam, W(100) planes kept their circular shape [2], while W(110) surfaces acquired a hexagonal shape and the channeled W(211) planes enlarged anisotropically due to anisotropic diffusion [2, 6]. Part of the observed anisotropy is due to differences in the adsorption energies of the vicinal faces for W adatoms. For instance, for passage from (211) to (111), an activation energy of 2.86 eV was calculated in [2, 7], while the corresponding value for migration from (211) to (100) was 3.23 eV. Therefore, more adatoms will be trapped on the edge of (111) than on the edge of (100). This explains the right-left asymmetry (parallel-antiparallel to the channels) of the collars mentioned above. Directional diffusion occurs along the channels without any concentration gradient

References for 3.2.1 to 3.2.3 on pp. 99/100

present, the bias being due to the different trapping probabilities at the two ends of the channels [2]. The passage from (110) to (321) requires an activation energy ~0.4 eV lower than along other directions [2].

The following table summarizes the activation energies (in eV) determined by migration experiments with FEM [1]:

		migration from							
		(011)	(112)	(123)	(001)	(012)	(013)	(122)	(111)
towards	(011)			>2.1		>2		>2	
	(112)			>3.6					>2.2
	(123)	1.8±0.2	4±0.4					>2.2	
	(012)	1.8±0.2		3.2±0.2	3.8±0.2		4.4±02		
	(013)					>4.2			
	(122)	1.8±0.2		2±0.2					
	(111)		2.05±0.2						
	(113)		3±0.2		3.8±0.2		4.1±0.3		
	(016)						4.4±0.2		

The experimentally determined activation energies Q generally agree with energies theoretically derived for atom hopping off the edges of (hkl) surfaces. A significant difference between calculated and measured Q was found only for the migration from (112) towards (111), where the experimentally found value of ~2.1±0.2 eV is ~0.8 eV lower than the calculated one (presumably because the tip was contaminated in this tip region) [1]. Earlier, Q=4.62±0.35 eV, 3.47±0.35 eV, and 1.21±0.09 eV had been determined for migration off the (110) surface and attributed to different migration processes. Similarly, 3.69±0.87 eV and 1.95±0.26 eV had been found for the (100) surface [3]; see also the value of ~1.95 eV for (100) in [8]. In agreement with the observation, the calculated Q values varied along the edges of an (hkl) plane [1]. These activation energies determine the residence time τ of the W adatoms. At 1000 K, τ is ~10^{11} s on W(111) but only ~1 s on W(211) [2].

A combined LEED and work function study of the first stages of epitaxy suggested that any W atom deposited on a clean W(110) single-crystal surface diffuses to neighboring terrace edges at 950 K (study at 7×10^{-9} Pa from ca. 300 K to 950 K) [9].

3.2.2 Grain-Boundary Grooving, Smoothing of Scratches, and Thermo- and Electrotransport Studies

Grain-Boundary Grooving. The grain-boundary grooving was studied at 1600 to 2300 °C on W bars preannealed at 2300 °C and having grain sizes of 2 to 5 mm. The surface diffusion coefficient D was determined in [10] from the grain-boundary grooving and smoothing of scratches. The surface energies required for the data evaluation were linearly inter- and extrapolated from the values 2.9 J/m² [11] and 2.3 J/m² [12] reported for 2030 K [11] and 3653 K [12], respectively. The dependence of D on the surface type (hkl) and the direction of the mass transport was illustrated by stereograms [10]. The following table summarizes the D values (in 10^{-7} cm²/s) [10]:

temperature in °C	1600	1800	2000	2200	2300
D (multiple-scratch smoothing)	1.164	5.48	13.04	36.05	63.6
D (grain-boundary grooving)	1.075	5.63	12.63	22.24	30.8

The grain-boundary grooving data at 2200 and 2300 °C fall somewhat below the straight line in the Arrhenius plot. But at the lower temperatures there is good agreement with the multiple-scratch data for which Arrhenius plots yield $D_0=0.2$ cm²/s and Q=2.31 eV [10]. For the surface self-diffusion on W sheets with preferred (001) orientation, the Arrhenius parameters $D_0=0.63\pm0.04$ cm²/s and $Q=2.74\pm0.02$ eV were determined from grain-boundary grooving in 10^{-5} Torr or in 1 atm Ar between 1700 and 3000 °C [13], while $D_0=8.5$ cm²/s and Q=3.4 eV had been given earlier by the same author in a study between 2600 and 3000 °C [14]; these data according to [15] were falsified by a calibration error in measuring the grooving angles.

Smoothing of Scratches. Surface diffusion was studied by the scratching-smoothing method on polycrystalline W samples which were given a 1 h anneal at 2500 °C to remove residual stresses and other artifacts produced by previous polishing. The grain sizes ranged up to 1 mm². The measurements involved 5 and 25 h long anneals at 2500 °C and 2.5 and 10 h long anneals at 2650 °C. Values for the product of surface energy and diffusion coefficient, $\sigma\cdot D$, ranged from 0.006 to 0.02 erg/s at 2500 °C and from 0.01 to 0.042 erg/s at 2650 °C (the variation is attributed to the effect of orientation mainly on D). Assuming the rather high value $\sigma=3$ J/m² for the surface energy (compare pp. 77/8), mean values for D (in 10^{-6} cm²/s) are 3.6 at 2500 °C and 7.0 at 2650 °C [16].

The anisotropy in the surface self-diffusion was studied between 2300 and 2600 K at $\sim2\times10^{-7}$ to 2×10^{-6} Torr on single crystals, cut to within ±1/3° parallel to (110), (211), (100), and (111) surfaces [17]; see also [18]. The anisotropy was studied between ~1850 and ~2500 K by measuring the rate of flattening of sinusoidal ripples or single scratches (and also by the rate of thermal grooving) [19]. Sinusoidal grooves were made, both as single grooves and as arrays thereof. The cold-worked layer was partially removed by electropolishing and anneals. The smoothing of the crimp profile (h = depth of the crimp profile) on annealing was followed as a function of time t at various λ, the wavelength of the crimp profile. On W(111) at 2500 K, the slope of the $\log(t^{-1}\cdot\log[h(0)/h(t)])$ vs. $\log(\lambda)$ curve indicates that the smoothing of the crimp profile occurs by surface diffusion at least up to 0.7 T_{melt} [17].

The diffusion on W(110) and W(211) is anisotropic consistent with the 2-fold symmetry of these surfaces [17, 18]. The W(111) and W(100) surfaces with 3- and 4-fold symmetries, respectively, did not show any anisotropy of D [18, 19]; see also [20] for W(100). The smoothing of scratches on clean W(100) single-crystal surfaces was studied between 2560 and 3150 K at $p<3\times10^{-9}$ Torr and the diffusion was along $\langle110\rangle$. The values $D_0=76000$ cm²/s and Q=5.55 eV were obtained for W(100) [21]. On W(100) and W(111), $D_0=4.2\pm0.4$ and 3.2 ± 0.6 cm²/s and $Q=3.42\pm0.03$ and 3.15 ± 0.05 eV, respectively, were determined between 2300 and 2600 K [17].

For W(110) the diffusion coefficient traces an ellipse in the polar diagram with the major axis along $[\bar{1}10]$ and the minor axis along [001], the axial ratio being 1.9 at 2500 K [19]. Between 2300 and 2600 K on W(110) $D_0=6.0\pm0.9$ and 7.0 ± 1.0 cm²/s and $Q=3.56\pm0.09$ and 3.63 ± 0.09 eV were determined for diffusion along $\langle110\rangle$ and $\langle100\rangle$, respectively [17]. On W(110) the crests and bottoms of the wells became flat, and the profile was no longer sinusoidal after extended anneals; terraces evolved by faceting [17].

On W(112) the ratio of the diffusion coefficients along $\langle111\rangle$- and $\langle110\rangle$-type directions, $D_{\langle111\rangle}:D_{\langle110\rangle}$, is 4.5 at 2500 K, the diffusion being easier along than across the $[11\bar{1}]$-oriented channels [19]. $D_0=1.9\pm0.2$ and 63 ± 13 cm²/s and $Q=2.88\pm0.03$ and 3.92 ± 0.06 eV for diffusion on (112) along $[11\bar{1}]$ and $[\bar{1}10]$, respectively, were found in [17]. In support of these data, the temperature dependence of the diffusion anisotropy on W(211) is consistent with $Q_{\langle111\rangle}$ being ~1 eV higher than $Q_{\langle110\rangle}$ [19].

References for 3.2.1 to 3.2.3 on pp. 99/100

Marker Displacements in Temperature Gradients and Electric Fields. The thermotransport on a W surface is from hot to cold [20, 22, 23], and the electrotransport of W is towards the cathode [20, 22] based on the displacement of indentations observed in a temperature gradient and in an electric field, respectively. The bulk effect has been separated from the surface effect, since the former depends on the radius of the indentations [20]. The activation energy for thermotransport is denoted Q_t and the heat of transport Q^*.

A W rod extending along $[1\bar{1}0]$ and exposing a (110) face was studied at 10^{-8} Torr and in 1 atm H_2. Displacements were observed with the indentations made on the (110) surface when both ends of the rod were kept at 273 K and its center was heated to 3250 K. The displacement of shallow marks was controlled by surface self-diffusion as evidenced by the displacement velocity [22]; see also the results for ∇T along $\langle 110 \rangle$ and along $\langle 111 \rangle$ on W(110) [23]. At 2600 to 3000 K, the velocity of the shallow marks was close to the growth rate of steps developing on the surface (deep marks were considerably slower, their transport properties being controlled by bulk diffusion) [22]. With $\nabla T \parallel \langle 111 \rangle$, marker displacements both along and across ∇T were observed in these studies. No significant differences in Q_t were found for diffusion along $\langle 110 \rangle$ parallel to ∇T and diffusion along $\langle 111 \rangle$ both parallel and transverse to ∇T. $Q_t = 3.3 \pm 0.4$, 3.4 ± 0.5, and 3.6 ± 0.7 eV were obtained for the marker displacements $\parallel \langle 110 \rangle$, $\parallel \langle 111 \rangle$, and $\perp \langle 111 \rangle$, respectively [23], obviously revising the somewhat higher value of 4 ± 0.4 eV given earlier in [22]. For transport $\parallel \langle 100 \rangle$ at 2300 to 2500 K, the same Q_t value of 3.3 ± 0.4 eV as for transport $\parallel \langle 110 \rangle$ was determined [20]. For thermotransport at 2350 to 2600 K on W(211) and W(100), Q_t values of 4.2 ± 0.4 and 2.4 ± 0.3 eV were determined for diffusion $\parallel \langle 110 \rangle$ and $\parallel \langle 111 \rangle$ on W(211), respectively, and $Q_t = 3.6 \pm 0.4$ eV for W(100) [20]. The heat of transport, Q^*, determined in that study tended to decrease with increasing temperature. In view of the rather large error limits (30 to 40%), the heat of transport was found to be isotropic and to be the same on all faces studied; the average Q^* value was 4.2 eV. The anisotropy of the mass transport is governed by the anisotropy of the surface self-diffusion coefficient (for D values, see a table in the paper) [20].

From **electrotransport** measurements, an effective charge of 0.75 elementary charges was deduced [20] revising the values of [22].

3.2.3 Blunting of W Tips. Ring Rate Studies

Blunting of Field Emitter Tips; Solid Drop Formation. The formation of constrictions and solid drops due to capillarity-induced surface diffusion was predicted by Nichols and Mullins [24]. Their theory was extended considering in addition the effect of evaporation by [25]. The predicted effect was confirmed experimentally by an SEM study of W tips heated to 2300 or 2600 K [26]. Pseudo-stationary tip profiles (well-suited for determining the surface self-diffusion coefficients) resulted in calculations for cone half-angles $\alpha \geq 3°$ [25]; see also [27]. Radii of curvature around 0.25 μm have been obtained for conical W tips at 3200 K in the absence of a field, while radii as low as 0.07 to 0.1 μm were achieved at this temperature in the presence of a field [28]. A thermo-field treatment in fields below 100 MV/cm sharpens the tips, while they are blunted in fields exceeding that value [29]. The SEM study [30] on tips with initial cone angles of $\alpha_0 = 1°$ to 15° heated between 1800 and 3000 K showed that for $\alpha_0 < 3°$ solid drops form at the tip apex [30] (in agreement with the calculations of [25]), while characteristic steady-state shapes form for $\alpha_0 > 3°$. Near 2900 K, the tip radius R increased with time t according to a $t^{1/4}$ law [30]. The increase of R with concomitant recession of the tip apex is called blunting. The $t^{1/4}$ rate law is expected from the Nichols-Mullins theory [24] mentioned above (see [31] for the different power laws pertinent for the various reshaping mechanisms). The validity of the $t^{1/4}$ law in the ranges 2100 to 2850 K [32] and 2500 to 2800 K [33] proved surface self-diffusion to be the dominating mechanism for tip blunting at these temperatures [32, 33]. The same conclusion holds for W points with radii of curvature between 3 and 30 μm studied between 2600 and 2900 K at 10^{-5} Torr [34]. W tips with R around 1 μm increased their radii by more than ~0.2 μm, when they were heated to above 2600 K (to remove oxygen or carbon) [35]. The rate of W removal from the apex of W tips ranged from one atomic layer per h to one layer per s between 1850 and 2700 K [3]. Much higher rates were found in [36], where the removal, however, was studied in a field of $\sim 2\times10^8$ V/cm. The removal started at 900 to 1000 °C with a rate of ~4 layers per s; the rate increased to ~12 and ~500 layers per s at 1100 and 2000 °C, respectively [36]. Dislocations are important for the W removal [36, 37]. The step heights of the layers removed from (110) were 4.6 ± 1.2 Å and multiples thereof with field-evaporated tips, while step heights of 2.5 ± 0.7 Å could be observed on thermal endforms [37].

Arrhenius plots of the annealing time required to reshape a built-up endform tip into a thermal endform tip (Sokol'skaya's method [38]) yielded $Q=3.1\pm0.1$ eV between 1600 and 2100 K [39, 40], and $Q=3.1\pm0.2$ eV and $D_0=0.9\cdot e^{\pm1}$ cm²/s were obtained between 2100 and 2850 K at 10^{-11} Torr [32]. $Q=3.2\pm0.2$ eV was found at 1180 to 1510 K by [38] and $Q\approx3.2$ eV with $D_0=0.1\times5^{\pm1}$ cm²/s were obtained in the range ~1740 to 2670 K [33]. The diffusion studies of [39, 40] have been extended in [41] to lower temperatures (1400 to 1800 K) including an anisotropy study of atom migration. For migration of atoms from the (210) regions towards the adjacent (110) and (100) faces, the diffusion parameters $D_0=34$ cm²/s and $Q=3.8$ eV were found. For the migration from regions around (311) and (611) towards the adjacent (211) and (100) faces, the values $D_0\approx0.3$ cm²/s and $Q=3.1$ eV were obtained.

While carbon coverages up to $\Theta=0.1$ monolayers did not alter Q, larger amounts ($\Theta\approx0.5$ to 1) increased Q from 3.1 ± 0.4 eV to 8.7 eV [39] or from 3.0 to 8.5 eV according to [42]. This may explain the high value of $Q=4.62\pm0.35$ eV found between ~2015 and ~2500 K in an earlier study [3].

The pseudo-stationary tip profiles formed by the simultaneous action of evaporation and self-diffusion were studied under 10^{-10} Torr on W tips having cone half-angles such as $\alpha=8°$ [43]. The need to correct the measured dR/dt rates for evaporation (or chemically induced volatilization, e.g., by oxygen) had been shown by [32]. Arrhenius plots of D measured between ~1900 and ~3450 K are strongly bent upward for $T>0.75\cdot T_{melt}$ ($T_{melt}\cong$ 3600 K) [43]. This extends the investigations to lower temperatures than those used in [44] where the limiting shapes of W tips heated to between ~3100 and ~3600 K were evaluated assuming a surface energy value of 2.5 J/m². $D_0=9.5\times10^7$ cm²/s and $Q=8.2$ eV resulted [44]. The large difference to the parameters $D_0=1$ cm²/s and $Q=3.1$ eV found by [32] in the range 2100 to 2800 K (confirmed in [43]) suggests that different mechanisms are operative above and below ~3000 K. At lower temperatures atom jumps to neighboring sites dominate, whereas at higher temperatures the diffusing atoms appear to be no longer in well-localized sites and seem to migrate over large distances as single atoms or clusters [44]. Plots of D vs. 1/T indicate that for $T/T_{melt}<0.75$ the D_0 equals 0.24 cm²/s and Q is 2.85 eV, while for $T/T_{melt}>0.75$ or 0.8, $D_0=10800$ cm²/s and $Q=5.57$ eV are found [45].

Anomalous dR/dt rates were observed above characteristic temperatures T* with built-up tips consisting only of {110}, {211}, and {100} surfaces ("rounded" regions being virtually absent). Usual rates dR/dt were observed for annealing below T*, but the blunting was 100 to 500 times faster when the tips were annealed above T*. T* strongly depends on R: T*(R) = 1750 and 2000 K for R = 0.3 and 0.5 μm, respectively [46].

References for 3.2.1 to 3.2.3 on pp. 99/100

Ring Rate Studies. Pulsed field emission allows one to image the edges of terraces on the weakly emitting (110) surfaces of a W field emitter tip. Due to the low duty cycles used (3×10^{-5}), the tips can be imaged even at temperatures as high as 3000 K without producing the shape changes usually observed under thermo-field treatments (for so-called build-up, see p. 100). A terrace edge is imaged as a dim ring that shrinks upon heating. Near 2000 K, for instance, the ring disappears within a few seconds and after a short period a new, shrinking ring appears. The repetition rate of this process is known as ring rate. It equals (in units of atomic layers) the rate at which the tip apex recesses due to capillarity-induced diffusion of the edge atoms, which migrate off the terraces and towards the shank of the tip [33]; see also [11, 47].

This diffusion of atoms from the apex towards the shank can be counteracted by a static electric field [33]. The determination of the balancing field by the ring rate stopping method is a check for the interpretation of the phenomena observed. It yielded for the surface energy at 2000 K the reasonable value of 2.5 ± 0.7 J/m^2 [33] in agreement with 2.9 ± 0.2 J/m^2 at 2030 K [11] and 2.8 ± 0.28 J/m^2 at 2100 K [48] obtained later by this method.

The ring rate depends strongly on T, but hardly on the radius of the tip, R, if the ratio R: length of the tip is close to zero (studied for R=1 to 10 μm) [49].

(110)-oriented field emitter tips have usually been used. Hence, the recession rates dz/dt at which the tip apex receded was taken as a measure for the readiness with which W atoms diffuse on W(110) planes. But in [48] diffusion coefficients were determined at $<10^{-12}$ Torr by this method for (110) and (100) surfaces. Between 2000 and 2600 K, $Q = 2.95 \pm 0.10$ eV and $D_0 = 0.1$ to 2.5 cm^2/s were found for W(110) and $Q = 2.7 \pm 0.3$ eV and $D_0 = 0.06$ to 1.5 cm^2/s for W(100) [48]. The following table lists diffusion parameters determined by the ring rate method from the apex recession rate:

Q in eV	D_0 in cm^2/s	T in K	p in Torr	Ref.
2.9 to 3.2	–	~1850 to ~2175	$<10^{-12}$	[48]
3.14 ± 0.08	4	~1750 to ~2700	$\sim 10^{-12}$	[11]
3.23 ± 0.1	$1.92 \cdot e^{\pm 0.3}$	1900 to 2000	10^{-11}	[50] (see [32])

The activation energy of Q=3.05 eV found with the ring rate method [42] is much larger than the value of 0.9 to 1.0 eV found for diffusion of single atoms on W(110) (see [51]) and similar values for W dimer dissociation on this surface (see [52]). This suggests that the diffusion rate is limited mainly by bonding of atoms at kink sites [42].

⟨110⟩-oriented W tips were studied at $p < 10^{-10}$ Torr by the probe-hole FEM technique. The tip was heated for 4 s at temperatures between 1100 and 1700 K. For fixed annealing temperatures, the plots of emission current vs. cumulated number of heating cycles has a sawtooth-like appearance for (110) and (100), the periods decreasing with increasing temperature. No periodic character was seen for (211) and rather irregular patterns resulted for (111). From Arrhenius plots of the observed periods, the low activation energies $Q = 1.19 \pm 0.11$ and 1.35 ± 0.08 eV were obtained for W(110) and W(100), respectively [53]. $Q = 1.3 \pm 0.2$ eV was derived for W(110) from the number of heating pulses required for a cycle. Each pulse lasted 5 s and the number per cycle decreased fom ~47 at 1200 K to ~10 pulses at 1400 K [54]; see also [55].

Changes in the steady-state bias field F as well as changes in the temperature change the ring rate to a new value. The transient time required for this change is not related to the thermal inertia of the tip [11]; rather, the time lag observed with the transients reflects the time constant for mass transport [49]. Anisotropies in the surface energy [11, 56] or the temperature dependence of the anisotropy [49] are perhaps responsible for this effect [11, 49, 56]. Quite extended transients (several min) were observed with W(110) but not with W(100) on tips, where the thermal time constant was at most a few s [49].

References for 3.2.1 to 3.2.3:

[1] Drechsler, M.; Vanselow, R. (Z. Kristallogr. **107** [1956] 161/81).
[2] Drechsler, M. Z. (Elektrochem. **58** [1954] 334/9).
[3] Müller, E. W. (Z. Phys. **126** [1949] 642/65).
[4] Golubev, O. L.; Shaikhin, B. M.; Shrednik, V. N. (Izv. Akad. Nauk SSSR Ser. Fiz. **40** [1976] 1599/604; Bull. Acad. Sci. USSR Phys. Ser. [Engl. Transl.] **40** No. 8 [1976] 40/3).
[5] Golubev, O. L.; Shrednik, V. N.; Pavlov, V. G. (Poverkhnost **1985** No. 2, pp. 50/7).
[6] Honigmann, B.; Müller, E. W.; Stranski, I. N. (Z. Phys. Chem. [Leipzig] **196** [1950/51] 6/11).
[7] Drechsler, M. (Z. Elektrochem. **58** [1954] 327/34).
[8] Pautov, D. M.; Shepilov, M. P. (Pis'ma Zh. Eksp. Teor. Fiz. **12** [1970] 393/6; JETP Lett. [Engl. Transl.] **12** [1970] 271/3).
[9] Hahn, P.; Clabes, J.; Henzler, M. (J. Appl. Phys. **51** [1980] 2079/84).
[10] Lai, Ho-yi (Jernkontorets Ann. **151** [1967] 801/21; C.A. **68** [1968] No. 98116).

[11] Barbour, J. P.; Charbonnier, F. M.; Dolan, W. W.; Dyke, W. P.; Martin, E. E.; Trolan, J. K. (Phys. Rev. [2] **117** [1960] 1452/9).
[12] Calverly, A. (Proc. Phys. Soc. [London] B **70** [1957] 1040/4).
[13] Allen, B. C. (Trans. Metall. Soc. AIME **245** [1969] 1621/32).
[14] Allen, B. C. (Trans. Metall. Soc. AIME **236** [1966] 915/24).
[15] Allen, B. C. (J. Less-Common Met. **29** [1972] 263/82).
[16] den Broeder, F. J. A. (High Temp.-High Pressures **10** [1978] 131/3).
[17] Zakurdaev, I. V.; Liberova, G. I. (Fiz. Tverd. Tela [Leningrad] **19** [1977] 1969/74; Sov. Phys.-Solid State [Engl. Transl.] **19** [1977] 1152/5).
[18] Zakurdaev, I. V.; Liberova, G. I.; Chernyak, E. Ya. (Izv. Akad. Nauk SSSR Ser. Fiz. **43** [1979] 1817/22; Bull. Acad. Sci. USSR Phys. Ser. [Engl. Transl.] **43** No. 9 [1979] 22/7).
[19] Mel'nik, G. I. (Poverkhnost **1985** No.11, pp. 140/2; Surf. Phys. Chem. Mech. **4** [1987] 3501/6).
[20] Zakurdaev, I. V.; Liberova, G. I.; Yakovlev, V. N. (Fiz. Tverd. Tela [Leningrad] **21** [1979] 456/62; Sov. Phys.-Solid State [Engl. Transl.] **21** [1979] 270/4).

[21] Bowden, F. P.; Singer, K. E. (Nature **222** [1969] 977/9).
[22] Zakurdaev, I. V.; Tsyganov, A. D. (Fiz. Tverd. Tela [Leningrad] **15** [1973] 1657/60; Sov. Phys.-Solid State [Engl. Transl.] **15** [1973] 1115/6).
[23] Zakurdaev, I. V.; Shuppe, G. N.; Yakovlev, V. N. (Fiz. Tverd. Tela [Leningrad] **18** [1976] 2784/6; Sov. Phys.-Solid State [Engl. Transl.] **18** [1976] 1624/5).
[24] Nichols, F. A.; Mullins, W. W. (J. Appl. Phys. **36** [1965] 1826/35).
[25] Vu, Thien Binh; Uzan, R. (Surf. Sci. **179** [1987] 540/60).
[26] Drechsler, M.; Piquet, A.; Uzan, R.; Vu, Thien Binh (Surf. Sci. **14** [1969] 457/60).
[27] Vu, Thien Binh; Drechsler, M. (J. Phys. Colloq. [Paris] **45** [1984] C 9-29/C 9-37).
[28] Vu, Thien Binh (J. Microsc. [Oxford] **152** [1988] 355/61).
[29] Shrednik, V. N.; Pavlov, V. G.; Rabinovich, A. A.; Shaikhin, B. M. (Izv. Akad. Nauk Ser. Fiz. **38** [1974] 296/301; Bull. Acad. Sci. USSR Phys. Ser. [Engl. Transl.] **38** No. 2 [1974] 104/10).
[30] Vu, Thien Binh; Piquet, A.; Roux, H.; Uzan, R.; Drechsler, M. (Surf. Sci. **25** [1971] 348/56).

[31] Herring, C. (J. Appl. Phys. **21** [1950] 301/3).
[32] Piquet, A.; Roux, H.; Vu, Thien Binh; Uzan, R.; Drechsler, M. (Surf. Sci. **44** [1974] 575/84).
[33] Dyke, W. P.; Dolan, W. W. (Adv. Electron. Electron Phys. **8** [1956] 89/185, 148/63).
[34] Boling, J. L.; Dolan, W. W. (J. Appl. Phys. **29** [1958] 556/9).
[35] Drechsler, M. (J. Phys. Colloq. [Paris] **49** [1988] C 6-87/C 6-91).
[36] Drechsler, M.; Pankow, G.; Vanselow, R. (Z. Phys. Chem. [Munich] **4** [1955] 249/63).
[37] Drechsler, M. (Z. Phys. Chem. [Munich] **6** [1956] 272/85).
[38] Sokol'skaya ["Sokolovskaia"], I. L. (Zh. Tekh. Fiz. **26** [1956] 1177/84; Sov. Phys.-Tech. Phys. [Engl. Transl.] **1** [1956] 1147/54).
[39] Pichaud, M.; Drechsler, M. (Surf. Sci. **32** [1972] 341/8, 342).
[40] Pichaud, M.; Drechsler, M. (Surf. Sci. **36** [1973] 813/6).

[41] Bardon, J.; Drechsler, M. (Rev. Phys. Appl. **9** [1974] 989/93).
[42] Bettler, P. C.; Bennum, D. H.; Case, C. M. (Surf. Sci. **44** [1974] 360/76).
[43] Vu, Thien Binh; Heyde, P. (NATO ASI Ser. B **86** [1983] 558/60).
[44] Vu, Thien Binh; Uzan, R.; Drechsler, M. (J. Phys. Lett. **39** [1978] L-385/L-388).
[45] Vu, Thien Binh; Melinon, P. (Surf. Sci. **161** [1985] 234/44).
[46] Ventova, I. D.; Fursei, G. N. (Zh. Tekh. Fiz. **47** [1977] 857/9; Sov. Phys.-Tech. Phys. [Engl. Transl.] **22** [1977] 513/4).
[47] Trolan, J. K.; Barbour, J. P.; Martin, E. E.; Dyke, W. P. (Phys. Rev. [2] **100** [1955] 1646/9).
[48] Bettler, P. C.; Barnes, G. (Surf. Sci. **10** [1968] 165/76).
[49] Bettler, P. C.; Love, G. G. (J. Appl. Phys. **38** [1967] 3835/7).
[50] Warner, H. R. (Diss. Carnegie Inst. Techn., Pittsburgh 1967 cited from [32]).

[51] Ehrlich, G.; Hudda, F. G. (J. Chem. Phys. **44** [1966] 1039/49).
[52] Tsong, T. T. (Phys. Rev. [3] B **6** [1972] 417/26).
[53] Stępień, Z. M.; Kukułka, J. (Surf. Sci. **216** [1989] 505/8).
[54] Stępień, Z. M.; Lenkow, W.; Kukułka, J. (Acta Univ. Wratislav. Mat. Fiz. Astron. No. 52 [1988] 323/8; C.A. **110** [1989] No. 146120).
[55] Stępień, Z. M.; Kukułka, J.; Lenkow, W. (J. Phys. Colloq. [Paris] **47** [1986] C 7-165/C 7-168).
[56] Bettler, P. C. (AD-660129 [1967] 1/7 [10 pp.]).

3.2.4 Build-Up of Emitter Tips and Formation of Protrusions

3.2.4.1 Build-Up

Local-field gradients drive diffusing atoms towards tip regions with smaller radii of curvature. As a consequence, the apex-to-shank direction of the matter transport observed without an electric field can be reversed in the presence of sufficiently strong fields. As more and more of the small-diameter upper terraces are "dissolved", the tip in that region will be covered by increasingly larger terraces, formerly buried under atoms which now are deposited near the edges of the deeper layers. This process continues until two adjacent close-packed $\{110\}$, $\{211\}$, or $\{100\}$ facets meet at a sharp edge ideally formed by a single row of atoms [1]. This process is commonly known as build-up. For variants of this main process, see "Types of Build-Up" below. Typical experimental conditions for the thermo-field treatment used are 20 MV/cm applied for 200 s at 1950 K and $p < 10^{-7}$ Pa [2]; fields of 21 to 28 MV/cm were applied at 1400 to 1900 K in [3]. Tips in the final stage of build-up are polyhedral and consist only of $\{110\}$, $\{211\}$, and $\{100\}$ faces [2].

W diffusion above 1170 K was noted, and build-up had been observed already in early FEM studies [4 to 6]. Nonetheless, the shaping of field emitters by both heat and thermo-field treatment as described in [7] is still actively studied; see for instance paper [8] describing ⟨111⟩-, ⟨310⟩-, and ⟨611⟩-oriented, built-up FEM tips for use as high-brilliance electron sources [8] or the papers [9, 10] describing the preparation and *in situ* regeneration of microtips for scanning tunneling microscopy.

W tips with radii of curvature near R = 0.15 μm have been built-up at 1600 K under 6.6 kV [11] and tips with R < 0.1 μm under $p < 10^{-9}$ Torr at 1440, 1520, and 1730 K under applied fields of ~38.6, 46.3, 54, and 61.7 MV/cm [12]. Three stages of build-up were observed by Sugata et al. [11] and two in his earlier paper [12]. At 1600 K the first two stages lasted for ~10 min each [11]. The first stage, where the central (110) plane and the {211} surfaces enlarged [11, 12], was due to migration of atoms diffusing outward from the central parts of these surfaces. This stage is indicated by a plateau in the time dependence of the field emission (FE) characteristics [12]. During the second stage in addition build-up around the {100} planes begins [11], the atoms now diffusing from the shank towards the tip apex [12]. The third stage is characterized by a build-up near {210} and further enlargement of the {211} and {100} surfaces [11]. The time dependence of the FE current at 1440 K under ~54 MV/cm suggested that the first stage (which was reached after an ~10 s induction period) lasted for ~1 min. The transient to the second stage started after ~100 s and was completed after ~3 min [12].

The table on p. 102 lists activation energies determined without (Q_0) and with (Q_F) an applied field. According to [2] the values do not depend on the tip radii R (ranging from 0.2 to 1 μm in [2]). But there is a marked dependence of Q_F on the residual gas pressure. Q_F for {100} build-up increased with pressure [3] in agreement with numerous literature data (see [3, p. 248] for a critical compilation of Q values from build-up data reported between 1956 and 1962). But for build-up of the ⟨111⟩ zone, the activation energies seem to decrease with increasing pressure [3]. The Q values for W diffusion around (110) and (100) were found to vary strongly with field ($0 \leq F \leq 50$ MV/cm) [13] (perhaps because of a field-dependent contamination). Q for the {111} faces decreased from 2.36 to as low as 0.72 eV when ~10^{-8} Torr O_2 were admitted to a system at $p \approx 10^{-10}$ Torr [14]. An up to ~10^{-7} Torr residual gas pressure increased the activation energy of surface self-diffusion for the average tip. This conclusion was drawn by [3] after reviewing literature data and fresh data by [15]. The value of Q = 3.5 eV found in [16] for the spreading of W adlayers on (110) at ~10^{-8} Torr and the values 1.6 to 2 eV deduced near 10^{-11} Torr by [17] for the migration of W adlayers off the (110) plane are explained analogously. But the very low Q value of 1.1 eV obtained at ~10^{-5} Torr (H_2O, O_2) by [16] is attributed to the formation of an oxide layer [3].

During build-up, the applied voltage required to draw a preset FE current continually decreases due to higher field enhancement by the build-up. Arrhenius plots of the times required to achieve a preselected degree of build-up indicate that the surface migration under 20 MV/cm is activated by $Q_F = 2.4$ eV whereas a value of $Q_0 = 3.3$ eV was quoted for the migration in zero field [2]. This is in conflict with the conclusion drawn by Bettler and Charbonnier [1] that build-up is *not* the reversal of tip blunting (for which $Q = 3.14 \pm 0.08$ eV had been derived by [22]). Their apparently obsolete conclusion was based on their result, $Q_F = 2.44 \pm 0.05$ eV and $Q_0 = 2.79 \pm 0.08$ eV [1].

A field of ~40 MV/cm enhanced the W atom migration on {211} and neighboring regions to such an extent that observable effects resulted within 1 h even at 800 K. The rate of change, which did not depend on the sign of the field, rapidly increased with the field, perhaps with the square or the cube of the field strength [23]. At 1000 K the restructuring

References for 3.2.4 on pp. 104/5

Q_0 in eV	Q_F in eV	F in MV/cm	T in K	p in 10^{-10} Torr	Ref.
3.3 ± 0.1	2.4 ± 0.1	20			[2]
~2.77 for (111)	2.36 ± 0.05	66	1460 to 1600	~1	[14]
~2.49 for (310)	2.10 ± 0.05	62	1400 to 1560		
	2.36 ± 0.1	~30 to ~50[a)]	1260 to 1640	<10	[18, 21]
	2.3 (±1?)	70	1470 to 1610	<1	[19]
	1.9 ± 0.2	100			
	1.96				[20]
	1.6 to 2			0.1	[17]
	1.9 ± 0.1	21 to 22	1400 to 1900	1	[3][b)]
	2.1 ± 0.2	21 to 22	1400 to 1900	10	
	2.3 ± 0.2	21 to 22	1400 to 1900	100	

[a)] The values 80 to 130 MV/cm given in [21] should be revised, see [3, p. 248]. – [b)] Q values for {100} build-up.

was significantly enhanced by applying a 45 MV/cm field. At 1400 K and 41 MV/cm, pronounced FEM pattern changes occurred within 1 min and within 20 min the {211} planes did grow to such an extent that they met the {110} planes [24]. The action of an ~40 MV/cm field applied for 1 h resulted in a surface that consisted to >50% of planes. The {211} planes almost met the {110} planes, and {111} and {310} developed. For F>40 MV/cm and at 1500 to 1600 K, most of the tip surface developed into planes [23].

The field evaporation of a built-up tip was studied under ~10^{-3} Torr He at 63 K. These studies showed that the surface layers in the built-up regions are disordered [12]. Prior to field evaporation, the atoms migrated from a stable site to a position where the local field was very high and where they were more easily field-evaporated. Disorder at the steps of the {211} surfaces was considered to be due to surface migration of step atoms [25].

In a field vs. temperature plot, a "beginning of migration" curve can be drawn which separates conditions where no W migration was observed from conditions that allow for migration. This curve drops from high fields at low temperatures to moderate fields at high temperatures [26] and lies below the "beginning of build-up" curve given for ~1130 to ~1320 K as a voltage vs. temperature plot [27, p. 135] (this means that at constant temperatures higher fields are required to initiate build-up than are required to enable migration). After the initial strong drop, the curve slopes down moderately above 1150 K [27].

Above some maximum temperature T_{max}, hysteresis occurred in the thermal field emission current density j which is no longer reproducible [26]. T_{max} is about 1350 K, and sudden changes in the emission characteristics indicate occurrence of faceting [28]. Apparently, changes in the atomic order occurring at high temperatures were frozen-in on cooling and gave rise to hysteresis [26]. The surface atom migration caused stepwise noise with {100}-built-up emitters during thermal field emission above 1900 K [29].

Types of Build-Up. Polyhedral tips were obtained by treating tips having radii of curvature, R, between 0.1 and 0.2 μm under $p < 10^{-10}$ Torr at $T < 1800$ K and $F < 50$ MV/cm [30]. Various types of build-up were reported. Two types, which differ significantly only in the latest stages of the process, have been described in [1] and are characterized by filling-in of either the initially dark {211} regions or by filling-in of the initially dark {100} regions [1]. {310}-built-up tips were produced by flashing to ~2700 K for 1 s and applying an electrical field to the emitter heated at 1800 K [29]. Tips with large {110} and {211} faces and extremely small {100} faces were produced by [31]. Build-up of {211} faces was the only mode observed at the lowest pressures (several 10^{-10} Torr) and low fields (21 to 22 MV/cm). {100} faces appeared with increasing pressures at fields ranging from 23 to 28 MV/cm [3]. {100} build-up was also produced by an oxygen-heat treatment of {310}-built-up tips [29]. Built-up tips consisting of {211} and {111} planes formed at 10^{-8} Pa and 2100 K under 40 MV/cm [29].

3.2.4.2 Protrusions

Macroprotrusions. Heating W tips (R = 0.1 to 0.2 μm) at $p < 10^{-10}$ Torr in fields exceeding 30 MV/cm resulted in appreciable overall growth of the tips and in outgrowths which were comparable in size with R ("macroprotrusions"). They formed on several close-packed faces, especially on {110} [30, 32]. The macroprotrusions left had bases comparable in size with the tip diameters and grew perpendicularly out of {110}, {100}, and {111} faces, essentially extending the lattice. FIM showed that the outgrowths form steps of {110} planes and the centers of the steps at various heights do not coincide [30]. At 1850 K the growth of large protrusions on the central (110) face did not always occur exactly normally to the (110) surface. Rather, "oblique" growth was sometimes observed suggesting that a dislocation mechanism is operative [33].

On the central (110) plane, several macroprotrusions were found simultaneously existing. They had wide (110) faces and rounded edges between them. The edges belonged to ⟨110⟩ zones and included {211}-type faces [32]. On tips with large radius R, *weaker* fields were required to produce macroscopic outgrowths than required at smaller R values [30]; see also [32]. For R = 1.8 μm the threshold field was 32 MV/cm at 2450 K, while it was 60 MV/cm at 2300 K with 0.45 μm [30]. In addition to the macroprotrusions, very sharp microprotrusions appeared [30, 32]; see also [34, 35]. Microprotrusions were found only in the immediate neighborhood of {110} surfaces and not *on* these. They were removed (without removal of the macroprotrusions) by briefly heating without a field, e.g., for 1 s at 2450 K [30, 32].

Microprotusions. Microscopic protrusions growing out normally off the central (110) plane reached 0.08 to 0.1 μm [33]; see also [36]. W tips with radii R ranging from 0.13 to 1 μm were built up under $p < 10^{-9}$ Torr at 2000 K by applying fields of F = 20 to 28 MV/cm for 30 s. For the final stage of remolding, the relationship $F = 8 \cdot \sqrt{\pi\sigma/R}$ is expected to hold, where σ denotes the surface energy. The nearly equal sizes of {110}, {211}, and {100} faces would be consistent with nearly equal σ values for these faces. The stability limit was exceeded when (with R = 0.7 μm for F = 28 MV/cm) brightly emitting spots appeared in the FEM image at the junction between {110} and {100} faces. This was associated with an accentuation of the corners, where the central (011) surface and the {001} and {112} faces meet [37]. Heated W tips in high fields strongly rearrange and microprotrusions form. The field at which this rearrangement occurred (near 1800 K presumably) decreased from ~60 MV/cm for R ≈ 0.15 μm to ~15 MV/cm for R > 1.2 μm (saturation) [35]. Local growth of the central (110) face was observed at 1900 K at 27 to 46 MV/cm. At the initial stage, "collars" appear in the FEM images. Collar-like rings appearing bright around the dark FEM images of the (310) faces developed during the build-up of ⟨310⟩-oriented W tips at $p < 10^{-10}$ Torr under 50 to 100 MV/cm applied at 1400 to 1600 K. In the final stage a bright spot appeared on (310), while the collar around (310) disappeared or strongly decreased its brightness [19]. Outgrowths have been observed also at the corner where

References for 3.2.4 on pp. 104/5

the central (011) face meets the (001) and (010) faces [2]. On polyhedral tips, consisting essentially of {110} and {100} faces, the regions of strong field emission were restricted to the corners at {111} regions or to microprotrusions seemingly randomly distributed over the tip [38]. Field desorption microscopy of hot tips (1200 K$<$T$<$2300 K and F$>$150 MV/cm) showed that microprotrusions moved along the edges and steps on the tip, and even the edges themselves moved [39].

Microprotrusions formed under ion bombardment on ⟨110⟩-oriented tips. These microprotrusions were distributed over the whole tip area, up to 50° off the emitter axis, but form rarely near the emitter apex. On ⟨111⟩-oriented tips, most of them occurred in an angular region between 4° and 10° off the emitter axis [40]. Protrusions having base diameters ranging between ca. 100 and 700 Å formed on ⟨111⟩-oriented field emitter tips heated at $\sim$0.3 T_m (T_m: melting point) under He^+ ion bombardment [41]. FEM in an atmosphere of 10^{-5} to 10^{-2} Torr H_2 exposes the tip to strong ion bombardment. Protrusions formed near the {111} faces, apparently due to ion-impact-enhanced surface diffusion. But they did not form near the central (011) surface and only rarely grew around {012} [42].

⟨110⟩-, ⟨210⟩-, and ⟨111⟩-oriented W tips were studied for use in high brilliance (He^+, etc.) ion sources. Tips with radii of a few hundred nm were annealed for at least 10 min at 2500 to 2800 K. After the temperature was lowered to between 1000 and 700 K, an electric field of 30 to 40 MV/cm was applied in $\sim 10^{-4}$ mbar He. The $\sim$0.5 μA field electron current drawn ionized the He. The resulting ion bombardment activated surface diffusion and enabled formation of microprotrusions. The results sensitively depend on the contamination level present, since heavier ions (Ne^+, Ar^+, etc.) sputter more efficiently than do He^+ ions [40]. The delicate balance between protrusion formation and sputtering may explain the somewhat different results reported by others ([43]) [40]. Depending on tip temperature and gas purity, the formation of protrusions takes 10 s to a few min after initiation of the electron emission. The protrusions have typically radii of 40 nm and withstand annealing at up to 1500 K and also withstand fields of at least 350 MV/cm [40].

References for 3.2.4:

[1] Bettler, P. C.; Charbonnier, F. M. (Phys. Rev. [2] **119** [1960] 85/93).
[2] Gorbatyi, N. A.; Pulatova, S.; Reshetnikova, L. V. (Poverkhnost **1986** No. 12, pp. 39/43).
[3] Vanselow, R.; Wang, J. L. (J. Less-Common Met. **35** [1974] 245/58).
[4] Benjamin, M.; Jenkins, R. O. (Proc. R. Soc. [London] A **176** [1940] 262/79, 267/70).
[5] Ashworth, F. (Adv. Electron. **3** [1951] 1/42).
[6] Dyke, W. P.; Barbour, J. P.; Martin, E. E.; Trolan, J. K. (Phys. Rev. [2] **99** [1955] 1192/5).
[7] Dyke, W. P.; Charbonnier, F. M.; Strayer, R. W.; Floyd, R. L.; Barbour, J. P.; Trolan, J. K. (J. Appl. Phys. **31** [1960] 790/805).
[8] Denizart, M.; Hadef, B.; Balladore, J. L. (J. Phys. Colloq. [Paris] **49** [1988] C 6-93/C 6-97).
[9] Vu, Thien Binh (J. Microsc. **152** [1988] 355/61).
[10] Vu, Thien Binh; Marien, J. (Surf. Sci. **202** [1988] L 539/L 549).
[11] Sugata, E.; Mun, D.; Yokoya, H. (J. Electron Microsc. **17** [1968] 221/8).
[12] Sugata, E.; Izeki, K. (J. Electron Microsc. **14** [1965] 189/95).
[13] Pautov, D. M.; Shepilov, M. P. (Pis'ma Zh. Eksp. Teor. Fiz. **12** [1970] 393/6; JETP Lett. [Engl. Transl.] **12** [1970] 271/3).
[14] Mun, D.; Sugata, E. (Jpn. J. Appl. Phys. **10** [1971] 1163/70).
[15] Pichaud, M.; Drechsler, M. (Surf. Sci. **32** [1972] 341/8, 342).
[16] Müller, E. W. (Z. Phys. **126** [1949] 642/65).
[17] Drechsler, M.; Vanselow, R. (Z. Kristallogr. **107** [1956] 161/81).

[18] Sokol'skaya, I. L. (Izv. Akad. Nauk SSSR Ser. Fiz. **20** [1956] 1151/2; Bull. Acad. Sci. USSR Phys. Ser. [Engl. Transl.] **20** [1956] 1044/5).
[19] Kim, H.; Inoue, T.; Hanawa, T.; Sugata, E. (Proc. 2nd Int. Conf. Solid Surf., Kyoto 1974, pp. 59/62).
[20] Matsuo, Y. (Wakayama Daigaku Gakugeigakubu Kiyo Shizenkagaku No. 12 [1962] 61/7 from C.A. **61** [1964] 10074).

[21] Sokol'skaya ["Sokolovskaia"], I. L. (Zh. Tekh. Fiz. **26** [1956] 1177/84; Sov. Phys.-Tech. Phys. [Engl. Transl.] **1** [1956] 1147/54).
[22] Barbour, J. P.; Charbonnier, F. M.; Dolan, W. W.; Dyke, W. P.; Martin, E. E.; Trolan, J. K. (Phys. Rev. [2] **117** [1960] 1452/9).
[23] Becker, J. A. (Bell Syst. Tech. J. **30** [1951] 907/32, 907/23).
[24] Becker, J. A. (Phys. Rev. [2] **82** [1951] 575/6).
[25] Kuroda, T.; Yagi, H.; Nakamura, S. (Mem. Inst. Sci. Ind. Res. Osaka Univ. **26** [1969] 101/14; C.A. **71** [1969] No. 74504).
[26] Wysocki, J. K. (Surf. Sci. **137** [1984] 506/14).
[27] Wysocki, J. (Acta Phys. Pol. A **42** [1972] 129/45).
[28] Swanson, L. W. (Surf. Sci. **70** [1978] 165/80).
[29] Tamura, N. (Scanning Electron Microsc. **1979** No. 1, pp. 31/8; C.A. **91** [1979] No. 112044).
[30] Shrednik, V. N.; Pavlov, V. G.; Rabinovich, A. A.; Shaikhin, B. M. (Izv. Akad. Nauk Ser. Fiz. **38** [1974] 296/301; Bull. Acad. Sci. USSR Phys. Ser. [Engl. Transl.] **38** No. 2 [1974] 104/10).

[31] Ptitsyn, V. E.; Fursei, G. N. (Izv. Akad. Nauk SSSR Ser. Fiz. **52** [1988] 1513/7; Bull. Acad. Sci. USSR Phys. Ser. [Engl. Transl.] **52** No. 8 [1988] 49/52).
[32] Shrednik, V. N.; Pavlov, V. G.; Rabinovich, A. A.; Shaikhin, B. M. (Phys. Status Solidi A **23** [1974] 373/81).
[33] Ventova, I. D.; Fursei, G. N.; Polezhaev, S. A. (Zh. Tekh. Fiz. **47** [1977] 849/56; Sov. Phys.-Tech. Phys. [Engl. Transl.] **22** [1977] 509/12).
[34] Pavlov, V. G.; Rabinovich, A. A.; Shrednik, V. N. (Pis'ma Zh. Eksp. Teor. Fiz. **17** [1973] 247/50; JETP Lett. [Engl. Transl.] **17** [1973] 177/9).
[35] Grishanov, B. I.; Zhukov, V. M.; Polezhaev, S. A.; Fursei, G. N.; Chibukov, Yu. Ya.; Yasnov, G. I. (Radiotekh. Elektron. [Moscow] **23** [1978] 575/83; Radio Eng. Electron. Phys. [Engl. Transl.] **23** [1978] 84/90).
[36] Drechsler, M. (Z. Elektrochem. **61** [1957] 48/55).
[37] Ventova, I. D.; Fursei, G. N. (Zh. Tekh. Fiz. **47** [1977] 844/8; Sov. Phys.-Tech. Phys. [Engl. Transl.] **22** [1977] 506/8).
[38] Pavlov, V. G.; Rabinovich, A. A.; Shrednik, V. N. (Zh. Tekh. Fiz. **45** [1975] 2126/34; Sov. Phys.-Tech. Phys. [Engl. Transl.] **20** [1975/76] 1337/41).
[39] Vlasov, Yu. A.; Pavlov, V. G.; Shrednik, V. N. (Pis'ma Zh. Tekh. Fiz. **12** [1986] 548/52; Sov. Tech. Phys. Lett. [Engl. Transl.] **12** [1986] 224/5).
[40] Jousten, K.; Böhringer, K.; Börret, R.; Kalbitzer, S. (Ultramicroscopy **26** [1988] 301/11).

[41] Schwoebel, P. R. (Surf. Sci. **181** [1987] 154/6).
[42] Cavaillé, J. Y.; Drechsler, M. (Rev. Phys. Appl. **12** [1977] 1631/9).
[43] Dranova, Zh. I.; Mikhailovskii, I. M. (Fiz. Tverd. Tela [Leningrad] **12** [1970] 132/7; Sov. Phys.-Solid State [Engl. Transl.] **12** [1970] 104/8).

3.2.5 Ion-Bombardment-Induced Surface Diffusion

The enhancement of surface diffusion by ion bombardment has already been addressed at the end of the foregoing section.

References for 3.2.5 on pp. 106

The low-temperature surface migration of W on W tips and its enhancement by 0.5 to 2.5 keV He^+ ion bombardment was studied by FEM and FIM at 0.005 Torr He (base pressure $<10^{-7}$ Torr) in the presence of strong fields. The shape of a W needle (shown as shadowgraph in [1]) subjected to $\sim$1 mA/cm^2 of 2 keV He^+ ions from a glow discharge indicated that ion bombardment and thermal annealing have similar effects on W points [1]. A particular change in an FEM tip profile under ion bombardment was brought about always by the same ion dose [2]. The He^+ ion impact-induced surface self-diffusion on a field-desorbed tip was studied by FEM. With 8.5 keV ions, a 10^{17} ions/cm^2 dose (apparently at $\sim$300 K) was required to have a blunting effect equivalent to that produced by heating 80 s at 1330 K [2]. This result indicates that the ion-bombardment-induced diffusion coefficient D_i for an He^+ impingement is $\sim$6 orders of magnitude lower (at least for 8.5 keV ions) than the D_i reported for 2 keV ions by [1]. The 8.5 keV He^+ ions were found to be $\sim$2 orders of magnitude more efficient than the 0.6 keV H_2^+ ions studied in [3]. D_i is proportional to the ion flux [2]. The values determined using FEM methods in [2, 3] are given in the following table:

D_i in cm^2/s	ion flux in cm$^{-2}\cdot$s^{-1}	ion species	Ref.
1.5×10^{-12}	1.2×10^{15}	8.5 keV He^+	[2]
1.6×10^{-11}	$\sim10^{18}$	0.6 keV H_2^+	[3]

Electron field emission in 10^{-6} to 10^{-3} Torr noble gases (He, Ne) results in bombardment by 0.5 to 5 keV ions produced by the ionization of the gas atoms. Near 300 K, the bombardment by $\sim$1 keV Ne ions at ion current densities between 0.5 and 5 μA/cm^2 resulted in effective surface diffusion coefficients of (2.8 to 30)$\times10^{-10}$ cm^2/s. Under thermal activation this would correspond to 1550 to 1700 K [4].

Interstitial W atoms were produced by He ion bombardment, and their diffusion from the bulk to the surface was studied by FIM at 21 to 78 K under a residual gas pressure of $\sim10^{-8}$ Torr and p(He)=(1 to 5)$\times10^{-3}$ Torr. A field-evaporated endform tip (R=0.01 to 1 μm) was bombarded by (1 to 7 keV) He ions while keeping a field of 300 to 550 MV/cm in order to field-desorb contaminants. The He ion beam impinged at 90°±5° with respect to the (110) surface normal. During the ion bombardment and up to 10 s thereafter, interstitial W atoms emerged from the bulk. Single atoms and dimers were observed in the FIM image. Less than 1% of the atoms appeared on the (110) surface which seemed to possess a reflecting barrier (of 29 meV) for the interstitials in addition to the barrier for their migration in the bulk (50 to 80 meV). 8.2% of the interstitials appeared in regions considered to be in the shadow of the ion beam [5]; see also the earlier paper [6].

References for 3.2.5:

[1] Dranova, Zh. I.; Mikhailovskii, I. M. (Fiz. Tverd. Tela [Leningrad] **12** [1970] 132/7; Sov. Phys.-Solid State [Engl. Transl.] **12** [1970] 104/8).
[2] Drechsler, M.; Junack, M.; Meclewski, R. (Surf. Sci. **97** [1980] 111/8).
[3] Cavaillé, J. Y.; Drechsler, M. (Surf. Sci. **75** [1978] 342/54).
[4] Dranova, Zh. I.; Mikhailovskii, I. M. (Fiz. Tverd. Tela [Leningrad] **13** [1971] 553/7; Sov. Phys.-Solid State [Engl. Transl.] **13** [1971] 443/6).
[5] Gerasimenko, V. I.; Dranova, Zh. I.; Mikhailovskii, I. M. (Fiz. Tverd. Tela [Leningrad] **25** [1983] 2456/61; Sov. Phys.-Solid State [Engl. Transl.] **25** [1983] 1410/3).
[6] Dranova, Zh. I.; Mikhailovskii, I. M. (KFTI-81-7 [1981] 104 pp., pp. 55/7; INIS Atomindex **13** [1982] No. 705326).

4 Work Function

4.1 Survey. Reviews

Thermionic emission is comprehensively reviewed in [1]. Particular emphasis is placed on the problems of patches and patch fields. The emission parameters A (Richardson constant, see p. 132) and Φ (work function) determined by thermionic emission measurements are critically compared with the parameters derived from photoelectric studies and the Φ data from electron beam and contact potential (Kelvin) experiments [1]. Methods for measuring Φ (thermionic and photoelectric emission, field emission, contact potential, and surface ionization) and results thereby achieved were comprehensively reviewed by [2]. The apparent work function Φ^* derived from Richardson plots for uniform surfaces is related to the true work function Φ approximately by $\Phi \approx \Phi^* + T \cdot (d\Phi/dT)$. The quantity actually derived from Richardson plots for a composite surface usually is Φ^{**}, the apparent work function of a patchy surface. For problems associated with patchiness and the distinction to be made between the "strong field" and the "weak field" cases, see [2]. At that time (1969) the opinion about the value of $\Phi(110)$ was divided. The high values usually found with probe-hole field emission microscopy (FEM) have to be discarded in view of the good agreement between the contact potential measurements by [3] and various thermionic and surface ionization studies. The weakest point in the FEM method appears to be the uncertainty about the field enhancement factor for the (110) face [2]. This was borne out later by Todd and Rhodin [4] who strongly favor macroscopic Φ determinations over FE methods. In addition, they point out that the Fowler-Nordheim theory describes the electron tunneling through the surface barrier by just two parameters, namely field and work function [4]. Early FEM studies that yielded lower $\Phi(110)$ values than the later probe-hole determinations suffer from spurious signals that tend to lower the Φ values for the surfaces with high work function, especially $\Phi(110)$ [2]. Rivière [2] seems to be rather reluctant with respect to any work prior to the mid-fifties. Though good vacuum conditions were sometimes achieved even in pre-war papers, it was then not realized that contaminants like carbon cannot be flashed off from W surfaces. A very critical selection of Φ values published up to the mid-sixties is given in [2].

The work functions of clean metal surfaces, including numerous data for W, are compiled in [5 to 9]. The work function of stepped surfaces has been reviewed in [10, 11]. Studies of work function changes brought about by adsorption are reviewed in [5 to 7, 9, 12].

Theoretical approaches to Φ of clean metal surfaces are briefly reviewed within the framework of the jellium model in [13]; extensions of this model that include lattice effects are also given [13].

References for 4.1:

[1] Herring, C.; Nichols, M. H. (Rev. Mod. Phys. **21** [1949] 185/270).

[2] Rivière, J. C. (Solid State Surf. Sci. **1** [1969] 179/289).

[3] Hopkins, B. J.; Pender, K. R. (Br. J. Appl. Phys. **17** [1966] 281/2).

[4] Todd, C. J.; Rhodin, T. N. (Surf. Sci. **36** [1973] 353/69).

[5] Fomenko, V. S.; Podchernyaeva, I. A. (Emissionnye i Adsorptionnye Svoistva Veshchestv i Materialov, Spravochnik, Atomizdat, Moscow 1975, pp. 1/320; C.A. **83** [1975] No. 121265).

[6] Fomenko, V. S. (Emissionnye Svoistva Materialov, Spravochnik, Naukova Dumka, Kiev 1970, 148 pp.; C.A. **76** [1972] No. 118932).

[7] Fomenko, V. S. (Handbook of Thermionic Properties, Plenum, New York 1966, 151 pp.).

[8] Michaelson, H. B. (J. Appl. Phys. **48** [1977] 4729/33).

[9] Hölzl, J.; Schulte, F. K. (Springer Tracts Mod. Phys. **85** [1979] 1/150).
[10] Wagner, H. (Springer Tracts Mod. Phys. **85** [1979] 151/221, 182/7).
[11] Rhead, G. E. (Surf. Sci. **68** [1977] 20/38, 21/2).
[12] Eberhagen, A. (Fortschr. Phys. **8** [1960] 245/94).
[13] Kiejna, A.; Wojciechowski, K. F. (Prog. Surf. Sci. **11** [1981] 293/338).

4.2 Experimental Work Function Data

4.2.1 Single-Crystal Surfaces

4.2.1.1 W(110)

Very controversial results have been reported for the work function of W(110). Measurements on macroscopic specimens cluster around $\Phi \approx 5.3$ eV (see, e.g., the tables given in [1 to 3]), but much higher values have often been obtained by field emission on point cathodes. While the low Φ(110) values obtained by FE in [4 to 6] were early recognized to be due to spurious signals [7], the reasons for the excursions to high values remained obscure for a long time. Todd and Rhodin's paper [2] is a careful investigation into the reliability of Φ values from Fowler-Nordheim (F-N) data and is especially devoted to the Φ(110) problem. This problem stems from insufficient control on and lack of knowledge about the field enhancement factor. Todd and Rhodin [2] conclude that F-N data are not well suited for Φ(110) determinations. Also, a combination of F-N data with energy distribution (ED) results (i. e., Young's method [8]) is unsatisfactory in that it presumes free electron behavior, a condition seldom fulfilled with transition metals [2]. For instance, from F-N and ED measurements at 77 K, Φ(110) was found at 7.1 eV for field-evaporated tips and at 8.8 eV for tips annealed 1 min near 1000 K [8]. The FIM study [9] discussing the anomalous behavior of W(110) suggested $\Phi(110) \geq 7$ eV [9]. On an emitter annealed 5 min at 1000 K, Φ(110) was determined from F-N plots to be 5.79 eV and from combined F-N slope and ED results to be 6.40 eV [10]. At 78 K, Φ(110) determined from F-N plots ranged between 6.28 ± 0.08 and 5.32 ± 0.10 eV [11]. At 15 K, F-N plots indicated $\Phi(110) = 5.83$ eV, and ED data indicated $\Phi(110) = 5.76$ eV [12]. From F-N plots, work function values of 6.30, 5.74, and 4.82 eV were obtained for the central part of the (110) plane and for edge regions facing (100) and (111), respectively [13]. Other high Φ(110) values (in eV) obtained from FE studies are: 6.5 [14], 6.0 [15 to 17], 5.96 [18], 5.92 [19], 5.80 ± 0.05 [20], 5.7 ± 0.15 [21], 5.7 ± 0.08 [22], 5.69 ± 0.05 [23], 5.6 to 5.9 [24], 5.54 [25], 5.5 [26], 5.5 ± 0.2 [28]. The values 5.3 eV at 78 and 350 K [29] and 5.2 ± 0.1 eV [30, 31] obtained from FE studies are consistent with the Φ(110) values known from macroscopic investigations but they are about 0.5 to 1 eV lower than most other FE-derived Φ(110) values. It is clear that the large range of Φ(110) values derived from FE cannot be explained by the usual experimental uncertainties.

Todd and Rhodin [2] used W tips which had been annealed at 1500 K. At 78 K and $p < 10^{-10}$ Torr, the tips were investigated by FE and gradually field-desorbed between the various measuring runs. Φ(110) values, derived by F-N plots from annealed tips, ranged between 5.9 and 6.0 eV. The onset of field evaporation is reflected in **Fig. 23** by a sharp drop to Φ(110) values between 5.2 and 5.4 eV [2], the range of Φ(110) values usually found by macroscopic methods; see below. The behavior shown in Fig. 23 [2] is reminiscent of observations by Müller [7] who found Φ(110) to be dependent on the annealing temperature. Annealing the tip at temperatures between ~1300 °C and ~2300 °C resulted in Φ values which systematically ranged from 5.99 to 5.70 eV, the higher Φ values being associated with the lower annealing temperatures, i. e., with conditions tending to increase the (110) plane [7]. Similarly, $\Phi(110) = 5.70$ eV was obtained after flashing the tip to >2200

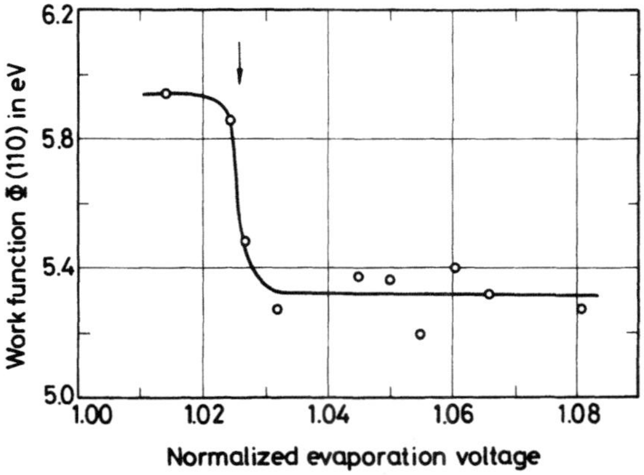

Fig. 23. The work function of W(110) after field evaporation in vacuum at the indicated voltage. The voltage is normalized in terms of the (lower) voltage required for field evaporation in helium. The arrow indicates the onset of field evaporation in vacuum [2].

K, while annealing at 1050 K for several min resulted in a Φ value as high as 6.30 eV (this change was attributed to quenched-in atomic disorder and/or to shape changes) [32]. This variation of (apparent) Φ values is due to changes in the field enhancement factor which is very hard to determine reliably for W(110) [2]. Correcting for the drop of the field from the (110) plane edge to the (110) plane center, Todd and Rhodin [2] found Φ(110) values in the range 5.38 to 5.47 eV for thermally annealed tips, and the range of corrected Φ(110) values, encompassing both field-evaporated and thermal endforms, was 5.2 to 5.47 eV. In addition, they showed that patch field corrections (see [33]) are negligible for the field emission problem with W(110) [2].

A combined FEM/FIM study [34] indicated a significant dependence of the emission yield on the atomic perfection of the W(110) surface. If attributable primarily to Φ changes, this would explain some of the discrepancies between Φ(110) values reported from determinations using different methods (using atomically perfect surfaces vs. imperfect surfaces) [34]. A calculation by [35] suggests that the spreading of the electron gas by the applied field might increase the (apparent) work function. For applied fields of $F \approx 34$ MV/cm, a work function shift of $\Delta\Phi(110) = \Phi(F \approx 0) - \Phi(F \approx 34\ \text{MV/cm}) \approx 0.45$ eV was calculated by a local density formalism [35]. However, a subsequent calculation showed that the magnitude of this effect should be much lower [36].

Some more recent FE studies from the eighties are discussed in the following. By a combined probe-hole FIM/FEM study, the work function at the center of a (110) plane was determined from F-N plots to be 5.18 ± 0.04 eV and 4.91 ± 0.01 eV off-center (the probe hole used had a diameter about as large as the image of an Re atom adsorbed on the W(110) plane). The decrease in Φ on moving off-center is attributed to spurious probe-hole current contributions from the more strongly emitting (110) vicinals [37]. In an extension of that study (using probe holes with projected radii of ≤2.7 Å), again $\Phi(110) \approx 5.1$ eV was found for a tip radius of ~220 Å and a (110) plane radius of ~32 Å. Under these conditions, with smaller plane radii (prepared by controlled field evaporation of the sharp tips formed by Ne sputtering), again Φ values were found to be as low as 4.9 eV [38]. $\Phi(110) = 5.43$ eV was found by [3] assuming $\Phi = 4.5$ eV for the average tip. Controlled field evaporation allowed the study (and elimination) of edge effects for W(110); patch field effects were shown to be negligible [3]. A similar study at the same laboratory yielded $\Phi(110) = 5.49$ eV [39].

References for 4.2.1 on pp. 118/21

While the average amounts to 5.46±0.03 eV, a value of 5.47 eV is considered to be the best estimate [3]. Combining photofield and FE data resulted in Φ(110)=5.34±0.02 eV [40].

Φ(110)=5.25±0.02 eV was determined at room temperature at ~10^{-10} Torr with the FE electron beam method on samples cleaned by flashing to 2100 K [41]; this value is recommended by [42]. A contact potential difference (CPD) of 0.50±0.02 eV has been determined between W(110) and a W foil, whereby Φ(polycr.) was assumed to be 4.55 eV (see [43]) leading to Φ(110)=5.05±0.02 eV at 300 K [44]. Prior to the measuring runs the crystal had been heat-treated at 2500 K for 100 h and in addition heated at 3000 K for 3 h. When the crystal was subjected to an additional oxygen-heat treatment (10^{-6} Torr at 2500 K for 20 h) after the cleaning run, Φ(110) increased from 5.05 to 5.15 eV and some etching occurred. Therefore, it is not clear whether the etching or any removal of carbon caused this increase [44]; a value of 5.11±0.02 eV was given in another study [45] by the same authors. A more recent CPD study with the electron beam method indicated an increase of Φ(110) from 5.08 eV to 5.30±0.02 eV after removing carbon impurities from the surface [122]. Under p≈10^{-11} Torr, Kelvin CPD measurements (referring to the Φ(polycr.)=4.55 eV recommended by [43]) yielded Φ(110)=5.11±0.01 eV [46]. Under ~10^{-10} Torr, Φ(110)≈ 5.35 eV was measured at room temperature using the thermionic electron beam method [47]. Using Shelton's method (see [48]), the work function difference ΔΦ=Φ(110)−Φ(100)= 0.65±0.02 eV was determined between surfaces which were given a prolonged treatment at ~2200 K in ~4×10^{-6} Torr oxygen to remove carbon and cleaned by repeated flashing up to 2500 K. Assuming Φ(100)=4.6 eV yielded Φ(110)=5.25 eV [49]. With the same assumption, Φ(110)≈5.3 eV was obtained from CPD measurements using the electron beam method [50]. The work function difference Φ(110)−Φ(111)=0.71 eV at 2000 K was obtained from the energy distribution of thermionic emission at 1.5×10^{-9} Torr, and Φ(110)=5.09 eV was estimated for the same temperature [51].

The value 5.30±0.12 eV is considered to be the best estimate for the thermionic work function Φ* of W(110) [1] based on a careful examination of the measurements by [52 to 57]. Φ*=5.35±0.05 eV was recommended by [58, 59]. More recently, Φ*=5.75 eV was derived from Richardson plots between ~2350 and 2800 K [60] and compared with earlier Φ* values obtained from Richardson plots at lower temperatures: 5.40±0.05 eV (~2050 to 2350 K) [52], 5.33±0.03 eV (~2070 to 2350 K) [53, 61], 5.30±0.06 eV (~1920 to 2330 K, Φ* value in agreement with surface ionization data of Na) [54], 5.18±0.08 eV (see below) [55], 5.0±0.2 eV [62]. All of these papers indicate a negative temperature coefficient dΦ/dT in agreement with [60], but |dΦ/dT| in these papers is smaller than the value of 230±90 µeV/K found by [60]; see p. 128. The rapid change of the true work function and its temperature coefficient in the neighborhood of a W(110) plane suggest that Φ*(110) is higher than previously published values indicate [60].

Thermionic measurements in the range 1850 to 2450 K yielded an effective work function of 5.22 eV for electropolished W(110) surfaces [63]. Φ*(110)=5.2 eV was obtained using a cylindrical electron projection tube [64]. Φ*(110)=5.26 eV was estimated by [56, p. 305]. Richardson plots from the W(110) surface, studied between 1400 and 2500 K under 10^{-10} Torr, suggest that there are two different regions, namely for T<1850 K, Φ*(110)=5.1± 0.04 eV and for T>1850 K, Φ*(110)=5.7±0.04 eV according to [65]. When a W surface parallel to (110) within ~2° was subjected under 10^{-9} Torr to an extended heat treatment at 2300 K, its thermionic work function decreased from 5.36 to 5.24 eV [66].

The thermionic Φ*(110) was measured between ~1860 and ~2260 K at p<10^{-10} Torr. Richardson plots indicated Φ*(110)=5.18±0.08 eV. In addition, Φ* was also determined by surface ionization (SI) of an NaCl molecular beam impinging on the same W(110) surface used in the thermionic study. From the data recorded between 1845 and 2136 K, Φ*(110)=

4.87 ± 0.10 eV was obtained. The average of these two independent determinations is Φ*(110) = 5.03 ± 0.15 eV [55]. Etching a W single crystal wire with [110] along the wire axis resulted in a faceted structure. Φ* = 5.25 ± 0.05 eV was obtained from Schottky plots (~1490 K ≤ T ≤ ~1880 K) differing only insignificantly for the surface normals [$\bar{1}$10], [112], and [11$\bar{2}$]. Φ* = 5.25 eV also resulted from a Saha-Langmuir evaluation of the SI of Na [57]. For a critique of this paper, see [67]. W(110) surfaces were studied by surface ionization of Sr and La [68], extending the former SI study [69] where Sr and Ca were used. The new value obtained with Sr, Φ*(110) = 5.52 ± 0.05 eV [68], revises the former one, 5.41 ± 0.04 eV [69], which had been obtained on a slightly misoriented crystal. Small amounts of oxygen slowly attack the hot (110) surface, thereby lowering Φ*(110) from 5.52 to 5.30 eV. SI studies involving La resulted in Φ*(110) = 5.44 ± 0.05 eV [68].

From photoemission spectra at 300 K, Φ(110) = 5.25 eV was obtained by [70] in good agreement with 5.24 ± 0.03 eV, similarly found by [71].

4.2.1.2 W(211)

The Φ values reported for this surface scatter considerably, and no recommended value is given in the data compilations [1, 42]. Φ(211) values in the range 4.65 to 4.88 eV were obtained, apparently at room temperature, under $p < 10^{-12}$ Torr of chemically active gases with the probe-hole FE technique. These Φ values involved an FEM tip that had been annealed at temperatures between ~1300 °C and ~2300 °C, the higher work function being associated with the lower annealing temperatures and vice versa. Some disorder or some other defects were suspected to be frozen in after discontinuing the anneals [7]. But Φ(211) values ranging from 4.7 to 4.90 eV were also obtained from field evaporation sequences applied to annealed emitter tips (just as described for W(110) on p. 108) [2]. Field emission energy distribution (ED) measurements at 78 and 295 K yielded a weighted mean Φ value of 4.89 ± 0.15 eV, considered to be reliable for the W(211) surface [72]. Φ values were determined by [10] from the slope of F-N plots (Φ_{FN}) and from the combined F-N slope and ED results (Φ_{ED}). Prior to each measurement the emitter was either annealed for 5 min at 1000 K or flashed to 1800 K and quenched to 77 K, the temperature of the Φ measurements. Fair agreement was found between Φ_{FN} and Φ_{ED}. The Φ_{ED} value was larger for the annealed emitter (5.05 ± 0.05 eV) than for the flashed and quenched tip (4.84 ± 0.06 eV) [10]. At 15 K the values Φ_{FN} = 5.04 eV and Φ_{ED} = 5.02 eV were determined for W(211) [12]. FE measurements of the temperature dependence of Φ resulted in Φ(211) = 5.0 eV (at 78 and 340 K), 4.92 eV (at 600 K), and 5.08 eV (at 700 K) [29].

Other Φ(211) values (in eV) derived from FE measurements are: 4.6 [31], 4.82 [25], 4.85 ± 0.1 [30], 4.87 ± 0.02 [16], 4.88 [19], 4.9 [24, 28, 73], 4.91 [26], 5.0 ± 0.05 [23], 5.05 ± 0.05 [20], 5.05 ± 0.08 [22], 5.12 ± 0.07 [11], 5.18 [13], 5.3 [14].

Correcting for a numerical error in the original paper [74], a value of Φ*(211) = 4.69 eV was given for the thermionic work function in [75]. Φ*(211) = 4.65 eV was quoted in another early paper [56]. Using a cylindrical electron projection tube, the thermionic Φ*(211) = 4.7 ± 0.1 eV was obtained under $p \approx 10^{-10}$ Torr [62]. At $p < 10^{-9}$ Torr a thermionic value of Φ*(211) = 4.76 ± 0.06 eV was found [61]. Similarly, Φ*(211) = 4.80 ± 0.05 eV resulted from a study on a cylindrical W single crystal [52]; see also [58, 59]. Very high values, 5.3 eV [64] and 5.25 eV [57], have also been reported for Φ*(211).

Φ(211) = 4.71 ± 0.05 eV was measured at 300 K with the (thermionic) electron beam method [76] in good agreement with the average of the thermionic [52, 56, 75, 77] and FE [7, 10, 78] data in the literature [76]. Φ(211) ≈ 4.7 eV resulted from CPD measurements using

References for 4.2.1 on pp. 118/21

the electron beam method and assuming Φ(100) = 4.6 eV [50]. A much higher value, Φ(211) = 4.90 eV [79] obtained by the FE retarding potential method, is cited in [80].

A single crystal wire, kept at ~1400 °C under 2×10^{-8} Torr during photoelectric measurements and flashed at 2200 °C in between, yielded Φ(211) = 4.50 eV [81].

4.2.1.3 W(100)

Φ(100) = 4.63 ± 0.02 eV was determined (at $p\approx10^{-10}$ Torr, obviously at room temperature) with the FE electron beam method on samples cleaned by heating them to 2100 K [41] (this Φ value is recommended by [42]). The Kelvin CPD between an aged W foil and W(100) under $p\leq5\times10^{-10}$ Torr gave Φ(100) = 4.65 ± 0.04 eV [45]. Φ(100) = 4.64 ± 0.01 eV was determined from Kelvin CPD measurements under $p\approx10^{-11}$ Torr, assuming Φ = 4.55 eV (see [43]) for the W foil used as reference [46]. Φ(100) = 4.60 ± 0.05 eV was obtained at 300 K by using the (thermionic) electron beam method [76]. This value agrees with the average of literature data from thermionic emission [52, 56, 75, 77, 82 to 84] and FE [78] measurements [76]. Under the conditions used by [76], a value of Φ(100) ≈ 4.52 eV was obtained [47], significantly lower than the other CPD results, but well in the range of other thermionic Φ*(100) values (see below). From photoemission spectra taken at 300 K, Φ = 4.62 ± 0.04 eV was extracted [71].

The value Φ* = 4.58 ± 0.08 eV is quoted to be the best estimate for the thermionic work function of W(100) [1, p. 209] based on the data obtained by [52, 54, 56, 82], while Φ*(100) = 4.60 ± 0.05 eV is recommended by [58, 59]. The thermionic value Φ*(100) = 4.66 ± 0.06 eV resulted from Richardson plots recorded between ~1950 and 2400 K at $p<10^{-8}$ Torr [54]. This value is higher than the Φ*(100) values usually reported, but it agrees well with the value calculated from the thermionic yield and also from SI studies with Na by the same authors [54]. Φ*(100) = 4.6 ± 0.1 eV was determined under $\sim10^{-10}$ Torr using a projection tube [62]. Φ* = 4.57 eV was obtained between 1900 and 2200 K under $p\approx3\times10^{-10}$ Torr [85]. Richardson plots gave Φ*(100) = 4.61 ± 0.05 eV for a surface with an average dislocation density of $(0.6\ \text{to}\ 1)\times10^{6}\ cm^{-2}$ [86]. Φ*(100) = 4.56 eV was obtained [75] after correcting a numerical error in the original thermionic emission study [74]. Other Φ*(100) values (in eV) derived from thermionic emission measurements are: 4.515 ± 0.062 [87], 4.52 (± ~0.5%) [56], 4.53 ± 0.05 [61], 4.55 ± 0.05 [52], 4.59 ± 0.02 [82], and 4.6 [64]. A single crystal ribbon was investigated at $p\leq10^{-10}$ Torr with Shelton's method (see [48]). The exposed surfaces were parallel to (100) within 1° yielding Φ*(100) = 4.55 ± 0.04 eV from Richardson plots between 1672.8 and 1818.6 K [83]. Φ*(100) = 4.52 ± 0.07 eV was measured between 1150 and 2200 K by [88] on ribbons outgassed at 2400 K for 15 to 18 h. This treatment resulted in recrystallization yielding surfaces exposing 95 to 96% (100) planes with the remainder being mainly (111) planes [88]. Under $p\approx10^{-7}$ Torr and between 1850 and 2450 K, electropolished W(100) surfaces gave an effective work function of 4.54 eV [63].

The thermionic emission from W(100) surfaces (aligned to within 0.5°) measured under $p<10^{-9}$ Torr between 1675 and 2150 K showed a discontinuity in the Richardson plots at 1820 K, attributed to a change in the emission parameters: Φ*(100) = 4.4 ± 0.2 and 4.76 ± 0.04 eV below and above 1820 K [89].

W(100) surfaces were investigated with surface ionization (SI) of Sr and La. Above 2500 K Saha-Langmuir plots yielded Φ*(100) = 4.56 ± 0.04 eV [68].

Φ(100) ≈ 4.64 eV was obtained from an energy-resolved FE study [90]. The field dependence of the energy distribution of field-emitted electrons was investigated at 295 K giving Φ(100) = 4.57 ± 0.14 eV [72]. This value agrees with Φ(100) = 4.59 ± 0.02 eV derived from F-N plots in a probe-hole FE study at 77 K with annealed tips [10]. Note, however, that the result of this study is later cited by [41] to be $\Phi(100) = 4.70 \pm 0.05$ eV.

An FEM H_2-adsorption study on W at 10^{-10} Torr yielded $\Phi(100) = 4.65 \pm 0.1$ eV [30]. $\Phi(100) = 4.7$ eV was derived from F-N data at 78 K [73]; for values at higher temperatures, see the paper [73]. At liquid N_2 temperature and $p < 10^{-11}$ Torr, $\Phi(100) = 4.70 \pm 0.05$ eV was obtained from F-N plots (assuming Φ(average tip) = 4.50 eV) and also from single-point measurements [16]. But another F-N study [20] indicated a much higher value, $\Phi(100) = 4.90 \pm 0.05$ eV (apparently at T < 100 K and $p < 10^{-8}$ Torr) [20]. Also, from FE measurements at 15 K and $p \ll 10^{-8}$ Torr, $\Phi(100) = 4.89$ eV was obtained [12]. F-N plots measured at 10^{-10} Torr (apparently at 80 K) were evaluated assuming Φ(average tip) = 4.54 eV and yielded $\Phi(100) = 5.1$ eV [14]. Other $\Phi(100)$ values (in eV) obtained from FE studies are: 4.6 ± 0.1 [28], 4.65 [17], 4.69 [91], ~4.7 [24], 4.71 [26], 4.83 ± 0.02 [22], 4.90 [13], 4.93 ± 0.06 [11], 4.95 ± 0.05 [23], 5.2 [92].

Reconstructed vs. Unreconstructed Surfaces (see p. 1). $\Phi(100)$ was measured with the retarding potential technique. On cooling from ~400 to 210 K, the CPD increased due to formation of a c(2×2) superstructure by ~35 meV, ~10 meV of which may be due to hydrogen adsorption. $\Delta\Phi = 20 \pm 10$ meV was assumed for the CPD between a reconstructed (low Φ) and a simple surface (high Φ) [93]. Similarly, the work function of W(100) decreased by 50 ± 5 meV on cooling from ~550 to 120 K and increased by the same amount on heating [94]. The review [95] on surface reconstruction of W(100) states good agreement between the T dependences of $\Delta\Phi$ [93, 94] and LEED results. A much faster cooling rate (15 s from 2000 to 120 K) was used in [94] than in [93], which minimized hydrogen adsorption [95].

4.2.1.4 W(111)

The value $\Phi^*(111) = 4.40 \pm 0.03$ eV is considered by [1] to be the best estimate based on thermionic measurements [52, 53, 56], and $\Phi^*(111) = 4.40 \pm 0.02$ eV is recommended by [58, 59]. The thermionic value $\Phi^*(111) = 4.39$ eV is given in the review [75], correcting a numerical error in the original paper [74]. The average value 4.42 ± 0.03 eV is given in [76] as a mean of thermionic [53, 56, 75, 77, 84], FE [7, 10, 78, 96], and SI [69] data.

$\Phi^*(111) = 4.38 \pm 0.01$ eV was derived from the thermionic emission of a W single-crystal wire studied at $p \approx 1.5 \times 10^{-9}$ Torr [56]. For a cylindrical W single crystal at $p \approx 10^{-9}$ Torr, $\Phi^*(111) = 4.42 \pm 0.03$ eV [52]. $\Phi^*(111) = 4.40$ eV was measured between 1900 and 2200 K at $p \approx 3 \times 10^{-10}$ Torr [85]. Using a cylindrical electron projection tube, $\Phi^*(111) = 4.4 \pm 0.1$ eV was obtained at $p \approx 10^{-10}$ Torr [62], superseding the value 4.2 eV previously found by a similar set-up [64]. Other $\Phi^*(111)$ values (in eV) are: 4.40 ± 0.03 [53, 61], 4.40 ± 0.02 [97], 4.43 ± 0.05 [98]. With electropolished W(111) surfaces at $p \approx 10^{-7}$ Torr, an effective work function of 4.54 eV was found by thermionic measurements between 1850 and 2450 K [63]. When a W surface parallel to (111) within ~2° was subjected at 10^{-9} Torr to an extended heat treatment at 2300 K, the thermionic work function increased from 4.46 to 4.70 eV [66].

Surface ionization of Sr and Ca was studied on single-crystal filaments exposing (111) surfaces giving $\Phi^*(111) = 4.49 \pm 0.04$ eV [69].

$\Phi(111) = 4.47 \pm 0.02$ eV ($p \approx 10^{-10}$ Torr, obviously at room temperature) was obtained by the retarding field method using an FE electron beam on samples cleaned by heating them to 2100 K [41] (this value is recommended by [42]). Kelvin CPD measurements at 300 K yielded $\Phi(111) = 4.4$ eV assuming the work function of the Pt reference electrode to be 5.0 eV [99]. $\Phi(111) \approx 4.5$ eV resulted from CPD measurements using the electron beam method and assuming $\Phi(100) = 4.6$ eV [50].

References for 4.2.1 on pp. 118/21

Photoelectric and thermionic Schottky deviations indicate a photoelectric value of $\Phi(111) = 4.32$ eV [100].

The values $\Phi(111) = 4.415$ and 4.404 eV were determined at 418 and 695 K, respectively, from the Boltzmann tail of the field emission energy distribution [101]. Formerly, $\Phi(111)$ was determined at 15 K and $p \leq 10^{-8}$ Torr from F-N plots (Φ_{FN}) and from electron energy distributions (Φ_{ED}) resulting in $\Phi_{FN} = 4.68$ eV and $\Phi_{ED} = 4.78$ eV (with $\Phi = 4.41$ eV being determined for the average tip) [12]. Similarly, $\Phi_{FN} = 4.49 \pm 0.02$ eV and $\Phi_{ED} = 4.80 \pm 0.03$ eV were determined on annealed tips at 77 K [10]. The F-N result is later cited by [41] to be 4.45 ± 0.03 eV. The field dependence of the energy distribution of field-emitted electrons was studied at 295 K leading to $\Phi(111) = 4.83 \pm 0.38$ eV [72].

At 78 K and $p < 10^{-10}$ Torr, W tips annealed at 1500 K were gradually field-evaporated and investigated by FE. $\Phi(111)$ values ranging from 4.45 to 4.55 eV were found from F-N plots [2]. Similarly, F-N plots and single-point measurements gave $\Phi(111) = 4.43 \pm 0.03$ eV at liquid N_2 temperature and $p < 10^{-11}$ Torr [16]. Other $\Phi(111)$ values (in eV) derived from FE measurements are: 4.81 ± 0.05 (?) [23], 4.60 ± 0.05 [20], 4.6 [14], 4.56 [13], ~4.5 at 77 K [24], 4.47 ± 0.03 [22], 4.46 [103], 4.45 ± 0.05 at 78 K [11], 4.45 [91], 4.43 [102], 4.42 [19], 4.40 ± 0.05 [30], 4.40 (at 295 K [17], 78 K [96]) [17, 25, 92, 96], 4.39 [7], 4.37 ± 0.03 [29], 4.32 ± 0.11 (?) [6], 4.2 ± 0.1 (?) [28].

Photofield and FE data were combined by [40] and gave $\Phi(\bar{1}11) = 4.55 \pm 0.01$ eV [40].

4.2.1.5 W(311)

The $\Phi(311)$ value recommended by [42] is the thermionic value $\Phi^* = 4.18 \pm 0.04$ eV which actually was obtained on a ribbon, the surface plane of which was 3° off (311) and 7.5° off (211) [104]. This value probably is much too low [1]. A surface ionization study on polycrystalline W filaments, which after recrystallization on heating exposed mainly (311) surfaces, yielded Φ^* values ranging from 4.75 to 4.88 eV with La and from 4.49 to 4.60 eV with Sr atomic beams [68].

$\Phi(311) = 4.55 \pm 0.05$ eV was obtained at 300 K with the (thermionic) electron beam method [76] in agreement with literature data, viz., the thermionic $\Phi^*(311) \approx 4.56$ eV by [105] and the photoelectric $\Phi(311) = 4.53$ eV by [106]. At $p \approx 10^{-11}$ Torr, Kelvin CPD measurements (referring to Φ(polycr.) = 4.55 eV recommended by [43]) yielded $\Phi(311) = 4.59 \pm 0.01$ eV [46]. The values $\Phi = 4.66 \pm 0.03$ and 4.71 ± 0.01 eV are cited in [1, p. 218] for annealed and additionally oxygen-treated W(311) surfaces, respectively. The values stem from unpublished Kelvin CPD measurements by the authors of paper [44].

Photoemission from a W ribbon exposing (311) surfaces at $\sim 10^{-10}$ Torr gave $\Phi(311) = 4.54$ eV [107]; see also [108].

$\Phi(311) = 4.50$ eV was obtained from F-N plots taken at room temperature at $\sim 5 \times 10^{-10}$ Torr [92]. F-N plots recorded at 78 K indicated $\Phi(311) = 4.46 \pm 0.05$ eV [11].

4.2.1.6 W(611)

The value $\Phi^*(611) = 4.32 \pm 0.02$ eV, derived from thermionic measurements, is recommended by [58, 59]. $\Phi^*(611) = 4.30 \pm 0.03$ eV [53, 61] is recommended by [1, 42].

Correcting for a numerical error in the thermionic emission study [74], $\Phi^*(611) = 4.39$ eV resulted [75]. Using a cylindrical electron projection tube, the thermionic $\Phi^*(611) = 4.3 \pm$

0.1 eV was found under $\sim 10^{-10}$ Torr by [62] and $\Phi^*(611) = 4.29$ eV ($\pm \sim 0.5\%$) under $p \approx 1.5 \times 10^{-9}$ Torr [56].

$\Phi(611) = 4.36 \pm 0.05$ eV was obtained at 300 K with the (thermionic) electron beam method [76]. This value agrees with the literature average, 4.30 ± 0.064 eV, from thermionic data ([53, 56, 75, 77]) and FE studies ([7, 10, 78, 96]) [76].

$\Phi(611)$ was determined by probe-hole FE from the slope of F-N plots (Φ_{FN}) and from combined F-N slope and energy distribution data (Φ_{ED}). For tips, annealed 5 min at 1000 K prior to the measurement, $\Phi_{ED} = 4.41 \pm 0.04$ eV and $\Phi_{FN} = 4.32$ eV resulted at 77 K [10]. F-N plots recorded at 78 K yielded $\Phi(611) = 4.32 \pm 0.06$ eV [11]. From FE measurements, $\Phi(611) = 4.30$ eV was obtained at room temperature [7, 92] and at 80 K (assuming Φ(average tip) = 4.54 eV) [14]. $\Phi(611) = 4.26$ eV was obtained assuming $\Phi(111) = 4.38$ eV [26]. Assuming Φ(average tip) = 4.54 eV gave $\Phi(611) = 4.26$ eV [13] (obviously superseding an earlier value of 4.20 eV obtained assuming Φ(average tip) = 4.50 eV [96]).

4.2.1.7 W(310)

The photoelectric value $\Phi(310) = 4.35$ eV was determined using a single-crystal wire kept at ~1400 °C during the measurements and flashed intermittently at 2200 °C (residual pressure: 2×10^{-8} Torr) [81]. Thermionic and CPD measurements on a surface parallel to (310) within 1° indicated $\Phi^* = 4.31 \pm 0.07$ eV [109]. The SI of Sr was studied on a filament which had recrystallized to an orientation close to (310). It yielded the high Φ^* value of 4.61 ± 0.03 eV [69].

$\Phi(310)$ was determined by probe-hole FE from the slope of F-N plots (Φ_{FN}) and from combined F-N slope and energy distribution data (Φ_{ED}). Prior to each measurement the emitter was either annealed for 5 min at 1000 K or flashed to 1800 K and quenched to 77 K. Φ_{FN} was slightly larger for the annealed emitters than for the flashed and quenched tips. $\Phi_{ED} = 4.16 \pm 0.03$ eV and $\Phi_{FN} = 4.28$ eV were obtained for the annealed tip [10]. $\Phi(310) =$ 4.31 eV was determined apparently at room temperature at $p < 10^{-12}$ Torr [7]. F-N plots recorded at 78 K suggested $\Phi(310) = 4.34 \pm 0.04$ eV [11]. $\Phi(310) = 4.35$ eV was obtained from FE data and assuming $\Phi(111) = 4.38$ eV [26]. This value also resulted from F-N plots taken at $\sim 5 \times 10^{-10}$ Torr and room temperature [92]. From F-N plots, $\Phi(310) = 4.48 \pm 0.03$ eV at 550 and 650 K [29]. An FEM study of Cs adsorption on W [24] indicated $\Phi(310) \approx 4.5$ eV at 77 K [24]. The very high value $\Phi(310) = 5.19 \pm 0.16$ eV, obtained by [72] from field emission energy distribution measurements, was considered by the authors to be unreliable.

4.2.1.8 Other Surfaces

This section includes a discussion of the variation of Φ between (hkl) poles and of its variation on the vicinal regions.

The work functions of various W planes other than (110), (112), (100), (111), (311), (611), and (310) were predominantly measured by field emission methods and are summarized in Table 1. The table includes data obtained by [76] using the thermionic electron beam method on a drawn W wire after recrystallization. These data were calibrated by a thermionic emission measurement on a W(411) surface [76].

References for 4.2.1 on pp. 118/21

Table 1
Work Functions of High-Index W Surfaces.

surface	Φ in eV [76][a)]	[11][b)]	[7][a)]	others
(210)		4.36 ± 0.05	4.34	4.32 [110], 4.33 ± 0.05 [23], 4.35 [25], 4.46 ± 0.05 [20]
(510)				4.43 ± 0.01 [40]
(610)		4.43 ± 0.04		4.31 [103]
(320)		4.58 ± 0.06		4.29 [25], 4.4 [14]
(520)	4.55 ± 0.05			
(411)	4.40 ± 0.05	4.42 ± 0.05		4.40[c)], 4.48[d)] [13]
(511)	4.35 ± 0.05			~4.2 [24]
(911)	4.56 ± 0.05			
(221)		4.30 ± 0.04	4.35	4.37 ± 0.02 [16], 4.4 [111]
(321)		4.50 ± 0.06	4.52	4.56[c)], 4.61[c)] [13]
(421)		4.35 ± 0.05		
(331)		4.68 ± 0.07		
(431)		4.74 ± 0.07		
(441)				5.15 (?) [17]
(322)		4.70 ± 0.05		
(722)	4.43 ± 0.05			
(922)	4.34 ± 0.05			
(332)		4.41 ± 0.05	4.46	4.79 ± 0.05 (?) [23]
(433)		4.62 ± 0.04		4.20 [17]

a) At room temperature. — b) At 78 K. — c) Annealed tip. — d) Field-evaporated tip.

The thermionic Φ and dΦ/dT data of (stepped) vicinals of (110) were measured between 2200 and 2800 K under 10^{-6} to 10^{-7} Pa; the carbon Auger signal of the samples was below the detection limit. All of the vicinal faces studied consisted of (110) terraces. The surfaces (10 9 0), (650), and (750) have $(1\bar{1}0)$-type steps with a density (in 10^6 cm^{-1}) of 2.35, 4.05, and 7.36, respectively. The following Φ and Φ* values (in eV) are found [60]:

surface	(110)	(10 9 0)	(650)	(750)
Φ at 2300 K	5.22 ± 0.02	5.10 ± 0.02	5.02 ± 0.02	4.89 ± 0.02
$\Phi^* = \Phi(T) - T(d\Phi/dT)$	5.75 ± 0.02	5.56 ± 0.02	5.30 ± 0.02	5.13 ± 0.02

The results from this set of vicinal surfaces indicate that Φ decreases linerarly with the density of $(1\bar{1}0)$ steps at a rate of ~0.3 eV per 6×10^6 cm^{-1}. The vicinals (10 9 1) and (15 15 2) studied in addition have step densities (in 10^6 cm^{-1}) of 4.05 and 4.20, respectively, about equal to that of (650) but differing in the type of steps, these being (101) and (112), respectively. The values Φ (at 2300 K) = 5.08 ± 0.02 eV and Φ* = 5.54 ± 0.02 eV were obtained for (10 9 1); the corresponding values for (15 15 2) are Φ = 5.05 ± 0.02 eV and Φ* = 5.44 ± 0.02 eV [60]. For a comparison of these results with those on stepped Au and Pt surfaces, see [112]. For a characterization of these stepped surfaces by LEED, see [113].

Contact potential differences ΔΦ were measured as a function of the azimuthal angle φ using the (thermionic) beam method on a cylindrical W single crystal (axis along ⟨110⟩) at a residual pressure below 5×10^{-11} Torr. The Φ vs. φ curve shown in **Fig. 24** was calibrated by assuming Φ(100) = 4.6 eV. The curve is in general agreement with the prediction by

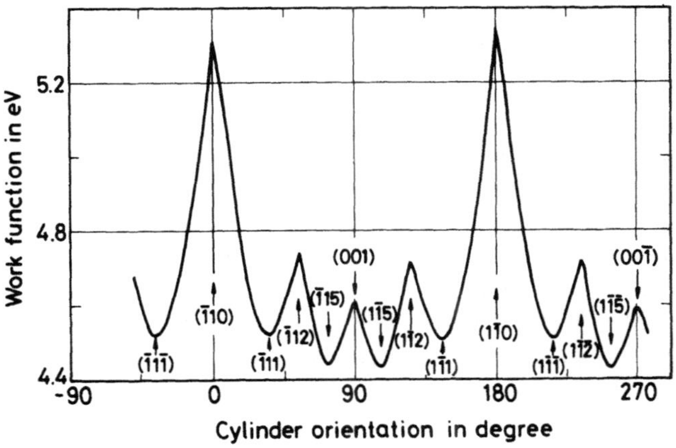

Fig. 24. Work function versus orientation for the ⟨110⟩ zone of W. The work functions are referenced against Φ(100) for which a value of 4.6 eV is assumed [50].

[114] that smoother surfaces have higher Φ values. Within a region of $\Delta\varphi \approx \pm 5°$ about the low-index plane maxima, the measured ΔΦ decreases linearly with increasing misorientation. This is attributed to an increase in the step density [50].

According to [115] the paper [60] indicates that in one zone the dependence of Φ on the tilt angle ψ can be described with high accuracy by $\Phi(\psi) = \Phi_s - D \cdot \psi$. Here Φ_s is the work function of a plane with simple Miller indices (densely packed plane) in this particular zone and ψ is the angle by which the actual surface is tilted away from the densely packed plane. D is a constant the value of which depends on the particular zone. For the ⟨100⟩ zone, D = 35 meV/degree is derived by [115] from the results of [60]. The results of [50] indicate D = 23 meV/degree, when the surface plane tilts away from (110) towards (111), and D = 19 meV/degree for the (001) towards (101) direction close to (001), see **Fig. 25** [115].

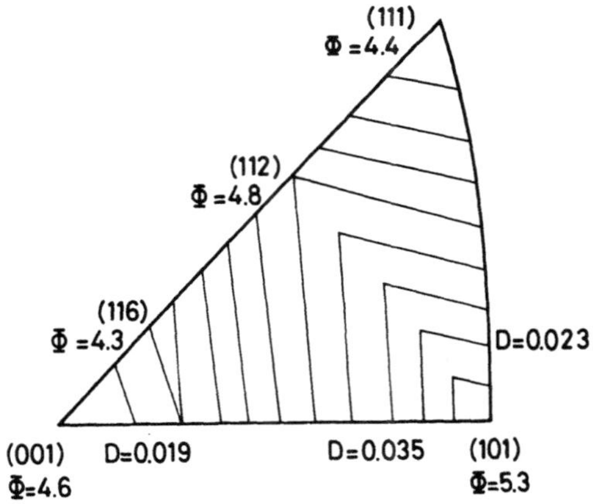

Fig. 25. Calculated contour map of the thermionic work function of W. The work functions Φ are given in eV for the principle orientations. Adjacent contour lines differ by 0.1 eV. The D values (in eV/degree) give the dependence of Φ on the orientation within a zone by the relation $\Phi(\psi) = \Phi_s - D \cdot \psi$, with Φ_s the work function of the orientation with simple Miller indices and ψ the angular deviation [115].

References for 4.2.1 on pp. 118/21

The effective work function of non-uniform W surfaces with ⟨110⟩ fiber texture was calculated and compared in [115] with experimental data from [116, 117].

The work function of rolled W ribbons was measured by thermionic emission and surface ionization of K (the ionization potential of which was assumed to be 4.32 eV) [119]. The rolled ribbon had a texture with W(100) tilted by ~5° off the surface and W⟨110⟩ at ±(13±5)° off the rolling direction (see [118]). The effective work function for these ribbons was determined by K surface ionization at 1574 K to be $\Phi_{eff} = 4.54 \pm 0.02$ eV. The thermionic emission was measured between 1150 and 1650 K. Assuming a linear dependence on the tilt angle ψ (in rad), $\Phi = \Phi_{100}(1 - k_b \cdot |\psi|)$, $\Phi_{100} = 4.65 \pm 0.03$ eV, and $k_b = 0.64 \times (1 \pm 0.19)$ were obtained [119]. A parabolic ψ dependence was discussed by [119, 120].

The Kelvin CPD of nominal W(110) and W(100) surfaces (the actual surface orientations were (13 12 1) and (19 1 0)) has been measured at $p < 10^{-10}$ Torr and 300 K assuming a work function of 5.0 eV for the Pt electrode used as reference; Φ("110") = 5.9 eV and Φ("100") = 4.7 eV were found [99].

Paper [109] extends the earlier thermionic emission and CPD study [104] on W ribbons (outgassed at 2800 K) where the surface plane was oriented 6° off (311) and 7.5° off (211) (ribbon I) and 3° off (311) and 7.5° off (211) (ribbon II). The Φ^* values 4.235±0.04 and 4.18±0.04 eV were found for ribbon I and II respectively [104]. Φ^* was measured in [109] on W ribbons at $\sim 2 \times 10^{-9}$ Torr, after the ribbons had been flashed to above 2700 K. $\Phi^* = 4.15 \pm 0.07$ eV was obtained from Richardson plots for a surface tilted 5° off the (311) orientation. The other surface studied was parallel to (310) within 1° and gave $\Phi^* = 4.31 \pm 0.07$ eV [109].

The electron emission from stepped surfaces was studied by [121] by comparing FEM and FIM images of W tips. According to these observations, terraces wider than about three atomic diameters and separated by monoatomic steps from their neighboring terraces behave as "patchy" surfaces. If the terraces are narrower than ~3 atomic diameters, the emission from these planes is essentially an emission from a rough, low-Φ surface [121].

References for 4.2.1:

[1] Rivière, J. C. (Solid State Surf. Sci. **1** [1969] 179/289).
[2] Todd, C. J.; Rhodin, T. N. (Surf. Sci. **36** [1973] 353/69).
[3] Polizzotti, R. S.; Ehrlich, G. (Surf. Sci. **91** [1980] 24/36).
[4] Wilkinson, M. K. (J. Appl. Phys. **24** [1953] 1203/9).
[5] Dyke, W. P.; Trolan, J. K.; Dolan, W. W.; Grundhauser, F. J. (J. Appl. Phys. **25** [1954] 106/12).
[6] Andreev, I. S. (Zh. Tekh. Fiz. **22** [1952] 1428/41).
[7] Müller, E. W. (J. Appl. Phys. **26** [1955] 732/7).
[8] Young, R. D.; Clark, H. E. (Appl. Phys. Lett. **9** [1966] 265/8).
[9] Plummer, E. W.; Rhodin, T. N. (J. Chem. Phys. **49** [1968] 3479/96).
[10] Swanson, L. W.; Crouser, L. C. (Phys. Rev. [2] **163** [1967] 622/41).

[11] Hellwig, S.; Block, J. H. (Z. Phys. Chem. [Munich] **83** [1973] 269/86).
[12] Lea, C.; Gomer, R. (J. Chem. Phys. **54** [1971] 3349/59).
[13] van Oostrom, A. G. J. (Philips Res. Rep. Suppl. **1966** No. 1, 102+6 pp.).
[14] Nikliborc, J.; Dworecki, Z. (Acta Phys. Pol. **32** [1967] 1023/4).
[15] Müller, E. W. (Z. Phys. **120** [1943] 261/9).
[16] Schmidt, L. D. (J. Chem. Phys. **46** [1967] 3830/41).
[17] van Oostrom, A. (J. Chem. Phys. **47** [1967] 761/9).

[18] Young, R. D.; Müller, E. W. (J. Appl. Phys. **33** [1962] 91/5).
[19] Polański, J.; Sidorski, Z. (Surf. Sci. **40** [1973] 282/94).
[20] Engel, T.; Gomer, R. (J. Chem. Phys. **50** [1969] 2428/37).

[21] Drechsler, M.; Müller, E. W. (Z. Phys. **134** [1952/53] 208/21).
[22] Chrzanowski, E. (Acta Phys. Pol. A **44** [1973] 711/29).
[23] Sanchez Avedillo, M.; de Segovia, J. L. (Electrón Fis. Apl. **19** [1976] 39/40).
[24] Sidorski, Z.; Pelly, I.; Gomer, R. (J. Chem. Phys. **50** [1969] 2382/91).
[25] Burmistrova, O. P.; Vladimirov, G. G.; Shakirova, S. A. (Fiz. Tverd. Tela [Leningrad] **18** [1976] 2712/5; Sov. Phys.-Solid State [Engl. Transl.] **18** [1976] 1581/3).
[26] Houston, J. M. (private communication to [27]; c. f. Diss. M.I.T., Cambridge, Mass., 1955 cited by [1, 99]).
[27] Dyke, W. P.; Dolan, W. W. (Adv. Electron. Electron Phys. **8** [1956] 89/185, 111/8).
[28] Smirnov, B. G.; Shuppe, G. N. (Zh. Tekh. Fiz. **22** [1952] 973/80).
[29] Jones, J. P.; Roberts, E. W. (Surf. Sci. **69** [1977] 185/204).
[30] Collins, R. A.; Blott, B. H. (J. Appl. Phys. **40** [1969] 5390/1).

[31] Gavrilyuk, V. M.; Medvedev, V. K. (Fiz. Tverd. Tela [Leningrad] **8** [1966] 1811/8; Sov. Phys.-Solid State [Engl. Transl.] **8** [1966] 1439/44).
[32] Swanson, L. W.; Strayer, R. W. (J. Chem. Phys. **48** [1968] 2421/42, 2435).
[33] Young, R. D.; Clark, H. E. (Phys. Rev. Lett. **17** [1966] 351/3).
[34] Plummer, E. W.; Rhodin, T. N. (Appl. Phys. Lett. **11** [1967] 194/6).
[35] Smith, J. R. (Phys. Rev. Lett. **25** [1970] 1023/6).
[36] Theophilou, A. K.; Modinos, A. (Phys. Rev. [3] B **6** [1972] 801/12, 808).
[37] Ernst, N.; Ehrlich, G. (J. Phys. Coll. [Paris] **45** [1984] C 9-293/C 9-296).
[38] Ernst, N.; Ehrlich, G. (J. Phys. Coll. [Paris] **47** [1986] C 2-47/C 2-51).
[39] Liu, R. (unpublished results, from [3]).
[40] Gaudin, G. A.; Lee, M. J. G. (Surf. Sci. **185** [1987] 283/98).

[41] Strayer, R. W.; Mackie, W.; Swanson, L. W. (Surf. Sci. **34** [1973] 225/48).
[42] Michaelson, H. B. (J. Appl. Phys. **48** [1977] 4729/33).
[43] Hopkins, B. J.; Rivière, J. C. (Proc. Phys. Soc. [London] **81** [1963] 590/2; Br. J. Appl. Phys. **15** [1964] 941/6).
[44] Hopkins, B. J.; Pender, K. R. (Br. J. Appl. Phys. **17** [1966] 281/2).
[45] Hopkins, B. J.; Pender, K. R. (Surf. Sci. **5** [1966] 155/9).
[46] Lea, C.; Mee, C. H. B. (J. Appl. Phys. **39** [1968] 5890/6).
[47] Zingerman, Ya. P.; Ishchuk, V. A.; Krutilina, T. A. (Fiz. Tverd. Tela **7** [Leningrad] [1965] 2569/71; Sov. Phys.-Solid State [Engl. Transl.] **7** [1965/66] 2078/9).
[48] Shelton, H. (Phys. Rev. [2] **107** [1957] 1553/7).
[49] Madey, T. E.; Yates, J. T., Jr. (Nuovo Cimento Suppl. [1] **5** [1967] 483/505).
[50] Gardiner, T. M.; Kramer, H. M.; Bauer, E. (Surf. Sci. **112** [1981] 181/96).

[51] Hutson, A. R. (Phys. Rev. [2] **98** [1955] 889/901).
[52] Azizov, U. V.; Shuppe, G. N. (Fiz. Tverd. Tela [Leningrad] **7** [1965] 1970/3; Sov. Phys.-Solid State [Engl. Transl.] **7** [1965/66] 1591/4).
[53] Sultanov, V. M. (Radiotekh. Elektron. [Moscow] **9** [1964] 317/20; Radio Eng. Electron. Phys. [Engl. Transl.] **9** [1964] 252/4).
[54] Sytaya, E. P.; Smorodinova, M. I.; Imangulova, N. I. (Fiz. Tverd. Tela [Leningrad] **4** [1962] 1016/20; Sov. Phys.-Solid State [Engl. Transl.] **4** [1962] 750/3).
[55] Fine, J.; Madey, T. E.; Scheer, M. D. (Surf. Sci. **3** [1965] 227/33).
[56] Smith, G. F. (Phys. Rev. [2] **94** [1954] 295/308).
[57] Hughes, F. L.; Levinstein, H.; Kaplan, R. (Phys. Rev. [2] **113** [1959] 1023/8).

[58] Protopopov, O. D.; Mikheeva, E. V.; Sheinberg, B. N.; Shuppe, G. N. (Fiz. Tverd. Tela [Leningrad] **8** [1966] 1140/6; Sov. Phys.-Solid State [Engl. Transl.] **8** [1966] 909/14).
[59] Shuppe, G. N. (Izv. Akad. Nauk SSSR Ser. Fiz. **30** [1966] 1935/41; Bull. Acad. Sci. USSR Phys. Ser. **30** [1966] 2016/23).
[60] Krahl-Urban, B.; Niekisch, E. A.; Wagner, H. (Surf. Sci. **64** [1977] 52/68).

[61] Sultanov, V. M.; Shuppe, G. N. (Izv. Akad. Nauk UzSSR Ser. Fiz.-Mat. Nauk **1965** No. 5, pp. 49/53).
[62] Gardner, F. M.; Girouard, F. E.; Boeck, W. L.; Coomes, E. A. (Surf. Sci. **26** [1971] 605/23).
[63] van Someren, L. (Surf. Sci. **20** [1970] 221/34).
[64] Gienapp, H. (Z. Angew. Phys. **12** [1960] 254/7).
[65] Azizov, E. E.; Mashtakova, V. A.; Shishkin, B. B. (Vestnik Mosk. Univ. Ser. III Fiz. Astron. **42** No. 6 [1987] 39/42; Moscow Univ. Phys. Bull. [Engl. Transl.] **42** No. 6 [1987] 42/5).
[66] Vakhabova, M. A.; Sytaya, E. P. (Izv. Akad. Nauk UzSSR Ser. Fiz.-Mat. Nauk **1980** No. 5, pp. 52/4; cf. Izv. Akad. Nauk UzSSR Ser. Fiz.-Mat. Nauk **1984** No. 5, pp. 91/2).
[67] Trischka, J. W. (J. Appl. Phys. **37** [1966] 455/6).
[68] Reynolds, F. L. (Surf. Sci. **14** [1969] 327/39).
[69] Reynolds, F. L. (J. Chem. Phys. **39** [1963] 1107/14).
[70] Weng, S.-L. (Phys. Rev. [3] B **25** [1982] 6188/94).

[71] Plummer, E. W.; Waclawski, B. J.; Vorburger, T. V.; Kuyatt, C. E. (Prog. Surf. Sci. **7** [1976] 149/82, 154).
[72] Vorburger, T. V.; Penn, D.; Plummer, E. W. (Surf. Sci. **48** [1975] 417/31).
[73] Jones, J. P.; Roberts, E. W. (Thin Solid Films **48** [1978] 215/28).
[74] Nichols, M. H. (Phys. Rev. [2] **57** [1940] 297/306).
[75] Herring, C.; Nichols, M. H. (Rev. Mod. Phys. **21** [1949] 185/270, 202).
[76] Haas, G. E.; Thomas, R. E. (J. Appl. Phys. **40** [1969] 3919/24).
[77] Coomes, E. A.; Girouard, F. E. (Proc. Thermionic Conv. Spec. Conf., Cleveland OH, 1964, pp. 47/52 from [76, 98]).
[78] Becker, J. A. (Bell Syst. Tech. J. **30** [1951] 907/32, 932).
[79] Strayer, R. W.; Swanson, L. W. (unpublished result, cited in [80]).
[80] Plummer, E. W.; Bell, A. E. (J. Vac. Sci. Technol. **9** [1972] 583/90).

[81] Mendenhall, C. E.; DeVoe, C. F. (Phys. Rev. [2] **51** [1937] 346/9).
[82] Brown, A. A.; Neelands, L. J.; Farnsworth, H. E. (J. Appl. Phys. **21** [1950] 1/4).
[83] Abey, A. E. (J. Appl. Phys. **39** [1968] 120/7).
[84] Engelmaier, W.; Stickney, R. E. (Proc. 26th Annu. Conf. Phys. Electron., Cambridge, Mass., 1966, p. 260 from [76]).
[85] Engelmaier, W.; Stickney, R. E. (Surf. Sci. **11** [1968] 370/94, 377).
[86] Chaikovskii, E. F.; Taran, A. A. (Pis'ma Zh. Tekh. Fiz. **5** [1979] 920/3; Sov. Tech. Phys. Lett. [Engl. Transl.] **5** [1979] 381/2).
[87] Yackob, Y. A. (Diss. Univ. Cincinnatti 1965, 79 pp. from Diss. Abstr. **26** [1965/66] 4764).
[88] Wilson, R. G. (J. Appl. Phys. **37** [1966] 3170/2).
[89] Mashtakova, V. A.; Shishkin, B. B. (Vestnik Mosk. Univ. Ser. III Fiz. Astron. **39** No. 2 [1984] 76/8; Moscow Univ. Phys. Bull. [Engl. Transl.] **39** No. 2 [1984] 88/90).
[90] Billington, R. L.; Rhodin, T. N. (Phys. Rev. Lett. **41** [1978] 1602/5).

[91] Gong, Yun-Ming; Leng, Rong-Hua (Solid State Commun. **84** [1992] 1085/8).
[92] Holscher, A. A. (J. Chem. Phys. **41** [1964] 579/80).
[93] Debe, M. K.; King, D. A. (Surf. Sci. **81** [1979] 193/237).
[94] Heilmann, P.; Heinz, K.; Müller, K. (Surf. Sci. **89** [1979] 84/94).

[95] King, D. A. (Phys. Scr. T **4** [1983] 34/43).
[96] van Oostrom, A. (Phys. Lett. **4** [1963] 34/6).
[97] Dikova, L. N.; Smorodinova, M. I.; Sytaya, E. P. (Dokl. Akad. Nauk UzSSR No. 3 **23** [1966] 19/21 from C.A. **65** [1966] 11469).
[98] Coggins, J. L.; Stickney, R. E. (Surf. Sci. **11** [1968] 355/69).
[99] Delchar, T. A.; Ehrlich, G. (J. Chem. Phys. **42** [1965] 2686/702).
[100] Stafford, D. F.; Weber, A. H. (J. Appl. Phys. **34** [1963] 2667/70).

[101] Wysocki, J. K. (Surf. Sci. **137** [1984] 506/14).
[102] Wysocki, J. K. (Surf. Sci. **104** [1981] 463/77).
[103] Oguri, T. (J. Phys. Soc. Jpn. **19** [1964] 83/91).
[104] Love, H. M.; Dyer, G. L. (Can. J. Phys. **40** [1962] 1837/40).
[105] Allen, F. G. (J. Phys. Chem. Solids **8** [1959] 119/21).
[106] Eisinger, J. (J. Chem. Phys. **29** [1958] 1154/60).
[107] Eisinger, J. (J. Chem. Phys. **27** [1957] 1206/7).
[108] Eisinger, J. (J. Chem. Phys. **28** [1958] 165/6).
[109] Love, H. M.; Wilson, J. R. (Can. J. Phys. **45** [1967] 225/7).
[110] Wysocki, J. K. (Phys. Rev. [3] B **28** [1983] 834/41).

[111] Wysocki, J. (Acta Phys. Pol. **35** [1969] 195/8).
[112] Besocke, K.; Krahl-Urban, B.; Wagner, H. (Surf. Sci. **68** [1977] 39/46).
[113] Besocke, K.; Wagner, H. (Surf. Sci. **52** [1975] 653/63).
[114] Smoluchowski, R. (Phys. Rev. [2] **60** [1941] 661/74).
[115] Gubbels, G. H. M.; Metselaar, R. (Surf. Sci. **226** [1990] 407/11).
[116] Samstadt, G. I.; Danko, J. C.; Case, J. M. (CR-1898 [1971] 144 pp. from [115], see also C.A. **75** [1971] No. 145115).
[117] Yang, L.; Hudson, R. G. (Proc. Thermionic Conv. Specialist Conf., Houston 1966, pp. 395 ff. from [115]).
[118] Körner, W. (Z. Metallkd. **64** [1973] 509/10).
[119] Körner, W. (Phys. Status Solidi A **22** [1974] 523/34).
[120] Körner ["Korner"], W. (Vide No. 167 [1973] 174/5).

[121] Knor, Z. (Collect. Czech. Chem. Commun. **44** [1979] 3434/40).
[122] Varaksin, B. P.; Titkov, A. S.; Silant'ev, V. I.; Shevchenko, N. A. (Poverkhnost **1991** No. 11, pp. 125/31; Phys. Chem. Mech. Surf. **7** [1992] 2820/35).

4.2.2 Polycrystal Surfaces and Average Tip Studies

An extensive compilation of the numerous data reported through 1974 for the work function of polycrystalline W gives experimental values in the range 4.2 to >5 eV [1]; see also the earlier compilation [2]. The Φ values definitely cluster in the range 4.5 to 4.6 eV and the value $\Phi = 4.54$ eV is recommended by [1] for the work function of polycrystalline W. In the following some selected data are discussed.

An unweighted average value of $\Phi^{**} = 4.54 \pm 0.02$ eV is cited in [3] for the thermionically determined Φ^{**} values from well-aged polycrystalline W wires. This value, derived from studies on patchy wires under high accelerating fields, is biased towards low work functions [3]. It is the average of 4.52 eV [4], 4.53 eV [5 to 7], and 4.55 eV [8, 9]. The data of [10] and [11] indicate that the emission from (100) and (311) surfaces is nearly the same. This justifies to average the Φ(100) data (4.52 eV [11], 4.56 eV [10], and 4.59 eV [12] including the value 4.49 eV found by [13] for a surface consisting mainly of (100) planes) with the Φ(311) data (4.53 eV from photoelectric studies on recrystallized W ribbons said to expose

References for 4.2.2 on pp. 123/4

mainly (311) planes [14] and ~4.57 eV from thermionic data measured between 1600 and 2400 K on a single-crystal ribbon [15]). The unweighted mean of these values is 4.543 ± 0.050 eV which is considered an approximate area average (pertinent for CPD data) over the various surfaces present on a polycrystal. The recommended value for the true work function of well-aged W polycrystals at 300 K is 4.545 eV [3]. Upon recrystallization, a drawn W wire may be expected to show surface planes evenly distributed along the ⟨110⟩ zonal axis. If, however, the wire is rolled into a ribbon and then recrystallized, the study [16] shows that there is a preferential growth of those planes that lie between the ⟨001⟩ and the ⟨112⟩ directions of that zone. The work function changes rapidly in the region of the most probable orientations (⟨115⟩ to ⟨113⟩ directions for the recrystallized ribbons); hence, there may be differences between different ribbons even in the second decimal place of Φ. The average Φ value of 4.46 eV for W ribbons found in this study is not recommended as a reference value [16] (this remark slightly criticizes [2, 17]) because of the poor reproducibility of the recrystallization patterns of W ribbons [16].

A Richardson plot (~1315 to ~2045 K) for a W filament gave $\Phi^{**}=4.58$ eV in an early paper [18], and $\Phi^{**}=4.54\pm0.07$ eV was measured at 2×10^{-9} Torr between ~1530 and 2545 K on a W wire by [19]. Thermionic measurements on a W ribbon in the range 1080 to 1400 K yielded work function values of 4.55 and 4.47 eV (±0.01 eV) under weak-field and high-field conditions, respectively [20]. The thermionic emission from W was measured between 915 and 965 K, using a cathode wire (W) in an Ar-filled Geiger-Müller counter. $\Phi^{**}=4.62$ eV was obtained from Richardson plots [21]. From the heat of evaporation of electrons emitted from W polycrystals, the values $\Phi=4.565$ and 4.600 eV were obtained at 2100 and 2700 K, respectively. Extrapolating back to 0 K yielded $\Phi=4.44$ eV in the early paper [22].

Polycrystalline W wires were "dc-etched" at 2400 K in vacuum better than 10^{-9} Torr. The thermionic Φ^{**} passed a minimum, typically after ~70 h of heat treatment, and eventually saturated between 4.8 and 5.1 eV depending on the specimen. The wires showed a pronounced (110) surface orientation after the prolonged heat treatment [23]. The electron emission from a sintered W strip was measured under laser and resistive heating conditions. $\Phi^{**}=4.56$ and 4.38 eV were obtained, respectively [24]. The average work function value 4.47 eV resulted with the measured value of the Richardson constant A (considering patchiness) in the thermionic emission study [25] using a cylindrical electron projection tube. Assuming for A the theoretical value 120 $A\cdot cm^{-2}\cdot K^{-2}$ gave the lower value 4.38 eV [25].

Well-aged W wires, recrystallized at 2700 K, were studied at $\sim10^{-7}$ Torr for their effective work function with respect to thermionic electron emission and surface ionization (SI). The thermionic Φ^{**} is heavily weighted towards low-Φ patches present, while Φ^{**}_{SI} at least for the hard-to-ionize species is weighted towards the high-Φ limit. The W wires used had a thermionic Φ^{**} of 4.58 ± 0.03 eV. The SI of In atoms measured between ~2100 and ~2610 K indicated $\Phi^{**}_{SI}=5.14\pm0.03$ eV [26]; see also the SI studies [27, 28], the latter study using easy-to-ionize Na atoms and also yielding $\Phi^{**}_{SI}\approx5.14$ eV for the (110) face of a recrystallized W wire. The Saha-Langmuir evaluation of SI data from potassium on a W foil between ~1130 and 2000 K gave $\Phi^{**}_{SI}=4.67\pm0.09$ eV [29]. SI of Cu and Ag on a W filament resulted in $\Phi^{**}_{SI}=4.66$ and 4.65 eV (assuming ionization potentials of 7.72 eV for Cu and 7.57 eV for Ag) [30]. The SI of sodium and potassium halides on W polycrystals indicated an average Φ^{**}_{SI} close to 4.69 eV, substantially higher than the accepted value of 4.54 eV for W polycrystals [31].

W films, vacuum-deposited on glass substrates (used as reference in [17, 32]), gave $\Phi=4.55\pm0.02$ eV (assuming $\Phi=4.54$ to 4.55 eV for aged W polycrystals) [17]. The photoemission from W films, condensed on glass at 90 K and ordered by annealing for 2 h at 403 K, yielded $\Phi = 4.53$ eV at 90 K and 4.55 eV at 293 K [33].

From the threshold wavelength for photoemission, $\Phi = 4.57 \pm 0.1$ eV was obtained on a W foil [29]. The photoelectric Φ of a W foil measured after flashing the foil to 2300 °C was 4.50 eV [34]. W foils, high-temperature treated at $\sim 10^{-7}$ Torr, showed a photoelectric Φ of 4.62 eV; after admission of Ar $\Phi = 4.6$ eV resulted [35]. From the threshold for the emission of secondary electrons, the work function of a W foil was determined to be 4.51 eV [36].

Average (Total) Tip Studies. From Fowler-Nordheim (F-N) plots of annealed ($\sim$2600 K) W tips, the Φ value for the average tip at $p < 5 \times 10^{-9}$ Torr is 4.54 eV [37]; see also [38]. But the determination of tip radii by Drechsler's and Henkel's method (see [39]) is reliable only for ⟨110⟩-oriented emitter tips (these tips gave an "average" Φ value of 4.55 eV). Tips with ⟨100⟩ and ⟨111⟩ orientation had been prepared by [40] in addition to ⟨110⟩-oriented tips by etching small cylinders cut out from a large single crystal in the required orientation. For the ⟨100⟩- and ⟨111⟩-oriented tips, unrealistically low values of 3.6 ± 0.05 and 3.40 to 3.45 eV, respectively, were obtained for the average-tip work function. Therefore, not too much significance should be attached to the "average" Φ value 4.5 eV, usually given for ⟨110⟩-oriented tips [40].

Atomic and ionic desorption of Cs from a field emitter tip was studied in [41] in an extension of a previous, similar study of K desorption [42]. The Φ measured by this method is the value of those places where ionic desorption occurs, i.e., surfaces with high work functions. The Φ value thus found is 5.3 eV [41, 42].

References for 4.2.2:

[1] Fomenko, V. S.; Podchernyaeva, I. A. (Emissionnye i Adsorptionnye Svoistva Veshchestv i Materialov, Spravochnik, Atomizdat, Moscow 1975, pp. 1/320; C.A. **83** [1975] No. 121265).
[2] Fomenko, V. S. (Handbook of Thermionic Properties, Plenum, New York 1966, 151 pp.).
[3] Hopkins, B. J.; Rivière, J. C. (Proc. Phys. Soc. [London] **81** [1963] 590/2).
[4] Reimann, A. L. (Proc. R. Soc. [London] A **163** [1937] 499/510).
[5] Nichols, M. H. (Phys. Rev. [2] **78** [1950] 158/61, 159).
[6] Dushman, S.; Rowe, H. N.; Ewald, J.; Kidner, C. A. (Phys. Rev. [2] **25** [1925] 338/60).
[7] Freitag, H.; Krüger, F. (Ann. Phys. [5] **21** [1934] 697/742).
[8] Wahlin, H. B.; Whitney, L. V. (Phys. Rev. [2] **50** [1936] 735/8).
[9] Rauh, E. G.; Thorn, R. J. (J. Chem. Phys. **31** [1959] 1481/5).
[10] Nichols, M. H. (Phys. Rev. [2] **57** [1940] 297/306).

[11] Smith, G. F. (Phys. Rev. [2] **94** [1954] 295/308).
[12] Brown, A. A.; Neelands, L. J.; Farnsworth, H. E. (J. Appl. Phys. **21** [1950] 1/4).
[13] Apker, L.; Taft, E.; Dickey, J. (Phys. Rev. [2] **73** [13] [1948] 46/50).
[14] Eisinger, J. (J. Chem. Phys. **29** [1958] 1154/60).
[15] Allen, F. G. (J. Phys. Chem. Solids **8** [1959] 119/21).
[16] Haas, G. E.; Thomas, R. E. (J. Appl. Phys. **40** [1969] 3919/24).
[17] Hopkins, B. J.; Rivière, J. C. (Br. J. Appl. Phys. **15** [1964] 941/6).
[18] Ahearn, A. J. (Phys. Rev. [2] **44** [1933] 277/86).
[19] Zubenko, Yu. V.; Sokol'skaya, I. L. (Zh. Tekh. Fiz. **32** [1962] 378/80).
[20] Camp, M.; Lecchini, S. M. A. (Proc. Phys. Soc. [London] **85** [1965] 815/7).

[21] Sastri, R. C.; Chatterjee, S. D. (Indian J. Phys. **41** [1967] 793/802).
[22] Krüger, F.; Stabenow, G. (Ann. Phys. [Leipzig] [5] **22** [1935] 713/34).

[23] Sytaya, E. P.; Shuppe, N. G. (Izv. Akad. Nauk SSSR Ser. Fiz. **26** [1962] 1349/53; Bull. Acad. Sci. USSR Phys. Ser. [Engl. Transl.] **26** [1962] 1372/5).
[24] Hiernaut, J.-P.; Magill, J.; Ohse, R. W.; Tetenbaum, M. (High Temp.-High Pressures **17** [1985] 633/8).
[25] Gardner, F. M.; Girouard, F. E.; Boeck, W. L.; Coomes, E. A. (Surf. Sci. **26** [1971] 605/23).
[26] Zandberg, E. Ya.; Tontegode, A. Ya. (Zh. Tekh. Fiz. **35** [1965] 1501/3; Sov. Phys.-Tech. Phys. [Engl. Transl.] **10** [1965/66] 1162/3).
[27] Zandberg, E. Ya. (Zh. Tekh. Fiz. **30** [1960] 1215/21).
[28] Shuppe, G. N.; Sytaya, E. P.; Kadyrov, R. M. (Izv. Akad. Nauk SSSR Ser. Fiz. **20** [1955] 1142/50; Bull. Acad. Sci. USSR Phys. Ser. [Engl. Transl.] **20** [1955] 1035/43).
[29] Schroen, W. (Z. Phys. **176** [1963] 237/52).
[30] Weiershausen, W. (Adv. Mass Spectrom. **1** [1959] 120/4).

[31] Zemel, J. (J. Chem. Phys. **28** [1958] 410/3).
[32] Rivière, J. C. (Proc. Phys. Soc. [London] B **70** [1957] 676/86).
[33] Suhrmann, R.; Mata Arjona, A.; Wedler, G. (Z. Elektrochem. **65** [1961] 786/8).
[34] Seddig, M.; Haase, G. (Z. Angew. Phys. **4** [1952] 105/8).
[35] Chistyakov, P. N.; Milovanova, R. A.; Lytkin, V. V. (Zh. Tekh. Fiz. **44** [1974] 228/30; Sov. Phys.-Tech. Phys. [Engl. Transl.] **19** [1974] 148/9).
[36] Alov, N. V.; Dadayan, K. A. (Pis'ma Zh. Tekh. Fiz. **13** [1987] 290/3; Sov. Tech. Phys. Lett. [Engl. Transl.] **13** No. 3 [1987] 119/20).
[37] Jones, J. P. (Proc. R. Soc. [London] A **284** [1965] 469/87).
[38] Jones, J. P. (Surf. Sci. **32** [1972] 29/44).
[39] Drechsler, M.; Henkel, E. (Z. Angew. Phys. **6** [1954] 341/6).
[40] Gorbatyi, N. A.; Gofman, I. I. (Radiotekh. Elektron. [Moscow] **8** [1963]; Radio Eng. Electron. Phys. [Engl. Transl.] **8** [1963] 1798/802).

[41] Schmidt, L. D.; Gomer, R. (J. Chem. Phys. **43** [1965] 2055/63).
[42] Schmidt, L.; Gomer, R. (J. Chem. Phys. **42** [1965] 3573/98).

4.3 Calculated Work Function Data

A theoretical treatment for the anisotropy of Φ has been given early by [1]. The smoothness of a surface is very important with respect to its Φ. The correct sequence of Φ(hkl) values was deduced to be $\Phi(110) > \Phi(211) > \Phi(100) > \Phi(111)$ [1]. A semiempirical theory of the (hkl) dependence of Φ is given in [2]. It rests on the observation that the measured Φ(hkl) values (cited from literature) linearly increase with the surface density. In this respect, only W(100) behaves anomalously (perhaps also (310) for which the then available experimental evidence was very controversial). The discrepancy with W(100) appears to be attributable to band structure effects ("Swanson hump"). A value of 3.955 eV is assumed for the isotropic part of Φ, while the anisotropic part of Φ(hkl) is due to the surface dipole layer [2]:

(hkl)	(110)	(211)	(100)	(111)	(321)	(210)	(332)	(221)	(611)
Φ(hkl) in eV	5.20	4.811	4.963	4.562	4.511	4.404	4.378	4.259	4.250

Slightly better agreement with measured data was obtained, when the difference $\Phi(211) - \Phi(111)$ was fixed on the experimental value 0.3 eV [2]. Values of 5.52, 4.66, and 4.47 eV were calculated for the work function of (110), (100), and (111) surfaces, respectively [3] using the empirical approach of [4]. According to a calculation using a molecular orbital approach, the work functions and stabilities of W surfaces decrease in the order (110),

(100), (211), and (111) [5]. The magnitude of Φ and its (hkl) dependence depend on the occupancy of s- and d-electron states of the surface atoms. Tight binding calculations give Φ(110) > Φ(111) [6]; see also [7]. The difference Φ(110) − Φ(100) = 0.48 eV resulted from the calculated dipole barriers for (110) and (100) surfaces [8], but the authors of [9] claim that the dipole barriers D given by [8] are actually twice the correct D values [9].

The work function of W(110) was calculated to be 5.62 eV by an ab initio study using a Green's function technique based on the linear-muffin-tin-orbitals method [10]; see also [11]. Earlier, a Φ(110) value of 6.70 eV had been calculated by an extension of the local density formalism [12].

Φ = 4.5 ± 0.2 eV was obtained from a self-consistent energy band study of the unreconstructed W(100) surface [13]. Φ(100) for unreconstructed surfaces was obtained from linearized augmented plane wave (LAPW) local-density calculations performed on films n = 1, 3, 5, 7, 9, (and 8) layers thick. For n ≥ 5 the work function amounted to 4.60 ± 0.05 eV [14]. A higher value, 4.77 eV, found earlier for a five-layer slab [15, 16] is attributed by [14] to a coarser grid in $\vec{k}$ space and a simpler technique used in the previous work. The more recent value Φ(100) = 5.4 eV from a 3-layer LAPW calculation [17] is considered by the authors [17] to be inferior to that by [14].

Φ was calculated for stepped surfaces by a semiempirical procedure based on the concept of electronegativity for the surface atoms [18]. Fair agreement was achieved in [18] (revising [19]) with the experimental Φ data published by [20] (see p. 116) for stepped (110) surfaces.

A rather simple jellium model for Φ was applied assuming five valency electrons for W. A value of Φ = 4.78 eV resulted for a polycrystalline surface 95% of which were assumed to consist of (100) surfaces [21] (obviously superseding the value Φ = 3.91 eV calculated by [22], who also used a jellium model). For (obviously polycrystalline) W the work function Φ = 4.68 eV was calculated using a simple model which considers the contribution of the free electrons to Φ [23].

An average Φ was calculated from the first four ionization potentials of the W atom by a simple model that ignores contributions from surface dipoles. For a correlation between Φ and the ionization potential for various metals, see [24]. For correlations between Φ and the electronegativity, see "Tungsten" Suppl. Vol. A 2, 1987, pp. 83/4.

References for 4.3:

[1] Smoluchowski, R. (Phys. Rev. [2] **60** [1941] 661/74).
[2] Modinos, A. (Surf. Sci. **75** [1978] 327/41).
[3] Hölzl, J.; Schulte, F. K. (Springer Tracts Mod. Phys. **85** [1979] 1/150, 28).
[4] Steiner, D.; Gyftopoulos, E. P. (Proc. 27th Conf. Phys. Electron., Cambridge, Mass., 1967, pp. 160/8 from [3]).
[5] Pancir, J.; Haslingerová, I. (Chem. Phys. Lett. **107** [1984] 598/602).
[6] Goryachev, Yu. M.; Podchernyaeva, I. A. (Phys. Status Solidi B **56** [1973] 443/51).
[7] Goryachev, Yu. M.; Podchernyaeva, I. A. (Zh. Fiz. Khim. **48** [1974] 813/7; Russ. J. Phys. Chem. [Engl. Transl.] **48** [1974] 469/71).
[8] Fritsche, L.; Gollisch, H.; Noffke, J. (Phys. Status Solidi B **89** [1978] 367/73).
[9] Taut, M.; Eschrig, H.; Schubert, M. (Phys. Status Solidi B **100** [1980] 243/50).
[10] Skriver, H. L.; Rosengaard, N. M. (Phys. Rev. [3] B **46** [1992] 7157/68).
[11] Skriver, H. L.; Rosengaard, N. M. (Phys. Rev. [3] B **45** [1992] 9410/2).
[12] Smith, J. R. (Phys. Rev. Lett. **25** [1970] 1023/6).

[13] Posternak, M.; Krakauer, H.; Freeman, A. J.; Koelling, D. D. (Phys. Rev. [3] B **21** [1980] 5601/12).
[14] Ohnishi, S.; Freeman, A. J.; Wimmer, E. (Phys. Rev. [3] B **29** [1984] 5267/78).
[15] Wimmer, E.; Freeman, A. J.; Weinert, M.; Krakauer, H.; Hiskes, J. R.; Karo, A. M (Phys. Rev. Lett. **48** [1982] 1128/31).
[16] Wimmer, E.; Freeman, A. J.; Hiskes, J. R.; Karo, A. M. (Phys. Rev. [3] B **28** [1983] 3074/91).
[17] Wu, R.; Chen, K.; Wang, D.; Wang, N. (Phys. Rev. [3] B **38** [1988] 3180/8).
[18] Rudnitskii, L. A. (Zh. Tekh. Fiz. **50** [1980] 355/61; Sov. Phys.-Tech. Phys. [Engl. Transl.] **25** [1980] 210/3).
[19] Rudnitskii, L. A. (Zh. Tekh. Fiz. **48** [1978] 2590/2; Sov. Phys.-Tech. Phys. [Engl. Transl.] **23** [1978] 1483/5).
[20] Krahl-Urban, B.; Niekisch, E. A.; Wagner, H. (Surf. Sci. **64** [1977] 52/68).

[21] Albrecht, H. E. (Phys. Status Solidi A **6** [1971] 135/42).
[22] Smith, J. R. (Phys. Rev. [2] **181** [1969] 522/9, 527).
[23] Mehrotra, R.; Mahanty, J. (J. Phys. C **11** [1978] 2061/4).
[24] Rose, A. (Solid State Commun. **45** [1983] 859/64).

4.4 Influence of Defects on the Work Function

The effect of W adatoms and voids on the work function of a W(110) surface was considered theoretically in [1]. For the coverage dependence of Φ in the W-on-W(110) adsorption system, see the calculation [2]. Another calculation showed that both W vacancies and W adatoms reduce the work function of (110), (211), and (100) W surfaces [3]. For an extension to adatom pairs and vacancy pairs, see [4].

The effect of size on Φ is considered for small metal particles having various forms, namely spheres and highly symmetrical polyeders. A decrease in size causes Φ to decrease. Depending on the shape of the particles, $-\Delta\Phi$ values ranging between 0.4 and 0.6 eV were calculated for W particles of 15 Å effective diameter (when compared with "infinitely large spheres or polyeders of the same kind"). An exception is the octaeder for which $-\Delta\Phi$ was calculated to be as large as 0.7 to 0.8 eV [5].

A combined field electron/field ion microscopic study of W(110) showed the field-emitted current to increase when depositing W atoms, indicating an apparent decrease of the work function [6]. Contact potential difference measurements during W adsorption on W(110) surfaces at 2×10^{-10} Torr indicate that Φ decreases by as much as 0.6 eV [7]. Φ reductions as large as ~0.5 eV were observed during the first stages of epitaxial growth of W on W(110) [8]. The experimental results of [7] on W cluster formation on W(110) were compared by [9] with a model relating cluster size and $\Delta\Phi$ vs. coverage as a function of substrate temperature [9]; see also [10]. The $\Delta\Phi$ caused by W adatoms deposited on the (110) terraces of stepped W surfaces was further studied on single crystals, whose surfaces were oriented close to (750), (650), and (40 37 1). Total $\Delta\Phi$ values of about 0, -0.2, and -0.4 eV were found, respectively, the work function drop thus increasing with the terrace width [11]. For the variation of the work function of these stepped W surfaces with step density and step orientation, see p. 116. For calculated dipole moments of edge adatoms on stepped W surfaces and a comparison with measured values derived from Φ data with stepped surfaces, see [12].

The thermal roughening of stepped surfaces has been studied by FEM on the (011)-to-(112) region. Φ data, apparently derived from F-N plots, are given for temperatures up to 1200 K for W(257) being at the edge of the (011) region, for W(123) being in-between (011) and (112), and for W(235) being at the edge of (112). For (257) there was little change in Φ below 900 K. But above 900 K, Φ decreased on heating, fairly steeply between 900 and 1000 K and less so for T increasing to 1200 K. This result indicates thermal roughening of the surface near 900 K [13, 14].

From Richardson plots and the total thermionic emission, $\Phi^* = 4.61 \pm 0.05$ eV was found for a W(100) surface having an average dislocation density of (0.6 to 1) $\times 10^6$ cm^{-2}. Regions with dislocation clusters (densities of 10^8 cm^{-2} or higher, produced by thermal treatment of the crystal) had $\Phi^* = 4.53$ eV [15]. The effect of irradiation of W(110) and W(111) surfaces with $\sim 10^7$ rad of 1.24 MeV γ radiation and subsequent annealing below 2500 K on the work function was studied by [16]. The formation of a hill-and-valley structure on the surfaces of annealed samples caused a significant downward shift of Φ from 5.3 ± 0.05 eV to values around 4.95 eV for (110) and from 4.5 ± 0.06 eV to values around 4.38 eV for (111). For histograms of Φ distributions on the annealed surfaces, see paper [16]; see also [17].

The photoemission from W films condensed on glass at 90 K under 10^{-10} Torr was studied at 90 K on a fresh, disordered film and at 90 and at 293 K on ordered films formed by annealing them 2 h at 403 K. A value of $\Phi = 4.34$ eV was found for the disordered film at 90 K. $\Phi = 4.53$ eV at 90 K and 4.55 eV at 293 K were obtained for an ordered film [18]. This result is compared in [19] with $\Phi = 4.36$ eV for a disordered film and $\Phi = 4.63$ eV for an ordered film obtained by annealing at 438 K by [20].

In the presence of Ar, the work function of W polycrystals was measured to be 4.6 eV. After an Ar glow-discharge treatment in cathode falls of 0.18 to 0.25 keV, Φ decreased to 4.35 eV, apparently due to Ar ion implantation [21].

References for 4.4:

[1] Rudnitskii, L. A. (Zh. Fiz. Khim. **57** [1983] 1227/33; Russ. J. Phys. Chem. [Engl. Transl.] **57** [1983] 739/42).
[2] Rudnitskii, L. A. (Dokl. Akad. Nauk SSSR **286** [1986] 146/50; Dokl. Phys. Chem. [Engl. Transl.] **286/291** [1986] 22/6).
[3] Reznik, A. I.; Rudenko, N. V. (Izv. Akad. Nauk SSSR Ser. Fiz. **46** [1982] 1282/7; Bull. Acad. Sci. USSR Phys. Ser. [Engl. Transl.] **46** No. 7 [1982] 61/6).
[4] Reznik, A. I.; Rudenko, N. V. (Zh. Fiz. Khim. **58** [1984] 1196/9; Russ. J. Phys. Chem. [Engl. Transl.] **58** [1984] 725/8).
[5] Rudnitskii, L. A.; Martinyuk, E. N.; Reznik, A. I. (Zh. Tekh. Fiz. **52** [1982] 1175/81; Sov. Phys.-Tech. Phys. [Engl. Transl.] **27** [1982] 711/5).
[6] Plummer, E. W.; Rhodin,T. N. (Appl. Phys. Lett. **11** [1967] 194/6).
[7] Besocke, K.; Wagner, H. (Phys. Rev. [3] B **8** [1973] 4597/600).
[8] Hahn, P.; Clabes, J.; Henzler, M. (J. Appl. Phys. **51** [1980] 2079/84).
[9] Albano, E. V.; Heras, J. M.; Schrammen, P.; Mann, M.; Hölzl, J. (Surf. Sci. **129** [1983] 137/54).
[10] Albano, E. V.; Heras, J. M.; Schrammen, P.; Mann, M.; Hölzl, J. (Symp. Surf. Sci. **1983** 213/7).
[11] Besocke, K.; Wagner, H. (Surf. Sci. **53** [1975] 351/8).
[12] Krahl-Urban, B.; Wagner, H.; Butz, R. (Surf. Sci. **93** [1980] 423/30).
[13] Gong, Y. M.; Gomer, R. (J. Chem. Phys. **88** [1988] 1359/69).
[14] Gong, Y. M.; Gomer, R. (J. Phys. Coll. [Paris] **48** [1987] C 6-15/C 6-20).
[15] Chaikovskii, E. F.; Taran, A. A. (Pis'ma Zh. Tekh. Fiz. **5** [1979] 920/3; Sov. Tech. Phys. Lett. [Engl. Transl.] **5** [1979] 381/2).

[16] Salomatin, A. A.; Mashtakova, V. A.; Shishkin, B. B.; Chernov, I. P.; Mamontov, A. P.; Nikitenkov, N. N. (At. Energ. **66** No. 1 [1989] 53/5; Sov. At. Energy [Engl. Transl.] **66** No. 1 [1989] 70/4).
[17] Mashtakova, V. A.; Salomatin, A. A.; Shishkin, B. B.; Mamontov, A. P.; Nikitenko, N. N.; Chernov, I. P. (Vestnik Mosk. Univ. Ser. III Fiz. Astron. **43** No. 5 [1988] 86/7; Moscow Univ. Phys. Bull. [Engl. Transl.] **43** No. 5 [1988] 95/6).
[18] Suhrmann, R.; Mata Arjona, A.; Wedler, G. (Z. Elektrochem. **65** [1961] 786/8).
[19] Suhrmann, R.; Wedler, G. (Z. Angew. Phys. **14** [1962] 70/4).
[20] Hahndorff, A. (Diss. Univ. Hannover 1960 cited by [7]).
[21] Chistyakov, P. N.; Milovanova, R. A.; Lytkin, V. V. (Zh. Tekh. Fiz. **44** [1974] 228/30; Sov. Phys.-Tech. Phys. [Engl. Transl.] **19** [1974] 148/9).

4.5 Temperature Coefficient of the Work Function

Survey. There is considerable disagreement in the literature about the work function temperature coefficient $d\Phi/dT$. For instance, a value of -230 ± 9 μeV/K was obtained in a thermionic emission study on W(110) [1]. This value has been compared in [1] with the values found by [2 to 6]. In agreement with [1] all of these papers indicate $d\Phi/dT$ to be negative, but always $|d\Phi/dT| < 230$ μeV/K. $d\Phi/dT$ values reported (in μeV/K) are -60 [2] and (calculated by [1] from data in these papers) -140 [3], -160 ± 10 [4], -100 [5], and < -150 [6]. These values were obtained from the apparent thermionic emission constant A^* (see p. 132) assuming that any deviations of A^* from the Richardson constant A_R are solely due to the temperature coefficient of the work function. More recently, $d\Phi/dT = -133 \pm 5$ μeV/K was found by [7] using Shelton's method (see [8]). A Kelvin CPD study yielded $d\Phi/dT = 63 \pm 6$ μeV/K [9]. These controversial $d\Phi/dT$ data for W(110) reflect the state of the art; simple explanations like bad vacuums or ill-defined surface conditions are unsatisfactory.

Theories contributed only little to the subject: According to [10] the thermal expansion of W ($\alpha = 7.3 \times 10^{-6}$ K^{-1}) contributes an amount of 34 μeV/K to $d\Phi/dT$ [10], while [11] using the jellium model estimated a total value of $d\Phi/dT = -14.4$ to -15.8 μeV/K [11]. For an attempt to calculate the contribution of surface point defects on $d\Phi/dT$, see [12].

In the following, investigations on polycrystals will be reported first (papers from the thirties compare favorably with more recent work). Field emission (FE) results for an average tip are given next. Then, FE data on various (hkl) surfaces follow. The section is concluded with a review on $d\Phi/dT$ results from macroscopic single-crystal surfaces, i.e., results from CPD, thermionic, and surface ionization (SI) studies.

Polycrystals and Average-Tip Data from Field Emission (FE). An early paper quoted $d\Phi/dT = 60$ μeV/K (± 20 to 30% estimated accuracy) from a measurement of the heat of evaporation of electrons emitted at 2100 to 2700 K [13]. A CPD study of a W foil found a strictly linear T dependence for Φ between room temperature and ≤ 1300 K at $\sim 10^{-10}$ Torr. This led to $d\Phi/dT = 110 \pm 10$ μeV/K [9]. The CPD between two W filaments, of which one was heated to between 300 and 400 K to 1070 K, yielded $d\Phi/dT = 70 \pm 5$ μeV/K [14]. $d\Phi/dT = 63$ μeV/K was obtained at 300 to 935 K from the CPD between two filaments measured at $\sim 10^{-8}$ Torr [15] confirming the value of 60 μeV/K found previously [16]. The SI data of various alkali metal atoms on a W filament between ~ 1400 and ~ 3000 K [18] are consistent with Φ of well-aged W increasing from 4.65 to 4.72 eV between 1600 and 2500 K as suggested by [17]. (For a critique of paper [18], see [19].) The negative value $d\Phi/dT = -43$ μeV/K from thermionic emission of a W filament at $< 10^{-8}$ Torr [20] (for a similar result, see the CPD study [21]) appears to be superseded.

A low-current FE study indicated $d\Phi/dT \approx 75$ µeV/K near 1000 K for the average tip [22]. Low-temperature FE measurements indicated $d\Phi/dT < 5$ µeV/K for the range below room temperature [23].

(hkl)-Resolved FE Studies. By probe-hole FE between 77 and 900 K [24] and 78 to 293 K [25], Φ(hkl) was determined from the slope of F-N plots recorded on annealed tips. The following dΦ/dT values (in µeV/K) were given [24, 25]:

surface	(110)	(211)	(100)	(111)	(310)	(611)	
dΦ/dT	−170	−143	−109	35	−32	50	[24]
dΦ/dT	-	-	-	6±3	-	57±3	[25]

$d\Phi/dT = -93 \pm 9$, 19 ± 20, and 144 ± 30 µeV/K for W(100), W(110), and W(210), respectively were found by FE on W tips, laser-heated from room temperature to ~440 K (reliable results were obtained for $\Delta T > 40$ K) [26]. Φ was measured vs. T for W(211), W(100), W(111), and W(510) by energy-resolved FE below 10^{-10} Torr; see **Fig. 26** [27]. dΦ/dT was determined by [27] over those T ranges where the Φ vs. T curves are approximately linear. For W(211), $d\Phi/dT = 6 \pm 3$ and 23 ± 5 µeV/K in the temperature ranges 300 to 550 and 650 to 900 K, respectively. For W(100), $d\Phi/dT = -92 \pm 3$ and 8 ± 6 µeV/K at 300 to 400 and 550 to 750 K, respectively. For W(111), $d\Phi/dT = -25 \pm 3$ µeV/K at 300 to 650 K [27] (to be compared with $d\Phi/dT < 10$ µeV/K given by [28]). From thermal FE yields and total energy distributions, studied between 78 K and ~800 K on (210), $d\Phi/dT \approx 190$ (110 to 250) µeV/K was obtained [29]. For W(510) $d\Phi/dT = 50 \pm 3$ and 67 ± 4 µeV/K at 300 to 550 and 650 to 900 K, respectively [27]. According to [27], there is fair to good agreement with dΦ/dT values determined by field or photofield emission energy distributions [26, 29, 30]. In view of the CPD results obtained between 100 and 600 K by [31], the temperature variation of Φ(100) (see Fig. 26)

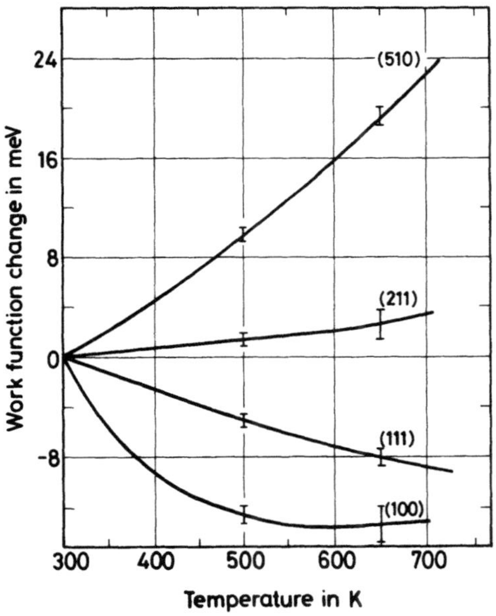

Fig. 26. Work function change as a function of temperature for (100), (111), (211), and (510) tungsten surfaces. Each curve is the best estimate obtained by averaging the results of several measurements. The error bars represent the uncertainty in the best-estimate work function changes [27].

References for 4.5 on pp. 130/1

was thought to be the high-temperature tail of a strongly nonlinear T dependence of Φ [27]. This explanation, however, required a different sign for dΦ/dT.

From total yields in photofield emission (PFE) in fields of $\sim 2.2 \times 10^9$ V/m, dΦ/dT ≈ 18 ± 14, 2 ± 10, and 87 ± 14 μeV/K was obtained for W(211), W(111), and W(310), respectively [32]. PFE spectra yielded Φ(110) vs. T curves linear up to 1100 K and indicated a dΦ/dT value as low as −2 ± 3 μeV/K [33].

Macroscopic Single-Crystal Surfaces. dΦ/dT was measured by Shelton's method (see [8]) at $\sim 10^{-10}$ Torr on ribbons spark-cut from a single crystal. The ribbons exposed surfaces parallel to W(110) and W(111) within 0.5°, and no carbon was seen by AES. The knees in the thermionic current-voltage characteristics as a function of temperature indicated dΦ/dT = −133 ± 5 and 82 ± 12 μeV/K for (110) and (111) [7]. Smith [34] reanalyzed the thermionic data given by [35]. For (211), (100), (111), and (611) he recalculated dΦ/dT = −67, −1, 49, and 49 μeV/K, respectively [34]. dΦ(110)/dT = −160 ± 10 μeV/K was obtained from the temperature dependence of the knee occurring in retarding potential plots [4].

Richardson plots recorded between ~1900 and 2400 K at $p < 10^{-8}$ Torr gave dΦ/dT ≈ −60 μeV/K for both (110) and (100) [2]. dΦ/dT = −60 and −1 μeV/K for (211) and (100) [36]. Using Shelton's method (see [8]), |dΦ/dT| was estimated to be <20 μeV/K for either surface, and dΦ/dT for (110) and (100) was believed to be of equal sign and similar magnitude [37]. Two W surfaces parallel within 0.5° to W(111) were kept at a temperature difference of 713 ± 10 K. From the saturation current densities determined for 6 temperatures between 1523 and 1973 K, a value of dΦ/dT = 175 μeV/K was obtained for W(111) [38]. Between 1400 and 2500 K two different regions were indicated by Richardson plots for W(110), namely Φ*(110) = 5.1 ± 0.04 eV for T < 1850 K and Φ*(110) = 5.7 ± 0.04 eV for T > 1850 K. For dΦ/dT of (110), (100), and (111) for both high- and low-temperature regions, see the paper [39].

The following dΦ/dT values (in μeV/K) were obtained in a thermionic emission study of stepped W(110) vicinals (see p. 116) [1]:

(hkl)	(110)	(10 9 0)	(650)	(750)	(10 9 1)	(15 15 2)
dΦ/dT	−230 ± 9	−200 ± 10	−120 ± 7	−106 ± 7	−202 ± 6	−168 ± 10

dΦ/dT = −130 ± 20 μeV/K was determined for W(110) from the SI of an NaCl molecular beam [4]. The average of the thermionically determined value dΦ/dT = −160 ± 10 μeV/K and the SI-derived value −130 ± 20 μeV/K is dΦ/dT = −140 ± 20 μeV/K [4].

W(110) and W(100) were investigated with the Kelvin CPD method between <100 K and 1300 K at $\sim 10^{-10}$ Torr. From the strictly linear T dependence found for Φ, the dΦ/dT values 63 ± 6 and 91 ± 4 μeV/K were obtained for W(110) and W(100) [9].

References for 4.5:

[1] Krahl-Urban, B.; Niekisch, E. A.; Wagner, H. (Surf. Sci. **64** [1977] 52/68).

[2] Sytaya, E. P.; Smorodinova, M. I.; Imangulova, N. I. (Fiz. Tverd. Tela [Leningrad] **4** [1962] 1016/20; Sov. Phys.-Solid State [Engl. Transl.] **4** [1962] 750/3).

[3] Sultanov, V. M. (Radiotekh. Elektron. [Moscow] **9** [1964] 317/20; Radio Eng. Electron. Phys. [Engl. Transl.] **9** [1964] 252/4).

[4] Fine, J.; Madey, T. E.; Scheer, M. D. (Surf. Sci. **3** [1965] 227/33).

[5] Azizov, U. V.; Shuppe, G. N. (Fiz. Tverd. Tela [Leningrad] **7** [1965] 1970/3; Sov. Phys.-Solid State [Engl. Transl.] **7** [1965/66] 1591/4).

[6] Gardner, F. M.; Girouard, F. E.; Boeck, W. L.; Coomes, E. A. (Surf. Sci. **26** [1971] 605/23).

[7] Azizova, D. Kh.; Gorbatyi, N. A. (Dokl. Akad. Nauk UzSSR **1983** No. 10, pp. 25/6).
[8] Shelton, H. (Phys. Rev. [2] **107** [1957] 1553/7).
[9] Hopkins, B. J.; Lee, T. J.; Williams, C. B. (J. Appl. Phys. **40** [1969] 1728/32).
[10] Seely, S. (Phys. Rev. [2] **59** [1941] 75/8).

[11] Kiejna, A.; Wojciechowski, K. F.; Zebrowski, J. (J. Phys. F **9** [1979] 1361/6; Corrigendum: J. Phys. F **11** [1981] 2495).
[12] Azizova, D.; Pazdzerskii, V. A.; Tsipinyuk, B. A. (Poverkhnost' **1984** No. 9, pp. 96/9; Phys. Chem. Mech. Surf. **3** [1985] 2707/15).
[13] Krüger, F.; Stabenow, G. (Ann. Phys. [Leipzig] [5] **22** [1935] 713/34).
[14] Reimann, A. L. (Proc. R. Soc. [London] A **163** [1937] 499/510).
[15] Potter, J. G. (Phys. Rev. [2] **58** [1940] 623/32).
[16] Waterman, A. T.; Potter, J. G. (Phys. Rev. [2] **51** [1937] 63).
[17] Reimann, A. L. (Phys. Rev. [2] **45** [1934] 898).
[18] Datz, S.; Taylor, E. H. (J. Chem. Phys. **25** [1956] 389/94).
[19] Trischka, J. W. (J. Appl. Phys. **37** [1966] 455/6).
[20] Nottingham, W. B. (Phys. Rev. [2] **49** [1936] 78/97).

[21] Langmuir, D. B. (Phys. Rev. [2] **49** [1936] 428/35).
[22] Drechsler, M. (Z. Phys. **167** [1962] 558/62).
[23] Young, R. D.; Müller, E. W. (J. Appl. Phys. **33** [1962] 91/5).
[24] Swanson, L. W.; Crouser, L. C. (Phys. Rev. [2] **163** [1967] 622/41).
[25] van Oostrom, A. (Phys. Lett. **4** [1963] 34/6).
[26] Gao, Y.; Reifenberger, R. (J. Vac. Sci. Technol. [2] A **4** [1986] 1289/93).
[27] Gaudin, G. A.; Lee, M. J. G. (Surf. Sci. **201** [1988] 540/58).
[28] Wysocki, J. K. (Surf. Sci. **104** [1981] 463/77).
[29] Wysocki, J. K. (Phys. Rev. [3] B **28** [1983] 834/41).
[30] Donders, P. J. (Diss. Univ. Toronto 1988 from Diss. Abstr. B **49** [1988] 1193).

[31] Heilmann, P.; Heinz, K.; Müller, K. (Surf. Sci. **89** [1979] 84/94).
[32] Donders, P. J.; Lee, M. J. G. (Surf. Sci. **167** [1986] 141/9).
[33] Lopinski, G. P.; Donders, P. J.; Lee, M. J. G. (Surf. Sci. **201** [1988] 294/310).
[34] Smith, G. F. (Phys. Rev. [2] **100** [1955] 1115/6).
[35] Hutson, A. R. (Phys. Rev. [2] **98** [1955] 889/901, 900).
[36] Sultanov, V. M.; Shuppe, G. N. (Izv. Akad. Nauk UzSSR Ser. Fiz.-Mat. Nauk **1965** No. 5, pp. 49/53).
[37] Madey, T. E.; Yates, J. T., Jr. (Nuovo Cimento Suppl. [1] **5** [1967] 483/505).
[38] Bulyga, A. V.; Solonovich, V. K. (Surf. Sci. **223** [1989] 578/84, Vysokochist. Veshchestva **4** No. 3 [1990] 74/9; High-Purity Subst. [Engl. Transl.] **4** [1990] 462/7).
[39] Azizov, E. E.; Mashtakova, V. A.; Shishkin, B. B. (Vestn. Mosk. Univ. Ser. III Fiz. Astron. **42** No. 6 [1987] 39/42; Moscow Univ. Phys. Bull. [Engl. Transl.] **42** No. 6 [1987] 42/5).

5 Electron Emission

5.1 Thermionic Emission

5.1.1 Introduction

General reviews about thermionic electron emission from metal surfaces are given in the comprehensive papers [1 to 4]. If no field is applied, the current density j of thermionic electrons emitted from a uniform surface at temperature T is described by the Richardson (-Laue-Dushman) equation

$$j = A\,T^2 \exp\left[-\Phi/(k_B T)\right],$$

where Φ is the work function of the metal, $A = A_R \cdot t$, A_R the Richardson constant ($= 4\pi\, e m_e k_B^2/h^3 = 120\ A \cdot cm^{-2} \cdot K^{-2}$), and t an average transmission coefficient that describes the probability of the electrons to be transmitted through the surface potential barrier. This transmission coefficient depends on energy, momentum, temperature, and surface structure in a complex manner (see e.g. [4]). In general, weak electric fields F are applied in thermionic emission measurements; they lower the potential barrier by $\Delta\Phi = (e^3 F)^{1/2}$, giving the well-known Schottky formula [5]

$$j = A\,T^2 \exp\left[-(\Phi - (e^3 F)^{1/2})/(k_B T)\right].$$

In strong electric fields, a transition from thermionic to field emission takes place. Thermal field and pure field emissions are discussed in Chapter 5.3, pp. 156/8, and Chapter 5.2, pp. 142/55, respectively.

The irradiation of W surfaces by laser beams can provoke the emission of thermionic electrons as a result of locally heating the metal surface to high temperatures. Laser-induced electron emission from W surfaces is discussed in detail in Chapter 5.6, pp. 197/201.

5.1.2 The Richardson Constant

Thermionic emission constants were estimated from plots of $\ln(j_0/T^2)$ versus 1/T (Richardson plot), where j_0 is the linearly extrapolated value at zero field from a plot of $\ln(j_0/T^2)$ versus $F^{1/2}$ (Schottky-plot). Since the work function is temperature-dependent (see Chapter 4.5, pp. 128/30), Richardson plots do not yield the true constants A and Φ but the "apparent emission constant" A^* ($= A \exp[-(1/k_B)(d\Phi/dT)]$ for temperature-independent t) and the "apparent work function" Φ^*; see e.g. [1, 4, 6]. In general, the surface of an experimental emitter is not uniform. It consists of an unknown number of patches of unknown area with different work functions. So experimentally one obtains values A^{**} and Φ^{**}, which are a complicated combination of the true emission constants of the different patches [1, 6].

In the original form of the Richardson equation, the transmission coefficient does not appear [7 to 9]. Using this simple form and disregarding the effect of patchiness, strongly varying Richardson constants A^{**} ranging from 22 [10] to 100 $A \cdot cm^{-2} \cdot K^{-2}$ [11] were reported from measurements on polycrystalline W wires [10 to 19]. The deviation from A_R was explained solely with a linear temperature dependence of the work function in early papers [10, 16, 20]. For a summary of the earliest work on thermionic emission from W, see "Wolfram" 1933, pp. 77/8. For a detailed discussion of the work function and its temperature dependence, see Chapter 4, pp. 107/31.

The possibility that electron reflection at the surface potential barrier may explain the low values of the experimentally determined Richardson constants was taken into account for the first time by Nottingham [21]. He introduced an empirical reflection factor to describe thermal emission data from polycrystalline W [21].

In a study on W single crystals, it was found that thermionic emission strongly depends on the crystallographic orientation [22]. This observation gives a simple explanation for the large variations in the experimentally determined Richardson constants of polycrystalline W. It was shown that the values obtained from the usual measurements on polycrystalline wires are characterized by the weighted average of properties of the various crystal faces occurring [23]. Average emission constants for polycrystalline W, estimated from single-crystal measurements on different surfaces, are 40 $A \cdot cm^{-2} \cdot K^{-2}$ [18] and 62 $A \cdot cm^{-2} \cdot K^{-2}$ [24]. The following emission constants were reported for the different crystal faces:

crystal plane	thermionic emission constant in $A \cdot cm^{-2} \cdot K^{-2}$							
	A*	A*	A*	A*	A*	A	A	A
(111)	35	52	70[1)]	50	128	101	92	130
(112)	125	120	140[1)]	180	–	61	55	70
(116)	53	40	–	35	–	103	94	–
(100)	117	105	130[1)]	130	200	113	104	127
Ref.	[23]	[25]	[26]	[27]	[28, 29]	[30]	[31]	[26]

[1)] Recalculated from data in [26].

In [26] the measured A* values (not written down in [26]) were converted to the emission constants A with the aid of temperature coefficients of the work functions taken from [30, 32]. The values of [30] were obtained by correcting the data from [25] and [23] for the temperature variation of the work function and for the special character of the energy distribution introducing an energy-dependent reflection coefficient [30]. In [31] these values were criticized and calculated again assuming that, due to bad analyzer resolution, a wrong energy distribution had been used in [30]. Averaging the values for the [111], [116], and [100] directions, an energy-independent reflection coefficient of 0.2 follows [31]. The high values reported in [28, 29] were explained with negative temperature coefficients of the work function [29].

For the W(110) surface, only low-precision A* values were obtained which are generally very large and vary strongly. Examples are $A^* = 320 \pm 200$ [27], 400 [26], or 1733 ± 180 $A \cdot cm^{-2} \cdot K^{-2}$ [33]. To explain these values, large negative temperature coefficients were assumed. It is remarkable that at the same time the work function of the (110) surface is the highest of all W surfaces (cf. Section 4.2.1, pp. 108/18). This effect of compensation was clearly observed for a number of different W surfaces in a study of Gardner et al. [24], who measured the thermionic electron emission as a function of the crystallographic orientation for the [110] zone of a W filament single crystal with an angular resolution of 1°. Before taking point-by-point emission data, the wire had been recrystallized by twisting, polishing, cleaning, and annealing at approximately 2000 K for 15 h. The coefficients Φ^* and A* estimated from the experimental data are shown in **Fig. 27**, p. 134 [24]. The data reveal an obvious relation between the minima and maxima of Φ^* and A*. A mathematical analysis shows a proportionality between Φ^* and ln A*, which is discussed in [27] and [34]. In the latter work it was found that this functional relation (named compensation effect therein) also holds for thermal emission data from single-crystal surfaces which are altered by adsorption of impurity atoms or by external electric fields [34].

In the older studies [23] and [25], very low emission constants (15 and 8 $A \cdot cm^{-2} \cdot K^{-2}$, respectively) were reported for the (110) surface of recrystallized W wires. The discussion of these low values yielded no satisfactory explanation. In the [110] direction, the emission

References for 5.1 on pp. 139/41

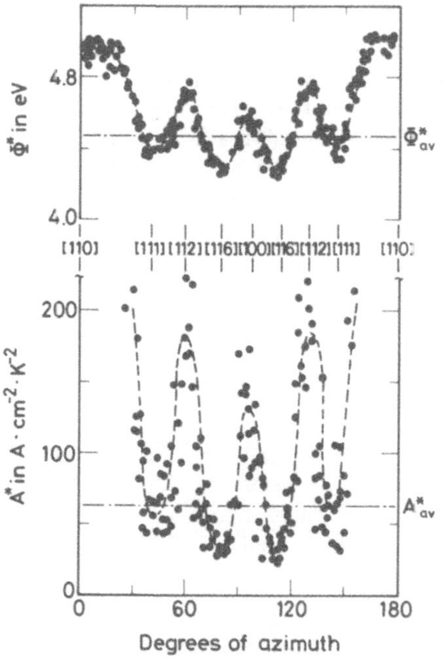

Fig. 27. Richardson constant A* and work function Φ* for recrystallized W filaments as a function of the crystal orientation. The values were measured on W wires with the axis parallel to [110]. Φ^*_{av} and A^*_{av} mark the average values [24].

intensity in dependence on the azimuth showed a strong minimum [22, 23, 25], and the (110) surface was found to have an especially rough [23] or shingle [25] structure. Polycrystalline W wires tend to line up the [110] direction of the crystallites along the axis of the wire after recrystallization (see "Tungsten" Suppl. Vol. A 3, 1989, p. 47). Since in addition the work function associated with this direction is especially high, the currents measured in [23] and [25] probably result from secondary effects as was suggested already in [25].

5.1.3 Influence of Surface Nonuniformity

As was just mentioned above, the surface of a real crystal generally is not uniform, but consists of many regions or **patches** with different work functions. The effect of patches on the emission properties of W single-crystal surfaces was investigated in [1, 24, 25, 27, 30, 35]. A detailed patch theory is presented in [1]. Since the work function as well as the thermionic emission constant depend on the crystallographic orientation, such patches produce a nonuniform electron emission. The effect of patchiness was discussed in [25] and [30] as an additional possibility (besides reflection effects and the temperature dependence of the work function) to explain the emission constant data differing from 120 $A \cdot cm^{-2} \cdot K^{-2}$ in W. From the observations in [25], the author concluded that large- and medium-scale patchiness can be ruled out, leaving only the existence of small-scale patches (some 10 Å in diameter). Reflection effects were stated to be a much more likely explanation for the electron emission properties of W than patches [30].

Introducing the term of the fundamental patch (which is the smallest independently emitting surface area), a refined patch theory was presented in [27]. The sum of fundamental patch areas with the same work function were called a micropatch, and it was concluded from a computer analysis of emission data from W that the distribution of micropatches has Gaussian form. Assuming such a distribution, the observed linear relationship between the work function and the logarithm of the emission constant could be explained [24, 27].

Since patches always reduce the electron emission current, an attempt was made to produce thermionic emitters with a more uniform emission by chemical vapor deposition [36]. W filaments were prepared by hydrogen reduction of tungsten hexafluoride and exhibited a high degree of uniaxial crystal orientation with the preferred axis centered in the direction of growth. The surface essentially was a single-crystal plane, resulting in an optimized efficiency of electron emission [36].

The existence of **clusters** of atoms on their own solid surface was shown to influence the thermionic electron emission properties [37]. The clusters were presumed to act as separate entities, although their atoms are bound to the solid. Atoms in clusters behave as if much more dilute than lattice atoms. The author considered that the valence electrons of such a cluster obey Maxwell-Boltzmann statistics, because they are less dense than in the bulk, whereas Fermi-Dirac statistics is expected to be valid for bulk electrons. Based on this idea he concluded that the Richardson and Schottky equations cited above must be altered for the case of electron emission from clusters in such a manner that the factor T^2 is replaced by $T^{1/2}$. Applying these assumptions to analyze some very old emission data for tungsten [38] and assuming that the area of all clusters is about 50% of the total surface area, it was possible to completely eliminate the temperature dependence attributed to the tungsten work function. This gave an electron number density of about $4 \times 10^{19}/cm^3$ for the clusters, which is 1500 times lower than the anticipated lattice electron number density of about $6 \times 10^{22}/cm^3$ [37].

The effect of **dislocations** on the thermionic emission current of a W(100) single-crystal surface was studied in [39]. Electrolytic etching in a 2% solution of NaOH and subsequent heating-cooling cycles of the samples brought about surface regions with dislocation densities higher than $10^8/cm^2$ at 1900 K. In these dislocation regions, the emission current was increased by a factor of 1.5 to 1.6 and the work function was lowered by 0.08 eV [39].

Radiation defects on a polycrystalline W surface, produced by 3 μs pulses of electron irradiation with energies above 1.25 MeV, were studied measuring the time dependence of the thermionic emission current. A mechanism involving the change in the concentration of equilibrium defects as a result of absorbed thermal energy was described. Enhanced relaxation times were found as the radiation dose was increased [40].

5.1.4 Energy Distribution

There are conflicting results in the literature on the energy distribution of the emitted thermionic electrons from W. In [31, 41 to 43] pure Maxwell distributions were found, whereas deviations from this distribution were observed in [21, 30, 44]. This point is of special interest, because the Richardson and Schottky equations were developed on the assumption that the electrons are emitted according to the Maxwell law.

Theoretically, a pure Maxwell energy distribution of thermionic electrons occurs only if electrons with energies above the top of the potential barrier contribute to the emission current, i.e., if the transmission coefficient is exactly 0 for electrons with energies below the top of the barrier and 1 for the others. Reflection of electrons by the barrier top may

References for 5.1 on pp. 139/41

result in a deficiency of low-energy electrons. This effect can be calculated with sufficient accuracy [21, 30]. A further source for deviations from the Maxwell distribution is the non-free-electron character of the emitter. Since this effect depends on specific properties of the emitting surface, it is more difficult to be accounted for [4]. Inelastic scattering processes of electrons by Coulomb collisions leading to a distortion of the electron energy distribution in thermionic emission were discussed in [45]. Theoretical calculations showed that the deviations of the transmission coefficient from unity for electrons above the barrier is no more than a few percent [45].

Experimentally, the energy distribution of thermionic electrons from a W filament was investigated for the first time in [41], where current-voltage characteristics were measured at eight different temperatures ranging from 1440 to 2475 K. According to this study, the energy distribution was purely Maxwellian at all temperatures. In [21], a deficiency of slow electrons was detected in measurements at 1160 K, but in another study investigating especially the low-energy range down to 0.03 eV with an improved cylindrical condensor at 1750 and 2000 K, these deviations could not be verified [42]. An increase of low-energy electrons and a broadening of the energy distribution compared to the Maxwellian form was reported from electron emission measurements in [44].

Deviations were found once more in a study on a single-crystal W filament using a magnetic velocity analyzer tube [30]. Measurements were made for various crystallographic orientations at 1700 and 2000 K. **Fig. 28** [30] shows a superposition of energy distributions obtained for five different crystallographic directions. To superpose these five distribution curves, they had been normalized and shifted along the energy axis yielding the differences of the true work functions as an additional result. The solid curve represents the theoretical

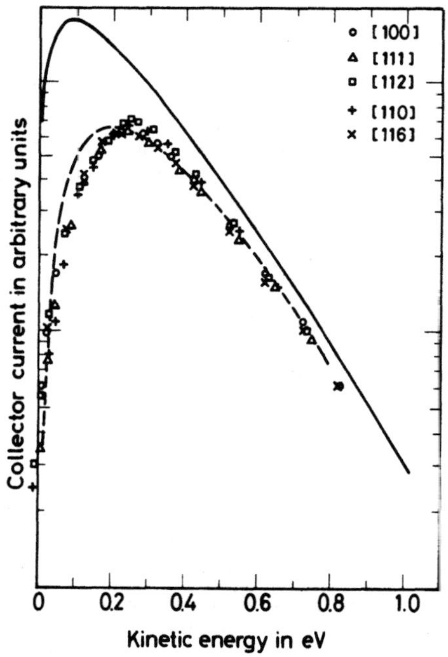

Fig. 28. Superposition of the energy distributions obtained from the directions of maximum and minimum emission (cf. Fig. 27) at T = 2000 K with an anode potential of 1000 V. A pure Maxwell distribution (solid line) and a Maxwell distribution modified by a reflection coefficient according to [21] (dashed line) are shown for comparison [30].

Maxwell distribution, and the dashed line corresponds to a Maxwell distribution modified by a reflection coefficient according to [21]. Obviously, the energy distribution has the same shape for all crystallographic directions, revealing a deficiency of slow electrons. The effect of patchiness to explain the non-Maxwellian character of the energy distribution was ruled out, since it would seem extremely unlikely that the various surfaces of a single crystal of W would be patchy in just such a manner as to yield identical energy distributions. Band structure effects were ruled out as well [30].

Dealing with the experimental results of [30], it was suggested that the true energy distribution nevertheless might be Maxwellian and that the observed deficiency of slow electrons could be attributed to analyzer resolution. Assuming a resolution a factor of ten poorer than in [30], a good fit to the Maxwell distribution was obtained [31].

Using a W whisker with its axis parallel to the [110] direction, the energy distribution from different crystal faces was investigated at 1890 and 2100 K [43]. As had been also found in [30], the shape of the energy distribution was the same for all crystallographic directions studied. However, the assumption of a reflection coefficient as in [21] and [30] yielded poor results. In the low-energy range, the experimental curve coincided with the pure Maxwell distribution. Deviations were found in the high-energy range, but this effect was explained by the neglect of the electron-energy dependence of the energy resolution [43].

5.1.5 Schottky Deviations

According to the Schottky equation, at constant temperature a linear relationship should exist between the logarithm of the emitted current and the square root of the field strength, giving the so-called Schottky line. However, careful measurements of the electron emission from polycrystalline W [46 to 49] and from single-crystal planes of W [25, 50] revealed that there exist small periodic deviations from the Schottky line. These deviations soon were recognized to be a quantum-mechanical effect [51]; since then, several theoretical attempts to explain the phenomenon were made [1, 52 to 59].

Periodic deviations were detected experimentally for the first time in thermionic emission measurements on a polycrystalline W wire at temperatures ranging from 1610 to 1940 K [46]. No appreciable effect of temperature upon the position of the peaks was found [46 to 48]. For the temperature dependence of the deviation amplitudes, different results were reported. Decreasing [46] as well as increasing [47] and temperature-independent [48] data were found when the temperature was increased. Oscillation amplitude and period increase with increasing field strength [46 to 48].

Theoretically, the periodic deviations from the Schottky line were explained to result from the interference between electron waves reflected from the potential barrier maximum and those reflected from the surface itself [51 to 59]. The interference resulting from this double reflection depends on the field strength, since the position of the maximum of the potential barrier is inversely proportional to $F^{1/2}$. The first theoretical calculations [52] were modified several times using improved models for the surface potential barrier [53 to 59]. Taking into account exchange and correlation forces, good agreement between theoretical and experimental results for W was achieved [58]. In [59] it was pointed out that a quantitative study on periodic Schottky deviations must be based on measurements of emission currents from single-crystal planes because of the distinct transmission properties of the different crystal faces. Up to now, the only available single-crystal data on W are the electron emission measurements of [50] on W(111) surfaces. Although there was clearly evidence of

References for 5.1 on pp. 139/41

periodic deviations on the Schottky lines from (100), (111), (112), and (116) data, no attempt was made to study these Schottky deviations due to limitations of the tube used [25].

According to theoretical calculations of [60], strong electric fields may set free electrons from the surface which tunnel through the top of the potential barrier and give an additional contribution to the amplitude of the periodic Schottky deviations. This increase leads to a better agreement with experiments at large fields [60].

Besides thermionic deviations, photoelectric deviations from the Schottky line were observed (see Section 5.5.4, p. 197). In [50] photoelectric and thermionic Schottky deviations were compared in a dual experiment on a single-crystal W(111) surface. The photoelectric deviations were found to be in fair agreement in period and phase with the thermionic deviations (see **Fig. 29**) [50].

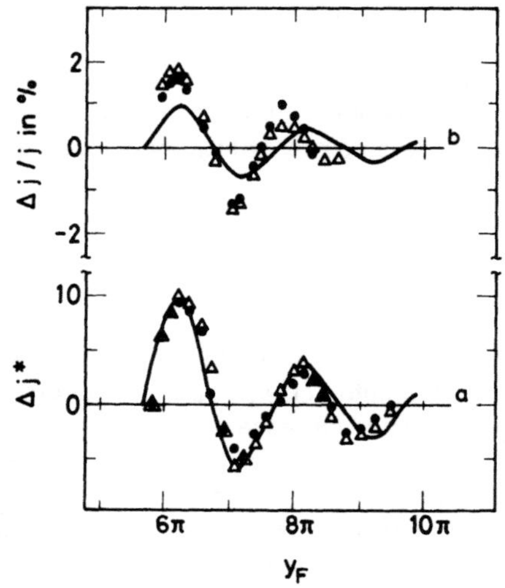

Fig. 29. Thermionic (a) and photoelectric (b) periodic deviations from the straight Schottky line obtained from the (111) surface of W. $y_F \equiv 357.1\ F^{-1/4}$ with F in V/cm. $\Delta j^* \equiv 10^3\ [\ln(j/j(F=0)) - (e^3F)^{1/2}/k_BT]/\ln 10$. The circles and triangles correspond to experimental points of two different experimental runs; the solid lines were calculated according to the theory in [55, 56]; [50].

5.1.6 Emission Current Anomalies Near T_M

When W wires were heated up to temperatures above 3000 K by large current pulses of some 10^6 A/cm^2, anomalously high thermionic emission currents were observed [61 to 63]. In [64] it was shown that this anomaly only appears for temperatures near the melting point. The authors presume a connection between anomalies of heat capacity and electron emission. A contamination effect is ruled out [64]. The assumption is made that the high emission currents are due to the appearance of lattice defects, particularly to transitions of metal atoms from crystal lattice sites to interstitial sites [65].

Anomalously high emission currents were also observed for a W arc cathode in an Ar atmosphere at low pressures. Current densities were measured a factor of 20 to 55 higher than the values calculated from the Richardson equation for the corresponding cathode temperatures [66]. The dependence of the current density on Ar pressure was calculated in [67].

A study of heated W wires exploding in a vacuum of 2×10^{-6} Torr showed that a discharge existed in a low-pressure metal vapor outside the wire and that the initial current carriers were thermionic electrons [68].

Anomalies in the current-voltage characteristics of plane-parallel diodes with single-crystal W cathodes were observed above a certain characteristic temperature Θ_{te} ($=1810 \pm 5$ K [69], 1860 ± 5 K [70], and 1850 K [29] for W(100), W(111), and W(110), respectively). At $T = \Theta_{te}$ the Richardson straight line exhibits a discontinuity, and for temperatures above this critical value the Richardson emission parameters are strongly increased [28, 29]. It was assumed that the effect is caused by the emission of electrons formed during the recombination of Frenkel pairs [71]. This recombination-emission current from thermal defects is superimposed on the normal thermal emission, and hence the Richardson equation is no longer valid for $T > \Theta_{te}$. In the energy distribution a peak corresponding to electrons with an average energy of ca. 2.5 eV appeared in addition to the spectrum of Maxwell electrons [70]. Assuming that the excess energy transferred to an electron in a recombination process is twice the formation energy E_f of a vacancy ($E_f = 3.15$ eV [72], 3.6 eV [73], see also "Tungsten" Suppl. Vol. A 2, 1987, p. 184), electrons with energies 1 to 2 eV below the Fermi level can participate in such a process [69, 70]. Since the electronic density of states is significant for these energies and depends strongly on energy (d electrons), a narrow peak should appear in the energy distribution as was observed in [70]. From measurements of voltage-current characteristics for recombination-emission currents of W, the temperature dependence of the vacancy formation energy and thermal entropy was calculated in [73].

5.1.7 Potential and Temperature Gradients Inside the Crystal

Directional transport processes of electric charge or heat inside a W single crystal were found to affect the emission of thermionic electrons. At the same time, a step-like macrostructure was observed to develop at the crystal surface. It was assumed that the surface is restructured by electromechanical forces leading to a change of the work function and thus indirectly to a change of the thermionic emission current [74].

References for 5.1:

[1] Herring, C.; Nichols, M. H. (Rev. Mod. Phys. **21** [1949] 185/270).
[2] Nottingham, W. B. (in: Flügge, S.; Handbuch der Physik, Vol. 21, Springer, Berlin-Göttingen-Heidelberg 1956, pp. 1/175).
[3] Haas, G. A.; Thomas, R. E. (in: Bunshah, R. F.; Techniques of Metals Research Vol. VI, Pt. 1, Wiley, New York 1972, pp. 91/262).
[4] Modinos, A. (Field, Thermionic, and Secondary Electron Emission Spectroscopy, Plenum, New York 1984, pp. 1/42, 295/325).
[5] Schottky, W. (Phys. Z. **15** [1914] 872/8, **20** [1919] 220/8; Z. Phys. **14** [1923] 63/106).
[6] Rivière, J. C. (in: Green, M.; Solid State Surface Science, Vol. 1, Dekker, New York 1969, pp. 179/289, 182/6).
[7] Richardson, O. W. (Philos. Mag. [6] **23** [1912] 594/627).
[8] Von Laue, M. (Jahrb. Radioakt. Elektron. **15** [1918] 205/56, 257/70).
[9] Dushman, S. (Phys. Rev. [2] **21** [1923] 623/36).
[10] Freitag, H.; Krüger, F. (Ann. Phys. [Leipzig] [5] **21** [1934] 697/742, 736).

[11] Zwikker, C. (Proc. K. Ned. Akad. Wet. **29** [1926] 792/802).
[12] Dushman, S.; Rowe, H. N.; Ewald, J.; Kidner, C. A. (Phys. Rev. [2] **25** [1925] 338/60).
[13] Ahearn, A. J. (Phys. Rev. [2] **44** [1933] 277/86).
[14] Nottingham, W. B. (Phys. Rev. [2] **47** [1935] 806).
[15] Wahlin, H. B.; Whitney, L. V. (Phys. Rev. [2] **50** [1936] 735/8).
[16] Reimann, A. L. (Proc. R. Soc. London A **163** [1937] 499/510).
[17] Reimann, A. L. (Philos. Mag. [7] **25** [1938] 834/48).
[18] Nichols, M. H. (Phys. Rev. [2] **59** [1941] 944).
[19] Nichols, M. H. (Phys. Rev. [2] **78** [1950] 158/61).
[20] Krüger, F.; Stabenow, G. (Ann. Phys. [Leipzig] [5] **22** [1935] 713/34).

[21] Nottingham, W. B. (Phys. Rev. [2] **49** [1936] 78/97).
[22] Martin, S. T. (Phys. Rev. [2] **56** [1939] 947/59).
[23] Nichols, M. H. (Phys. Rev. [2] **57** [1940] 297/306).
[24] Gardner, F. M.; Girouard, F. E.; Boeck, W. L.; Coomes, E. A. (Surf. Sci. **26** [1971] 605/23).
[25] Smith, G. F. (Phys. Rev. [2] **94** [1954] 295/308).
[26] Azizov, U. V.; Shuppe, G. N. (Fiz. Tverd. Tela [Leningrad] **7** [1965] 1970/3; Sov. Phys.-Solid State [Engl. Transl.] **7** [1966] 1591/4).
[27] Coomes, E. A.; Petrauskas, A. A. (AD-664892 [1967] 86 pp.; C.A. **69** [1968] No. 47158); see also Coomes, E. A.; Boeck, W. L. (AED-CONF-65-324-23 [1965] 201/5; C.A. **66** [1967] No. 69964).
[28] Mashtakova, V. A.; Naumova, L. Yu.; Shishkin, B. B. (Poverkhnost **1984** No. 2, pp. 74/8; Phys. Chem. Mech. Surf. [Engl. Transl.] **3** [1985] 455/64).
[29] Azizov, E. E.; Mashtakova, V. A.; Shishkin, B. B. (Vestn. Mosk. Univ. Ser. Fiz. Astron. **42** No. 6 [1987] 39/42; Moscow Univ. Phys. Bull. [Engl. Transl.] **42** No. 6 [1987] 42/5).
[30] Hutson, A. R. (Phys. Rev. [2] **98** [1955] 889/901).

[31] Smith, G. F. (Phys. Rev. [2] **100** [1955] 1115/6).
[32] Sytaya, E. P.; Smorodinova, M. I.; Imangulova, N. I. (Fiz. Tverd. Tela [Leningrad] **4** [1962] 1016/20; Sov. Phys.-Solid State [Engl. Transl.] **4** [1962] 750/3).
[33] Krahl-Urban, B.; Niekisch, E. A.; Wagner, H. (Surf. Sci. **64** [1977] 52/68).
[34] Vanselow, R. (Surf. Sci. **149** [1985] 381/93).
[35] Herring, C. (Phys. Rev. [2] **59** [1941] 889/96).
[36] Weissman, I.; Kinter, M. L. (J. Appl. Phys. **34** [1963] 3187/94).
[37] Biblarz, O. (J. Vac. Sci. Technol. [2] A **8** [1990] 127/33).
[38] Jones, H. A.; Langmuir, I. (Gen. Electr. Rev. **30** [1927] 310/9, 354/61, 408/12).
[39] Chaikovskii, E. F.; Taran, A. A. (Pis'ma Zh. Tekh. Fiz. **5** [1979] 920/3; Sov. Tech. Phys. Lett. [Engl. Transl.] **5** [1979] 381/2).
[40] Rudenko, N. V.; Golovanov, Yu. N.; Abdullin, A. G.; Takibaev, Zh. S. (Fiz.-Khim. Probl. Krist. No. 2 [1971] 206/12; C.A. **80** [1974] No. 101367).

[41] Germer, L. H. (Phys. Rev. [2] **25** [1925] 795/807).
[42] Dobretsov, L. N.; Uvarov, P. Ya. (Tr. 6th Soveshch. Katodn. Elektron., Kiev 1951 [1952], pp. 43/54; C.A. **1956** 7573).
[43] Gorbatyi, N. A.; Chekina, V. A. (Izv. Akad. Nauk SSSR Ser. Fiz. **38** [1974] 230/3; Bull. Acad. Sci. USSR Phys. Ser. [Engl. Transl.] **38** No. 2 [1974] 43/6), Chekina, V. A.; Gorbatyi, N. A. (Emiss. Elektron. Mater. 2nd Semin. Probl. Rab. Vykhoda Elektronov Adsorbts. Met. Zavisimosti Kristallogr. Napravlenii, Ryazan, USSR, 1971 [1974], pp. 85/92; C.A. **85** [1976] No. 135512).
[44] Boersch, H. (Naturwissenschaften **40** [1953] 267/8).
[45] Moizhes, B. Ya.; Naumov, V. N.; Nemchinskii, V. A. (Zh. Tekh. Fiz. **48** [1978] 2120/7; Sov. Phys.-Tech. Phys. [Engl. Transl.] **23** [1978] 1211/5).

[46] Seifert, R. L. E.; Phipps, T. E. (Phys. Rev. [2] **56** [1939] 652/63; **53** [1938] 493).
[47] Turnbull, D.; Phipps, T. E. (Phys. Rev. [2] **56** [1939] 663/7).
[48] Nottingham, W. B. (Phys. Rev. [2] **57** [1940] 935).
[49] Houde, A. L. (Diss. Univ. Notre Dame 1952, from [56]).
[50] Stafford, D. F.; Weber, A. H. (J. Appl. Phys. **34** [1963] 2667/70), Stafford, D. F. (Diss. St. Louis Univ. 1962, 111 pp.; Diss. Abstr. **24** [1963] 4261).

[51] Mott-Smith, H. M. (Phys. Rev. [2] **56** [1939] 668/9).
[52] Guth, E.; Mullin, C. J. (Phys. Rev. [2] **59** [1941] 575/84).
[53] Juenker, D. W.; Colladay, G. S.; Coomes, E. A. (Phys. Rev. [2] **90** [1953] 772/8).
[54] Miller, S. C., Jr.; Good, R. H., Jr. (Phys. Rev. [2] **92** [1953] 1367/72).
[55] Juenker, D. W. (Phys. Rev. [2] **99** [1955] 1155/60).
[56] Cutler, P. H.; Gibbons, J. J. (Phys. Rev. [2] **111** [1958] 394/402).
[57] Belford, G. G.; Kuppermann, A. A.; Phipps, T. E. (Phys. Rev. [2] **128** [1962] 524/31).
[58] Yang, J.-J.; Yang, T.-T. (Phys. Rev. [3] B **1** [1970] 3614/23).
[59] Modinos, A. (Surf. Sci. **115** [1982] 469/500).
[60] Guth, E.; Mullin, C. J. (Phys. Rev. [2] **61** [1942] 339/48).

[61] Lebedev, S. V. (Zh. Eksp. Teor. Fiz. **27** [1954] 487/500; Phys. Abhandl. Sowjetunion [Ger. Transl.] **10** [1957] 115/32; SCL-T-**257** [1954] 1/28; N.S.A. **13** [1959] No. 17089).
[62] Lebedev, S. V.; Khaikin, S. E. (Zh. Eksp. Teor. Fiz. **26** [1954] 723/35).
[63] Borodovskaia, L. N.; Lebedev, S. V. (Zh. Eksp. Teor. Fiz. **28** [1955] 96/110; Sov. Phys.-JETP [Engl. Transl.] **1** [1955] 71/83).
[64] Ivantsova, I. L.; Lebedev, S. V. (Teplofiz. Vys. Temp. **9** [1971] 1168/72; High Temp. [Engl. Transl.] **9** [1971] 1077/81).
[65] Lebedev, S. V. (Teplofiz. Vys. Temp. **8** [1970] 252/9; High Temp. [Engl. Transl.] **8** [1970] 240/6).
[66] Dorodnov, A. M.; Kozlov, N. P.; Pomelov, Ya. A. (Teplofiz. Vys. Temp. **12** [1974] 10/6; High Temp. [Engl. Transl.] **12** [1974] 9/14).
[67] Nemchinskii, V. A. (Zh. Tekh. Fiz. **44** [1974] 2548/50; Sov. Phys.-Tech. Phys. [Engl. Transl.] **19** [1974] 1577/8).
[68] Holmström, I.; Händel, S. K.; Stenerhag, B. (J. Appl. Phys. **39** [1968] 2998/3002).
[69] Mashtakova, V. A.; Shishkin, B. B. (Pis'ma Zh. Eksp. Teor. Fiz. **30** [1979] 99/102; JETP Lett. [Engl. Transl.] **30** [1979] 89/92).
[70] Mashtakova, V. A.; Pirogov, Y. A.; Kovalev, S. V.; Goloskokova, L. Y.; Babaev, V. G. (Izv. Vyssh. Uchebn. Zaved. Radiofiz. **28** [1985] 372/8; Radiophys. Quantum Electron. [Engl. Transl.] **28** [1985] 253/8).

[71] Mashtakova, V. A.; Shishkin, B. B. (Vestn. Mosk. Univ. Ser. Fiz. Astron. **39** No. 2 [1984] 76/8; Moscow Univ. Phys. Bull. [Engl. Transl.] **39** No. 2 [1984] 88/90).
[72] Kraftmakher, Ya. A.; Strelkov, P. G. (Fiz. Tverd. Tela [Leningrad] **8** [1966] 1049/52, **4** [1962] 2271/4; Sov. Phys.-Solid State [Engl. Transl.] **8** [1966] 838/41, **4** [1962] 1662/4).
[73] Goloskokova, L. Yu.; Shishkin, B. B. (Vestn. Mosk. Univ. Ser. Fiz. Astron. **43** No. 2 [1988] 63/8; Moscow Univ. Phys. Bull. [Engl. Transl.] **43** No. 2 [1988] 64/8).
[74] Zakurdaev, I. V.; Chernyak, E. Ya.; Shuppe, G. N.; Yakovlev, V. N. (Fiz. Tverd. Tela [Leningrad] **14** [1972] 3540/5; Sov. Phys.-Solid State [Engl. Transl.] **14** [1972] 2977/80; Izv. Akad. Nauk SSSR Ser. Fiz. **37** [1973] 2547/52; Bull. Acad. Sci. USSR Phys. Ser. [Engl. Transl.] **37** No. 12 [1973] 67/71).

5.2 Field Emission

5.2.1 Reviews

The early work on field emission (FE) by both experimentalists and theorists is reviewed in [1, pp. 177/88]. The experimental work during the period from 1936 to 1943, much of which is devoted to tungsten and field emission microscopy (FEM), is treated in [1, pp. 188/95]. For early FE work, see [2, 3] and also [4, 5] where Fowler-Nordheim (F-N) and energy distribution (ED) curves are given and FEM and field ion microscopy (FIM) are reviewed [4, 5]. The significance of early FEM and FIM results (with emphasis on the work of Müller) in the light of more recent studies is discussed by [6]. Various aspects of FEM on metals were reviewed by [7]; see also the review of FE and FIM by [8]. Field emission with emphasis on total energy distributions (TEDs) of FE electrons is treated in [9]. Theoretical aspects of field emission spectroscopy are discussed in [10].

References for 5.2.1:

[1] Jenkins, R. O. (Rep. Prog. Phys. **9** [1942/43] 177/97).
[2] Ashworth, F. (Adv. Electron. **3** [1951] 1/42).
[3] Dyke, W. P.; Dolan, W. W. (Adv. Electron. Electron Phys. **8** [1956] 89/185).
[4] Müller, E. W. (Ergeb. Exakt. Naturwiss. **27** [1953] 290/360).
[5] Good, R. H., Jr.; Müller, E. W. (in: Flügge, S.; Handbuch der Physik, Vol. 21, Springer, Berlin-Göttingen-Heidelberg 1956, pp. 176/231).
[6] Drechsler, M. (Surf. Sci. **70** [1978] 1/18).
[7] Swanson, L. W.; Bell, A. E. (Adv. Electron. Electron Phys. **32** [1973] 193/309).
[8] Melmed, A. J. (Adv. Mater. Res. **1** [1967] 103/43).
[9] Gadzuk, J. W.; Plummer, E. W (Rev. Mod. Phys. **45** [1973] 487/548).
[10] Modinos, A. (Surf. Sci. **70** [1978] 52/91, 67/85).

5.2.2 Total Energy Distributions

Total energy distributions (TEDs) of field-emitted electrons at low and elevated temperatures are reviewed in this section for various surfaces of W. To emphasize deviations from free-electron behavior, enhancement factors R(ε) are derived from TEDs by dividing the measured energy distribution j(ε) by the distribution $j_0(\varepsilon)$ expected for free electrons ($\varepsilon = E - E_F$ with E_F being the Fermi energy). Results concerning high- and low-energy tails in TEDs are summarized at the end of this section for all W surfaces studied.

W(110). TEDs of field-emitted electrons, measured between 77 and 1044 K at applied fields of F=0.37 V/Å, are in rough agreement with the predictions of the free-electron model [1]; see also [2]. TEDs were recorded at 10^{-10} Torr for various fields at a resolution of 0.19±0.03 and 0.1±0.02 eV, using retarding field and velocity selector analyzers, respectively [3]. The TED measured at 300 K under F=0.235 V/Å excellently agrees with the curve calculated for free-electron behavior and Φ=5.2 eV [3, 4]. TEDs at 405±10 K, measured at a field of 0.31 V/Å and a resolution of 0.07 eV using laser light for heating the emitter tip, are consistent with free-electron behavior [5]; see also the TEDs from W(110) measured at room temperature under F=0.18 V/Å [6] or F=0.31 V/Å [7], at 20 K [8], 15 K [9]; see also [10].

A feature in the TED at ε= −0.5 eV was found to be not very sensitive to contamination [11]. R(ε) curves derived from these TEDs (see also the TED from (110) plotted in [12]) are plotted in subsequent papers of this research group [13, 14] and, in a slightly extended version, in [15] (see **Fig. 30**). They show (compared with other surfaces) only weak structure.

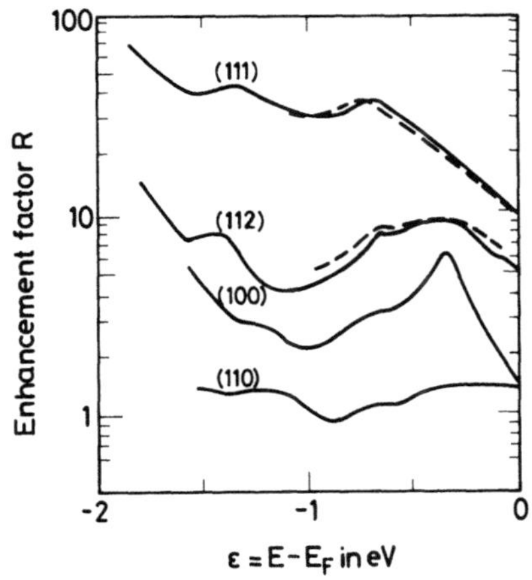

Fig. 30. The enhancement factor R(ε) for the low-index planes W(110), W(100), W(112), and W(111) at 78 K. The curves refer to free-electron distributions calculated assuming the respective work functions 5.25, 4.64, 4.90, and 4.45 eV and are arbitrarily shifted vertically in order to place all curves on one graph. The (110) and the (100) curves are for fields of ~0.33 and ~0.37 V/Å. Field effects are shown for the (112) and (111) curves: the dashed and solid (112) curves are for fields of ~0.29 and ~0.38 V/Å; the dashed and solid (111) curves are for fields of ~0.27 and ~0.33 V/Å [15].

An R(ε) curve derived from TEDs at 300 K shows no significant structure in the range $0 \geq \varepsilon \geq -0.6$ eV [16].

TEDs from W(110) were studied in [17] on a [110]-oriented field emitter between 300 and ~1435 K. The tip was heated by laser radiation focused ~15 μm off the apex at the shank on a spot ~4 μm in size. The enhancement factor appears to be temperature-independent up to ~1195 K (deviations observed at ~1435 K may be due to build-up and the resulting changes in the local field). There is a strong anomaly in the TED and the corresponding R(ε) curve from W(110): the enhancement factor sharply drops in the range 0.2 to 0.7 eV above E_F and remains approximately constant at higher ε values. TEDs at 1030±30 K at fields of 0.26 and 0.30 V/Å indicate this anomaly to be only weakly field-dependent. The anomalies observed are attributed to the band limit of a thermally populated, p-like electron band [17].

Free-electron-like behavior was predicted for W(110) by calculations [18, 19]; see also [20]. Extending these studies the TED was calculated for various bulk potentials and surface barriers. Calculated R(ε) curves are presented, some of which show considerable structure at energies above E_F [21]. A subsequent calculation using the bulk potential of [22] showed no structure in R vs. ε curves at energies below E_F [23].

W(211). TEDs from W(211) were measured between 78 and about 945 K [1] (see also [2]), at liquid H_2 temperatures [24] (see also [9]), at 78 K [13], [14, pp. 509/10] (for earlier results, see [25]). While the TEDs obtained by [1, 2, 24] agreed with the free-electron model over most of the energy range studied, considerable structure in the TEDs was observed by [11, 13, 14]. The corresponding R(ε) curves are plotted in [13, 14]. A subsequent study [15] of this research group showed some change in R(ε) upon variation of the applied field (see Fig. 30; note that two features at 0.69 and 1.50 eV below E_F are given explicitly in the earlier paper [11]). The R(ε) curve derived from TEDs at 300 K by [16] is in rough

References for 5.2.2 on pp. 147/9

agreement with the curve obtained by [13, 14]. In agreement with the results shown in Fig. 30, R(ε) curves derived from TEDs at 300 K at $p < 10^{-10}$ Torr and a field of 0.26 V/Å showed a shoulder at $\varepsilon = -0.62$ eV and a broad peak centered at -0.35 eV. At higher temperatures the shoulder shifts to higher ε values and smears into the broad peak [26].

W(100). The TED of field-emitted electrons from W(100) is shown in **Fig. 31** [11]. The so-called Swanson peak is a hump in the TED at $\varepsilon = -0.37$ eV [11] (-0.36 eV [27], -0.35 eV [1, 2, 8, 28, 29], -0.34 eV [5], or -0.33 eV [16] were reported in other studies; see also the TED in [25], that for 77 K under 0.403 V/Å in [30], or that for 15 K in [9]). The Swanson peak is extremely sensitive to contamination (by CO, N_2, H_2, O_2) and disappears at coverages between 1/2 and 1 monolayer [11]; see also the adsorption studies using Kr [9] and hydrogen [13]. It was first observed as a broad shoulder on the trailing edge of the TED with ⟨110⟩- and ⟨310⟩-oriented emitter tips between 77 and about 900 K and indicates deviations from free-electron behavior [1, 2]. The feature at $\varepsilon = -0.78$ eV (see Fig. 31) is much less sensitive to contamination [11]. R vs. ε plots derived from the TEDs of [11] are given in [13 to 15]; see Fig. 30, p. 143. R(ε) curves were derived from TEDs taken at 300 to 1025 K under $p < 10^{-10}$ Torr at a resolution of 0.06 eV. The FWHM of the peak at $\varepsilon = -0.37$ eV increases linearly with T at a rate of $(8.5 \pm 0.8) \times 10^{-5}$ eV/K which within the error limits coincides with the Boltzmann constant, suggesting thermal broadening [26].

Several workers [18, 19, 23, 31 to 39] calculated TEDs or R(ε) curves for W(100) and gave various explanations for the Swanson hump, e.g., band structure effect [18, 32] or surface resonance [19, 34, 35]. In contrast to earlier studies [18, 19, 35], all features in the experimental TED study of [11] could be well reproduced by the calculation of [23] using the bulk potential of [22]. The two high-energy peaks were attributed to bands of surface states/resonances, while the low-energy peak derives from bulk density of state effects [23].

The TED of field-emitted electrons from a W(100)-oriented, built-up emitter was measured at a resolution of ~0.05 eV between 78 and 1900 K and base pressures below 8×10^{-10} Torr [40]. The emission was confined to a cone with a half-angle of ~8° (for this emission characteristics, see [41, 42]). This TED did not show the Swanson hump. It did show some structure near $\varepsilon = -0.8$ eV [40] in agreement with the reports by [11] (see above). The lack of the Swanson hump may be due to the built-up tip being pyramidal with a very

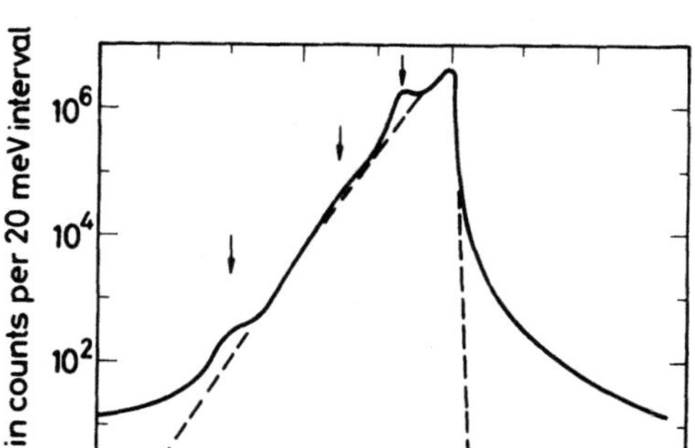

Fig. 31. Total energy distribution of field-emitted electrons from W(100) measured at 78 K (solid curve). Dashed lines represent the prediction of free-electron theory [11].

small topmost plane (<20 Å). For fields up to ~0.5 V/Å the FWHM of the TED measured at 84 K was close to that expected for free electrons, but at ~0.9 V/Å the FWHM was twice that value. Similar discrepancies occurred at higher temperatures, e.g. at 1685 K; see figures in the paper [40]; for earlier results see [43]. A combination of broadening due to the reduced lifetime of the tunneling electrons and collective Coulomb interactions has been tentatively invoked [40].

The TED data from [40] were analyzed by [44]; see also [45]. The current dependence of the energy broadening observed by [40] indicates electrostatic interactions in the high-intensity beam (the randomly emitted electrons experience random interactions that transform potential energy into kinetic energy, the so-called "Boersch effect") [44]. For the Boersch effect, see also [46].

TEDs from W(100) were measured at 1800 K at fields of 0.255 to 0.329 V/Å. Two humps at 0.9 and 2.4 eV above E_F were observed and interpreted in terms of band structure effects [69].

Interactions between s electrons and d holes and electron-phonon interactions are considered in the interpretation of the TED at 78 to 950 K from (100) surfaces [27].

W(111). TEDs from W(111) measured between 77 and 812 K could be well described by the free-electron model [1]. Over most of the energy range investigated, this behavior was also found in a study at liquid H_2 temperatures and above [24] (see also [9]) and in a study at 78 K [25]. A small bump is said to exist between 0.12 eV below E_F and E_F on the TEDs measured at 10 temperatures between 84 and 695 K over the energy range $-0.6\ \mathrm{eV} < \varepsilon < +0.35\ \mathrm{eV}$. Though very weak, this feature seems to be real, since it was not observed with other surfaces [47]; see also [48]. The TED measured at $\sim 3\times10^{-9}$ Torr using electrons emitted from a region ~30 Å in diameter showed a break in slope near $\varepsilon = -0.7$ eV. The contaminated tip had a much broader distribution and did not show this break [49]. Similarly, a break at $\varepsilon = -0.75$ eV was observed in TEDs at 77 K from a W(111) thermal end-form tip as well as from a built-up tip, the apex of which showed a three-atom microfacet [50].

Considerable structure in TEDs from W(111) was observed by [11] (see also the TED at 78 K plotted in [51, 52], [14, p. 539]). The $R(\varepsilon)$ curves derived from this TED are plotted in [13 to 15]; see Fig. 30, p. 143. The feature at about 0.7 eV below E_F was found to be not very sensitive to contamination [11]. The $R(\varepsilon)$ curve derived from TEDs at 300 K by [16] roughly agrees with the curve obtained by [13, 14]. The $R(\varepsilon)$ curve derived from TEDs taken at 300 K and $p < 10^{-10}$ Torr at a resolution of 0.06 eV showed a peak at $\varepsilon = -0.69$ eV which shifts to higher energy and decreases in height as the temperature is increased from 450 to 850 K [26]. For a TED and $R(\varepsilon)$ curve obtained at ~1190 K, see [17].

Field emission electron spectroscopy from "single-atom tips" was compared with emission from macroscopic tips having radii of curvature less than 100 nm [53]. The tips were ⟨111⟩-oriented W tips on which by thermo-field treatment single-atom protrusions formed. The macroscopic tip showed single-peaked TEDs which are known to be generated on clean ⟨111⟩ W tips (see [49] above). Tips with single-atom protrusions, on the other hand, showed doubly and triply peaked TEDs at about half the voltage applied on macroscopic tips. The peaks shift linearly to lower ε values with applied field emission voltage. The FWHM of these (single-atom tip) TED peaks is ~0.24 eV. The leading peak (at the Fermi edge) and the other peaks could *not* be fitted with the expression for the tunneling current from a free-electron metal through a triangular barrier. The peak shapes were, however, fitted by Lorentzians. A model is proposed according to which the emission occurs through

References for 5.2.2 on pp. 147/9

the broadened energy levels of the nonmetallic W atom on top of the tip [53]. For the controversy raised by this paper, see [70, 71].

W(210). The TED from (210) has been measured near and above 20 K at a resolution of 0.025 to 0.04 eV. Free-electron behavior was observed over most of the energy range studied. The experimental half width of the TED was 0.178 eV at 250 K, the predicted value being 0.168 eV [24]; see also [8]. Only small deviations from free-electron behavior have been noted between 78 and 950 K [54]. But the total currents observed between 140 and 760 K in the TED and the thermal FE currents at $p<1.5\times10^{-10}$ Torr were smaller than expected for free electrons [28]. For the TED from the (210) region of a ⟨110⟩-oriented tip at 78 K, see [55]. Old data at estimated 500 K are given in [56].

W(310). The TEDs from W(310) measured between 77 and 929 K at an applied field of F=0.306 V/Å can be well described by the free-electron model [1]; see also [2]. The intrinsic width of the TED at room temperature was measured to be 169±1 meV [57]. [11] found features in TEDs at $\varepsilon=-0.39$ and -1.45 eV. The R(ε) curve given in [16], derived from TEDs taken at 300 K, showed a peak at $\varepsilon\approx-0.35$ eV.

The FE currents from W(310) were smaller than those expected for free electrons [28]. Interactions between s electrons and d holes and also between electrons and phonons were considered to explain the TEDs at 78 to 950 K [27]. A TED was obtained at F=0.229 V/Å from the region around (310) [58]. For TEDs recorded at various fields, see [3].

Other Surfaces. The TEDs of FE and PFE electrons from the region around W(510) were determined at F=0.26 V/Å [59]. No anomalous behavior was noted at 300 and 665 K for W(510) in the TED study [26] performed under $p<10^{-10}$ Torr at a resolution of 0.06 eV [26]. The study [54] (confined to (210)) was extended to measurements of the TED at 78 to 950 K from (610) surfaces. Interactions between s electrons and d holes and between electrons and phonons were considered [27]. The FE currents from W(320) were smaller than those expected for free electrons [28]. TEDs from W(611) measured between 77 and 918 K are consistent with free electron behavior [1]; see also [2]. The R(ε) curves determined at 8 different positions along the [110] zone are compared in [14, p. 514].

"Average Tip" Studies. The TED from a tip slightly contaminated by carbon was measured at 1570 K and 10^{-11} Torr at fields F estimated to range from 0.08 to 0.4 V/Å [60]. The results closely agree with a free-electron theory. There are exponentially decaying tails at the low- and high-energy ends of the distribution, corresponding to the tunneling probability for electrons with energies near E_F and to the Boltzmann tail of the Fermi distribution, respectively. The intermediate region is consistent with the barrier model $V(z)=-e^2/(4z)-eFz$ [60].

The half width of the energy distribution of electrons emitted at 1500 K to 2100 K from a W whisker was wider than that expected for pure FE or pure thermionic emission. Half width vs. T curves measured at current densities of 53 and 150 A/cm² are compared with theoretical curves [61].

For early measurements of TEDs with a retarding potential analyzer, see [62]. A classical TED study from W tips at 21, 77, and 300 K, using an improved retarding potential analyzer capable of measuring the narrow width of the distribution [63], and the companion theoretical paper [64] showed that the observed quantity corresponds to total energy distributions and not to normal energy distributions as tacitly assumed in older studies.

High- and Low-Energy Tails in TEDs. High-energy tails extending to ε regions ($\varepsilon=E-E_F$), where free-electron theories predict negligible emission, have been observed by several workers in TEDs from W surfaces (see e.g. Fig. 31, p. 144). Various explanations have been given, but the origin of these tails seems to be not entirely clear yet.

The TED of field-emitted electrons was studied on W(310) and W(111) surfaces, obviously at room temperature [65]. In agreement with earlier studies [24, 51], the current density j(ε) in the tail was found to be proportional to the square of the total FE current density, J^2. For 0.4 eV<ε<4 eV, the data showed a $j \propto \varepsilon^{-2.6}$ behavior. At higher energies, there are deviations being the stronger the higher the energy is. Even at ε=16 eV, a significant signal was found. Measurements on W(111) showed that raising T raised j(ε): between 300 and 720 K j(3.18 eV) increased by 43±4% (the square of the total FE current increased by 40±1%). All these findings are in agreement with a Coulomb scattering mechanism, i.e., the tails arise from electron–electron interactions in the vacuum near the field emitter; internal mechanisms that have previously been proposed (e.g. by [51], see below) do not play a significant role [65]; see also [66]. For a discussion of the broadening of the TED observed at high electric field and also explained by this mechanism, see p. 145.

Hot hole-electron cascades have been invoked to explain the high-energy tail of the TED from W(111) at 78 K. The intensity of the high-energy tail of the FE current depends on J^2 for 0.1 eV<ε<4 eV. This tail was proposed to result from steady-state cascade processes occurring by electron-hole production in the conduction band. The holes decay by Auger-type processes [51].

The current in the high-energy tail at ε>0.1 eV after subtraction of the Boltzmann tail was found to increase nearly with the square of the total current in the room-temperature TED from W(111) [67]; see also [68]. The current density j(ε>0) and i*, the current in the high-energy tail of the TED, are proportional to I^2 for (211), (111), and (210), the three orientations investigated in [24] (I denotes the total probe-hole current). The I^2 dependence of i* indicates electron-electron scattering, but the low total currents used ($I<10^{-8}$ A) are considered to rule out electron–electron scattering on the vacuum side of the barrier. Scattering in the metal was thought to produce a pair of electrons with energies $E_F+\Delta E$ and $E_F-\Delta E$. Thus, a two-electron tunneling mechanism was proposed [24]. High-energy tails in the TEDs from W(100) [11] and W(210) [55] increased with increasing current density J.

The low-energy tails in TEDs (observed e.g. on W(100) [11], W(311) [65], and W(111) [51]) have been studied in much less detail. These tails also increase with current density J [11, 51]. The results obtained by [65] suggest that the Coulomb scattering mechanism is valid also for these tails.

References for 5.2.2:

[1] Swanson, L. W.; Crouser, L. C. (Phys. Rev. [2] **163** [1967] 622/41, 632/36).
[2] Swanson, L. W.; Crouser, L. C. (Phys. Rev. Lett. **16** [1966] 389/92).
[3] Haavig, D. L. (Diss. Purdue Univ. 1983, 176 pp.; Diss. Abstr. B **44** [1984] 3845).
[4] Haavig, D. L.; Reifenberger, R. (Surf. Sci. **151** [1985] 128/42).
[5] Gao, Y.; Reifenberger, R. (J. Vac. Sci. Technol. [2] A **4** [1986] 1289/93).
[6] Gao, Y.; Reifenberger, R.; Kramer, R. M. (J. Phys. E **18** [1985] 381/3).
[7] Gao, Y.; Reifenberger, R. (Phys. Rev. [3] B **35** [1987] 8301/7).
[8] Young, P. L.; Gomer, R. (Phys. Rev. Lett. **30** [1973] 955/8).
[9] Lea, C.; Gomer, R. (J. Chem. Phys. **54** [1971] 3349/59).
[10] Stepień, Z.; Lenkow, W. (Pr. Nauk Inst. Technol. Elektron. Politech. Wroclav. No. 18 [1979] 437/42; C.A. **92** [1980] No. 86701).
[11] Plummer, E. W.; Gadzuk, J. W. (Phys. Rev. Lett. **25** [1970] 1493/5).

[12] Ehrlich, C. D.; Plummer, E. W. (Phys. Rev. [3] B **18** [1978] 3767/71).
[13] Plummer, E. W.; Bell, A. E. (J. Vac. Sci. Technol. **9** [1972] 583/90).
[14] Gadzuk, J. W.; Plummer, E. W (Rev. Mod. Phys. **45** [1973] 487/548).
[15] Vorburger, T. V.; Penn, D.; Plummer, E. W. (Surf. Sci. **48** [1975] 417/31).
[16] Ivanov, V. A.; Kirsanova, T. S.; Tumareva, T. A. (Fiz. Tverd. Tela [Leningrad] **23** [1981] 664/8; Sov. Phys.-Solid State [Engl. Transl.] **23** [1981] 377/9).
[17] Donders, P. J.; Lee, M. J. G. (Surf. Sci. **175** [1986] 197/214).
[18] Nicolaou, N.; Modinos, A. (Phys. Rev. [3] B **11** [1975] 3687/96).
[19] Modinos, A.; Nicolaou, N. (Phys. Rev. [3] B **13** [1976] 1536/47).
[20] Modinos, A. (Surf. Sci. **70** [1978] 52/91, 67/85).

[21] Modinos, A.; Oxinos, G. (J. Phys. C **14** [1981] 1373/80).
[22] Posternak, M.; Krakauer, H.; Freeman, A. J.; Koelling, D. D. (Phys. Rev. [3] B **21** [1980] 5601/12).
[23] Modinos, A. (Field, Thermionic, and Secondary Electron Emission Spectroscopy, Plenum, New York 1984, pp. 141/8).
[24] Lea, C.; Gomer, R. (Phys. Rev. Lett. **25** [1970] 804/6).
[25] Plummer, E. W.; Young, R. D. (Phys. Rev. [3] B **1** [1970] 2088/109, 2100/2).
[26] Gaudin, G. A.; Lee, M. J. G. (Surf. Sci. **201** [1988] 540/58).
[27] Czyżewski, J. J. (Surf. Sci. **39** [1973] 1/20).
[28] Wysocki, J. K. (Phys. Rev. [3] B **28** [1983] 834/41).
[29] Gao, Y.; Reifenberger, R. (Phys. Rev. [3] B **35** [1987] 4284/90).
[30] Swanson, L. W.; Crouser, L. C. (Phys. Rev. Lett. **19** [1967] 1179/81).

[31] Gadzuk, J. W. (Phys. Rev. [2] **182** [1969] 416/26).
[32] Nagy, D.; Cutler, P. H. (Phys. Rev. [2] **186** [1969] 651/6).
[33] Duke, C. B.; Fauchier, J. (Surf. Sci. **32** [1972] 175/204).
[34] Gadzuk, J. W. (J. Vac. Sci. Technol. **9** [1972] 591/6).
[35] Nicolaou, N.; Modinos, A. (Surf. Sci. **60** [1976] 527/39).
[36] Kar, N.; Soven, P. (Solid State Commun. **20** [1976] 977/9).
[37] Soven, P.; Plummer, E. W.; Kar, N. (Chem. Phys. Solid Surf. **1977** 27/47).
[38] Kar, N. (Surf. Sci. **70** [1978] 101/13).
[39] Dalidchik, F. I. (Zh. Eksp. Teor. Fiz. **87** [1984] 1384/99; Sov. Phys.-JETP [Engl. Transl.] **60** [1984] 795/803).
[40] Bell, A. E.; Swanson, L. W. (Phys. Rev. [3] B **19** [1979] 3353/64).

[41] Swanson, L. W.; Martin, N. A. (J. Appl. Phys. **46** [1975] 2029/50, 2035).
[42] Swanson, L. W.; Crouser, L. C. (J. Appl. Phys. **40** [1969] 4741/9).
[43] Swanson, L. W. (J. Vac. Sci. Technol. **12** [1975] 1228/33, 1230).
[44] Knauer, W. (Optik [Stuttgart] **59** [1981] 335/54, 349).
[45] Knauer, W. (J. Vac. Sci. Technol. **16** [1979] 1676/9).
[46] Dietrich, W. (Z. Phys. **152** [1958] 306/13).
[47] Wysocki, J. K. (J. Phys. E **15** [1982] 1376/85, 1379/81).
[48] Wysocki, J. K. (Surf. Sci. **137** [1984] 506/14).
[49] Isaacson, M.; Gomer, R. (Appl. Phys. **15** [1978] 253/6).
[50] Vu Thien Binh; Marien, J. (Surf. Sci. **202** [1988] L 539/L 549).

[51] Gadzuk, J. W.; Plummer, E. W. (Phys. Rev. Lett. **26** [1971] 92/5).
[52] Kuyatt, C. E.; Plummer, E. W. (Rev. Sci. Instrum. **43** [1972] 108/11).
[53] Vu Thien Binh; Purcell, S. T.; Garcia, N.; Doglioni, J. (Phys. Rev. Lett. **69** [1992] 2527/30).
[54] Czyżewski, J. J. (Surf. Sci. **33** [1972] 589/606).
[55] Workowski, C. J. (J. Phys. E **13** [1980] 67/73).
[56] Müller, E. W. (Z. Phys. **120** [1943] 261/9).

[57] Venus, D.; Lee, M. J. G. (Rev. Sci. Instrum. **56** [1985] 1206/11).
[58] Lee, J. M. G. [M. J. G.] (Phys. Rev. Lett. **30** [1973] 1193/6).
[59] Reifenberger, R.; Goldberg, H. A.; Lee, M. J. G. (Surf. Sci. **83** [1979] 599/616, 613/5).
[60] Gadzuk, J. W.; Plummer, E. W. (Phys. Rev. [3] B **3** [1971] 2125/9).
[61] Zaima, S.; Sase, M.; Adachi, H.; Shibata, Y. (J. Phys. D **11** [1978] L 179/L 180).
[62] Henderson, J. E.; Dahlstrom, R. K. (Phys. Rev. [2] **55** [1939] 473/81).
[63] Young, R. D.; Müller, E. W. (Phys. Rev. [2] **113** [1959] 115/20).
[64] Young, R. D. (Phys. Rev. [2] **113** [1959] 110/4).
[65] Donders, P. J.; Lee, M. J. G. (Phys. Rev. [3] B **35** [1987] 6578/87).
[66] Donders, P. J.; Lee, M. J. G. (Phys. Rev. [3] B **41** [1990] 1781/8).
[67] Bakhtizin, R. Z.; Yumaguzin, Yu. M. (Phys. Status Solidi B **150** [1988] 103/8).
[68] Bakhtizin, R. Z.; Yumaguzin, Yu. M. (Poverkhnost' **1987** No. 7, pp. 51/5; C.A. **107** [1987] No. 145693).
[69] Yasuhari Hirai (Surf. Sci. **287/288** [1993] 605/8).
[70] Ernst, N.; Unger, J.; Fink, H.-W.; Grunze, M.; Müller, H. U.; Völkl, B.; Hofmann, M.; Wöll, C. (Phys. Rev. Lett. **70** [1993] 2503).
[71] Vu Thien Binh; Purcell, S. T.; Garcia, N. (Phys. Rev. Lett. **70** [1993] 2504).

5.2.3 Fowler-Nordheim Characteristics

The linearity of an $\ln(j/F^2)$ vs. 1/F plot (where j denotes the current density and F the electric field), also known as Fowler-Nordheim (F-N) plot, is commonly regarded as adequate proof that the emission is due to stable field emission. Such linear F-N plots have been experimentally obtained by many authors on various tungsten surfaces (see e.g [1, 2]), and in most cases work functions have been derived from the slope of these plots (see Chapter 4, p. 107). In this section some selected data as well as studies emphasizing the conditions under which deviations from Fowler-Nordheim behavior occur are discussed.

The absolute tunneling (FE) current density from W(110) was measured at room temperature and base pressures of 10^{-10} Torr using a probe-hole apparatus. The F-N data, $\ln(j/U^2)$ vs. 1/U plots, were analyzed in terms of a triangle potential and a triangle-plus-image potential. Good agreement between the measured data and free-electron behavior was found for the two models [3]; see also [4].

F-N characteristics have been recorded for an average tip and for various W faces [5, 6]. Nonlinearities in the F-N curves are noted for both the very low and very high current ends of the studied range covering 10^{-19} to 10^{-1} A. In this context, the problem was addressed of averaging exponentials (the F-N curve for the whole tip taken to be an average is not expected to follow a typical F-N dependence) [6]. Probe-hole currents from W(110) have been measured on W tips subjected to heating (at 2100 K)-quench-FE cycles. The current values observed scattered more than was compatible with the scatter in the Φ values determined from F-N plots. The result might suggest variations in the frozen-in disorder of the surface [7].

The F-N plots recorded at $<10^{-12}$ Torr of chemically active gases are straight lines up to current densities of typically (depending on the emitter used) $j=6\times10^6$ A/cm^2. At higher j values (e.g., $j>10^7$ A/cm^2 [8]) the data fell below the F-N line. This was attributed to space charge effects [8, 9]; see also [10] and the review [11]. The deviations from the F-N law were explained by the shape of the potential barrier to be tunneled through [12]. On the other hand, the results given in [8, 9] were criticized for being due to inappropriate electronic circuits and the measured data were found to fall above the F-N curve rather

References for 5.2.3 on pp. 150/1

than below it [13]. With respect to the results published by [8], only slight but significant discrepancies were noted with pulsed field emission at 10^{-11} Torr by [14] who found the F–N plots to be linear up to $j = (1 \text{ to } 3) \times 10^6$ A/cm²; at higher j values again a *downward* deviation from linearity was noticed [14]. In agreement with the results of [8], downward deviations from straight line behavior of log j vs. 1/U curves were observed at high current densities. The deviations showed an oscillatory behavior [25, 26] similar to that observed in thermionic emission (see p. 137) [26].

The F–N plots from (100) built-up tips are straight lines [15]. Similarly, linear F–N plots were obtained from (111) built-up tips [16]. On the other hand, microtips having a monoatomic apex, produced from built-up tips by an additional oxygen treatment, showed strong deviations from the straight F–N line (the current increase with applied field was smaller than that expected for F–N behavior) [16, 17]; see also [18]. These deviations were interpreted in terms of space charge effects [16, 18], but can also be attributed to the particular geometry of the tip [17]. F–N characteristics were studied on W tips remolded by thermal treatment in high electric fields [19, 20]. The remolding process replaces the smooth, round shape of the thermally annealed tip by sharp edges. The concomitant local field enhancement decreases the slope of the F–N curve [20]. The F–N plots for tips on which microprotrusions had formed at the edges of the tip during the remolding process indicated that high local current densities can be achieved (the average current density was $\sim 6 \times 10^6$ A/cm²). Linear F–N plots were obtained for emission from microprotrusions up to current densities of 5×10^8 A/cm² (deviations from linearity at higher j values may be due to space charge effects). Steady-state j values up to 10^{10} A/cm² were reached at the corners and outgrowths of remolded tips in the $\{311\}$ and $\{111\}$ regions. The emission is stable and no breakdown occurs. The removal of the microprotrusions by high j values does not destroy the tip. This makes it unlikely that thermofield-produced microprotrusions initiate breakdown in vacuum ($\sim 10^{-9}$ Torr) [19]. For F–N plots obtained with pulsed field emission (pulse duration: 2×10^{-8} to 1.1×10^{-3} s) on remolded tips see [21]. The FE current measured on many microneedles protruding from a 10 µm diameter W filament at $\sim 10^{-8}$ Torr obeyed the F–N relationship [22].

Plots of log j vs. 1/U are presented for emitters operated at $p < 10^{-9}$ Torr with voltage pulses applied for 7 and 10 ns. At high current densities ($j \approx 10^9$ A/cm²), the log j vs. 1/U curve bends downwards [23]. The maximum FE current density j a tip can stand without damage depends strongly on the tip radius r. With µs pulses and for r = 1 to 3 µm, $j_{max} \approx (1 \text{ to } 2) \times 10^7$ A/cm², while with ns pulses, j_{max} is about one order of magnitude greater. At high j values the FE is buffered by space charge effects, causing the measured values to deviate below the F–N curve [24].

References for 5.2.3:

[1] Müller, E. W. (J. Appl. Phys. **26** [1955] 732/7).

[2] Hellwig, S.; Block, J. H. (Z. Phys. Chem. [Munich] **83** [1973] 269/86).

[3] Ehrlich, C. D.; Plummer, E. W. (Phys. Rev. [3] B **18** [1978] 3767/71).

[4] Ehrlich, C. D. (Diss. Univ. Pennsylvania 1979, 154 pp.; Diss. Abstr. B **40** [1979] 2732).

[5] Gofman, I. I. (Dokl. Akad. Nauk Uzb. SSR **19** No. 6 [1962] 26/8; C.A. **58** [1963] 6284).

[6] Gofman, I. I. (Fiz. Tverd. Tela [Leningrad] **4** [1962] 2005/14; Sov. Phys.-Solid State [Engl. Transl.] **4** [1962/63] 1471/7).

[7] Stepień, Z. M.; Lenkow, W.; Kukułka, J. (Acta Univ. Wratislav. Mat. Fiz. Astron. No. 48 [1986] 207/11; C.A. **105** [1986] No. 70563).

[8] Dyke, W. P.; Trolan, J. K. (Phys. Rev. [2] **89** [1953] 799/808).

[9] Barbour, J. P.; Dolan, W. W.; Trolan, J. K.; Martin, E. E.; Dyke, W. P. (Phys. Rev. [2] **92** [1953] 45/51).

[10] Dyke, W. P.; Trolan, J. K.; Martin, E. E.; Barbour, J. P. (Phys. Rev. [2] **91** [1953] 1043/54).

[11] Dyke, W. P.; Dolan, W. W. (Adv. Electron. Electron Phys. **8** [1956] 89/185, 109/11).
[12] Lewis, T. J. (Proc. Phys. Soc. [London] B **68** [1955] 938/43).
[13] Gofman, I. I.; Protopopov, O. D.; Shuppe, G. N. (Fiz. Tverd. Tela [Leningrad] **2** [1960] 1323/7; Sov. Phys.-Solid State [Engl. Transl.] **2** [1960] 1203/8).
[14] Buder, E. E. (Diss. St. Louis Univ. 1974, 227 pp. from Diss. Abstr. B **36** [1975] 2881).
[15] Tamura, N. (Scanning Electron Microsc. **1979** No. 1, pp. 31/8; C.A. **91** [1979] No. 112044).
[16] Vu Thien Binh; Marien, J. (Surf. Sci. **202** [1988] L 539/L 549).
[17] Sáenz, J. J.; García, N.; de Raedt, H.; Vu Thien Binh (J. Phys. Colloq. [Paris] **50** [1989] C8-73/C8-78).
[18] Golubok, A. O.; Masalov, S. A.; Tarasov, N. A. (Ultramicroscopy B **42/44** [1992] 1574/9).
[19] Pavlov, V. G.; Rabinovich, A. A.; Shrednik, V. N. (Zh. Tekh. Fiz. **45** [1975] 2126/34; Sov. Phys.-Tech. Phys. [Engl. Transl.] **20** [1975/76] 1337/41).
[20] Golubev, O. L.; Shaikhin, B. M.; Shrednik, V. N. (Izv. Akad. Nauk SSSR Ser. Fiz. **40** [1976] 1599/604; Bull. Acad. Sci. USSR Phys. Ser. [Engl. Transl.] **40** No. 8 [1976] 40/3).

[21] Zhukov, V. M.; Polezhaev, S. A.; Varekha, L. M. (Radiotekh. Elektron. [Moscow] **33** [1988] 2153/62; Sov. J. Commun. Technol. Electron. **34** [1989] 121/8; C.A. **110** [1989] No. 17192).
[22] Okuyama, F.; Aoyagi, M.; Kitai, T.; Ishikawa, K. (J. Vac. Sci. Technol. **15** [1978] 98/102).
[23] Fursei, G. N.; Antonov, A. A.; Gulin, B. F. (Vestn. Leningr. Univ. Ser. 4 Fiz. Khim. **1971** No. 2, pp. 71/4; C.A. **75** [1971] No. 145128).
[24] Grishanov, B. I.; Zhukov, V. M.; Polezhaev, S. A.; Fursei, G. N.; Chibukov, Yu. Ya.; Yasnov, G. I. (Radiotekh. Elektron. [Moscow] **23** [1978] 575/83; Radio Eng. Electron. Phys. [Engl. Transl.] **23** [1978] 84/90).
[25] Sokol'skaya, I. L.; Fursei, G. N. (Radiotekh. Elektron. [Moscow] **7** [1962] 1474 ff.; Radio Eng. Electron. Phys. [Engl. Transl.] **7** [1962] 1395/403).
[26] Fursei, G. N. (Radiotekh. Elektron. [Moscow] **6** [1961] 298/302; Radio Eng. Electron. Phys. [Engl. Transl.] **6** [1961] 169/77).

5.2.4 Single-Electron vs. Multiple-Electron Tunneling

This section deals with studies aimed at detecting multiple-electron tunneling events in field emission from W; these have been proposed to explain unusual tails in energy distributions; see pp. 146/7. The topic has recently been reviewed by [1].

After improving experimentally a preliminary study [2], probe-hole measurements [3] on various W faces at 5×10^{-10} Torr and 77 to 1000 K gave no indication for multiple-electron tunneling events (pair tunneling etc.). However, the field emission statistics spectra did show multiple peaks after inserting auxiliary acceleration electrodes into the apparatus [3]. This indicates that earlier results [4] suggesting considerable contributions from multiple-electron events were falsified by secondary electron emission [3]. The statistics of FE from W has been studied at liquid He temperatures at $p < 10^{-12}$ Torr for various emission directions. Within an error limit of 0.1%, all emission processes were of the one-particle-at-a-time type [5]. Electron pairs did not significantly contribute to the tunneling current from W(100) surfaces at current densities of $j = 2 \times (10^{-6}$ to $10^{2})$ A/cm^2. The single-electron character of the FE statistics was established with an accuracy of 99% even at ~ 100 A/cm^2 [6]; see also [7].

References for 5.2.4 on p. 152

References for 5.2.4:

[1] Fursey [Fursei], G. N.; Kocheryzhenkov, A. V.; Maslov, V. I. (Surf. Sci. **246** [1991] 365/72, 371).
[2] Fursei, G. N.; Mokhasne, M. M.; Egorov, N. V.; Ponomarev, V. S.; Shchemelev, V. N. (Fiz. Tverd. Tela [Leningrad] **18** [1976] 631/2; Sov. Phys.-Solid State [Engl. Transl.] **18** [1976] 368/9).
[3] Fursei, G. N.; Egorov, N. V.; Kocheryzhenkov, A. V. (Pis'ma Zh. Tekh. Fiz. **7** [1981] 798/801; Sov. Tech. Phys. Lett. [Engl. Transl.] **7** [1981] 343/5).
[4] Gazier, C. (Phys. Lett. A **35** [1971] 243/4).
[5] Kocheryzhenkov, A. V.; Maslov, V. I.; Fursei, G. N. (Fiz. Tverd. Tela [Leningrad] **29** [1987] 2471/2; Sov. Phys.-Solid State [Engl. Transl.] **29** [1987] 1421/2).
[6] Maslov, V. I.; Fursei, G. N.; Kocheryzhenkov, A. V. (Zh. Tekh. Fiz. **59** No. 10 [1989] 164/7; Sov. Phys.-Tech. Phys. [Engl. Transl.] **34** [1989] 1192/4).
[7] Maslov, V. I.; Fursey [Fursei], G. N.; Kocheryshenkov [Kocheryzhenkov], A. V. (J. Phys. Colloq. [Paris] **50** [1989] C8-113/C8-116).

5.2.5 Energy Exchange Effects

Energy exchange phenomena associated with field electron emission have been reviewed by [1]. In an early study [2], the energy losses of a W cathode were measured with a Ta-W thermocouple for FE and thermionic emissions. While thermionic emission gave 4.46±0.09 eV per electron, no measurable energy loss was found for FE at room temperature consistent with the view that in FE the electrons tunnel through the barrier

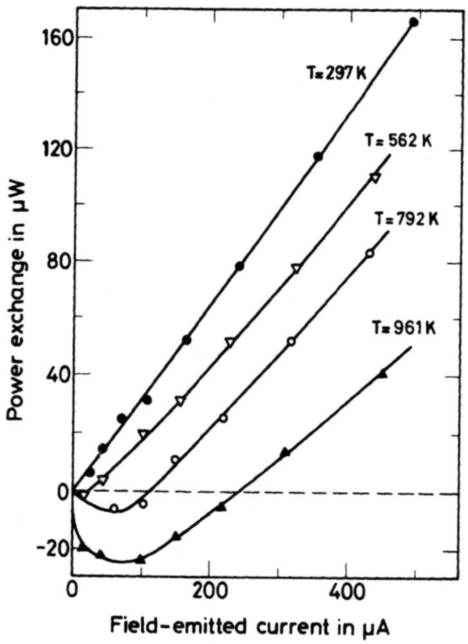

Fig. 32. The experimentally determined power exchange at the emitter as a function of the field-emitted current at four temperatures as indicated. Positive and negative power exchange values indicate emission heating and emission cooling of the tip, respectively [8].

and do not surmount the surface potential [2]. Several minor flaws in the theoretical expressions given in [2] were acknowledged in [3] in response to the criticism expressed by Nottingham [4]. The cooling and heating of thin W wires by FE was studied at 10^{-7} Torr. At a fixed field, the heating observed at low temperatures changes to cooling with increasing temperature (at the inversion temperature T*) [5] ("Nottingham effect", see [4]). There is fair agreement between the measured heats of electron evaporation (both positive and negative) [5] and the calculations by Levine [6]. T* was found to increase from 1092 K at 0.479 V/Å to 1360 K at 0.542 V/Å [7]. Much lower inversion temperatures (e.g. 550 K at 0.46 V/Å) were observed in a subsequent investigation [8]. **Fig. 32** shows that the Nottingham inversion (i.e., the zero heat effect) occurs at increasingly higher emitter currents (or fields) as the emitter temperature is increased. Cathode cooling occurs whenever the average energy $\langle E \rangle$ of the emitted electrons is higher than $\langle E' \rangle$, the average energy of the replacement electrons supplied from the Fermi sea [8] (see also [7]). A comparison of experimental and theoretical results indicates significant discrepancies which may be associated with the large contribution of holes to the electrical conduction of W [8]. The results of energy distribution measurements on various W surfaces were discussed in view of the Nottingham effect by [9, 10].

References for 5.2.5:

[1] Swanson, L. W.; Bell, A. E. (Adv. Electron. Electron Phys. **32** [1973] 193/309, 242/52).
[2] Fleming, G. M.; Henderson, J. E. (Phys. Rev. [2] **58** [1940] 887/94).
[3] Fleming, G. M.; Henderson, J. E. (Phys. Rev. [2] **59** [1941] 907/8).
[4] Nottingham, W. B. (Phys. Rev. [2] **59** [1941] 906/7).
[5] Drechsler, M.; Jähnig, G. (Z. Naturforsch. **18a** [1963] 1367/9).
[6] Levine, P. H. (J. Appl. Phys. **33** [1962] 582/7).
[7] Charbonnier, F. M.; Strayer, R. W.; Swanson, L. W.; Martin, E. E. (Phys. Rev. Lett. **13** [1964] 397/401).
[8] Swanson, L. W.; Crouser, L. C.; Charbonnier, F. M. (Phys. Rev. [2] **151** [1966] 327/40).
[9] Czyżewski, J. J. (Surf. Sci. **33** [1972] 589/606).
[10] Czyżewski, J. J. (Surf. Sci. **39** [1973] 1/20).

5.2.6 Field Emission in Magnetic Fields. Polarized Electron Emission

Very controversial results have been reported for the changes produced by an external magnetic field on field emission currents from tungsten. A steady decrease in the emission current was observed in longitudinal magnetic fields up to B = 1.5 T at 300 K. On cooling, oscillations in FE current vs. B plots developed, the amplitudes of which reached several % at 78 K. The F-N plots for FE in the presence of magnetic fields showed breaks [1]. This paper is criticized by [2] for reporting an artefact caused by adsorbed layers, since no changes in FE current were observed upon applying a magnetic field [2] (in agreement with the earlier paper [3]). However, the later study [4] also showed the FE current density j (above a critical field of about 0.2 T) to decrease with B together with oscillatory variations ($\sim$10%) up to B = 2 T. The very rich structure in j vs. B plots observed with W tips at He temperatures and below 10^{-10} Torr was reproducible for the same tip, but varied between tips [4].

In contrast to these results, the presence of longitudinal magnetic fields of B = 0.04 to 0.4 T was observed to *increase* the FE current density by up to an order of magnitude (at $p < 2 \times 10^{-10}$ Torr) in [5]. After switching off the B field, the tips showed a high-emission memory effect. This memory decayed at 800 K within tens of s and at 1800 to 2000 K within less than 1 s. The B field alone did not produce the high-emission state, since,

References for 5.2.6 on p. 154

in addition, an FE current must be emitted when the field is applied [5]. Similarly, the FE current density j(B) of a ⟨100⟩-oriented W tip was measured by pulsed FE at $p<5\times10^{-10}$ Torr [6, 7]. The ratio j(B)/j(B=0) peaked sharply near 0.125 T, the ratio j(0.125 T)/j(0) being ~10 and ~19 for 300 and 77 K, respectively [6]. The ratio j(B)/j(0) was measured between 77 and 850 K and peaked near B=0.1, 0.22, and 0.34 T. The ratio j(B=0.1 T)/j(0) decreased with temperature up to 400 K and remained nearly constant at higher temperatures [7]. Operation of the FEM in the B range, where j(B)/j(0) peaked, resulted in irreversible FEM pattern changes indicating sample deformation. Near the current peaks (j vs. B), current oscillations with time in the 1 to 100 kHz range have been observed. The oscillations started at current densities of $j=(1 \text{ to } 2)\times10^5$ A/cm^2 and stopped when j increased to $(3 \text{ to } 5)\times10^5$ A/cm^2 (simultaneously cutting off the FE current for the duration of the particular voltage pulse) [6].

FE from W in the absence of a magnetic field yields unpolarized electrons [8]. Polycrystalline and single crystalline emitter tips were used near 80 K at base pressures below 10^{-8} Torr, and pulsed B fields (B=0.3 to 2.5 T; pulse duration: 0.13 ms, repetition rate: 5 Hz) were applied parallel and antiparallel to the tip. Under these conditions, spin-polarized emission occurred as demonstrated by Mott scattering. The following table gives the maximum and most frequently observed degrees of polarization, P_{max} and P_{freq}, for various W surfaces [9]:

emission along	⟨110⟩	⟨100⟩	⟨111⟩	⟨polycr.⟩ [a]
P_{max} (±2.0%)	11.0	13.0	12.0	18.0
P_{freq} (±0.5%)	5.1	4.8	4.4	5.0

[a] Polycrystalline emitter, crystallites preferentially having ⟨110⟩ parallel to the wire axis.

References for 5.2.6:

[1] Buribaev, I.; Shishkin, B. B. (Fiz. Tverd. Tela [Leningrad] **12** [1970] 3309/11; Sov. Phys.-Solid State [Engl. Transl.] **12** [1970] 2678/9).
[2] Waites, R. F.; Schwettmann, H. A. (Phys. Rev. [3] B **8** [1973] 2420/5).
[3] Flood, D. J. (J. Phys. Chem. Solids **31** [1970] 1649/50).
[4] Kennedy, P. J.; Muir, A. Y. (Solid State Commun. **27** [1978] 279/81).
[5] Fursei, G. N.; Ptitsyn, V. E.; Egorov, N. V. (Pis'ma Zh. Tekh. Fiz. **5** [1979] 1161/4; Sov. Tech. Phys. Lett. [Engl. Transl.] **5** [1979] 486/7).
[6] Ptitsyn, V. E.; Fursei, G. N.; Egorov, N. V. (Pis'ma Zh. Eksp. Teor. Fiz. **31** [1980] 733/7; JETP Lett. [Engl. Transl.] **31** [1980] 692/5).
[7] Ptitsyn, V. E.; Fursei, G. N.; Egorov, N. V. (Pis'ma Zh. Tekh. Fiz. [1980] 619/22; Sov. Tech. Phys. Lett. [Engl. Transl.] **6** [1980] 269/70).
[8] Pimbley, W. T. (Diss. Pennsylvania State Univ. 1959, 60 pp.; Diss. Abstr. **20** [1959] 1835).
[9] Regenfus, G.; Sütsch, P. (Z. Phys. **266** [1974] 319/28).

5.2.7 Flicker Noise

The field emission current fluctuation (flicker noise) method is widely used for determining diffusion coefficients on single-crystal planes; see, e.g. [1, 2]. The effect of finite resolution on the determination of diffusion coefficients by the FE fluctuation method is treated in [3]. Surface self-diffusion coefficients obtained by this method on W are dealt with in Chapter 3.1, pp. 82/92.

Flicker noise was observed at frequencies below 10 Hz. The probe-hole current from regions comprising ~30 surface atoms was recorded on W(100) at temperatures between 77 and 900 K. There were weak flicker noise at 500 K and more distinct noise at higher temperatures due to increased, thermally initiated atomic displacements; see the probe-hole current vs. field curves shown in [4]. With the exception of (310), the flicker noise from W surfaces studied increases with decreasing surface density, see the comparison of atomic density (in 10^{14} atoms/cm^2) with the current fluctuation $\Delta I/I_{probe}$ in the following table [4]:

surface	(110)	(100)	(211)	(310)	(111)	(611)
atomic density	14.14	10.0	8.16	6.32	5.77	3.2
$\Delta I/I_{probe}$	0	1.1%	2.4%	8.3%	2.6%	2.8%

Current fluctuations were measured in the frequency range 50 Hz to 1 MHz with FE from (211), (100), and (310) surfaces using ⟨100⟩- and ⟨310⟩-oriented tips at pressures of active gases below 5×10^{-10} Torr. Noise "power" (rms current fluctuation) vs. T curves are given at 2, 5, and 10 kHz for T increasing from ~300 to ~1350 K. The frequency, current, and temperature dependence of the current fluctuations observed are consistent with the generation of mobile atoms on net-plane terraces [5].

The flicker noise of the FE current from (311) and (210) regions of W tips, built-up by thermo-field treatment at 1500 K, was analyzed in the frequency range f=2 Hz to 20 kHz (base pressure: ~2×10^{-10} Torr) [6]. The spectral density W(f) has been recorded while decreasing the temperature in steps from 1490 to 500 K and from 1430 to 600 K for (210) and (311) faces, respectively. At certain temperatures the log W vs. log f curves show straight-line behavior, while at other temperatures there are one or even two breaks in the curves. Plots of the frequencies at which the breaks occur vs. 1/T show Arrhenius-type behavior [6]. The W(111) data have been recorded on a smooth tip heated stepwise from 545 to 1490 K. Plotting log W vs. log f yielded a family of straight lines at various temperatures. Keeping the frequencies fixed at f=0.1, 1, and 10 kHz, the log W vs. 1/T curves indicate a superposition of two Arrhenius curves for any f [6]. A small fraction of the W atoms on a clean (111) surface seems to flip-flop from one position to another, resulting in small emission changes that give rise to the noise in the FE from W(111) between ~77 and ~590 K [7].

The relative noise amplitude was measured on a W(100) built-up tip at temperatures up to 1800 K at f=2, 5, and 10 kHz (residual pressure: 5×10^{-10} Torr). The onset of noise was observed at about 300 K, and the noise followed an approximate 1/f dependence [8]. The noise threshold temperature of the (124) region of built-up W tips was observed to be 700 K [9].

For current fluctuations probably associated with surface diffusion of adsorbed residual gas species on W tips, see [10, 11].

References for 5.2.7:

[1] Gomer, R. (Surf. Sci. **38** [1973] 373/93).
[2] Naumovets, A. G.; Vedula, Yu. S. (Surf. Sci. Rep. **4** [1985] 365/434, 386/9).
[3] Gomer, R.; Auerbach, A. (Surf. Sci. **167** [1986] 493/501).
[4] Swanson, L. W.; Crouser, L. C. (Phys. Rev. [2] **163** [1967] 622/41, 631/3).
[5] Swanson, L. W. (Surf. Sci. **70** [1978] 165/80).
[6] Biernat, T.; Kleint, C. (Appl. Surf. Sci. **27** [1987] 411/22).
[7] Dharmadhikari, C.; Gomer, R. (Surf. Sci. **143** [1984] 223/42, 225/7, 240/1).

[8] Swanson, L. W. (J. Vac. Sci. Technol. **12** [1975] 1228/33).
[9] Biernat, T.; Kleint, C. (Acta Univ. Wratislav. Mat. Fiz. Astron. No. 40 [1981] 13/22 from C.A. **95** [1981] No. 213795).
[10] Vasin, V. A.; Nevrovskii, V. A. (Radiotekh. Elektron. [Moscow] **28** [1983] 1163/8; Radio Eng. Electron. Phys. [Engl. Transl.] **28** No. 6 [1983] 92/7).
[11] Yamamoto, S.; Fukuhara, S.; Saito, N.; Okano, H. (Surf. Sci. **61** [1976] 535/49).

5.3 Thermal Field Emission

Electron emission in a range of temperature and applied field intermediate between those typical for field emission and thermionic emission is referred to as thermal field emission (TFE). The use of this term in the literature is not uniform, since several sub-regimes can be discerned and defined in different ways (compare e.g. [1] and [2]), and the term TFE is used either for the whole region or for a certain sub-regime. This subject has been theoretically approached by [3 to 5] and more recently by Christov [6 to 9]. According to Christov's theory, the extended field emission region is followed by an intermediate emission region (the TFE region is a sub-regime of this region in Christov's terminology) reached at $T > 5/6\ T_c$ [8, 9]. Numerical values for the critical temperature T_c as a function of work function Φ and field F derived from this theory are given in [10]. For a metal with a work function of 4.5 eV (typical for W), T_c increases from 803 K at a field of 0.15 V/Å to 1839 K at 0.35 V/Å. T_c decreases with increasing work function [10]. For a different definition of the thermal field emission region, see e.g. [1]. This section focuses on studies the results of which were discussed in view of the theories mentioned above. Electron energy distributions measured at elevated temperatures are included in Section 5.2.2, pp. 142/7.

Older Studies. The theory by [3] was used for data evaluation and discussion of results by [11] who studied the transition between field and thermionic emission on a W single-crystal wire [11]. For thermal FE results obtained between 500 and 2000 K in fields ranging from 0.258 to 0.667 V/Å, see the paper [12]. The FE from W point cathodes was studied at a base pressure of 10^{-8} Torr for current densities of $j = 2 \times (10^{-11}$ to 1) A/cm^2 [13]. Plots of log i vs. 1/U were linear between 295 and 880 K. At higher temperatures, thermionic emission became appreciable. At the highest temperatures, the emission was thermionic at low fields and FE at high fields [13]; poor agreement was stated by [13] comparing his results with predictions by the Murphy-Good theory [5]. However, the calculations by [13] were criticized by [7, p. 765].

According to the theory of Murphy and Good [5], the FE current density j varies with temperature T according to $j(T) = j_0 \cdot (\pi kT/d)/\sin(\pi kT/d)$, where d is a parameter depending on the work function and the field [5]. This was experimentally verified at $\sim 2 \times 10^{-10}$ Torr between 4.2 and 400 K [14]. The temperature dependence of the FE was measured between 300 and 1350 K at very low current densities [15]. Comparing the results obtained with the theory of [5] indicated that the T dependence of j may be partly due to the T dependence of the work function [15]. The theory by [6] is compared in [7] with the experimental results of [11 to 15]. Except for paper [13], there was no contradiction to published experimental data [7].

TFE from Built-Up Emitter Tips. The TFE from ⟨310⟩ and ⟨111⟩ built-up emitters was studied (for stability and brilliance) at typically 2×10^{-8} Torr and temperatures ranging between ~1100 and ~1600 °C by [16]. Fowler-Nordheim (F-N) plots are straight for cold emission from (100) built-up tips, but bend downward at high temperatures (1350 and 1810 K). The (310) built-up tip at 1810 K also yielded F-N plots that were bent downward [17].

(hkl)-Resolved TFE Studies. Good agreement with Christov's theory of electron emission ([6]) was found in a probe-hole study [10] performed between 300 and 1500 K below 10^{-10} Torr and low current densities using a tip prepared by heating it to 2500 K for 10 s. Current density j vs. T curves for the (310) surface have been recorded for fields of F = 0.155, 0.17, 0.183, and 0.2 V/Å; see **Fig. 33**. If these fields were 10% larger (which is in the range of experimental uncertainty), then the experimental data (dashed curve in Fig. 33) would fall into the range of curves calculated for F = 0.155, 0.185, and 0.215 V/Å using Christov's theory [10]. Similar results as those obtained with (310) are presented for (211) and (100) faces [10]. In previous papers of this research group, plots of the current density vs. T are given for 300 to 1500 K at fields of 0.16 to 0.22 V/Å on W(211) [18] and at 0.16 to 0.24 V/Å on W(111) and W(310) [19]. Depending on the field emission conditions used, the cathode was either heated or cooled by the emission [18] in fair agreement with the theory of Levine [20].

Thermal field emission was studied between room temperature and ~1100 K on (211), (111), (210), (221), (321), and (421) faces of W tips. Plots of the relative current [i(T) − i(295 K)]/i(295 K) vs. T are given for various fields ranging from 0.24 to 0.37 V/Å [21]; see also the preliminary investigation [22] confined to (111), (221), (321), and (421) faces. The experimental studies were extended to high temperatures on W(310) by [23].

FE and TFE are quite satisfactorily described by the theoretical approach of [6] as shown by a comparison with the experimental results on emission from W(310) published by [10] and [23] and from W(111) published by [21 to 23]; see [9] and also the earlier paper [8].

TFE results on W(111) obtained at temperatures up to 1442 K were interpreted by the theories of [5] and [6] in the low- and high-temperature regions, respectively [2]. In a

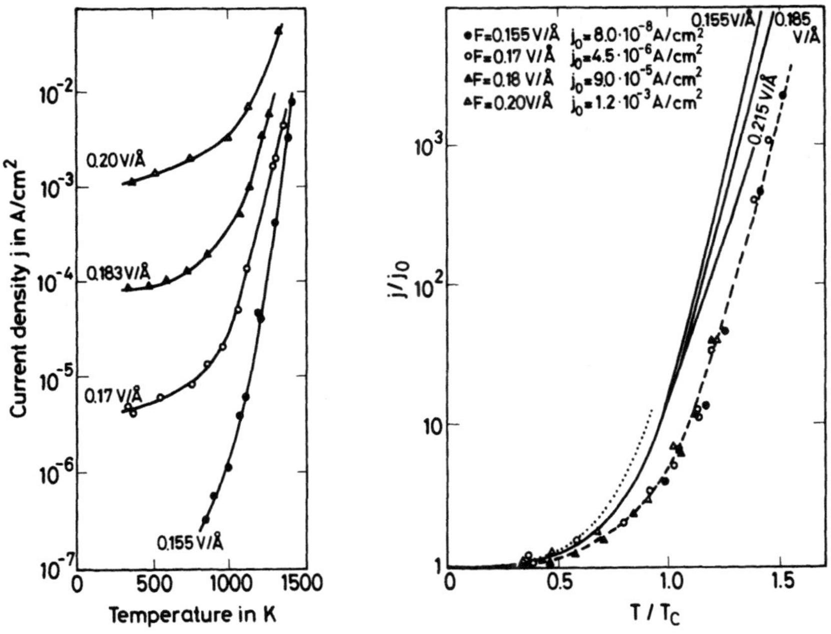

Fig. 33. Left panel: measured current density versus temperature at four different fields for the W(310) surface [10]. Right panel: the same data compared with the theories of Guth and Mullin [3] (dotted line) and of Christov [6, 8] (solid lines). The dashed line is a guide to the eye for the experimental data [10].

References for 5.3 on p. 158

study performed by [24] at $p<1.5\times10^{-10}$ Torr in the range 78 to 800 K, the TFE current i(T) was smaller than expected for free electrons on W(210), W(310), and W(320). A minimum in i(T) was found near room temperature. The results were interpreted in terms of the temperature dependence of the work function (see p. 129) [24]. TFE results at 1300 and 1500 K were used to determine the local field strength for various surfaces of a W field emitter [25].

References for 5.3:

[1] Swanson, L. W.; Bell, A. E. (Adv. Electron. Electron Phys. **32** [1973] 193/309, 296/304).
[2] Wysocki, J. K. (Surf. Sci. **104** [1981] 463/77).
[3] Guth, E.; Mullin, C. J. (Phys. Rev. [2] **61** [1942] 339/48).
[4] Dolan, W. W.; Dyke, W. P. (Phys. Rev. [2] **95** [1954] 327/32).
[5] Murphy, E. L.; Good, R. H., Jr. (Phys. Rev. [2] **102** [1956] 1464/73).
[6] Christov, S. G. (Phys. Status Solidi **17** [1966] 11/26).
[7] Christov, S. G.; Vodenicharov, C. M. (Solid State Electron. **11** [1968] 757/66).
[8] Christov, S. G. (Contemp. Phys. **13** [1972] 199/222).
[9] Christov, S. G. (Surf. Sci. **70** [1978] 32/51, 35/8).
[10] Bermond, J. M.; Lenoir, M.; Prulhière, J. P.; Drechsler, M. (Surf. Sci. **42** [1974] 306/23).

[11] Andreev, I. S. (Zh. Tekh. Fiz. **22** [1952] 1428/41; Phys. Abstr. **56** [1953] No. 6080).
[12] Dyke, W. P.; Barbour, J. P.; Martin, E. E.; Trolan, J. K. (Phys. Rev. [2] **99** [1955] 1192/5).
[13] Haag, H. W. (Z. Phys. **178** [1964] 365/74).
[14] Klein, R.; Leder, L. B. (Phys. Rev. [2] **124** [1961] 1046/9).
[15] Drechsler, M. (Z. Phys. **167** [1962] 558/62).
[16] Shimizu, R.; Kataoka, Y.; Hashimoto, H. (Technol. Rep. Osaka Univ. **23** [1973] 403/9 [jap.: 1121/54]; C.A. **81** [1974] No. 69929).
[17] Tamura, N. (Scanning Electron Microsc. **1979** No. 1, pp. 31/8; C.A. **91** [1979] No. 112044).
[18] Drechsler, M. (RGE Rev. Gen. Electr. **80** [1971] 568/70; C.A. **76** [1972] No. 28847).
[19] Drechsler, M.; Bermond, J.-M.; Prulhière, J.-P. (C. R. Séances Acad. Sci. B **269** [1969] 1267/70).
[20] Levine, P. H. (J. Appl. Phys. **33** [1962] 582/7).

[21] Wysocki, J. (Acta Phys. Pol. A **42** [1972] 129/45).
[22] Wysocki, J. (Acta Phys. Pol. **35** [1969] 195/8).
[23] Wysocki, J. K. (Acta Univ. Wratislav. Mat. Fiz. Astron. No. 29 [1977] 99/105).
[24] Wysocki, J. K. (Phys. Rev. [3] B **28** [1983] 834/41).
[25] Bermond, J. M. (Surf. Sci. **50** [1975] 311/28).

5.4 Photofield Emission

5.4.1 Introduction

Photofield emission (PFE) is field emission (FE) under excitation by photons with an energy less than the work function Φ. The interest in PFE stems from the fact that it may be used to explore the electronic band structure of the emitter material between the Fermi level and the vacuum level. PFE is discussed in the early theoretical papers [1 to 4]. A more recent model for PFE that takes interactions in the electron gas into account is given in [5]: visible-light photons penetrate some 200 Å into the bulk and create electron-hole pairs at several mean free path lengths beneath the surface; Auger-type deexcitations generate cascades of electron-hole pairs. A PFE model with a noninteracting electron gas, on the other hand, only considers electrons tunneling out of the metal with energies which they acquired as final energies in the primary photoexcitation process. Experimentally obtained energy distributions of photofield-emitted electrons on W(310), W(211), and W(111) are compared with these model calculations [5]; see Section 5.4.2 below. No indication was found for multiphoton excitation in PFE under irradiation below 0.1 MW/cm^2 [6, 7]; see Section 5.4.3 below. For a discussion of multiphoton excitation in laser-induced electron emission, see Chapter 5.6, pp. 197/201.

5.4.2 Photofield Emission Energy Distributions (PFEEDs)

The first measurements of energy distributions of photofield-emitted electrons from W were performed by Lee [8] on W(310) and by Liu et al. [9] on W(111). These were followed by numerous studies on PFEEDs from W single-crystal surfaces especially by Lee and coworkers [5, 10 to 17] and Reifenberger and coworkers [10, 18 to 23].

When p-polarized radiation is used, PFE is dominated by photoexcitation of electrons from occupied crystal states near the surface to free-electron-like tunneling states above the Fermi level [13]. Deviations from free-electron behavior due to direct transitions in the bulk were observed with s-polarized radiation where surface excitation is minimized [14]. PFEEDs were recorded on W(110), (211), (100), (111), and (510) at base pressures in the 10^{-11} Torr range. The tip was irradiated by a tunable Kr^+ laser with photons of 1.916 to 3.536 eV and power densities below 0.1 MW/cm^2 at normal incidence (s polarization) [14] or at grazing incidence (p polarization) [13]. The field at the surface was $F=0.25$ V/Å [13, 14]. The irradiation heated the tip typically less than 100 K (never by more than 200 K). With 3.536 eV photons (excitation to above the barrier), no longer only normal components of the electron waves contribute to the emitted current, but also electrons having a significant parallel momentum. This makes the interpretation of the data difficult. A large number of features were observed in plots of the electron energy vs. the enhancement factor R(E) which characterizes the departures from free-electron behavior. Extrema and changes in the slope of the R vs. E curves indicate extrema or changes in the slope of the one-dimensional density-of-states curves along the respective surface normals [14]. The results are compared with the bulk band structure calculated by Christensen and Feuerbacher [24]. One of the PFEED features on the (100) surface shifting its position as a function of the photon energy was associated with the "Swanson hump" which for dark FE occurs around 0.35 eV below E_F. Its observation under s-polarized irradiation may be due to some misalignment or indicates small contributions from antisymmetric wave functions (the intensity observed with s polarization is some 350 times weaker than with p-polarized radiation) [14].

Features observed with p-polarized light at initial-state energies $E-E_F-h\nu=\varepsilon-h\nu$ in the R(E) curves are summarized in the following table taken from [13]:

surface	initial-state energy $E-E_F-h\nu$ in eV
(110)	−0.89, −0.79, −0.58, −0.47
(211)	−1.44, −1.20, −0.794(34), −0.635(21), −0.338(14)
(100)	−0.563(13), −0.358(5), −0.234(13)
(111)	−1.56, −1.30, −0.925(31), −0.688(16)
(510)	−0.711(30), −0.354(22)

The pronounced peak observed on the (100) plane at 0.358 eV below E_F corresponds to the Swanson hump surface state at 0.35 eV below E_F. In the curve of (510), a broad peak was seen at an initial energy around 0.35 eV below E_F. It is likely due to the Swanson hump, since (100) and (510) are separated by only 11° [13]. In a preceding paper describing

References for 5.4 on pp. 166/7

the experimental apparatus, a bump in the PFEED of W(111) was observed with p-polarized light at an initial energy of 0.64 eV below E_F [12].

Features observed in plots of the enhancement factors as a function of the initial-state energy of different W surfaces were discussed in [17]. A weak shoulder at 0.37 eV below E_F was observed in the R(E) curve of W(110). It disappeared as the temperature was raised from 330 K to 860 K. This feature could not be explained on the basis of the bulk band structure. In the PFEED of W(211), an anomalous feature occurred at an initial energy of 0.66 eV below E_F [17]. In the study mentioned above [13], a feature at a similar value was observed (0.635 eV below E_F; see table above). The enhancement factor measured on W(111) at different temperatures revealed a feature around 0.75 eV below E_F, attributed to initial states lying about 0.72 eV below E_F. In the corresponding PFEED, an anomalous feature occurred at 0.69 eV below E_F [17] in good agreement with the value of 0.688 eV below E_F given above [13]. On W(510) no feature was found by [17].

For W(100) the well-known surface state at 0.35 eV below E_F was clearly resolved in PFEEDs measured with radiation at 2.41 to 3.54 eV (F=0.29 V/Å, $p<6\times10^{-11}$ Torr, Ar^+ laser power densities near 10 kW/cm^2) in [21]; see also [20]. The measured PFE signals were deconvoluted into contributions of free-electron behavior and emission from a surface state [20]. There is a strong dependence of the emitted intensity on the polarization of the illumination. The total photofield and surface state signals show a $\cos^2\psi$ dependence on the angle ψ between the surface normal and polarization vector. The experimentally observed polarization dependence is consistent with a d_{z^2} character of the surface state [20, 21]. The enhancement factor R(E) measured on W(100) peaks at an initial-state energy of 0.37 eV below E_F and $\ln R(E, T=920\ K)-\ln R(E, T=330\ K)$ shows a dip at this energy value [17].

For $\varepsilon=E-E_F=0.5$ to 3.5 eV, essentially free-electron behavior was observed for W(110), W(211), W(100), W(111), and W(310). The free-electron behavior is modified by electron cascades formed upon the relaxation of photoexcited electrons. At final energies close to E_F, strong departures from free-electron behavior were found. PFEEDs at normal incidence exhibit considerable fine structure which for (110), (211), and (310) strongly depends on the polarization. It weakens on heating, presumably due to final-state scattering by thermal phonons [25]; see also [5].

Deviations from free-electron behavior were also found by [22] on W(110) and W(111): With photon energies between 2.41 and 3.54 eV and a field of 0.3 V/Å, PFEEDs were recorded on W(110) at initial-state energies ranging from 1.5 eV below to 0.5 eV above E_F. A p-polarized beam was incident at 70°, the plane of incidence contained the [001] direction. Significant deviations from free-electron behavior occurred at initial-state energies ranging from 1.2 to 0.4 eV below E_F. The measured energy distributions were compared with theoretical curves, considering various direct transitions in the bulk. On W(111) a PFEED was recorded with F=0.24 V/Å and p-polarized light of 3.54 eV incident at an angle of 35° in a plane that contained the [$\bar{1}\bar{1}2$] direction. This PFEED showed deviations from free electron behavior between about 0.7 and 1.4 eV below E_F. Around 0.7 eV below E_F, a band gap occurs in the calculated band structure [22].

PFEEDs from W(110) [18, 19, 26], W(100), and W(310) [26] of subthreshold photo-emitted electrons were studied at $\sim10^{-10}$ Torr, using a mixture of about 20 mW of 3.531 eV and 10 mW of 3.408 eV radiation from an Ar^+ laser [18, 19, 26]. Despite the polarization being nearly parallel to the shank of the tip, the irradiation raised the temperature by about 100 K [19]. The PFEEDs measured with F=0.23 V/Å on W(110) and W(310) are in good agreement with theoretical curves assuming free-electron-like final states [18, 19, 26]. Disagreement was observed for fields below 0.12 V/Å [26]. The model calculations for W(100) disagreed with the experimental data for all fields applied [26].

PFEEDs from W(110) between 300 and 1400 K showed a discontinuity in the slope at an initial energy around 0.5 eV above E_F under near-UV illumination. This anomaly was observed with s-polarized light at normal incidence for [$\bar{1}$10] polarization, but vanished when the polarization vector was parallel to [001]. It was also observed with p-polarized light. These anomalies observed with both FE and PFE from W(110) are attributed to the band limit of a thermally populated p-like electron band (responsible for the N-centered hole ellipsoids on the Fermi surface; cf. Section 1.2.3.3, pp. 64/5 and "Tungsten" Suppl. Vol. A 2, 1987, p. 124) [15, 25].

Anomalous high-energy tails that deviate from the exponential decrease expected were observed in PFEEDs measured from W(111) and W(310). The effect is attributed to Coulomb scattering of electrons in the vacuum near the field emitter. The photocurrent excited by the 351 nm (3.54 eV) radiation from a Kr^+ laser was maximized when using p-polarized radiation incident at $\alpha \approx 60°$. The temperature increase due to radiation was $\Delta T = 14 \pm 3$ K for an ambient temperature of ~300 K (estimated by analyzing the high-energy tails in the TED with and without laser heating). The TEDs from FE and PFE show high-energy tails which extend to initial energies where free-electron theories predict negligible emission [16]. The TED of electrons emitted from W(310) is interpreted as a superposition of contributions from thermally populated states (decreasing according to exp $[-a \cdot \varepsilon]$, where $\varepsilon = E - E_F$) and a Coulomb-scattering term (identified by its intensity being proportional to the square of the total FE current) exhibiting an $\varepsilon^{-2.6}$ dependence. The Coulomb tail dominates the dark TED for $\varepsilon > 1$ eV. The results for $0 < \varepsilon < h\nu/2$ favor the interacting electron gas model which assumes that the PFE electrons result from Auger-type deexcitations in the solid. The enhancement factor R for irradiation of W(111) by s-polarized 351 nm (3.53 eV) radiation is nearly unaffected by the applied field. This implies that the electrons detected at a given energy, ε_1, tunnel through the barrier at just that energy, ε_1, and that the low-energy tail in the PFEED arises from photoelectrons which lose energy as a result of inelastic scattering processes within the solid [5]. The results obtained with p-polarized light are compared with the predictions of a model [27] of surface photoexcitation for a noninteracting electron gas. Although significant deviations (especially at low energies) occur, the enhancement observed is much smaller than the corresponding enhancement found with s-polarized light [5].

There was no indication found for photoexcited electrons tunneling through the potential barrier [28]. Using Cd, Ar^+, and He-Ne lasers ($h\nu = 2.82$, 2.41, and 1.97 eV, respectively), the total energy distributions (TEDs) of the emitted electrons were determined with a resolution of ~0.04 eV at the high-energy threshold. The TEDs were close to Maxwell distributions and for $\lambda = 514.5$ nm showed a high-energy tail not present with dark emission. The extent of the tail varied with the voltage applied, and a temperature effect was noted [28].

5.4.3 Yield Studies

The quantum yield for tunneling of photoexcited electrons at $F \approx 0.2$ V/Å was measured with photons of 2.41 eV. At power densities below ~100 W/cm², the quantum yield was less than 10^{-10} electrons/photon for any polarization. Tunneling of the photoexcited electrons is less probable than their thermalization which converts much of the absorbed radiation into heat [28].

The relative yield, defined as the yield in PFE scaled to the yield in FE, is plotted vs. photon energy for p-polarized light with 2.41 to 3.54 eV incident at 70° and 35° on W(110)

References for 5.4 on pp. 166/7

and W(111) surfaces, respectively. The yield values were obtained from the areas under the energy distribution curves of FE and PFE. The relative yields increased from $\sim 2\times10^{-7}$ to $\sim 4\times10^{-6}$ on W(110) and from $\sim 10^{-6}$ to $\sim 2\times10^{-5}$ on W(111) for hν increasing from 2.41 to 3.54 eV [23].

The dependence of the photocurrent on the intensity of illumination was studied for $F\approx 0.24$ V/Å at $p<10^{-11}$ Torr using 476, 521, and 568 nm radiation (i.e., 2.60, 2.38, and 2.18 eV) from a Kr^+ laser. A linear increase was found at least below 3.5 kW/cm² for s- and p-polarized radiation incident on W(111) at ~26° and on (310) at ~70°. This dependence indicates one-photon excitation processes. The linear increase of the photocurrent means that the photoyield is independent of the irradiance intensity [7]. This result was confirmed on (111) when the measurements were extended to intensities of 100 kW/cm². Small nonlinearities found on (211) and (310) surfaces are attributed to work function changes due to laser heating [6]. In the PFE study [29] (extending the scope of the previous paper [7] to the polarization dependence of emission from (111), ($\bar{1}$11), (031), and (013) surfaces), the quantum yield was measured at 0.27 V/Å with 351 nm (3.53 eV) radiation. Depending on polarization and surface orientation, quantum yields in the range between 10^{-5} and 10^{-3} electrons/photon were obtained. Indirect bulk and surface photoeffects contribute to the photocurrent which is dominated by electrons with energies close to the peak of the potential barrier. For a field strength of 0.27 V/Å, this peak lies between 2 and 3 eV above E_F. Thus, the photocurrent at photon energies used in this study (2.2 to 2.6 eV) is due to transitions from initial states close to E_F, where the joint density of states for direct bulk transitions is only very small. The relative importance of the two effects depends on the angle of incidence of the light. For near-normal incidence, indirect bulk excitations dominate, but at large angles of incidence the emission stems primarily from the surface photoeffect. Surface-effect photoemission was observed even from the geometrical shadow of the tip. This emission is interpreted as arising from strong diffraction effects (Mie resonances) [29].

High quantum yields and strong PFE current pulses were obtained using an Nd:YAG laser operating at 1.17, 2.33, and 3.49 eV (1064, 532, and 355 nm, respectively) and producing pulses of 10 ns or 35 ps width with a power of some μJ per pulse [30, 31]. The values of the field strength are not given in the papers. Current densities as high as 10^{12} A/cm² (for the ps pulses) were measured. Only about 10^3 photons were required for the release of one electron [30, 31].

5.4.4 Current-Field Characteristics

Fowler-Nordheim (F-N) plots of data from FE give straight lines (cf. Section 5.2.3, pp. 149/50). In contrast, F-N plots of PFE data show deviations from the linear behavior at strong fields [8, 32 to 38]. Below a certain critical field value, electron emission occurs predominantly by tunneling through the surface potential barrier like in FE, but for high values of the electric field, emission from above the barrier can take place resulting in a weaker slope in the F-N plot; see e.g. [8]. The influence of the surface potential barrier on PFE, especially at higher fields where deviations from the F-N formula appear, was theoretically investigated in [39] using a scattering theory formalism and assuming a free-electron model. The experimental data of Lee and Reifenberger [34] are in poor agreement with the results calculated in [39].

PFE was studied with a W tip illuminated by $\sim 10^{12}$ photons/s (power density <1 kW/cm²) of 441.6 nm (2.81 eV) and 10^{11} photons/s of 325.0 nm (3.82 eV) from an He-Cd laser [36, 40]. The strongly sloping part in the F-N plot of the 2.81 eV photons corresponds to tunneling through the barrier, the weakly sloping part (high-field region) corresponds to emission from above the barrier (lowered by the Schottky effect). The measured portion of the 3.82 eV curve is entirely within the energy range where emission from above the barrier occurs [36]. In addition, a comparison of FEM patterns under FE and PFE conditions taken at 0.056 and 0.225 V/Å is given in [36] (see also the subsequent paper [41], where the dependence of the brightness of the FEM patterns on the angle between the polarization vector and the tip axis is discussed).

F-N plots were recorded at room temperature in the 10^{-10} Torr range with 2.4, 2.7, and 3.5 eV radiation from an Ar^+ laser irradiating a W(110) surface. Notably at $h\nu = 3.5$ eV, the F-N plot deviates from the theoretical line at high fields [37]. This effect was also observed in a combined FE and PFE study [38] at pressures below 10^{-10} Torr, where (051), (011), (111), and $(\bar{1}11)$ planes were illuminated at angles of incidence of $\alpha = 73°$, 58°, and 26°, respectively, using a Kr^+ laser beam whose polarization was parallel to the tip shank (to minimize surface heating by the radiation; see Section 5.4.5, pp. 165/6). The (111) surface was studied by p- and s-polarized radiation. F-N curves for PFE, excited by photons with 2.336 to 2.645 eV and by the closely spaced pair of UV lines of the Kr^+ laser averaging at $h\nu = 3.508$ eV, are plotted. Under UV photoexcitation, the electrons just pass over the image potential barrier of the $(\bar{1}11)$ surface for $F \gtrsim 0.072$ V/Å (studied between 0.044 and 0.079 V/Å), indicated by the slight downward bend of the F-N plots. Similarly, but more distinctly, the F-N curve from the (051) surface irradiated by 3.508 eV radiation is bent downward for $F \gtrsim 0.056$ V/Å (studied between 0.036 and 0.065 V/Å) [38].

F-N curves were compared for FE and PFE currents measured during ~1 W/cm² illumination by 2.27 and 2.15 eV radiation (filtered Hg lamp). Although the PFE currents were 10^5 to 10^7 times smaller than corresponding FE currents, the slopes of the curves were nearly identical [35].

In many PFE studies on W, especially those where band structure effects were investigated, the current-field characteristics were plotted in the form of ln j vs. F^{-1} instead of ln (j/F^2) vs. F^{-1}. Knees or shoulders occurring in these curves were interpreted in terms of direct and indirect optical transitions [35, 42 to 49].

An early study [42] already showed that band structure effects must be taken into account to explain the PFE currents measured on W. The experimental data were compared with model calculations excluding [3] and including [4] band structure effects. The best results were obtained when direct and indirect transitions were taken into account [42].

Shoulders on the PFE current-voltage curves and slope changes were observed on clean (111) and (310) surfaces excited by 2.409 and 2.707 eV and by 2.409, 2.540, and 2.707 eV, respectively, at base pressures in the range 10^{-12} to 10^{-11} Torr. They are attributed to direct optical transitions in the case of (310) and to direct and indirect optical transitions in the case of (111) [43] referring to the band structure calculations in [24].

PFE studies on the band structure near the Fermi surface were performed by Radoń and Jaskółka on W(111) [44], W(100) [45], and W(211) [46] using p- and s-polarized radiation at 1.96 to 2.707 eV. The fields applied ranged from about 0.38 to 0.5 V/Å. The pressure in the tube was around 10^{-11} Torr. Discontinuities in the slopes of the curves in the case of s-polarized illumination are attributed to transitions between filled and empty bulk bands below E_F. Discontinuities in the curves in the case of p-polarized illumination, which were not obtained with s-polarized light, were attributed to surface states [44 to 46]. With ~1 W/cm² radiation impinging at $\alpha \approx 65°$ on a W(111) surface, two surface states were found located at about 0.85 and 0.35 eV below E_F [44]. Similarly, when a W(100) surface was irradiated by a chopped (8.5 kHz, strongly suppressing the thermofield current signal), s-

References for 5.4 on pp. 166/7

and p-polarized, ~100 W/cm^2 beam incident at α=39°, two surface states were observed, one at 0.07 to 0.22 eV above E_F and one (the Swanson hump?) at 0.43 to 0.44 eV below E_F (the spread is due to dispersion) [45]. PFE current vs. 1/F curves for W(211) were measured by Radoń and Jaskółka [46] under the same experimental conditions as in [45]. The 100 W/cm^2 radiation used impinged under an incidence angle of ~40° onto W(211). The results of [46] for W(211) are consistent with the relativistic band structure model of [24] (as are those for W(111) in [44] and W(100) in [45]) [46].

PFE from W(211) was also studied at liquid nitrogen temperatures using chopped 632.8 nm (1.959 eV) radiation [47]. Apparently, s-polarized radiation incident at ~20° was used (for experimental conditions, see [43]). The logarithm of the photofield current plotted vs. 1/F gives a partially linear curve interrupted by two horizontal lines. The low-field (high 1/F) ends of these plateaus occur at a=0.596 and b=0.567 V/Å. These knees are associated with indirect transitions in the bulk. "a" and "b" correspond to transitions from $E_{initial}=E_F-0.04$ eV to $E_{final}=E_F+1.96$ eV and from $E_{initial}=E_F+0.02$ eV to $E_{final}=E_F+1.98$ eV, respectively [47].

According to [28], the knees in the I-V curves of the photofield current are not related to the band structure, since near threshold (the conditions prevailing in the experiments by [28]) only "general" metal properties are relevant and the detailed band structure is immaterial [28]. (Radoń [43] and Radoń et al. [44 to 46], however, adhere to the view that PFE characteristics allow one to test details of the electronic structure in the bulk and surface band structure even in the vicinity of the Fermi level; see above.)

To interpret the structure occurring in the current vs. reciprocal-field curves, only a set of final states (selected through momentum conservation) is assumed to contribute to the photocurrent [48, 49]. With increasing field, electron emission changes from tunneling to free transmission, when the barrier maximum passes the energy of a final state. The observed knees or shoulders are manifestations of such transitions. The knees in the experimental curves discussed in [48, 49] (the W(111) PFE data of Radoń [43]) occur close to but not exactly at the F values where the final-state energies equal the barrier maximum [48, 49].

For pulsed (100 ns) ruby laser radiation (1.78 eV) with intensities around some MW/cm^2, it was shown that the PFE current is primarily caused by nonequilibrium electrons excited above the potential barrier maximum. Plots of ln j vs. $U^{1/2}$ suggest a stepped nature of the energy spectrum of electrons inside the metal below the vacuum level [50].

Current Oscillations

The PFE current contains a component which is periodic with the electric field. A W tip was irradiated by ~2 W/cm^2 of chopped 546 nm radiation from an Hg lamp. The amplitude and phase of the extrema strongly depended on the tip conditions despite the base pressure being as low as 1.5×10^{-10} Torr [32, 33]. The periodic changes seen in the i_{PFE} vs. field plots from W tips illuminated by filtered Hg light (hν=2.15, 2.27, 2.85, 3.06, and 3.40 eV) were strongly contamination-dependent, yet rather insensitive to the photon energy used ($p<5\times10^{-10}$ Torr). At mean fields of F≈0.395 V/Å, the PFE current was periodic in F with a period length of 1.289×10^{-3} V/Å [51, 52].

When a W(510) surface was irradiated with 3.51 eV photons at ~80 K and electric fields applied in the range 0.05 to 0.3 V/Å, periodic current oscillations were observed with the phase proportional to $1/\sqrt{F}$ [34]. Paper [53] questioned these measurements. The oscillations found in [34] were thought to be an experimental artifact. Experiments performed under similar conditions showed that the oscillatory effect can be caused by magnetic deflection of the beam across the retarding mesh of the energy analyzer (deflection of decelerated electrons in the earth's field was sufficient to bring about the effects called into question) [53]. This result is in line with several theoretical studies which did not support the anomalies claimed to exist with PFE [27, 54 to 56].

The study [53] has, however, no conclusive argument against the results presented by [18, 37]. In [37] the PFE current from W(110) and W(111) was measured at base pressures in the 10^{-10} Torr range and with 2.4, 2.7, and 3.5 eV radiation in the field range 0.14 to 0.25 V/Å. With hν = 3.5 eV, an oscillatory component was observed, the amplitude of which amounted to ~5% of the background value in good agreement with the 4.6% given in [34]. The similar results for W(111) and W(110) indicate that the PFE oscillations do not strongly depend on the surface type [37]. The following PFE oscillation frequencies are given for W(110) in [18, 37, 57], for W(111) in [37, 57], for W(210) in [57], and for W(510) in [34]:

surface	(110)	(111)	(210)	(510)
hν in eV	3.53	2.71	3.53	3.53
frequency in $10^5\sqrt{V/cm}$	1.03 ± 0.09	0.98 ± 0.08	0.89	0.97 ± 0.003

These PFE oscillations are not related to the periodic Schottky deviations, since the latter have a periodicity of $F^{-1/4}$. In addition, there are many more PFE oscillations over the same fractional range of electric fields than there are periodic Schottky deviations. Finally, the resonances in the transmission responsible for Schottky deviations do not occur for fields above ~0.08 V/Å (far below the field strengths used in this study) [37].

In a subsequent paper [58], the authors of [18] and [37] tried to explain the periodic component of the photofield current, calling into question the static nature of the image charge potential and assuming an oscillating component in the surface potential barrier arising from the time dependence involved in the formation of the image charge potential [58]. According to another theoretical study [59], this model seems to be inadequate, since a study [60] of dynamical corrections to the static image potential shows that the effects postulated in [58] are practically nonexistent for the electrons (of very low kinetic energy) emitted in a PFE experiment. A tentative explanation is given in [59] for the rapid oscillations observed in PFE. Emission from a band of surface states (these states being final states in the photoexcitation process) on a plane of small area on the field emission tip is suggested to cause the observed oscillations. The experimental results of [34] appear to be in qualitative agreement with the predictions [59].

5.4.5 Temperature Dependence of PFE

The temperature rise ΔT caused by laser heating is consistently greater (by ~45%) for light polarized perpendicular to the tip axis than for light polarized parallel to the shank. Diffraction seems to have a major effect on ΔT. Thus, in order to keep heating by laser radiation at a level as low as possible, the plane of polarization has to be kept parallel to the tip shank [61]. The effect of laser heating on the surface temperature is dealt with in more detail in Chapter 5.6, pp. 197/201. In the following paragraphs, the dependence of the PFE currents and energy distributions on temperature is discussed.

PFEEDs of W(110), (111), (211), (100), and (310) surfaces were measured using light at normal incidence with energies from 1.833 to 3.536 eV. The structure observed in the PFEEDs weakens when the surfaces are heated from 300 to 900 K. The PFE current from W(110) under [$\bar{1}$10]-polarized near-UV illumination decreased by ~40% between 300 and

References for 5.4 on pp. 166/7

1000 K, while the emission from the (211), (100), (111), and (310) surfaces increased approximately linearly by as much as 20%. On W(110) at 300 K, the photoexcited electrons are scattered by phonons at a rate comparable with electron-hole pair creation processes [25]. When a W(110) surface was heated to between 300 and 1400 K to cause emission from thermally populated electronic states above the Fermi level, anomalous emission was observed at an energy of about 0.5 eV above E_F independent of temperature. This anomaly is attributed to the band limit of a p-like electron band (see p. 161) [15, 25].

The temperature dependence of the energy distribution of photofield-emitted electrons was studied on W(110), (211), (100), (111), and (510) with 2.603, 3.048, and 3.536 eV radiation for fields of F=0.2 to 0.3 V/Å at a base pressure of $\sim 5\times10^{-11}$ Torr. P-polarized light at nearly grazing incidence was used (typically $\alpha\approx75°$). The field emitter was heated by a second, 0.1 W laser focused to a ~4 μm-diameter spot. This resulted in a temperature increase of about 700 K. The data are presented by plotting the difference between the logarithms of the enhancement factors at two different temperatures. In addition to the effect of the temperature dependence of the work function, emission anomalies correlated with the structure in the PFEEDs were observed. This indicates that the deviations from free-electron behavior in p-polarization are characteristic for surface photoexcitation [17].

References for 5.4:

[1] Bagchi, A. (Phys. Rev. [3] B **10** [1974] 542/53).
[2] Guth, E.; Mullin, C. J. (Phys. Rev. [2] **59** [1941] 867/73).
[3] Neumann, H. (Physica [Amsterdam] **44** [1969] 587/94).
[4] Neumann, H. (Ann. Phys. [Leipzig] [7] **26** [1971] 89/93).
[5] Donders, P. J.; Lee, M. J. G. (Phys. Rev. [3] B **41** [1990] 1781/8).
[6] Donders, P. J.; Lee, M. J. G. (Surf. Sci. **167** [1986] 141/9).
[7] Venus, D.; Lee, M. J. G. (Surf. Sci. **116** [1982] 359/68).
[8] Lee, M. J. G. (Phys. Rev. Lett. **30** [1973] 1193/6).
[9] Liu, R.; Ehrlich, G.; Polizzotti, R. S. (J. Vac. Sci. Technol. **11** [1974] 276/7).
[10] Reifenberger, R.; Goldberg, H. A.; Lee, M. J. G. (Surf. Sci. **83** [1979] 599/616, 613/5).

[11] Venus, D.; Lee, M. J. G. (Phys. Rev. [3] B **28** [1983] 437/9).
[12] Venus, D.; Lee, M. J. G. (Rev. Sci. Instrum. **56** [1985] 1206/11).
[13] Venus, D.; Lee, M. J. G. (Surf. Sci. **172** [1986] 477/95).
[14] Venus, D.; Lee, M. J. G. (Phys. Rev. [3] B **34** [1986] 4449/63).
[15] Donders, P. J.; Lee, M. J. G. (Surf. Sci. **175** [1986] 197/214).
[16] Donders, P. J.; Lee, M. J. G. (Phys. Rev. [3] B **35** [1987] 6578/87).
[17] Lopinski, G. P.; Donders, P. J.; Lee, M. J. G. (Surf. Sci. **201** [1988] 294/310).
[18] Reifenberger, R.; Egert, C. M.; Haavig, D. L. (J. Vac. Sci. Technol. [2] A **2** [1984] 927/30).
[19] Haavig, D. L.; Reifenberger, R. (Surf. Sci. **151** [1985] 128/42).
[20] Gao, Y.; Reifenberger, R. (Phys. Rev. [3] B **32** [1985] 1380/2).

[21] Gao, Y.; Reifenberger, R. (Phys. Rev. [3] B **35** [1987] 4284/90).
[22] Gao, Y.; Reifenberger, R. (Phys. Rev. [3] B **35** [1987] 6627/36).
[23] Gao, Y.; Reifenberger, R. (Phys. Rev. [3] B **35** [1987] 8301/7).
[24] Christensen, N. E.; Feuerbacher, B. (Phys.Rev. [3] B **10** [1974] 2349/72).
[25] Donders, P. J. (Diss. Univ. Toronto 1988, VII+185 pp. [1/157, 159/86]; Diss. Abstr. B **49** [1988] 1193).
[26] Haavig, D. L. (Diss. Purdue Univ. 1983, 176 pp.; Diss. Abstr. B **44** [1984] 3845).
[27] Schwartz, C.; Cole, M. W. (Surf. Sci. **115** [1982] 290/300).
[28] Bakhtizin, R. Z.; Yumaguzin, Yu. M. (Poverkhnost **1984** No. 10, pp. 49/53; Phys. Chem. Mech. Surf. [Engl. Transl.] **3** [1985] 2928/36).

[29] Venus, D.; Lee, M. J. G. (Surf. Sci. **125** [1983] 452/72).
[30] Boussoukaya, M.; Bergeret, H.; Chehab, R.; Leblond, B. (BNL-52 090 [1987] 164/79).

[31] Boussoukaya, M.; Bergeret, H.; Chehab, R.; Leblond, B.; Le Duff, J. (Nucl. Instrum. Methods Phys. Res. A **279** [1989] 405/9).
[32] Wysocki, J.; Kleint, Ch. (Phys. Status Solidi A **20** [1973] K 57/K 59).
[33] Wysocki, J.; Kleint, Ch. (Acta Phys. Pol. A **48** [1975] 157/74).
[34] Lee, M. J. G.; Reifenberger, R. (Surf. Sci. **70** [1978] 114/30).
[35] Radoń, T.; Kleint, C. (Surf. Sci. **60** [1976] 540/60).
[36] Teisseyre, Y.; Haug, R.; Coelho, R. (Surf. Sci. **75** [1978] 592/608).
[37] Egert, C. M.; Reifenberger, R. (Surf. Sci. **145** [1984] 159/74).
[38] Gaudin, G. A.; Lee, M. J. G. (Surf. Sci. **185** [1987] 283/98).
[39] Taranko, E. (Acta Phys. Pol. A **53** [1978] 761/4).
[40] Teisseyre, Y.; Coelho, R.; Haug, R. (C. R. Séances Acad. Sci. B **274** [1972] 1202/5).

[41] Teisseyre, Y.; Haug, R.; Coelho, R. (Surf. Sci. **87** [1979] 549/60).
[42] Neumann, H.; Kleint, C. (Ann. Phys. [Leipzig] [7] **27** [1971] 237/47).
[43] Radoń, T. (Surf. Sci. **100** [1980] 353/67).
[44] Radoń, T.; Jaskółka, S. (Surf. Sci. **200** [1988] 199/208).
[45] Radoń, T.; Jaskółka, S. (Surf. Sci. **231** [1990] 160/4).
[46] Radoń, T.; Jaskółka, S. (Surf. Sci. **247** [1991] 106/10).
[47] Żebrowski, J.; Kleint, C. (Phys. Lett. A **121** [1987] 463/5).
[48] Kleint, Ch.; Modinos, A. (Surf. Sci. **177** [1986] 389/98).
[49] Modinos, A.; Kleint, Ch. (Phys. Lett. A **112** [1985] 414/6).
[50] Azizov, S. T.; Lugovskoi, V. B. (Zh. Tekh. Fiz. **51** [1981] 2554/9; Sov. Phys.-Tech. Phys. [Engl. Transl.] **26** [1981] 1509/12).

[51] Radoń, T.; Kleint, C. (Acta Univ. Wratislav. Mat. Fiz. Astron. **22** [1977] 35/45).
[52] Radoń, T.; Kleint, C. (Ann. Phys. [Leipzig] [7] **34** [1977] 239/52).
[53] Donders, P. J.; Lee, M. J. G. (Surf. Sci. **160** [1985] 280/92).
[54] Puri, A.; Schaich, W. L. (Phys. Rev. [3] B **28** [1983] 1781/4).
[55] Wu, J.-W.; Mahan, G. D. (Phys. Rev. [3] B **28** [1983] 4839/41).
[56] Young, R. A. (Solid State Commun. **45** [1983] 263/6).
[57] Egert, C. M. (Diss. Purdue Univ. 1983, 164 pp.; Diss. Abstr. B **44** [1984] 2801).
[58] Reifenberger, R.; Haavig, D. L.; Egert, C. M. (Surf. Sci. **109** [1981] 276/90).
[59] Modinos, A.; Kleint, C. (Solid State Commun. **50** [1984] 651/4).
[60] Harris, J.; Jones, R. O. (J. Phys. C **6** [1973] 3585/604; J. Phys. C **7** [1974] 3751/4).

[61] Lee, M. J. G.; Reifenberger, R.; Robins, E. S.; Lindenmayr, H. G. (J. Appl. Phys. **51** [1980] 4996/5006).

5.5 Photoelectric Emission

5.5.1 Photoelectric Yields

Among the many papers, which report ultraviolet photoemission spectra showing electron-energy-resolved photoelectric yields (intensity vs. E_{kin}, see the energy distribution curves in Section 5.5.2, pp. 169/85), virtually only one paper [1] seems to give an absolute yield (Y) value for a clean W surface (this value refers to the emission from a surface state and does not give the total yield). The photoelectric yield of the surface state near 0.4 eV below E_F was observed to peak around a photon energy of 16 eV with a value of $Y \approx 0.0013$ electrons per incident photon impinging under $\alpha = 45°$ on the sample [1]. This yield was obtained on a polycrystalline W sample recrystallized in a predominantly (100)

References for 5.5.1 on pp. 168/9

orientation during a 100 h heat treatment in oxygen applied to remove carbon. Prior to any measuring run, the sample was flashed for 5 s at ~2700 K. The pressure was $\sim 1.5 \times 10^{-10}$ Torr (mainly noble gas from the light source) [1].

With W polycrystals flashed for 15 s at 2200 K, yields of $Y \approx 0.035$, 0.03, 0.012, 0.0043, and 0.0005 electrons/photon were found for 21.2, 16.9, 11.8, 10.2, and 7.7 eV radiation, respectively [2]. This result supports an earlier estimate of $Y = 0.05$ made for clean W irradiated with photons of 21.2 eV in a study of the yield as a function of oxygen exposure [3].

For W exposed to air, a value of $Y \approx 0.112 \pm 0.016$ was obtained at 584 Å (21.2 eV); Y decreased from 0.1 near 850 Å to 0.01 near 1290 Å [4]. This is consistent with the measured quantum efficiency of a W photodiode (behind an MgF_2 window). The efficiency decreased exponentially with λ from ~1% near 1150 Å to ~0.01% near 1850 Å [5].

The photoelectric yield Y was determined from intensity vs. λ spectra taken under normal incidence with λ between 31.6 and 304 Å. The W polycrystals used had been exposed to ambient air prior to the measurements. Between 304 and 1216 Å such spectra were taken under 45° incidence [6]. Including the normal-incidence data of another study [7], an increase of the yield with the wavelength of the photons from about 0.02 electrons/photon near 50 Å to 0.08 electrons/photon at about 400 Å was obtained [6]. Y was measured in the range 473 to ~1000 Å (26.2 to ~12.4 eV) on polycrystalline W at $p \leq 10^{-4}$ Torr. Degassing under these conditions reduced the maximum yield found in this range from ~0.15 electrons/photon (contaminated) to a maximum of ~0.05 electrons/photon (degassed). The "clean" curve is much flatter than the curve from gas-covered W and ranges between ~0.02 and ~0.05 electrons/photon [8]. For heavily contaminated W surfaces studied between $\lambda \approx 400$ and 1200 Å (~31 to ~10 eV), see [9].

For W degassed in vacuo by rf heating, the measured λ dependence of Y is compared in [10] with the dependence calculated according to [11]. The threshold wavelength λ_0 is 3120 Å [10, pp. 282/6]. Studying the photoemission from vapor-deposited W at $p < 10^{-7}$ Torr, λ_0 ranged from 2800 to 2830 Å for W when freshly deposited at 90 K and decreased to 2780 to 2760 Å on warming to room temperature with a concomitant decrease of the yield from $\sim 33 \times 10^{-6}$ to $\sim 5 \times 10^{-6}$ electrons per 2447 Å photon. For W films deposited at 0 °C, $\lambda_0 \approx 2670$ Å was found [12].

The (X-ray) photoelectric yield of a W photocathode was measured at various angles of incidence. For incidence angles $\alpha \leq 75°$, the quantum yield κ increases proportionally with 1/cos(α) [13]; see also [14]. The κ for Ti-filtered X radiation (mean photon energy 4.4 keV) under normal incidence is $\kappa = 0.012$ electrons/photon (~4.4 keV), for $\alpha = 75°$ it is $\kappa = 0.03$ [14]. For $\alpha > 75°$ the yield curve deviates from the 1/cos(α) behavior to smaller values, apparently because of reflection of the X-rays. For $\alpha = 80°$ the quantum yields $\kappa =$ 0.018, 0.048, 0.059, 0.11, 0.054, 0.073, and 0.081 were obtained for $\lambda = 1.537$, 2.743, 3.351, 5.395, 7.111, 8.321, and 9.870 Å, respectively [13]. Quantum yields of κ (in %) = 3.6, 4.0, 2.2, and 0.94 have been measured with X-rays incident at $\alpha \approx 60°$ for the K lines of O, C, B, and Be, i.e., for $\lambda = 23.6$, 44, 67, and 113 Å, respectively. The nondegassed W surface was studied at 5×10^{-7} Torr [15]. On polished (emery cloth) W and for $\alpha \approx 60°$, the quantum yields κ (in %) = 0.36, 0.82, 0.60, 0.80, 1.50, 0.94, 1.70, and 2.3 were measured for the K_α lines of Cu, Cr, Ti, Ca, Cl, Al, Na, and the L_α line of Cu, i.e., for $\lambda = 1.537$, 2.278, 2.74, 3.35, 4.72, 8.32, 11.9, and 13.3 Å, respectively [16].

References for 5.5.1:

[1] Waclawski, B. J.; Plummer, E. W. (Phys. Rev. Lett. **29** [1972] 783/6).
[2] Waclawski, B. J.; Hughey, L. R. (J. Opt. Soc. Am. **59** [1969] 1494/5).

[3] Waclawski, B. J.; Hughey, L. R.; Madden, R. P. (Appl. Phys. Lett. **10** [1967] 305/7).
[4] Hinteregger, H. E.; Watanabe, K. (J. Opt. Soc. Am. **43** [1953] 604/8).
[5] Heath, D. F.; McElaney, J. H. (Appl. Opt. **7** [1968] 2049/52).
[6] Heroux, L.; Hinteregger, H. E.; Bedo, D. E.; McMahon, W. J.; Cohen, M.; Manson, J. E.; Newburgh, R. G.; Chagnon, C. W. (AD-648918 [1966] 86 pp.; C.A. **68** [1968] No. 99715).
[7] Cairns, R. B.; Samson, J. A. R. (GCA-TR-65-29-N [1965] cited from [6]).
[8] Walker, W. C.; Wainfan, N.; Weissler, G. L. (J. Appl. Phys. **26** [1955] 1366/71).
[9] Wheaton, J. E. G. (J. Opt. Soc. Am. **54** [1964] 1287/8).
[10] Dubois, E.; Bathier, M. (J. Phys. Radium [8] **23** [1962] 277/86).

[11] Fowler, R. H. (Phys. Rev. [2] **38** [1931] 45/56).
[12] Hermann, H. (Diss. TH Braunschweig 1955, pp. 1/60).
[13] Rumsh, M. A.; Lukirskii, A. P.; Shchemelev, V. N. (Dokl. Akad. Nauk SSSR **135** [1960] 55/7; Sov. Phys. Dokl. **5** [1960] 1231/3).
[14] Ganeev, A. S.; Israilev, I. M. (Zh. Tekh. Fiz. **31** [1961] 376/82).
[15] Lukirskii, A. P.; Rumsh, M. A.; Smirnov, L. A. (Opt. Spektrosk. **9** [1960] 511/5; Opt. Spectrosc. [Engl. Transl.] **9** [1960] 265/7).
[16] Lukirskii, A. P.; Rumsh, M. A.; Karpovich, I. A. (Opt. Spektrosk. **9** [1960] 653/7; Opt. Spectrosc. [Engl. Transl.] **9** [1960] 343/6).

5.5.2 Valence-Band Spectra

In this section emphasis is placed on surface electronic properties. For photoemission measurements on the bulk valence-band structure, see "Tungsten" Suppl. Vol. A 2, 1987, pp. 116/7.

5.5.2.1 Introduction

The increasingly easier access to synchrotron radiation is reflected in the literature on photoemission which expanded considerably during the seventies. In addition to the tunability of the wavelength, synchrotron radiation is most remarkable for its high degree of polarization. The basic information about photoemission from W, both from bulk and surface states, is nonetheless due to ultraviolet photoemission spectroscopy (UPS) studies using resonance radiation from glow discharges. The Swanson hump observed in field emission measurements (see p. 144), for instance, was identified as a surface state in UPS work with discharge lamps [1, 2].

For theories of UPS from metal surfaces and the phenomenological three-step model of UPS from the bulk, see the review [3]. UPS and electron states at clean surfaces have been comprehensively reviewed by [4]; see especially the review of UPS on W in [4, pp. 203/11]. The study of the electronic structure by angle-resolved UPS (ARUPS) is reviewed in [5]. Polarization-dependent ARUPS (PARUPS) spectra were found to be in excellent agreement with polarization-related selection rules [6, 7] (for these, see [8]). The photoexcitation is described by matrix elements $\langle f|\vec{A}\cdot\nabla|i\rangle$ containing explicitly the vector potential $\vec{A}$. Thus, ARUPS spectra depend on the orientation of $\vec{A}$ with respect to the symmetry elements of the crystal. PARUPS spectra give initial state symmetries for two geometries: emission along the surface normal and emission in a mirror plane [7, pp. 308/9]. An example for the polarization dependence of the emission in a mirror plane is given in Fig. 38, p. 179.

Feuerbacher and Christensen's extensive papers [9, 10] give the results of calculations of the W band structure and emission spectra [9] which were complemented by an experi-

 References for 5.5.2.1 on p. 171

mental photoemission study [10]. These papers are of fundamental importance for understanding the UPS spectra from W. Three emission types may be expected: (i) emission due to direct transitions, (ii) emission due to inelastic scattering, and (iii) surface emission. Electrons excited by direct transitions and leaving the crystal unscattered carry information on the bulk band structure along a specific symmetry line in $\vec{k}$ space. The signature for direct transitions in UPS spectra is a shift of the initial state energy as hν is varied. The inelastically scattered electrons carry information on the conduction-band density of final states. Surface photoemission is considered a one-step process in which a transition takes place from an initial state localized near the surface to a free-electron-like state outside the crystal. Electrons emitted due to such a process carry infomation on the surface density of initial states [10]. A peak in UPS spectra is assigned to a surface state or surface resonance if the following three criteria are fulfilled: (i) it is sensitive to surface contamination (preferably by hydrogen adsorption which has the least effect on final states); (ii) the initial state energy (E_i) is independent of the photon energy hν, because a continuum of final states is available; (iii) it should be attributable to a surface state or resonance from a theoretical point of view. A true surface state exists in a symmetry band gap of the bulk band structure along the momentum component normal to the surface. It is characterized by a wave function that decays in the bulk. A surface state therefore is trapped at the crystal/vacuum interface, i.e., it is a localized state. If the state couples to a bulk band, it is not entirely located at the surface and is called a surface resonance. The distinction between surface states and surface resonances on the basis of measured spectra is subtle and requires knowledge of the bulk band structure [11]; see also [4].

There are three meaningful ways to plot UPS data that describe the excitation of an electron at E_i by hν to the final energy E_f. In energy distribution curves (EDCs) the intensity is plotted vs. the initial-state energy relative to the Fermi level ($E_i - E_F = E_f - h\nu - E_F = E_{kin} + \Phi - h\nu$ with Φ denoting the work function and E_{kin} the kinetic energy of the emitted electrons). The kinetic energy of the photoelectrons is scanned at fixed hν yielding the

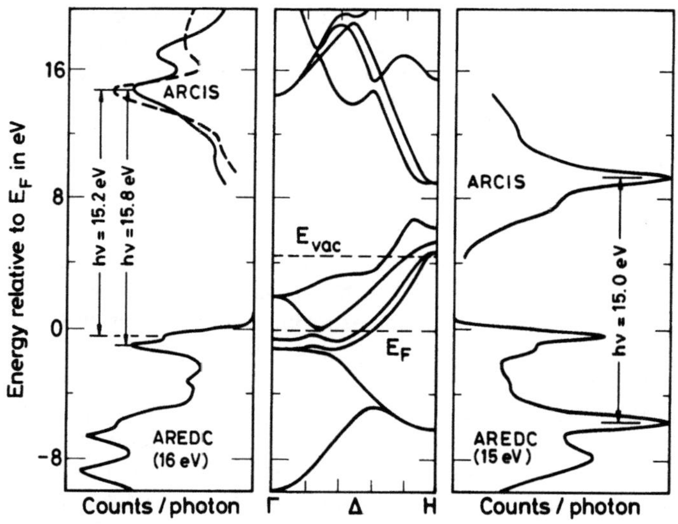

Fig. 34. Angle-resolved energy distribution curves and constant-initial-state curves (AREDC and ARCIS, respectively) for normal emission from W(100). The middle panel shows calculated bands (taken from Christensen and Feuerbacher [9]). The initial energy for each ARCIS spectrum is indicated by the horizontal line at the bottom of the double arrow [6, 12].

EDCs N(E_f; hν fixed). Constant final state (CFS) spectra are obtained scanning hν at a fixed E_f yielding N(hν, E_f fixed). To obtain constant initial state (CIS) spectra, E_f and hν are scanned synchronously yielding N(hν, E_i fixed). In ARUPS the EDCs and CIS spectra sample transitions on a plane in momentum space. ARCFS spectra, on the other hand, are taken at constant $k_{\parallel}$, the momentum component parallel to the surface [6, 12]. **Fig. 34** [6, 12] illustrates s-polarized AREDCs and ARCIS spectra from W(100) taken at normal emission, i.e., along the Δ line of the bulk. The measured densities of states are in good agreement with the bulk band structure calculation of [9] (note that the s polarization used strongly suppresses surface-related features) [6, 12].

References for 5.5.2.1:

[1] Waclawski, B. J.; Plummer, E. W. (Phys. Rev. Lett. **29** [1972] 783/6).
[2] Feuerbacher, B.; Fitton, B. (Phys. Rev. Lett. **29** [1972] 786/9).
[3] Glasser, M. L.; Bagchi, A. (Prog. Surf. Sci. **7** [1976] 113/47).
[4] Feuerbacher, B.; Willis, R. F. (J. Phys. C **9** [1976] 169/216).
[5] Smith, K. E.; Kevan, S. D. (Prog. Solid State Chem. **21** [1991] 49/131).
[6] Smith, R. J.; Lapeyre, G. J.; Anderson, J.; Williams, G. P. (Conf. Ser.-Inst. Phys. No. 39 [1978] 220/32).
[7] Lapeyre, G. J.; Anderson, J.; Smith, R. J. (Surf. Sci. **89** [1979] 304/18).
[8] Hermanson, J. (Solid State Commun. **22** [1977] 9/11).
[9] Christensen, N. E.; Feuerbacher, B. (Phys. Rev. [3] B **10** [1974] 2349/72).
[10] Feuerbacher, B.; Christensen, N. E. (Phys. Rev. [3] B **10** [1974] 2373/90).
[11] Holmes, M. W.; King, D. A. (Proc. R. Soc. London A **376** [1981] 565/83).
[12] Lapeyre, G. J.; Smith, R. J.; Anderson, J. (J. Vac. Sci. Technol. **14** [1977] 384/90).

5.5.2.2 Studies Comparing Different Surfaces W(hkl)

Much work on UPS from different W surfaces was done by the Feuerbacher group [1 to 7]. The comprehensive paper [1] is the experimental complement to the W band-structure calculation [2] which obviously supersedes the bulk band-structure calculation [8]. **Fig. 35**, p. 172, [3] gives EDCs from the (110), (100), and (111) surfaces excited by various photon energies. The strong peak at 0.4 eV below E_F (feature No. 1 in Fig. 35) hardly changes size or position (E_i) within the hν range 7.7 to 11.7 eV under study. It was assigned to emission from a surface state (Swanson hump) [4, 9]. Feature No. 2 at 4.5 eV below E_F is due to surface emission from a region with high DOS near the surface [5]. Feature No. 3, on the other hand, shifts to lower initial energies as hν increases. It is attributed to electrons emitted after direct interband transitions. Feature No. 4 (which does *not* coincide with No. 1) at 0.4 eV below E_F in the UPS spectra from W(110) gains intensity when hν increases from 7.7 to 11.7 eV. This intensity change and its moderate sensitivity to H_2 adsorption suggest that this peak seen on W(110) is neither a surface state nor a surface resonance [1, 3, 6]. The intensity of the peak from W(110) at 0.4 eV below E_F linearly decreases with increasing hydrogen exposure (at least up to 2.2 L H_2), while the W(100) feature at 0.4 eV below E_F is very much more adsorption-sensitive (virtually disappearing after an exposure to 0.4 L of H_2 [7], corresponding to a coverage of a small fraction of a monolayer [1]) [6]. Feature No. 5 coincides in the UPS spectra from W(110) and W(111). It was assumed that it is due to a critical point of the fourth band at Γ common to both faces [3]. A comparison of the EDCs from W(110) and W(100) taken at higher hν values shows a remarkable difference in general behavior: in the W(100) spectrum taken at hν = 21.2 eV it is the low-energy (i.e., the inelastically scattered) electrons that contribute the major intensity, while in the corresponding W(110) spectrum the intensity of inelastically scattered electrons is weak. This

References for 5.5.2.2 on p. 173

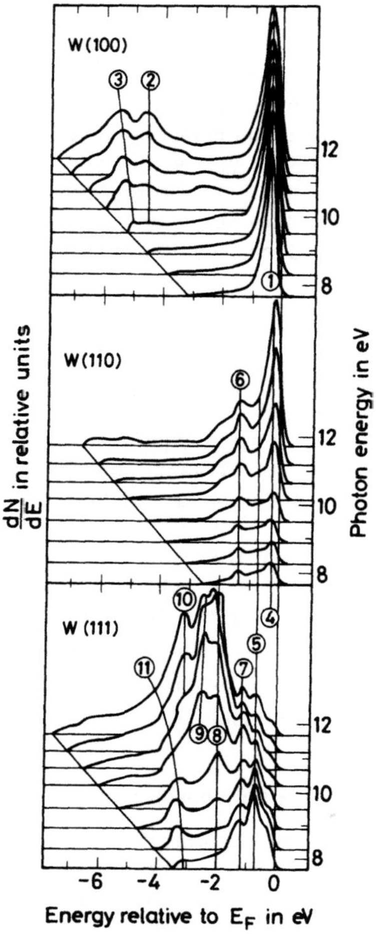

Fig. 35. Energy-distribution spectra of photoelectrons emitted normal to three single-crystal surfaces of W. Photon energies are given in the right-hand scale. The features labeled 1 to 11 are discussed in the text [3].

fact is attributed to a lack of available final states in the case of W(110). This surface shows, in addition to a conduction band gap, a large valence band gap which is not present on W(100) [1].

The one-dimensional surface DOS of W(100) obtained in a tight-binding calculation shows a strong peak in the spd hybridization gap at 2.3 eV below E_F [10]. In contrast to the conclusions drawn in [1], no narrowing of the gap width at the surface was found by [10]. Rather, the dangling-bond model used in [10] yields a strong reduction of the second moment of the band by removing the singularities at the band edges [10]. This calculation was extended to W(111) in [11]. Again, bulk band edge singularities are removed by confining the electrons to the surface layer, and a strong surface DOS peak emerged in the spd gap. One surface feature located at 1.36 eV below E_F is caused by a resonance, while a feature at 0.75 eV below E_F is brought about by a surface state [11]. These calculated features are identified with the experimental features at 1.2 eV and 0.7 eV below E_F found in the normal-emission UPS study [1, 3] (labeled "7" and "5", respectively, in Fig. 35 [3]). Inclusion of electron-electron scattering in the calculations [11] resulted in a broad DOS peak around 2.7 eV below E_F which may be associated with a prominent feature in the UPS spectrum measured by [1, 3]. The close correspondence of the calculated W(111) with the measured UPS results (by [3]) points to surface reconstruction (not included in the calculations) as a possible source for the poor correspondence in the case of W(100) between calculated DOS and measured UPS spectrum.

Indirect transitions are unimportant under the conditions of room-temperature UPS. This was shown experimentally by the weak emission in the range 1.5 to 4 eV below E_F for W(100) [1, p. 2380] and between 2.5 and 5 eV below E_F for W(110) [1, p. 2384]. The band structure calculation [2] indicates high densities of states in parts of the Brillouin zone other than the Δ line ([100]); hence, there would be more emission seen normal to W(100), if nondirect transitions were important [1, p. 2380]. The case of W(111) is less clear-cut [1, p. 2386].

UPS spectra recorded on W(110) and W(100) with 16.85 and 21.22 eV radiation are given in [12]. The measured EDCs for collection cones of ±45° and ±3.5° compare fairly well [12] with the band structures calculated by [2]. For W(100) the band gap centered at 8 eV above E_F occurred some 0.8 eV higher in energy than the calculations of [2] indicated, while for (110) the upper edge of the calculated gap may be ~0.5 eV too high [12]. On the other hand, an experimental study showed for W(110) an upward shift of the gap by nearly 1 eV [1]. The spectra from W(110) and W(100) taken with 16.85 eV radiation and collected at ±45° show a strong peak at 3.15 eV below E_F [12] (which was not seen with ±15° collection (see [9]), but which appeared on an EDC from W(110) in [13]). According to [12], most of the angular variations in spectra recorded with radiation energies $h\nu \geq 15$ eV are smoothed out for collection angles ≥20°.

References for 5.5.2.2:

[1] Feuerbacher, B.; Christensen, N. E. (Phys. Rev. [3] B **10** [1974] 2373/90).
[2] Christensen, N. E.; Feuerbacher, B. (Phys. Rev. [3] B **10** [1974] 2349/72).
[3] Feuerbacher, B.; Fitton, B. (Phys. Rev. Lett. **30** [1973] 923/6).
[4] Feuerbacher, B.; Fitton, B. (Phys. Rev. Lett. **29** [1972] 786/9).
[5] Feuerbacher, B. (Surf. Sci. **47** [1975] 115/23).
[6] Feuerbacher, B.; Fitton, B. (Surf. Sci. **37** [1973] 386/93; discussion on pp. 393/4).
[7] Feuerbacher, B.; Fitton, B. (Phys. Rev. [3] B **8** [1973] 4890/2).
[8] Petroff, I.; Viswanathan, C. R. (Phys. Rev. [3] B **4** [1971] 799/816).
[9] Waclawski, B. J.; Plummer, E. W. (Phys. Rev. Lett. **29** [1972] 783/6).
[10] Hermanson, J.; Kawajiri, M.; Schwalm, W. (Solid State Commun. **21** [1977] 327/9).
[11] Kawajiri, M.; Hermanson, J.; Schwalm, W. (Solid State Commun. **25** [1978] 303/5).
[12] Plummer, E. W.; Waclawski, B. J.; Vorburger, T. V.; Kuyatt, C. E. (Prog. Surf. Sci. **7** [1976] 149/82, 150/6).
[13] Baker, J. M.; Eastman, D. E. (J. Vac. Sci. Technol. **10** [1973] 223/6).

5.5.2.3 UPS Studies on W(110) Surfaces

The normal-emission ARUPS spectrum from 42 eV photons incident under an angle of 45° and with the electric field along $[1\bar{1}0]$ shows peaks at 1.0, 1.5, and 3.5 eV below E_F which are very sensitive to hydrogen adsorption. The corresponding spectrum for polarization along [001] shows a peak located at 1 eV below E_F with a shoulder near 0.5 eV

References for 5.5.2.3 on pp. 177/8

below E_F, the peak at 1.5 eV below E_F is strongly reduced or even missing, and the peak at 3.5 eV below E_F is absolutely missing [1]. The dispersion of the UPS features observed on W(110) is compared in [1] with a projection of the bulk bands (see [2]) onto the (110) surface. The observed polarization dependence indicates d_{xz} character for the features at 3.5 and 1.5 eV below E_F ([1$\bar{1}$0], [001], and [110] playing the roles of x, y, and z, respectively). The state corresponding to the shoulder at 0.5 eV below E_F has d_{yz} character. The peak at 1.0 eV below E_F appears in the EDCs for both polarizations, suggesting $d_{(3z^2-r^2)}$ or $d_{(x^2-y^2)}$ symmetry. The angle-of-incidence dependence of the intensity is compatible with d_{xz} symmetry; thus it appears that this state has both d_{xz} and $d_{(3z^2-r^2)}$ or $d_{(x^2-y^2)}$ character. Spectra for ten energies between 24 and 50 eV show considerable variations in the peak intensity, but at least for the major peaks the positions do not change with hν, indicative of electronic states located near the surface [1]. The leading peak of these spectra seems to correspond to feature No. 6 in Fig. 35, p. 172.

A strong peak observed by [3] near 3.4 eV below E_F in EDCs measured with photon energies of 16.8 eV was assigned to a direct transition, since it was found neither at higher nor lower photon energies [3]. The ARUPS spectra in the later paper [1], on the other hand, do show a significant peak near 3.7 eV below E_F for photon energies of 30 and 45 eV [1]. For excitation energies higher than about 9.5 eV, direct transitions occur, and above the threshold energy of 10.1 eV the intensity of peak No. 4 (see Fig. 35, p. 172) strongly increases. This threshold and the slope of the plot of intensity vs. hν [3] fairly agree with the prediction based on the band structure by [2]. The intensity of the surface features No. 4 and 6 in Fig. 35, p. 172, is fairly constant within the range of the calculated conduction band gap [3]. The pronounced scarcity of emission from initial states between about 2.5 and 5 eV below E_F indicates that emission of electrons that suffered umklapp processes is rather unimportant, although high densities of states are expected (according to [2]) elsewhere at 3 and 4.5 eV below E_F in reciprocal space. However, the EDC region of low emission does not exactly coincide with the calculated position of the band gap. The experimental gap is shifted by nearly 1 eV towards higher energies, apparently because of *d*-band narrowing at the surface [3].

Extending the earlier papers [4 to 6], ARUPS measurements from W(110) surfaces using photon energies of 16.8, 21.2, and 40.8 eV were reported in [7]. Spectra taken along the [001] azimuth of the (110) surface are along the $\bar{\Sigma}$ symmetry line of the surface Brillouin zone (SBZ), those along [1$\bar{1}$0] are along the $\bar{\Delta}$ symmetry line. These ARUPS studies, which include detection angles as large as 85°, were the first to reveal a surface state on W(110). The dispersion of this surface state correlates with the projected band gap between the third and fourth bands of the W bulk band structure [7].

In addition to ARUPS spectra along the $\bar{\Sigma}$ and $\bar{\Delta}$ symmetry lines (intersections of the SBZ with the ΓNP and ΓNH planes of the bulk BZ, respectively), other series of ARUPS spectra were recorded at various exit angles and along a range of azimuths spaced at 5° intervals between the $\bar{\Sigma}$ and the $\bar{\Delta}$ lines. These additional studies showed that the surface states found far away from the zone center on $\bar{\Sigma}$ and $\bar{\Delta}$ are the same state (which does not exist around $\bar{\Gamma}$) [6, 7]. The map derived from the ARUPS spectra agrees well with the calculated contour plot of the bottom of the uppermost absolute band gap (below E_F). This supports the assignment of the observed adsorption-sensitive structures in the UPS spectra to emission from an extended surface state in the band gap between the third and fourth bands of W [7]. The intensity map of the surface state near 0.5 eV below E_F, recorded with unpolarized radiation, shows that the emission maximum is not along the $\bar{\Delta}$ line, but occurs at an azimuth some 20° or 30° away [7]. This observation is consistent with the tight-binding model in [5], where the transferability of surface states was discussed.

For a large portion of the $\bar{\Sigma}$ line, the adsorption-sensitive feature near 0.5 eV below E_F is a true surface state. But close to the SBZ boundary, the state no longer exists in the band gap and turns into a surface resonance [7].

UPS studies along the $\bar{\Delta}$ line: EDCs taken with photon energies of 16.8, 21.2, and 40.8 eV are shown in **Fig. 36** [7]. The EDC obtained with photons of 21.2 eV [6, 7] at $\vartheta=0°$ (normal exit) agrees well with the corresponding EDC in [3]. The peak labeled "A", which appears at exit angles of $\vartheta \geq 55°$, has an unusually strong intensity and moves away from the Fermi level as ϑ increases. The attenuation of peak "A" under hydrogen exposure suggests that it is either due to a surface state or a resonance. "A" is not due to an initial state in the first SBZ, rather it involves an umklapp process. Feature "B" (arrowed) was first seen at $\vartheta=20°$ at 1.8 eV below E_F and moved towards E_F as ϑ increased to 35° [7]; see also [6]. The 16.8 eV EDC shows the adsorption-sensitive feature "D" which appears for $\vartheta=30°$ at 0.9 eV below E_F and disperses towards the Fermi level with increasing ϑ. It disappears for $\vartheta>45°$. Feature "E" in the EDC at $\vartheta=80°$ is also sensitive to gas adsorption. The feature "F" in the 40.8 eV EDC taken along the $\bar{\Delta}$ line is tentatively ascribed to emission from the surface. The adsorption-sensitive feature "G" appears at $\vartheta=37.5°$ and 0.3 eV below E_F and disperses at higher binding energies as ϑ increases (for $\vartheta=55°$ it appears at 1.5 eV below E_F). All features labeled "A" to "G" seem to correspond to emissions from the same (dispersing) surface state. Features "A", "E", and "G" are due to emissions involving surface umklapp processes, while features "B", "D", and (tentatively also) "F" correspond to direct emissions [7]. The surface states on W(110) found by UPS seem to

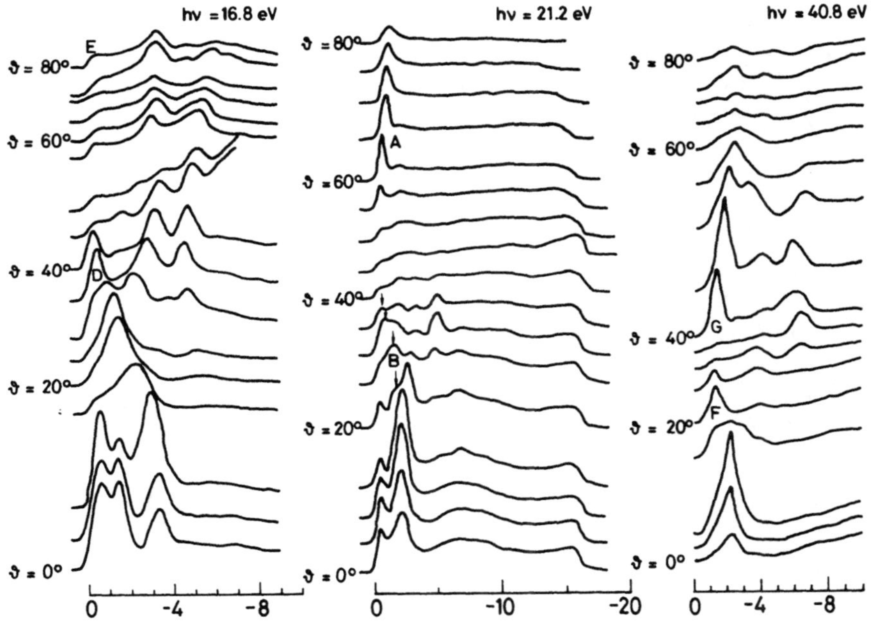

Fig. 36. Families of energy distribution curves for the $\bar{\Delta}$ symmetry line with electron detection angles ranging from 0° to 80° in 5° steps. Left, middle, and right panels refer to excitation by Ne^{I} (16.8 eV), He^{I} (21.2 eV), and He^{II} (40.8 eV) radiation, respectively, incident at 45° on W(110). The two upper curves in both the left and the right panel are scaled up by a factor 3.3. The features labeled "A" to "G" are discussed in the text [7].

References for 5.5.2.3 on pp. 177/8

be related with surface states on W(100) having the same orbital character. The fact that the surface states prominent in the W(110) UPS spectra were not seen in field emission spectra is explained in [6].

UPS studies along the $\bar{\Sigma}$ line: EDCs taken with photon energies of 16.8, 21.2, and 40.8 eV are given in **Fig. 37** [7]. The series of spectra obtained with photons of 16.8 eV in the left panel of Fig. 37 shows an adsorption-sensitive feature that appears first in the spectrum at $\vartheta = 45°$ (at 0.9 eV below E_F) and disperses towards higher energy as ϑ increases to 65°. This surface feature stems from the first SBZ. From the same surface state a small feature seems to originate which behaves similarly on the $\bar{\Sigma}$ symmetry line and which is said to occur with 21.2 eV radiation at 1.8 eV below E_F for $\vartheta = 30°$ and to disperse towards E_F as ϑ inreases to 50°. The series of EDCs taken with photons of 40.8 eV given in the right panel of Fig. 37 shows the two adsorption-sensitive features "J" and "M". Feature "J" appears in the $\vartheta = 25°$ spectrum at 0.9 eV below E_F, disperses towards the Fermi level as ϑ increases, and vanishes at $\vartheta = 30°$. It stems from the second SBZ. The small feature labeled "M" observed for $\vartheta \geq 55°$ is also attributed to indirect transitions, but involves more than one surface reciprocal lattice vector. It is associated with the third SBZ and emerges for $\vartheta = 55°$ at 0.5 eV below E_F. As ϑ increases it disperses to lower energies. With $\vartheta = 80°$ ($k_{\parallel} = 2.9$ Å^{-1}) "M" is observed at 1.8 eV below E_F [7].

At 1.18 ± 0.04 eV below E_F a surface resonance was found in a pseudogap opened by spin-orbit interaction. Near $\bar{\Gamma}$ this resonance occurs 0.27 ± 0.02 eV above the lower band

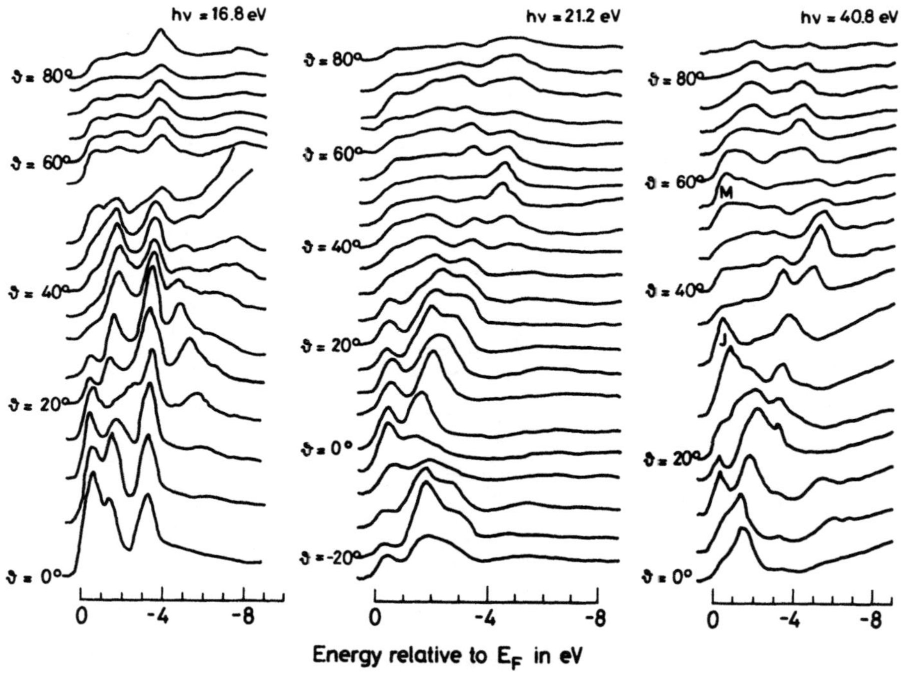

Fig. 37. Families of energy distribution curves for the $\bar{\Sigma}$ symmetry line with electron detection angles ranging from 0° to 80°, −20° to +80°, and 0° to 85° for Ne^{I} (16.8 eV), He^{I} (21.2 eV), and He^{II} (40.8 eV) radiation, respectively, incident at 45° on W(110). The Ne^{I} EDC for $\vartheta = 80°$ and the He^{I} EDCs for $\vartheta = 70°$ to 80° are scaled up by a factor 3.3. The features labeled "J" and "M" are discussed in the text [7].

edge and 0.47±0.02 eV below the upper band edge. Away from $\bar{\Gamma}$, surface states were studied along the $\bar{\Sigma}$ line for $k_{\parallel}=0.6\ Å^{-1}$ with photons of 39 eV. The peak at 1.18±0.04 eV below E_F seen in normal-emission EDCs taken with 20 eV radiation (with $\vec{A}$ at 45° in the (001) mirror plane) is found using 38 eV radiation at the same initial energy, even though with less intensity. The peak is quenched by exposure to 0.4 L hydrogen. These properties identify this peak as surface feature. It is suppressed in spectra taken with $\vec{A}$ in the (110) mirror plane and also suppressed in spectra taken as little as 3° off the surface normal [8]. In addition, a broad (composite) feature at about 0.3 eV below E_F was seen in ARUPS spectra obtained with photons of 39 eV. It was split but not suppressed on exposure to H_2. The low-energy component of this double peak shows a downward shift of 0.5 to 0.6 eV for monolayer adsorption of hydrogen [9 to 11]. ARUPS spectra taken with $\vec{A}$ parallel to $\bar{\Sigma}$ show that the asymmetric shape of this feature in the $k_{\parallel}=0.62\ Å^{-1}$ EDC is due to overlap of peaks at 0.31 and 0.20 eV below E_F. The former peak belongs to a state existing for $k_{\parallel}\approx 0.4$ to $0.8\ Å^{-1}$. Its behavior under hydrogen and oxygen exposure suggests that it is a surface state. Since hydrogen induces surface reconstruction on W(110), a connection between the peak at 0.3 eV below E_F and reconstruction is conjectured [9].

For radiation polarized with $\vec{A}\parallel\bar{\Delta}$ and $h\nu=20$ eV, ARUPS spectra taken along $\bar{\Delta}$ for $k_{\parallel}=0.5$ to $1.14\ Å^{-1}$ show a double peak near 1 eV below E_F for $k_{\parallel}=0.5\ Å^{-1}$ which splits when increasing $k_{\parallel}$. Both peaks disperse towards E_F; they cross the Fermi level at $0.84\ Å^{-1}$ and $1.1\ Å^{-1}$. Since they are sensitive to contamination and do not disperse with $k_{\perp}$ [12, 13], they are identified as surface resonances [12]. The resonance crossing E_F near $k_{\parallel}=1.10\ Å^{-1}$ is coupled to the bulk continuum by spin-orbit interaction [13]. For a discussion of the Fermi surface, see Section 1.2.3.3, pp. 64/5.

The anomalous broadening of ARUPS line widths has been studied on W(110) surfaces in paper [14] concentrating on the line labeled AB perpendicular to $\bar{\Sigma}$; see Fig. 21, p. 65 [12]. This line intersects the irregular electron pocket (centered on $\bar{\Gamma}$) and the ellipsoidal hole pocket located on the SBZ boundary. A family of AREDCs ($k_{\parallel\bar{\Sigma}}=1.01\ Å^{-1}$, various $k_{\parallel\bar{\Delta}}$) measured at $h\nu=20$ eV shows two features that cross the Fermi level on varying $k_{\parallel\bar{\Delta}}$. Both of them have a nonvanishing width at E_F. Two peaks and a shoulder are seen at 0.1 eV below E_F with the photoemission intensity plotted as a function of $k_{\perp\bar{\Sigma}}$. A least-squares fit of three Lorentzian profiles added to a linear background results in two peaks (at 0.27 and $0.47\ Å^{-1}$) having an FWHM of $0.08\ Å^{-1}$ each and a shoulder corresponding to a peak with an FWHM of $0.095\ Å^{-1}$. Along other lines in the SBZ, quantitatively similar results were obtained. Whenever the Fermi contour was resonant with the projected bulk continuum, a momentum width of $0.095\ Å^{-1}$ was observed, slightly larger than the $0.08\ Å^{-1}$ found for surface-state crossings. Apparently, the resonance contribution to the observed line widths is small. The observed broadening of the surface states on crossing E_F is attributed to phonon emission [14].

References for 5.5.2.3:

[1] Blanchet, G. B.; di Nardo, N. J.; Plummer, E. W. (Surf. Sci. **118** [1982] 496/512, 498/503).
[2] Christensen, N. E.; Feuerbacher, B. (Phys. Rev. [3] B **10** [1974] 2349/72).
[3] Feuerbacher, B.; Christensen, N. E. (Phys. Rev. [3] B **10** [1974] 2373/90).
[4] Holmes, M. W.; King, D. A. (Ned. Tijdschr. Vacuumtech. **16** [1978] 286/7).
[5] Holmes, M. W.; Inglesfield, J. E. (Surf. Sci. **89** [1979] 133/41).
[6] Holmes, M. W.; King, D. A.; Inglesfield, J. E. (Phys. Rev. Lett. **42** [1979] 394/7).
[7] Holmes, M. W.; King, D. A. (Proc. R. Soc. London A **376** [1981] 565/83).
[8] Gaylord, R. H.; Kevan, S. D. (Phys. Rev. [3] B **36** [1987] 9337/40).
[9] Gaylord, R. H.; Kevan, S. D. (Phys. Rev. [3] B **37** [1988] 8491/4).
[10] Kevan, S. D.; Gaylord, R. H. (J. Vac. Sci. Technol. A **6** [1988] 636/7).

[11] Jeong, K.; Gaylord, R. H.; Kevan, S. D. (J. Vac. Sci. Technol. A **7** [1989] 2199/202).
[12] Gaylord, R. H.; Jeong, K. H.; Kevan, S. D. (Phys. Rev. Lett. **62** [1989] 2036/9).
[13] Gaylord, R. H.; Jeong, K.; Dhar, S.; Kevan, S. D. (J. Vac. Sci. Technol. A **7** [1989] 2203/6).
[14] Dhar, S.; Kevan, S. D. (Phys. Rev. [3] B **41** [1990] 8516/8).

5.5.2.4 UPS Studies on W(100) Surfaces

Two surface resonances (at 0.4 and 4.2 eV below E_F) were found for normal exit ($\vartheta = 0°$), while three surface resonances (at 0.4, 0.8 to 0.9, and 4.2 eV below E_F) were seen at exit angles $\vartheta > 2°$ [1, 2]. The feature around 0.4 eV below E_F is related to the Swanson hump (see p. 144) and assigned to a surface resonance in the spin-orbit gap. It is fairly isotropic on the W(100) surface, but was detected only over ~20% of the surface Brillouin zone (SBZ) [3]. At the center of the SBZ, it is located at 0.3 [4, 5], 0.35 [6], 0.4 [1, 2, 7 to 11], 0.5 eV [12], or 0.6 eV below E_F [13]. The low-lying surface resonance was found at 4.2 [1], 4.3 [12], 4.4 [7], and 4.5 eV below E_F [14].

According to [1] the high-lying resonances (at 0.4 and 0.8 to 0.9 eV below E_F) show hardly any dispersion along the [011] azimuth ($=\bar{\Sigma}$ line) and an only weak shift towards E_F along $\bar{\Delta}$ [1]. Using better resolution, a fairly isotropic dispersion was found in [3, 8] for the topmost resonance which shifts slightly towards E_F for increasing values of ϑ [8]. For the dispersion along $\bar{\Sigma}$ of the two high-lying resonances, see also the papers [4, 6, 15] reviewed below in the section on reconstruction of W(100). Along the $\bar{\Sigma}$ azimuth the topmost resonance was observed only up to $k_{\parallel} = 0.69$ Å^{-1}; for $k_{\parallel} > 0.69$ Å^{-1} it was too weak to be observed [5]. The second high-lying surface state is a doublet state, absent at $\bar{\Gamma}$ and crossing the Fermi level near 1.2 Å^{-1} [5]. Selection rules (from [16]) show that this state has an even and an odd component which behave differently, especially near $k_{\parallel} = 0.7$ Å^{-1} [5].

A lot of information on the surface emission from W(100) was obtained from measurements of the polarization dependence of the peak intensities [1, 5, 9, 12, 17 to 19]. The paper [9] is cited in [17] as being the first to report strong s- versus p-polarization effects. These were confirmed in [12] showing more clearly the presence of a large peak at 0.3 eV below E_F and of a smaller one at 4.3 eV below E_F under p polarization and their absence under s polarization [12]. For normal emission and $\vec{A}$ normal to the (001) surface, only totally symmetric initial states were detected. Thus, these surface states have s-, p_z-, or d_{z^2}-orbital character [17].

Extending the normal incidence/normal emission studies [20, 21], PARUPS spectra were taken in [18] using ~88% s-polarized 13 to 20 eV radiation incident at 42.3° on a W(100) surface. The electron collection angle also was $\vartheta = 42.3°$. The s-polarized spectra depend on the angle enclosed between $\vec{A}$ and the crystal symmetry elements, i.e., (001) and (011) mirror planes perpendicular to the (100) surface. Some of the peaks in the UPS spectra observed for emissions from the (001) and (011) mirror planes show a strong polarization dependence [18]. For example, UPS spectra recorded with 20 eV radiation and electron exit in the [010] azimuth for $\vec{A}$ both parallel and normal to the (001) mirror plane show a peak at 5.4 eV below E_F which is strong with $\vec{A}$ in the plane spanned by the surface normal and the line of electron exit, but weak for $\vec{A}$ normal to that plane. A similar strong polarization dependence is observed for a peak at 3.5 eV below E_F in spectra taken with 16 eV photons in the [010] azimuth and for a peak located at 3.3 eV below E_F in spectra taken with 13 eV photons in the [0$\bar{1}$1] azimuth. These peaks are strong for $\vec{A}$ normal to the plane of electron exit and weak for $\vec{A}$ parallel to that plane [18]. The peaks fit into the band structure calculated by [22]. The polarization dependence is explained in terms of selection rules (see [16]) governing transitions between states of specific parity with respect to reflection in the mirror plane. The observations are consistent with the known parities of the W bands [18]. The initial state at 3.3 eV below E_F is an odd-parity state according to the $\vec{A}$ dependence of the corresponding peak in the UPS spectra [17]. While the corresponding CIS allocate the final state fairly unambiguously to a specific band in the band structure calculated by [22] merely by the final-state energy, the initial state corresponding to the $E_i = E_F - 3.3$ eV peak is much harder to identify [19]. Among the closely lying bands the correct assignment requires a symmetry analysis developed by [16]; for details, see [19]. Initial state symmetries studied on W(100) were reviewed in [17].

Fig. 38 [1] compares two AREDCs along $\bar{\Sigma}$ measured with different polarization on W(001). In the upper spectrum the $\vec{A}$ vector of the p-polarized radiation was confined to the same (110) mirror plane along which the electrons were detected [1]. The selection rules from [16] indicate that in this case the initial state has even parity. In the lower spectrum $\vec{A}$ was perpendicular to the plane of emission. Here, the second high-lying resonance is much more pronounced, indicating odd parity with respect to the (110) mirror plane. This conclusion is consistent with calculations where mainly $d_{zx,zy}$ and $d_{x^2-y^2}$ orbitals are attributed to that resonance [1]. Since this second feature seems to vanish for normal emission [1], it either does not exist for $k_{\parallel} = 0$ or it is a purely $d_{x^2-y^2}$-type state or resonance [17]. The different character of the two high-lying resonances is also indicated by the exit-angle dependence of the peak intensities. The intensity of the topmost resonance decreases with increasing $k_{\parallel}$, while the second resonance passes an intensity maximum at small but non-zero $k_{\parallel}$ values. In addition, the intensity ratio between the topmost and second resonances depends on the photon energy. The second resonance is less weak at high hν values (e.g., 29 eV) than at lower ones (e.g., 19 eV). Considering the reflection and refraction proper-

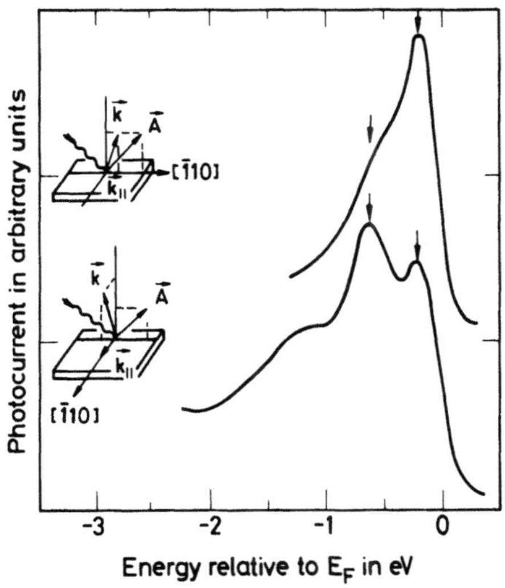

Fig. 38. Angle-resolved energy distribution curves from W(001) measured with 29 eV photons incident at 45° and electron collection with $k_{\parallel} = 0.26\ Å^{-1}$ in a (110) mirror plane. The arrows point to the high-lying resonances. For the upper spectrum, the vector potential $\vec{A}$ is in the same plane as the energy analyzer. For the lower spectrum, the $\vec{A}$ vector was rotated through 90° around the surface normal and, therefore, contains a component normal to the mirror plane of electron collection [1].

References for 5.5.2.4 on p. 183

ties of the p-polarized radiation, the angle-of-incidence dependence of the intensity reveals the symmetry of the initial states involved in the emission [1]. The low-lying surface state (for $\bar{\Gamma}$ located at 4.3 eV below E_F) shows significant dispersion along $\bar{\Sigma}$ [5] in contrast to the behavior along $\bar{\Delta}$ studied in [1]. The low-lying resonance (at 4.2 eV below E_F) has the same symmetry properties as the resonance at 0.4 eV below E_F (both of them have mainly d_{z^2} and/or s character) [1]; see also [5]. Similarly, a comparison of normal-emission EDCs (hν = 18 eV) from W(001) taken with s-polarized radiation and with mixed s and p polarization, suggests that the surface state near E_F derives from a d_{z^2} orbital [19]. This assignment is in accordance with an earlier comparison by [9] between p- and s-polarized normal-emission EDCs which showed a strong reduction of the surface resonance at 0.4 eV below E_F in the s-polarized spectrum (the observed suppression was compatible with a total suppression in view of the incomplete polarization and the divergence of 15° used) [9]. The assignment to a d_{z^2} orbital by [19] contradicts earlier calculations by [23, 24].

There is an intensity minimum in the region $E_f = E_F + 8$ eV in a series of normal-emission EDCs recorded with 10 to 16 eV [12] and 16.8 and 21.2 eV [14] radiation. This minimum is due to a lack of final states associated with the conduction band gap centered at $E_f = E_F + 8$ eV [12, 14]. It gives rise to a peak-like feature at 11.8 eV below E_F and a dip near 12.7 eV below E_F [7]. A peak seen in the 14 eV EDC in the final-state gap (but not in the other EDCs) could not be explained. Except for this feature and for identified surface states, all the major UPS features observed were rationalized in terms of $\vec{k}$-conserving, direct transitions [12]. This confirms the quantitative agreement (found at least up to 10 eV above E_F) between the experimental results obtained by [14] and the band structure calculation in the companion paper [22]. The sharp doublet at 1.5 eV below E_F and the broad and weaker peak at 6.2 eV below E_F in the 21.22 eV EDC are attributed to direct transitions [7] in agreement with the comparison between calculated DOS and measured EDCs given in [14, p. 2381].

Fig. 39 [1] shows the resonance-type variation with hν of the photoexcitation cross section for the resonances at 0.4 and 4.2 eV below E_F [1]; see also [2]. These data agree with those of [10], but are in poor agreement with those of [12] who reported a continuous

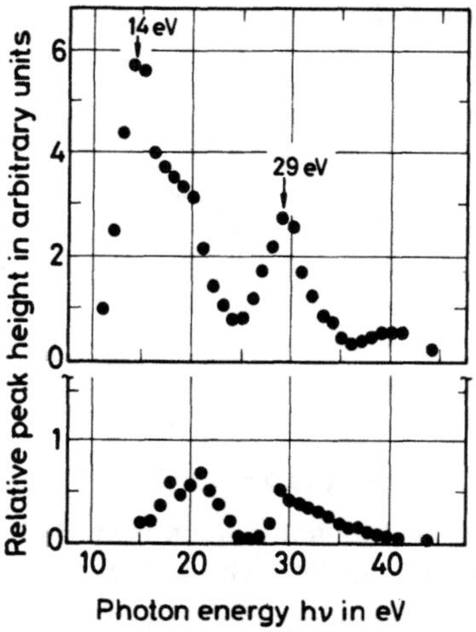

Fig. 39. Normal-exit photoemission intensity of the surface resonances on W(100) at 0.4 eV below E_F (upper panel) and at 4.2 eV below E_F (lower panel) as a function of the energy of the incident p-polarized photons [1].

intensity decrease of the state at 0.5 eV below E_F when hν was increased from 10 to 25 eV, i.e., beyond the plasmon energy ($E_{pl} \approx 23$ eV) [12]. For the state at 0.4 eV below E_F the photoexcitation cross section is strong at hν = 14 and 29 eV. In addition, Fig. 39 shows a pronounced shoulder at hν = 18 eV. Normal-exit EDCs taken with 13 to 35 eV radiation indicate that these features are largely due to final-state effects [1, 2]. In [12] the amplitude of the peak at 4.3 eV below E_F was found to be even larger around the plasmon energy E_{pl} than elsewhere in the hν range 13 to 27 eV [12] (again in disagreement with [1]; see Fig. 39). Despite the disagreement between [1] and [12], these results and those of the normal-emission study [25] using 10.2 to 40.8 eV photons refute the conclusion drawn in [26] that the emission from the surface resonance at 0.4 eV below E_F is quenched at $h\nu > E_{pl}$ [12, 25]. The hν-dependent intensity variation of the high-lying resonance can be explained without the many-body effects invoked by [25, 26]. Instead, all of the observed phenomena are understandable in terms of a one-electron model. The topmost resonance is mainly excited by the normal component of the electric field just outside the surface. The second resonance, by contrast, is excited by the component of the electric field parallel to the surface [1]. Strong suppression of normal photoemission is expected for hν close to E_{pl} because of the effect of refraction of p-polarized light [27, 28].

For $\vartheta = 0°$ the resonance at 0.4 eV below E_F remains symmetric and narrow (with respect to E_i) for all photon energies between 10 and 40 eV [1, 2]. An hν-independent intrinsic line width of 0.4 eV was found [1, 2] which agrees well with the value of 0.35 eV found in field emission spectra by [29]. The much wider width and the asymmetric shape of the peak reported earlier [25] (see also [10, 11]) are explained by the much larger acceptance angle used in those earlier studies and possibly by small misalignments of the surface (the effect of the latter is illustrated by a series of spectra taken at $\vartheta = 0°$, 3°, 6°, and 9° along the [100] azimuth = $\bar{\Delta}$ line) [1]. With a resolution of ~0.2 eV a width of ~0.5 eV resulted for the peak at 0.4 eV below E_F [8] (see also [11]), and with an acceptance cone of ~ ±7.5° a width of ~0.6 eV was found [10].

The emission intensity of the resonance at 0.4 eV below E_F as a function of the polar angle measured in the plane of incidence was found to be asymmetric with respect to $k_{\parallel} = 0$ for hν = 29 eV. Data taken at hν = 12 eV, on the other hand, are fairly symmetric in shape. This asymmetry is attributed to the fact that the square of the matrix element $|\langle f|\vec{A}\cdot\vec{p} + \vec{p}\cdot\vec{A}|i\rangle|^2$ contains an interference term that changes sign on reversal of $\vec{p}$ (i.e., on changing $\vec{k}_{\parallel}$ into $-\vec{k}_{\parallel}$) [1].

ARUPS Studies of W(100) Surface Reconstruction

The surface reconstruction on W(100) was discussed on the basis of ARUPS measurements of the Fermi surface; see the review [30, pp. 80/8]. Strong electron-phonon interaction is to be expected for those regions of the surface Brillouin zone (SBZ), where the electronic bands are nearly nondispersing and close to E_F [30]. A Fermi surface study [6] revealed such nonadiabatic regions. Surface reconstruction on W(100) may be understood in terms of a charge density wave (CDW) model. The nonadiabaticity indicates an inherent instability of the W(100)-(1 × 1) surface. Extended SBZ regions with nearly nondispersive electron bands (corresponding to localized electrons) located near E_F provide a link between CDW-based concepts and local bonding theories [30, pp. 80/8].

The reconstruction of the W(100) surface was studied by [5, 6] using synchrotron radiation; see also [15]. The ARUPS spectra taken by [6] are compatible with an incommensurate surface structure above room temperature. The results are consistent with a CDW model with respect to reconstruction, but also with local bonding of surface orbitals, i.e., with

References for 5.5.2.4 on p. 183

localized and delocalized interactions [6]. The ARUPS study [6] of the Fermi surface is summarized in [15]. A family of UPS spectra are shown taken along the $\bar{\Sigma}$ symmetry line with 20 eV photons impinging along the [011] ($=\bar{\Sigma}$) azimuth under an angle of incidence of 60°. The spectra are dominated by the intense emission from the surface state at 0.35 eV below E_F. As $k_{\parallel}$ increases along $\bar{\Sigma}$, the peak weakens and disperses towards E_F. For $k_{\parallel}=0.31\ Å^{-1}$ it is near 0.25 eV below E_F. There is no dispersion between $k_{\parallel}=0.31$ and 0.64 $Å^{-1}$, but the large error is consistent also with crossings of E_F in this range. On further increasing $k_{\parallel}$, the peak shifts downwards and reaches 0.6 eV below E_F near $k_{\parallel}=0.82\ Å^{-1}$ [15]. The minimum energy 0.67 eV below E_F is reached at $k_{\parallel}=0.79\ Å^{-1}$ [6]. For larger $k_{\parallel}$ values, the peak disperses back to E_F and crosses the Fermi level near 1.19 $Å^{-1}$ (in line with an earlier study [5] that extended well into the second SBZ with data collected for $k_{\parallel}$ up to 1.9 $Å^{-1}$). The other surface state, which appears to develop from a shoulder of the Swanson peak as $k_{\parallel}$ increases, reaches its minimum energy (0.94 eV below E_F) near $k_{\parallel}=0.38\ Å^{-1}$ and seems to merge with the Swanson hump state near 0.79 $Å^{-1}$ [6]. The Fermi surface (see Section 1.1.3.5, pp. 55/6) consists of well-defined hole pockets (around the $\bar{X}$ and $\bar{M}$ points) mimicking projections of the bulk Fermi surface onto a (100) plane and of areas (hatched in Fig. 18, p. 55 [6, 15]) with energies close to or equal to E_F (the experiments did not allow to decide whether the Fermi surface is crossed). But in any case, there will be a high density of electronic states allowing for strong interaction between electrons and phonons to occur [15]. The results presented by [6] differ from those given in [4] and reviewed by [31, p. 37]. There is also a difference in the dispersion curve measured by [5] who found the Swanson hump state to cross the Fermi level closely halfway of the $\bar{\Sigma}$ symmetry line, while [6] finds this state to be close to the Fermi level over a considerable portion of that line. For further results obtained away from $\bar{\Sigma}$, see [6].

The dispersion of the surface states (especially those occurring at 0.3 and 4.3 eV below E_F at $\bar{\Gamma}$) was studied in detail in [4] (extending the earlier paper [32]). Features seen at 0.6 and 1.4 eV below E_F in the spectra of the $(\sqrt{2}\times\sqrt{2})$R45°-reconstructed surface are due to umklapp processes of bulk features at $\bar{\Gamma}$ seen in normal emission spectra from the (1×1) surface [4]; see also [33]. Since these features are due to bulk transitions, the paper [4] offered no convincing driving mechanism for the surface reconstruction of W(100) [4].

Spin-Polarized Electron Emission from W(100)

Electron-spin-polarized UPS spectra from unreconstructed W(100) surfaces are reported for unpolarized 10.2 eV radiation at normal incidence. Both intensity and polarization of the photocurrent, emitted at $\vartheta=70°$ along the [011] azimuth of the (100) surface, show a pronounced structure when plotted as a function of the kinetic energy E_{kin} of the electrons [34]. In addition, there is an azimuthal dependence of intensity and polarization of photoelectrons at $E_{kin}=5$ eV emitted under an angle of $\vartheta=70°$ [34, 35]. The polarization profiles are explained by an appreciable spin-orbit-induced spin dependence of the transmission of photoelectrons through the surface (i.e., the polarization arises from the spin-dependent matching of eigenstates of the crystal (Bloch spinors) to eigenstates of the vacuum (free electron spinors)) [34].

The spin polarization spectrum for W(100) surfaces was calculated for normal electron emission excited by circularly polarized, normally incident radiation. Depending on the photon energy, 100% polarization is expected for hν near 10 eV; see a histogram in [36] covering the hν range 9.93 to 10.88 eV [36]. The spin polarization of the excited photoelectrons from a cesiated W(100) surface was measured in dependence on photon energy using circularly polarized light [37, 38].

References for 5.5.2.4:

[1] Weng, S.-L.; Plummer, E. W.; Gustafsson, T. (Phys. Rev. [3] B **18** [1978] 1718/40).
[2] Weng, S.-L.; Gustafsson, T.; Plummer, E. W. (Phys. Rev. Lett. **39** [1977] 822/5).
[3] Feuerbacher, B.; Willis, R. F. (J. Phys. C **9** [1976] 169/216).
[4] Campuzano, J. C.; Inglesfield, J. E.; King, D. A.; Somerton, C. (J. Phys. C **14** [1981] 3099/113).
[5] Holmes, M. W.; Gustafsson, T. (Phys. Rev. Lett. **47** [1981] 443/6).
[6] Smith, K. E.; Elliott, G. S.; Kevan, S. D. (Phys. Rev. [3] B **42** [1990] 5385/7).
[7] Mullins, D. R.; Lyman, P. F.; Overbury, S. H. (Surf. Sci. **277** [1992] 64/76, 70).
[8] Willis, R. F.; Feuerbacher, B.; Fitton, B. (Solid State Commun. **18** [1976] 1315/9).
[9] Feuerbacher, B.; Fitton, B. (Solid State Commun. **15** [1974] 295/9).
[10] Waclawski, B. J.; Plummer, E. W. (Phys. Rev. Lett. **29** [1972] 783/6).

[11] Feuerbacher, B.; Fitton, B. (Phys. Rev. Lett. **29** [1972] 786/9).
[12] Lapeyre, G. J.; Smith, R. J.; Anderson, J. (J. Vac. Sci. Technol. **14** [1977] 384/90).
[13] Murotani, T.; Fujiwara, K.; Nishijima, M. (Proc. 2nd Int. Conf. Solid Surf., Kyoto 1974, pp. 409/12).
[14] Feuerbacher, B.; Christensen, N. E. (Phys. Rev. [3] B **10** [1974] 2373/90).
[15] Smith, K. E.; Elliott, G. S.; Kevan, S. D. (Appl. Surf. Sci. **48/49** [1991] 135/8).
[16] Hermanson, J. (Solid State Commun. **22** [1977] 9/11).
[17] Lapeyre, G. J.; Anderson, J.; Smith, R. J. (Surf. Sci. **89** [1979] 304/18).
[18] Anderson, J.; Lapeyre, G. J.; Smith, R. J. (Phys. Rev. [3] B **17** [1978] 2436/45).
[19] Smith, R. J.; Lapeyre, G. J.; Anderson, J.; Williams, G. P. (Conf. Ser.-Inst. Phys. No. 39 [1978] 220/32).
[20] Smith, R. J.; Anderson, J.; Hermanson, J.; Lapeyre, G. J. (Solid State Commun. **19** [1976] 975/8).

[21] Lapeyre, G. J.; Anderson, J.; Smith, R. J. (Eur. Space Agency Spec. Publ. ESA SP-118 [1976] 165/9).
[22] Christensen, N. E.; Feuerbacher, B. (Phys. Rev. [3] B **10** [1974] 2349/72).
[23] Smith, N. V.; Mattheiss, L. F. (Phys. Rev. Lett. **37** [1976] 1494/7).
[24] Weng, S.-L. (Phys. Rev. Lett. **38** [1977] 434/7).
[25] Feuerbacher, B.; Willis, R. F. (Phys. Rev. Lett. **37** [1976] 446/50).
[26] Egelhoff, W. F., Jr.; Linnett, J. W.; Perry, D. L. (Phys. Rev. Lett. **36** [1976] 98/100).
[27] Bagchi, A.; Kar, N. (Phys. Rev. Lett. **40** [1978] 803/6).
[28] Bagchi, A.; Kar, N. (Phys. Rev. [3] B **18** [1978] 5240/7).
[29] Plummer, E. W.; Gadzuk, J. W. (Phys. Rev. Lett. **25** [1970] 1493/5).
[30] Smith, K. E.; Kevan, S. D. (Prog. Solid State Chem. **21** [1991] 49/131).

[31] King, D. A. (Phys. Scr. T **4** [1983] 34/43).
[32] Campuzano, J. C.; King, D. A.; Somerton, C.; Inglesfield, J. E. (Phys. Rev. Lett. **45** [1980] 1649/52).
[33] Anderson, J.; Lapeyre, G. J. (Phys. Rev. Lett. **36** [1976] 376/9).
[34] Feder, R.; Kirschner, J. (Solid State Commun. **40** [1981] 547/50).
[35] Kirschner, J.; Feder, R.; Wendelken, J. F. (Phys. Rev. Lett. **47** [1981] 614/7).
[36] Reyes, J.; Helman, J. S. (Phys. Rev. [3] B **16** [1977] 4283/7).
[37] Zürcher, P.; Meier, F.; Christensen, N. E. (Phys. Rev. Lett. **43** [1979] 54/7).
[38] Zürcher, P.; Meier, F. (J. Appl. Phys. **50** [1979] 2097/8).

5.5.2.5 UPS Studies on W(111) Surfaces

PARUPS spectra have been recorded on W(111) using 13, 16, 18, and 19 eV radiation. With normal emission the s-polarized spectrum shows two features at 0.5 and 2.0 eV below E_F, sensitive to hydrogen exposure. The p-polarized normal-emission spectra show a peak near 0.9 eV below E_F that has a shoulder near 0.4 eV below E_F; it splits into two well-resolved features when using off-normal (42.3°) detection [1]. The dipole-allowed initial-state symmetries for normal and mirror-plane emission have been given in the PARUPS study [2] using the symmetry selection rules formulated by [3]. A W ribbon was studied with >95% s-polarized radiation impinging at an angle of 42.3°. The acceptance angle also was 42.3° (±2°) off-normal. AREDCs were taken with 13 to 30 eV photons at 1 eV intervals for $\vec{k}_{\parallel}$ along $\bar{\Sigma}$, keeping $\vec{A}$ parallel to the mirror plane. The strongest surface feature by far is a flat band of states close to the Fermi level. At higher binding energies, several other flat bands, observed mostly near $\bar{\Gamma}$ [2, 4], were identified as surface features on account of their sensitivity to hydrogen adsorption [2]. The flatness of the bands indicates strongly localized wave functions, and it appears that the open structure of the W(111) surface causes the surface atoms to interact more with underlayer atoms than with other surface atoms [2]; see also the earlier angle-integrated and angle-resolved study [4]. With $\vec{A}$ perpendicular to the mirror plane, no feature was found that is sensitive to hydrogen adsorption [2].

EDCs taken with 7.7 to 11.7 eV photons show a rich structure; see Fig. 35, p. 172 [5, 6]. The two high-lying features at 0.7 and 1.2 eV below E_F (features Nos. 5 and 7 in Fig. 35) are attributed to surface emission because of their constant position in the spectra and their fairly hν-independent intensity. The two lowest-lying features (Nos. 10 and 11 in Fig. 35) vary slightly with the initial state energy E_i and are therefore attributed to bulk direct transitions. The two features Nos. 8 and 9 have intensities that vary with hν. Despite their constant E_i they are attributed to contributions from bulk and surface effects [5, 6]. They are shifted by ~1 eV from the bulk peak position towards higher initial-state energies by *d*-band narrowing at the surface. Comparing these spectra with others taken at photon energies of 13, 16.8, and 21.2 eV indicates changes in the character of the spectra. While surface emission accounts for ~85% of the intensity in the 10.2 eV spectrum, direct transitions are more important than surface emission in the 13 eV spectrum. Inelastically scattered electrons dominate the low-energy part of the 21.2 eV spectrum [5].

The spin polarization spectrum for W(111) surfaces was calculated for normal electron emission excited by circularly polarized, normally incident radiation. 100% spin polarization is expected for hν near 6.8 eV; see histograms in [7] covering the hν range 6.8 to 10.88 eV [7].

References for 5.5.2.5:

[1] Cerrina, F.; Williams, G. P.; Anderson, J.; Lapeyre, G. J.; Bisi, O.; Calandra, C. (Recent Dev. Condens. Matter Phys. Pap. 1st Gen. Conf., Antwerp 1980 [1981], Vol. 2, pp. 391/7; C.A. **96** [1982] No. 11953).

[2] Cerrina, F.; Anderson, J. R.; Lapeyre, G. J.; Bisi, O.; Calandra, C. (Phys. Rev. [3] B **25** [1982] 4949/61).

[3] Hermanson, J. (Solid State Commun. **22** [1977] 9/11).

[4] Cerrina, F.; Williams, G. P.; Anderson, J.; Lapeyre, G. J.; Bisi, O.; Calandra, C. (Vide Couches Minces No. 201 [1980] 1162/5; C.A. **93** [1980] No. 214196).

[5] Feuerbacher, B.; Christensen, N. E. (Phys. Rev. [3] B **10** [1974] 2373/90).

[6] Feuerbacher, B.; Fitton, B. (Phys. Rev. Lett. **30** [1973] 923/6).

[7] Reyes, J.; Helman, J. S. (Phys. Rev. [3] B **16** [1977] 4283/7).

5.5.2.6 UPS Studies on Surfaces of Tungsten Polycrystals

The UPS studies on polycrystalline W reported in the literature appear to have been done essentially on W(100) surfaces (because rolling or heat treatment recrystallized the samples used). The EDCs of photoelectrons from bulk W polycrystals were recorded using energies between 7.7 and 21.2 eV. DOS peaks above the vacuum level were found at 6.7 and 8.5 eV above E_F [1]. A 100 h heat treatment in oxygen, applied to remove carbon, tended to produce a predominant (100) orientation of the polycrystalline sample used by [2]. A photon energy of 7.7 eV was used mainly to study the sensitivity to hydrogen exposure of the surface state at 0.4 eV below E_F. For the absolute photoelectric yields measured by [2], see pp. 167/8. In [3] the EDC was measured for emission from rolled W ribbons whose rolling texture was essentially preserved through the aging process. Most of the ~50 μm-size crystallites were oriented with their (100) planes within 10° of the ribbon surface. On increasing the angle of incidence from 0° to 45°, the photocurrent increased by a factor of 5. Current-voltage characteristics (the derivatives of which are EDCs) excited by 4.89 and 5.80 eV radiation are given in [3].

Conduction band spectra were measured on sputter-deposited W films using photon energies of 21.2 and 40.8 eV. Since the microcrystalline samples were inhomogeneous and contaminated by at least 0.3 monolayers of oxygen, no details of the UPS spectra are given here [4].

References for 5.5.2.6:

[1] Farrell, H. J. (Diss. Univ. Notre Dame 1972, 116 pp. from Diss. Abstr. B **32** [1972] 6588).
[2] Waclawski, B. J.; Plummer, E. W. (Phys. Rev. Lett. **29** [1972] 783/6).
[3] Apker, L.; Taft, E.; Dickey, J. (Phys. Rev. [2] **73** [1948] 46/50).
[4] Penchina, C. M.; Sapp, E.; Tejeda, J.; Shevchik, N. (Phys. Rev. [3] B **10** [1974] 4187/94).

5.5.3 Core-Level Spectra

5.5.3.1 Introduction

The topic of surface core-level shifts is reviewed in [1, 2]. In addition to an extensive presentation of the theoretical background illustrated by many examples, experimental case studies are also given; for pure metals, see [1, pp. 326/37]. The microscopic model developed by [3] is compared in [2] with the thermodynamical model of [4, 5].

The lower coordination of the atoms at the surface of a solid shifts the binding energy of the core-level electrons of surface atoms towards values slightly different from those of the corresponding bulk-atom electrons. In the case of W surfaces, this surface core-level shift (SCLS) occurs towards lower binding energies, and a negative sign is used throughout the following text for shifts in this direction. The sign of the shift changes within the 5d transition series. The origin of this effect can be understood qualitatively by using a simple tight-binding model as follows (see e.g. [2, 3, 6, 7]): Due to the reduced coordination at the surface, the band width narrows. Assuming that there is no shift of the center of the surface band, a deficiency or an excess of electrons would occur depending on the position of E_F below or above the *d*-band center (i.e. depending on the number of *d* electrons). To satisfy layerwise charge neutrality, the center of the *d* band shifts in energy by ΔE. The surface core levels rigidly follow this shift of the *d*-band center. Band structure calculations give a change of sign of ΔE for a band filling slightly smaller than five electrons per atom (due to the asymmetry of the band), meaning a small negative SCLS for W in agreement with the experiments.

References for 5.5.3.1 on pp. 186/7

The surface sensitivity is discussed in regards to the contributions that core-hole lifetime, phonons, plasmons, electron-hole pairs, and electronic correlations have on the line shape. Synchrotron radiation sources allow one to tune the photon energy to the optimum surface sensitivity of the core-level spectra [2]. The surface-induced modification of the position, shape, and intensity of core-level lines is reviewed in [8].

Core-level spectroscopic studies on W surfaces are almost exclusively restricted to the W $4f_{7/2}$ level. In the bulk, the W $4f_{7/2}$ and W $4f_{5/2}$ core levels were found at 31.42±0.05 eV and 33.58±0.06 eV below E_F, respectively [9]. The W $4f_{7/2}$ bulk core-level values measured by [2, 10 to 13] amount to 31.4±0.1 eV below E_F; see also "Tungsten" Suppl. Vol. A 2, 1987, pp. 18 and 103/4. Apart from an extra lifetime broadening (additional ~53 meV), the W $4f_{5/2}$ levels are an exact replica of the W $4f_{7/2}$ level [9].

An example of a W $4f_{5/2}$ plus W $4f_{7/2}$ spectrum excited by 70 eV photons on W(110) is shown in [2, p. 27]. For a comparison of W $4f_{7/2}$ spectra of W(110), W(100), and W(111) surfaces, see [2, p. 33] and also the previous review [6]. Earlier work by the same group on W(110) [11, 14] and on W(100) [15] is reviewed in [7]. Core-level spectroscopic studies concerning the W(100) surface reconstruction are reviewed in [2, pp. 38/44].

The W $4f_{7/2}$ spectra of W(320) surfaces were presented to exemplify the core-level spectroscopy of stepped surfaces. The various components in the W $4f_{7/2}$ spectra of the stepped W(320) surface can be attributed to the various core-level shifts associated with the different types of surface atoms (differing in coordination) present on W(320) surfaces [2, pp. 44/9], [10, 16].

Angular variations in the core-level photoelectron intensity are discussed in terms of photoelectron diffraction, see e.g. [2, pp. 50/1].

For the variation of the photoelectron mean free path as a function of the kinetic energy, see e.g. [17] and a review of applications of electron spectroscopy (UPS, XPS, and AES) in surface science [18]. A chart showing the mean escape depth of electrons plotted vs. their energy is given in [18]. On a W foil the escape depth was determined to be $\lambda = 12.8$ Å (estimated error: ±20%) for ~1450 eV photoelectrons [19]. Using sputter-deposited Mo-on-W and W-on-Mo substrates, escape depths of 6 and 17 Å were found for 120 and 1736 eV Auger electrons. At kinetic energies of 48 and 350 eV, these are below 5.2 and 7.7 Å [20]. On W(110) surfaces the minimum escape depth for normal emission (and thus maximum sensitivity for emission from the surface layer) occurred at 75 eV photon energy [21].

References for 5.5.3.1:

[1] Egelhoff, W. F., Jr. (Surf. Sci. Rep. **6** [1987] 253/415).
[2] Spanjaard, D.; Guillot, C.; Desjonquères, M.-C.; Tréglia, G.; Lecante, J. (Surf. Sci. Rep. **5** [1985] 1/85).
[3] Desjonquères, M. C.; Spanjaard, D.; Lassailly, Y.; Guillot, C. (Solid State Commun. **34** [1980] 807/10).
[4] Johansson, B.; Mårtensson, N. (Phys. Rev. [3] B **21** [1980] 4427/57).
[5] Rosengren, A.; Johansson, B. (Phys. Rev. [3] B **22** [1980] 3706/9).
[6] Eastman, D. E.; Himpsel, F. J.; van der Veen, J. F. (J. Vac. Sci. Technol. **20** [1982] 609/16).
[7] Lecante, J.; Guillot, C. (Helv. Phys. Acta **56** [1983] 427/33).
[8] Tréglia, G.; Sébilleau, D.; Desjonquères, M. C.; Spanjaard, D.; Guillot, C.; Chauveau, D.; Lecante, J. (Proc. 10th Taniguchi Int. Symp., 1987 [1988], pp. 281/98).
[9] van der Veen, J. F.; Himpsel, F. J.; Eastman, D. E. (Phys. Rev. [3] B **25** [1982] 7388/97).
[10] Purcell, K. G.; Jupille, J.; King, D. A. (Surf. Sci. **208** [1989] 245/66).

[11] Tran Minh Duc; Guillot, C.; Lassailly, Y.; Lecante, J.; Jugnet, Y.; Vedrine, J. C. (Phys. Rev. Lett. **43** [1979] 789/92).
[12] Alnot, P.; Auerbach, D. J.; Behm, J.; Brundle, C. R.; Viescas, A. (Surf. Sci. **213** [1989] 1/24).
[13] Nyholm, R.; Berndtsson, A.; Mårtensson, N. (J. Phys. C **13** [1980] L 1091/L 1096).
[14] Tréglia, G.; Desjonquères, M. C.; Spanjaard, D.; Lassailly, Y.; Guillot, C.; Jugnet, Y.; Tran Minh Duc; Lecante, J. (J. Phys. C **14** [1981] 3463/73).
[15] Guillot, C.; Thuault, C.; Jugnet, Y.; Chauveau, D.; Hoogewijs, R.; Lecante, J.; Tran Minh Duc; Tréglia, G.; Desjonquères, M. C.; Spanjaard, D. (J. Phys. C **15** [1982] 4023/32).
[16] Chauveau, D.; Roubin, P.; Guillot, C.; Lecante, J.; Tréglia, G.; Desjonquères, M. C.; Spanjaard, D. (Solid State Commun. **52** [1984] 635/9).
[17] Brundle, C. R. (Surf. Sci. **48** [1975] 99/136, 100).
[18] Brundle, C. R. (J. Vac. Sci. Technol. **11** [1974] 212/24, 213/5).
[19] Carlson, T. A.; McGuire, G. E. (J. Electron Spectrosc. Relat. Phenom. **1** [1972] 161/8).
[20] Tarng, M. L.; Wehner, G. K. (J. Appl. Phys. **44** [1973] 1534/40).

[21] Purcell, K. G.; Derby, G. P.; King, D. A. (J. Phys. Condens. Matter **1** [1989] 1373/9).

5.5.3.2 Surface Core-Level Shifts

5.5.3.2.1 Experimental Data

W(110). The W(110) surface is the most densely packed bcc surface and exhibits only two, well-separated peaks. These peaks can be readily attributed to bulk and surface atoms. The (110) surface is the only one producing a clearly resolved surface peak [1].

Values for the W $4f_{7/2}$ SCLS on W(110) of -0.29 ± 0.01 eV at both 100 ± 20 K and at 290 ± 10 K [1], -0.30 eV for W $4f_{7/2}$ [2 to 5] and also for the W $4f_{5/2}$ level nearby [2], -0.32 eV [33], and -0.321 ± 0.001 eV [6] were measured. Spectra taken on both stepped and W(110) surfaces are consistent with a W(110) underlayer core-level shift amounting to more than 0.05 eV [1].

A shift of the core-level binding energy on W surfaces was first observed by [2]. The authors used p-polarized radiation with an energy of 53 to 140 eV impinging on a W(110) surface along the [001] azimuth. The intensity ratio of low-binding to high-binding energy peaks decreased from 1.3 to 0.6 when α changed from 70° to 27°. This change was accompanied by a reduction of the total intensity to about 30%. The low-binding-energy peak was identified as a surface core-level peak because of its strong dependence on exposure to hydrogen or oxygen. The polarization dependence of the intensity and also the $h\nu$-independent energy location of that peak follow the pattern known for surface states [2].

W(100). The SCLS in the (1 × 1) high-temperature (HT) phase of W(100) was investigated by [7 to 14]. For spectra covering the range of W $4f_{7/2}$ and W $4f_{5/2}$ emissions, see [7], [8, p. 11]. W $4f_{7/2}$ core-level spectra were taken on a W crystal cut to within 0.1° parallel to the (100) face. The contamination was below the 0.2% monolayer detection limit of AES. LEED indicated that the low density of terrace steps on the flat surfaces ensured a sharp phase transition between the HT phase and the c(2 × 2) reconstructed low-temperature (LT) phase. HT and LT phases could be studied in the absence of noticeable amounts of the other phase. The high-temperature spectra were taken at 500 K in order to reduce the possible influence of hydrogen adsorption from the residual gas [9]; see also [10]. There was only a subtle difference between the spectra from the c(2 × 2) and the (1 × 1) surfaces. A comparison of the HT- and LT-W $4f_{7/2}$ spectra indicates a small intensity change between the bulk and the slightly shifted surface peaks [9, 10]. For the HT and LT phase, values

References for 5.5.3.2 on pp. 193/4

for the SCLS of -0.37 ± 0.01 eV and -0.35 ± 0.01 eV, respectively, were measured [9]. (A value of -0.38 ± 0.01 eV is given for the HT phase in [10].) The spectra obtained by [11] at ~350 K and ~160 K reveal a more pronounced difference between the two phases (-0.40 eV and -0.35 eV for the HT and LT phase, respectively). The quality of the crystal surfaces used may be responsible for the different results according to [9].

Other values measured for the SCLS in the HT phase are -0.40 eV [12], -0.39 ± 0.02 eV [13], -0.36 ± 0.01 eV [7], -0.35 eV [14], and -0.27 ± 0.04 eV [8].

In the LT phase several measurements [9, 11, 15] yielded an identical SCLS of -0.35 eV.

Underlayer atom core-level shifts of -0.19 ± 0.02 eV [13], -0.16 eV [12], and -0.14 ± 0.01 eV [7, 9] for the HT phase and of -0.11 eV [15] for the LT phase are reported. The underlayer atom core-level shifts of the HT and LT phases differed by less than 0.020 eV according to [11].

W(111). $4f_{7/2}$ core-level spectra of a W(111) surface were decomposed into three peaks attributed to bulk, surface, and underlayer atoms. With respect to the bulk peak, the surface peak and the underlayer peak are shifted by -0.43 ± 0.01 and -0.10 ± 0.02 eV, respectively [16, 17]. A reanalysis of the data of [16] with more appropriate line shapes yielded shifts of -0.43 ± 0.01 and -0.11 ± 0.01 eV [7]. The decomposition of the W(111) core-level spectra into three peaks was questioned by Purcell et al. [18]. They analyzed their spectra using four components. The atoms at the selvedge give rise to three components. One of these, shifted by -0.46 ± 0.03 eV, is attributed to the topmost surface atoms, the other two, shifted by -0.36 ± 0.03 and -0.11 ± 0.02 eV, are attributed to two different underlayer sites [18]; see also [1]. The experimental data can also be fitted to a three-peak model (from [7]) which in addition to the bulk peak assumes a broadened surface peak and an underlayer peak. This model yields surface- and underlayer-atom core-level shifts of -0.42 ± 0.02 and -0.11 ± 0.02 eV, respectively [18] in good agreement with the values given in [7]. The aforementioned four-peak model is proposed not only because it fits the data well, but also because of structural considerations, suggestive of three different coordination types being important for the W(111) surface: (i) surface atoms having four nearest neighbors (nn) and three next-nearest neighbors (nnn), (ii) first-underlayer atoms (seven nn and three nnn), and (iii) second-underlayer atoms (seven nn and six nnn) [18]. Replying the critique of [18], the data given in [7] were analyzed again comparing improved fits of three and four lines to the spectra [19]. The authors [19] confirmed that the spectra can be fitted satisfactorily with four peaks, but found that the resulting line intensities are in conflict with the attenuation of signals originating from deeper layers. An improved three-line fit yielded shifts of -0.439 ± 0.009 and -0.112 ± 0.003 eV [19].

Stepped W Surfaces. Stepped surfaces provide a range of atom sites differing in coordination (wedge, terrace, and edge sites). Hence, stepped surfaces give an opportunity to investigate the role coordination plays for the core-level energies. Because of the different atomic sites present, core-level spectra from stepped surfaces are more complex than spectra from smooth surfaces.

Core-level shifts on stepped W surfaces were investigated by [1] and [20], [21, pp. 44/9], emphasizing the W(320) surface. **Fig. 40** shows a schematic representation of its atomic structure. It resembles the W(110) surface and corresponds to a $[6(110) \times (1\bar{1}0)]$ structure consisting of terraces with six parallel rows of atoms. The edge-row atoms have 4 first and 4 second neighbors (4 nn and 4 nnn coordination, respectively). The following three rows on the terrace consist of atoms with 6 nn and 4 nnn coordination. The terrace atom row adjacent to the step wedge consists of atoms with 6 nn and 5 nnn coordination, while the wedge row consists of atoms with 8 nn and 5 nnn coordination. The experimental core-

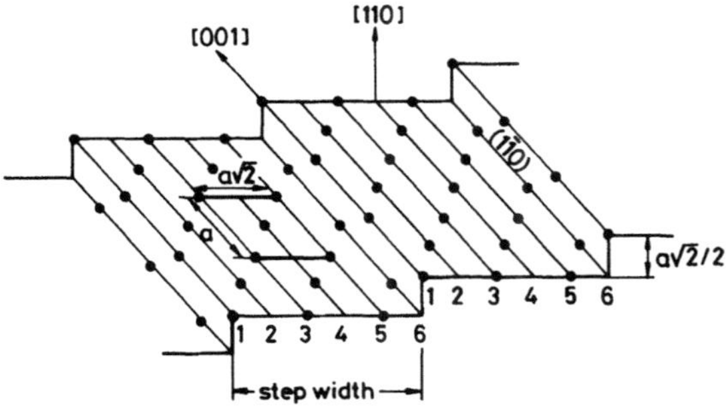

Fig. 40. Schematic representation of the atomic structure of stepped W(320) surfaces. Distances are given in units of the lattice parameter, a [20].

level shifts S_1, $S_2=S_3=S_4$, S_5, and S_6 were assigned to these atom sites [1, 20]. The analysis of the spectra of the W(320) plane performed by [20] yielded the core-level shifts $S_1=-0.58\pm0.01$ eV, $S_2=S_3=S_4=-0.30\pm0.01$ eV, $S_5=-0.44\pm0.02$ eV, and $S_6=-0.18\pm0.02$ eV. In [1] significantly smaller values, $S_1=-0.41\pm0.06$ eV, $S_2=S_3=S_4=-0.25\pm0.03$ eV, $S_5=-0.25\pm0.03$ eV, and $S_6=-0.10\pm0.02$ eV, were found. The spectra from stepped W(320) measured by [20] are compared in [21, p. 48] with spectra from a smooth W(110) surface. The analysis of the core-level spectra by [20] is questioned by Kim et al. [33] on grounds of the unrealistic bulk-peak intensity variations with photon energy obtained in [20].

In [1] detailed analyses are given for spectra from surfaces cut within 0.5° parallel to the stepped W(320), W(310), and W(610) planes. These three surfaces were selected to represent the following varieties: the (610) surface consists of (100)-type terraces and (110)-type steps; the (320) surface represents (110)-type terraces and (100)-type steps; the (310) surface contains equal numbers of (110)- and (100)-type steps. **Fig. 41**, p. 190, (from [1]) shows W $4f_{7/2}$ core-level spectra of W(110), W(100), W(111), W(320), W(310), and W(610) planes. The spectra were collected normal to the surface with an angular resolution of 4° half width and with an energy resolution estimated to be 0.16 eV. The 60 to 90 eV radiation impinged at an angle 70° off the surface normal. Similar shifts result for similar environments irrespective of the surface involved. All spectra from the stepped W(320), W(310), and W(610) surfaces could be decomposed into four lines. This result was expected for W(310), but for W(320) and W(610) five different peaks were expected (as was pointed out for W(320) above). Attempts to fit five lines to the spectra always resulted in a negligible intensity for one of the lines. It was concluded that two of the different types of atom sites must have very similar environments. The spectra of W(320), W(310), and W(610) surfaces could not be fitted to a three-peak model [1]. The core-level shifts for the various types of W atoms present on and just beneath the various surfaces are compared in Table 2, p. 191 (adapted in part from [1]), which also includes experimental and theoretical results of other authors. **Fig. 42**, p. 190, (taken from [1]) demonstrates the effect of the atomic coordination on the surface core-level shifts for different W surfaces. In order to include second- and third-nearest neighbor atoms, an effective coordination number, $N_{eff}=N_{nn}+(1/2)\,N_{nnn}+(1/8)\,N_{nnnn}$, was introduced where N_{nn}, N_{nnn}, and N_{nnnn} denote the numbers of nearest, next-nearest, and third-nearest neighbors, respectively. The two horizontal lines (connecting two values from W(320) and two values from W(610)) indicate that only four peaks were fitted to the spectra, although five types of atom sites with different coordination exist (cf. also Table 2, p. 191). The measured values for the SCLS increase fairly linearly with

References for 5.5.3.2 on pp. 193/4

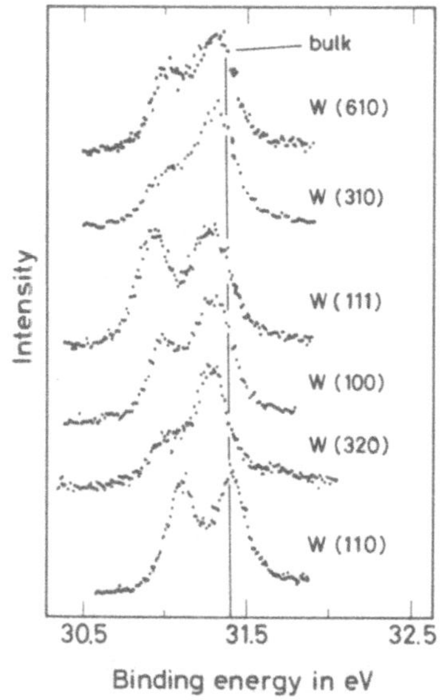

Fig. 41. W $4f_{7/2}$ core-level spectra collected at a photon energy of 75 eV for the low-index surfaces W(110), W(100), and W(111) and the stepped surfaces W(610), W(310), and W(320). The bulk core-level position is indicated by the solid line [1]

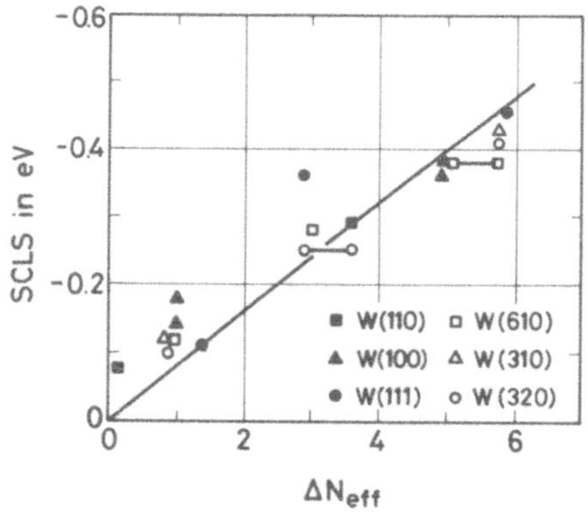

Fig. 42. Dependence of the surface core-level shifts (SCLS) on the difference in effective atomic coordination number (see text) between surface atom and bulk atom, $\Delta N_{eff} = N_{eff}(bulk) - N_{eff}(surface)$. The straight line refers to an extended bond-breaking model [1].

the difference between bulk and surface coordination in good agreement with the results of an extended bond-breaking model (cf. Section 5.5.3.2.2) represented by the straight solid line in Fig. 42 [1].

5.5.3.2.2 Comparison with Theoretical Results

Theoretically determined SCLS values are compared with some selected experimental values in Table 2 which was adapted in part from [1]. The top row indicates the type and

Table 2
Comparison of Experimental and Theoretical Values of the SCLS on W.

	$\{8,5,..\}U_{100}$	$\{7,6,9\}U_{111}(1)$	$\{7,3,9\}U_{111}(2)$	$(6,5,..\}S_{base}$	$\{6,4,7\}S_{110}$	$\{4,5,8\}S_{100}$	$\{4,4,6\}S_{step}$	$\{4,3,9\}S_{111}$	Ref.
experiments (selected values)									
W(110)					−0.29 ± 0.01				[1]
					−0.321 ± 0.001				[6]
W(100) un-	−0.16					−0.40			[12]
recon-	−0.14 ± 0.01					−0.36 ± 0.01			[7]
structed	−0.14 ± 0.01					−0.37 ± 0.01			[9]
W(111)		−0.11 ± 0.02	−0.36 ± 0.03					−0.46 ± 0.03	[18]
		−0.112 ± 0.003						−0.439 ± 0.009	[19]
W(610)	−0.12 ± 0.01			−0.28 ± 0.02		−0.38 ± 0.01	−0.38 ± 0.01		[1]
W(310)	−0.12 ± 0.01			−0.29 ± 0.02			−0.43 ± 0.01		[1]
W(320)	−0.18 ± 0.02			−0.44 ± 0.02	−0.30 ± 0.01		−0.58 ± 0.01		[20]
	−0.10 ± 0.02			−0.25 ± 0.03	−0.25 ± 0.03		−0.41 ± 0.06		[1]
theory (microscopic)									
W(110)					−0.27				[22]
					−0.0225				[26]
W(100) un-	−0.23					−0.55			[12]
recon-	−0.225					−0.570			[32]
structed	−0.15					−0.525			[26]
W(111)		−0.10	−0.20					−0.43	[21]
		−0.09	−0.18					−0.41	[26]
W(320)	−0.11			−0.29	−0.25		−0.34		[20]
theory (thermodynamic)									
RJM					−0.17	−0.40		−0.60	[25]
BBR	−0.11	−0.07	−0.24		(−0.30)	(−0.35)		(−0.43)	[27]
BBT(1)	−0.04		−0.18			−0.30		−0.39	[28]
BBT(2)	−0.06	−0.09	−0.18	−0.17	−0.23	−0.31	−0.36	−0.37	[1]
BBT(3)	−0.08	−0.12	−0.23	−0.21	−0.29	−0.39	−0.45	−0.47	[1]

References for 5.5.3.2 on pp. 193/4

the coordination number of the atoms; their core-level shifts are listed in the respective column. The symbol {nn, nnn, nnnn} gives the numbers of nearest, next-nearest, and third-nearest neighbors of the atoms, denoted by U for underlayer and S for surface atoms. For bulk W atoms, {nn, nnn, nnnn} = {8, 6, 12}. For U_{100}, nnnn = 8 on W(100) and 9 on W(610), W(310), and W(320). For S_{base}, nnnn = 8 on W(310) and W(610), and 9 on W(320). The hkl indices of U and S indicate that the coordination is that of U and S atoms on W(hkl) surfaces. There are two U_{111} species differing in the number of second neighbors. S_{base} and S_{step} indicate atoms at the base and at the edge (top) of a step formed by (100)- and (110)-type planes.

The theoretical results are divided into two classes. On the one hand, values for the SCLS were calculated according to microscopic models based on the tight-binding approximation. In these models it is assumed that the displacement of the core levels rigidly follows the displacement of the valence *d* bands; see e.g. [21, 22]. On the other hand, the SCLS was related to thermodynamical properties. Bulk W core-level binding energies were calculated assuming a fully screened final state and using the equivalent core approximation (Z+1 approximation) for the screening valence charge distribution around the core-ionized site [23 to 25].

In the thermodynamical model of [23 to 25] the SCLS is related to the surface heat of segregation of a substitutional (Z+1) impurity and depends on the surface crystal plane. The model obviously correctly reflects the trend of measured shifts for the variety of elements considered. For W both calculated and measured shifts are negative, but the absolute values of theory and experiment do not agree very well. Results of this thermodynamical model developed by Rosengren, Johansson, and Mårtensson are listed in the row labeled "RJM" in Table 2. In a more recent theoretical approach, surface segregation energies of impurities with atomic number Z+1 in a matrix of atomic number Z were calculated with the tight-binding method. No significant improvement of the SCLS values on W resulted [26]. Other theoretical approaches based on thermodynamical arguments are the bond-breaking models of Rosengren [27] (labeled "BBR" in Table 2) and of Tománek et al. [28] (labeled "BBT" in Table 2) in which the cohesive energy is written as a sum of pair interactions. The BBR model uses three experimental SCLS values (given in parentheses) for scaling [27]. The results of the BBT model, which considers only first and second neighbors, are given in the row BBT(1). Row BBT(2) gives the results for an extended model [1] which also considers third neighbors. To improve the fit with experimental data, the calculations were scaled with S_{110}; these scaled results are given in row BBT(3) [1]. For stepped surfaces, the core level shifts for the various types of W atoms present on and just beneath the various surfaces are compared in [1] including also experimental and theoretical results of other authors. The different lines in the measured spectra were assigned to particular coordination sites using the model by [28]. Good agreement was found with the results of the bond-breaking model of [27], while the simpler version by [28] appears less accurate, apparently because it neglects third neighbors.

According to the microscopical model of [22], the valence *d* bands are shifted at the surface as a result of the charge rearrangement at the surface due to the narrowing of the surface local density of states. The W $4f_{7/2}$ core levels reflect this displacement of the valence *d* bands. The model of [22] uses a tight-binding formalism (see also [29]). It is in fair agreement with the results obtained on W(110), W(111), and the stepped W(320) surfaces studied in [1]. This microscopic model gives, however, a questionable assignment for the atom row at the base of the steps on W(320) surfaces (see, e.g., the assignment in [21]), and there are large discrepancies with W(100) [1].

No surface core-level shift resulted from a seven-layer-film self-consistent calculation of the unreconstructed W(100) electronic surface structure. This LAPW calculation included all relativistic corrections except for the spin-orbit interaction [30]. An extended study found no significant change for the surface core-level shifts upon relaxing the surface layers [31]. Tight-binding quenched molecular dynamics calculations indicated that zigzag reconstruction is favored for W(100). This method was used to find the energy minimum with respect to alternate lateral displacements of atoms and involved the first four interlayer spacings. The lateral displacement decreased rapidly with the distance to the surface. This multilayer model of W(100) reconstruction yields significantly better agreement [32] with the experimental surface core-level shifts measured by [11] than the tight-binding model used in [11]. The SCLS values of HT and LT phases were calculated to differ by 0.07 eV. The corresponding core-level shift differences for underlayer atoms were calculated to be 0.01 eV [32].

References for 5.5.3.2:

[1] Purcell, K. G.; Jupille, J.; King, D. A. (Surf. Sci. **208** [1989] 245/66).
[2] Tran Minh Duc; Guillot, C.; Lassailly, Y.; Lecante, J.; Jugnet, Y.; Vedrine, J. C. (Phys. Rev. Lett. **43** [1979] 789/92).
[3] Jugnet, Y.; Prakash, N. S.; Porte, L.; Tran Minh Duc; Nguyen, T. T. A.; Cinti, R.; Poon, H. C.; Grenet, G. (Phys. Rev. [3] B **37** [1988] 8066/71).
[4] Jugnet, Y.; Prakash, N. S.; Tran Minh Duc; Poon, H. C.; Grenet, G.; Pendry, J. B. (Surf. Sci. **189/190** [1987] 782/7).
[5] Chauveau, D.; Guillot, C.; Villette, B.; Lecante, J.; Desjonquères, M. C.; Sébilleau, D.; Spanjaard, D.; Tréglia, G. (Solid State Commun. **69** [1989] 1015/8).
[6] Riffe, D. M.; Wertheim, G. K.; Citrin, P. H. (Phys. Rev. Lett. **63** [1989] 1976/9).
[7] Wertheim, G. K.; Citrin, P. H.; van der Veen, J. F. (Phys. Rev. [3] B **30** [1984] 4343/7).
[8] Alnot, P.; Auerbach, D. J.; Behm, J.; Brundle, C. R.; Viescas, A. (Surf. Sci. **213** [1989] 1/24).
[9] Jupille, J.; Purcell, K. G.; King, D. A. (Phys. Rev. [3] B **39** [1989] 6871/9).
[10] Jupille, J.; Purcell, K. G.; Derby, G.; Wendelken, J.; King, D. A. (Springer Ser. Surf. Sci. **11** [1988] 463/9).
[11] Guillot, C.; Desjonquères, M. C.; Chauveau, D.; Tréglia, G.; Lecante, J.; Spanjaard, D.; Tran Minh Duc (Solid State Commun. **50** [1984] 393/6).
[12] Guillot, C.; Thuault, C.; Jugnet, Y.; Chauveau, D.; Hoogewijs, R.; Lecante, J.; Tran Minh Duc; Tréglia, G.; Desjonquères, M. C.; Spanjaard, D. (J. Phys. C **15** [1982] 4023/32).
[13] Jupille, J.; Purcell, K. G.; King, D. A. (Solid State Commun. **58** [1986] 529/33).
[14] van der Veen, J. F.; Himpsel, F. J.; Eastman, D. E. (Solid State Commun. **40** [1981] 57/60).
[15] Mullins, D. R.; Lyman, P. F. (Surf. Sci. Lett. **285** [1993] L 473/L 478).
[16] van der Veen, J. F.; Heimann, P.; Himpsel, F. J.; Eastman, D. E. (Solid State Commun. **37** [1981] 555/9).
[17] van der Veen, J. F.; Himpsel, F. J.; Eastman, D. E. (Phys. Rev. [3] B **25** [1982] 7388/97).
[18] Purcell, K. G.; Jupille, J.; Derby, G. P.; King, D. A. (Phys. Rev. [3] B **36** [1987] 1288/91).
[19] Wertheim, G. K.; Citrin, P. H. (Phys. Rev. [3] B **38** [1988] 7820/3).
[20] Chauveau, D.; Roubin, P.; Guillot, C.; Lecante, J.; Tréglia, G.; Desjonquères, M. C.; Spanjaard, D. (Solid State Commun. **52** [1984] 635/9).
[21] Spanjaard, D.; Guillot, C.; Desjonquères, M. C.; Tréglia, G.; Lecante, J. (Surf. Sci. Rep. **5** [1985] 1/85).
[22] Desjonquères, M. C.; Spanjaard, D.; Lassailly, Y.; Guillot, C. (Solid State Commun. **34** [1980] 807/10).
[23] Johansson, B.; Mårtensson, N. (Phys. Rev. [3] B **21** [1980] 4427/57).

[24] Johansson, B.; Mårtensson, N. (Helv. Phys. Acta **56** [1983] 405/26).
[25] Rosengren, A.; Johansson, B. (Phys. Rev. [3] B **22** [1980] 3706/9).
[26] Said, M.; Desjonquères, M. C.; Spanjaard, D. (Phys. Rev. [3] B **47** [1993] 4722/9; Surf. Sci. **287/288** [1993] 780/4).
[27] Rosengren, A. (Phys. Rev. [3] B **24** [1981] 7393/6).
[28] Tománek, D.; Kumar, V.; Holloway, S.; Bennemann, K. H. (Solid State Commun. **41** [1982] 273/9).
[29] Tréglia, G.; Desjonquères, M. C.; Spanjaard, D.; Lassailly, Y.; Guillot, C.; Jugnet, Y.; Tran Minh Duc; Lecante, J. (J. Phys. C **14** [1981] 3463/73).
[30] Posternak, M.; Krakauer, H.; Freeman, A. J.; Koelling, D. D. (Phys. Rev. [3] B **21** [1980] 5601/12).
[31] Posternak, M.; Krakauer, H.; Freeman, A. J. (Phys. Rev. [3] B **25** [1982] 755/61).
[32] Legrand, B.; Tréglia, G.; Desjonquères, M. C.; Spanjaard, D. (J. Phys. C **19** [1986] 4463/72).
[33] Kim, B.; Chen, J.; Erskine, J. L.; Mei, W. N.; Wei, C. M. (Phys. Rev. [3] B **48** [1993] 4735/40).

5.5.3.3 Angle-Resolved Spectra (Photoelectron Diffraction)

The intensity data of photoelectron emission usually show strong spatial variations due to diffraction of photoelectrons by the surrounding atoms of the emitter. Photoelectron diffraction, specifically considering the W $4f_{7/2}$ core level spectra, was studied theoretically on W(110), W(100), and W(111) surfaces. The selection rules for dipole transitions restrict possible final states essentially to those with *d*-type symmetry ($l=2$), since at the electron energies considered ($E_{kin} \approx 30$ eV) *g*-type ($l=4$) final states may be neglected. Polar diagrams illustrate the azimuthal dependence of the calculated photoelectron intensity which strongly depends on the photon energy, the electron exit angle, and the angle between the polarization vector and surface normal. The theoretical study of W(110), W(100), and W(111) surfaces is complemented by experiments on W(100) [1]. The azimuthal dependence of angle-resolved core-level spectra was studied using p-polarized radiation (polarization vector fixed at 22.5° off the surface normal) with $h\nu = 65$ eV. The emitted electrons were detected with an energy resolution of 0.19 eV and at an exit angle 30° off the surface normal with an angular resolution of $<1°$. The surface was clean according to AES and was parallel to (100) to within $\pm 0.5°$. The high degree of order was verified by the LEED patterns showing $p(1 \times 1)$ and $c(2 \times 2)$ symmetry at room temperature and ~90 K, respectively. **Fig. 43** [1] shows the azimuthal dependence of the $4f_{7/2}$ intensities for the total, bulk, and surface emissions for both the unreconstructed and the reconstructed W(100) surface [1] (extending an earlier study [2] which was restricted to an unreconstructed W(100) surface). Conditions similar to those used by [1] also prevailed in a photoelectron diffraction study of W(110) surfaces by [3], where azimuthal scans were possible for rotations of $\varphi \leq 215°$ [3].

Angle-resolved photoelectron emission from surface and bulk atoms was studied on W(110) using ~50 to 125 eV radiation. At 70 eV the angular and energy resolutions were ~2° and 0.3 eV. In most of the experiments by [4, 5] the radiation impinged at an angle α of 70° and the electron exit angle ϑ was 22.5°. The polarization vector was 22.5° off the surface normal and within the incidence plane [4, 5]. The experimental intensity ratio for bulk and surface emission is plotted in [5] vs. photon energy and azimuthal angle φ, respectively. More recently, Kim et al. [9] found rather dissimilar results in corresponding experimental measurements and theoretical calculations [9]. The variations of the W $4f_{7/2}$ surface and bulk peaks are mainly due to final-state scattering of the photoelectrons by surrounding atoms [4, 5]. For the W(111) surface, the intensity ratio of the surface to bulk

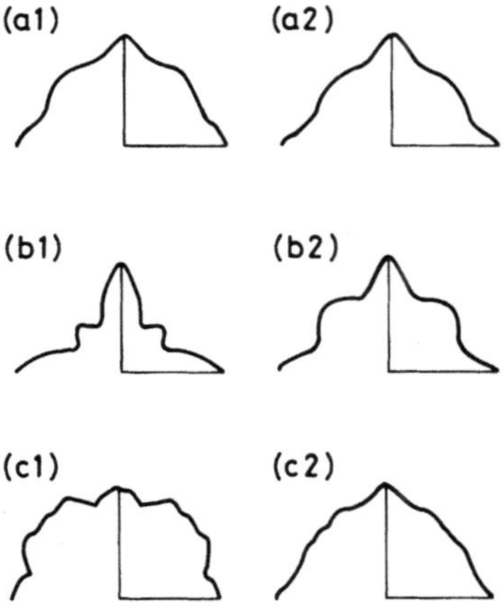

Fig. 43. Polar plots of the azimuthal dependence of $4f_{7/2}$ core-level photoemission from unreconstructed (left-hand curves) and reconstructed (right-hand curves) W(100) surfaces. The upper, middle, and bottom rows show total, bulk, and surface emission curves, respectively. The p-polarized radiation (hν = 65 eV) was incident at ~67.5°, the electron detection was at an angle of 30° [1].

photoelectron signals from the $4f_{7/2}$ core level was calculated as a function of the emission angle. For emissions 60° off the surface normal, ratios of 0.22 and 0.25 were found for Al K_α and Mg K_α radiation, respectively. For normal emission, the respective ratios were 0.10 and 0.11 [6].

For W(100) a single scattering model of photoelectron diffraction describes satisfactorily the azimuthal patterns, even when measured under conditions ($E_{kin} \approx 30$ eV) where bulk and surface emissions are well separated. This agreement, however, is achieved only when appropriate initial and final state wave functions are considered [7].

The semiempirical concept of an isotropic electron mean free path that varies with (hkl) ameliorates the agreement between single-scattering plane-wave theory [8] and the observed energy and angle dependences of the $4f_{7/2}$ core-level photoelectron intensities for W(110) [3] and W(100) [1]. While the experimental data in these papers ([1, 3]) were still evaluated with a constant, (hkl)-independent mean free path of 5 Å, a perfect match of experimental and calculated curves was now achieved by assuming a mean free path (for $4f_{7/2}$ electrons excited by 65 eV photons) of 8 and 5 Å for W(110) and W(100), respectively [8].

References for 5.5.3.3:

[1] Sébilleau, D.; Tréglia, G.; Desjonquères, M. C.; Spanjaard, D.; Guillot, C.; Chauveau, D.; Lecante, J. (J. Phys. [Paris] **49** [1988] 227/36).

[2] Sébilleau, D.; Desjonquères, M. C.; Chauveau, D.; Guillot, C.; Lecante, J.; Tréglia, G.; Spanjaard, D. (Surf. Sci. **185** [1987] L 527/L 533).

[3] Chauveau, D.; Guillot, C.; Villette, B.; Lecante, J.; Desjonquères, M. C.; Sébilleau, D.; Spanjaard, D.; Tréglia, G. (Solid State Commun. **69** [1989] 1015/8).

[4] Jugnet, Y.; Prakash, N. S.; Tran Minh Duc; Poon, H. C.; Grenet, G.; Pendry, J. B. (Surf. Sci. **189/190** [1987] 782/7).

[5] Jugnet, Y.; Prakash, N. S.; Porte, L.; Tran Minh Duc; Nguyen, T. T. A.; Cinti, R.; Poon, H. C.; Grenet, G. (Phys. Rev. [3] B **37** [1988] 8066/71).

[6] Chadwick, D.; Karolewski, M. A. (J. Electron Spectrosc. Relat. Phenom. **24** [1981] 181/7).
[7] Tréglia, G.; Sébilleau, D.; Desjonquères, M. C.; Spanjaard, D.; Guillot, C.; Chauveau, D.; Lecante, J. (Proc. 10th Taniguchi Int. Symp., 1987 [1988], pp. 281/98).
[8] Tréglia, G.; Desjonquères, M. C.; Spanjaard, D.; Sébilleau, D.; Guillot, C.; Chauveau, D.; Lecante, J. (J. Phys. Condens. Matter **1** [1989] 1879/88).
[9] Kim, B.; Chen, J.; Erskine, J. L.; Mei, W. N.; Wei, C. M. (Phys. Rev. [3] B **48** [1993] 4735/40).

5.5.3.4 Line Shapes and Widths. Background

Polycrystals. XPS measurements using Mg K_α (1253.6 eV) radiation were performed on polycrystalline W surfaces, clean at the 1% level (from XPS). The background intensity of the core-level emission appearing at the low-energy side of the lines is attributed to inelastically scattered electrons. This background falls off logarithmically with the energy away from the line (at least to ~150 eV below the line) [1].

The skew line shape of fast photoelectrons from the metal surfaces is attributed to interactions of the conduction electrons with the as-created electron holes. For an analytical expression, see [2].

W(110). Using photons of 65 to 95 eV energy impinging at $\alpha=70°$, normal-emission $4f_{7/2}$ core-level spectra were taken near 100, 290, and 700 K. The measured phonon broadening of the levels was 71 ± 25 and 92 ± 21 meV at 100 ± 20 and 290 ± 10 K, respectively [4] in agreement with the theoretical expectations of 60 and 90 meV FWHM broadening calculated for 77 and 300 K [5]. At 210 and 310 K the Gaussian widths of the bulk lines after correction for the instrumental width were 45.4 ± 7.6 and 57.3 ± 8.7 meV, respectively. From their temperature dependence these Gaussian widths are identified to be due to phonons excited during photoemission [3]. These measured widths are significantly smaller than the theoretical estimates given by [5] and [6]. Changes (with respect to the bulk) observed at the surface are due to changes in the natural line width and in the conduction-electron screening. There is no anomalous phonon broadening, no evidence for crystal-field splitting, and no need to invoke unresolved lines from subsurface layers. While the phonon broadening at the surface (48.4 ± 5.5 and 65.6 ± 4.2 eV at 210 and 310 K, respectively) is not much different from the bulk, the Lorentzian (lifetime) core-hole width at the surface (84 ± 3 meV) is distinctly larger than in the bulk (60 ± 3 meV) [3]. The bulk value is in good agreement with the earlier bulk values of 51 meV, 55 meV [7], and 61 meV [8]. The more recent values reported by [3, 4] obviously supersede the older data for the lifetime widths of 110 ± 20 meV for bulk and surface given in [9] and 190 and 220 meV for bulk and surface, respectively, given in [10]. The singularity index is 0.035 ± 0.003 and 0.063 ± 0.003 for the bulk and surface core-level lines, respectively [3].

W(100). The core-level lines from unreconstructed W(100) have asymmetric Lorentzian-Gaussian shapes (30% Lorentzian on the lower binding-energy side and 70% Lorentzian on the higher binding-energy side) [11]. The phonon broadening of 4f core-level lines and the influence of electron-hole pair formation on the lineshapes were studied theoretically by [5]. The calculated FWHM due to phonon broadening for atoms in the bulk is 60 and 90 meV at 77 and 300 K, respectively. For atoms on the W(100) surface, the corresponding values are 60 and 110 meV [5].

W(111). The broadening of the surface core-level line on W(111) is attributed to an increased natural linewidth [3]. Fitting four lines to the spectra, a Gaussian width of 150 ± 8 meV resulted; fits of three lines to the spectra resulted in Gaussian widths of 193 ± 8 and 191 ± 12 meV for constrained (for details, see the paper) and unconstrained fits, respectively [8].

References for 5.5.3.4:

[1] Tougaard, S.; Ignatiev, A. (Surf. Sci. **124** [1983] 451/60).
[2] Doniach, S.; Sunjic, M. (J. Phys. C **3** [1970] 285/91).
[3] Riffe, D. M.; Wertheim, G. K.; Citrin, P. H. (Phys. Rev. Lett. **63** [1989] 1976/9).
[4] Purcell, K. G.; Derby, G. P.; King, D. A. (J. Phys. Condens. Matter **1** [1989] 1373/9).
[5] Sébilleau, D.; Tréglia, G.; Desjonquères, M. C.; Guillot, C.; Chauveau, D.; Spanjaard, D. (J. Phys. C **20** [1987] 2647/57).
[6] Flynn, C. P. (Phys. Rev. Lett. **37** [1976] 1445/8).
[7] Wertheim, G. K.; Citrin, P. H.; van der Veen, J. F. (Phys. Rev. [3] B **30** [1984] 4343/7).
[8] Wertheim, G. K.; Citrin, P. H. (Phys. Rev. [3] B **38** [1988] 7820/3).
[9] van der Veen, J. F.; Himpsel, F. J.; Eastman, D. E. (Phys. Rev. [3] B **25** [1982] 7388/97).
[10] Tran Minh Duc; Guillot, C.; Lassailly, Y.; Lecante, J.; Jugnet, Y.; Vedrine, J. C. (Phys. Rev. Lett. **43** [1979] 789/92).
[11] Guillot, C.; Thuault, C.; Jugnet, Y.; Chauveau, D.; Hoogewijs, R.; Lecante, J.; Tran Minh Duc; Tréglia, G.; Desjonquères, M. C.; Spanjaard, D. (J. Phys. C **15** [1982] 4023/32).

5.5.4 Photoelectric Schottky Deviations

For Schottky deviations in thermionic emission, see p. 137. Oscillatory deviations from the zero-field current-voltage characteristics were predicted in a theoretical study of the field dependence of the surface photoeffect [1]. These photoelectric Schottky deviations were measured on a W single-crystal filament. The photoelectric current was sampled from a 30° region around the W(111) surface normal. Alternate photoelectric and thermionic Schottky runs were made, the thermionic one near 1300 K (since for that temperature the thermionic emission currents equaled the maximum photoelectric currents of 10^{-10} A). The field dependence of the photoelectric current is plotted in Fig. 29, p. 138 [2]. The photoelectric deviation amplitude is 1.7% at $\sqrt{F}=350$ $V^{1/2}/cm^{1/2}$ (i.e., at $y_F \approx 6.1\ \pi$; see curve b in Fig. 29). The periodic current deviations observed in [2] are in fair agreement with the theoretical prediction given in [3], while the thermionic data are excellently described by the model for the surface potential barrier and the periodic deviations given in [4]. The period and phase of the simultaneously determined photoelectric and thermionic deviations agree well for the higher fields, i.e., for the lower values of y_F; see Fig. 29. The disagreement at lower fields may be attributed to gas adsorption during the runs (made from high to low fields) [2]. Earlier, the emission had been measured on a polycrystalline W filament under 2537 Å irradiation for fields up to ~170 kV/cm. The measured deviations from the photoelectric Schottky line were in fair agreement with theoretical predictions; see the curves of $\Delta i/i_0$ vs. $\sqrt{F}$ (up to ~400 $V^{1/2}/cm^{1/2}$) given in [5].

References for 5.5.4:

[1] Guth, E.; Mullin, C. J. (Phys. Rev. [2] **59** [1941] 867/73).
[2] Stafford, D. F.; Weber, A. H. (J. Appl. Phys. **34** [1963] 2667/70).
[3] Juenker, D. W. (Phys. Rev. [2] **99** [1955] 1155/60).
[4] Cutler, P. H.; Gibbons, J. J. (Phys. Rev. [2] **111** [1958] 394/402).
[5] Buder, E. E.; Ruddick, J. J.; Weber, A. H. (Phys. Rev. [2] **91** [1953] 561/6).

5.6 Laser-Induced Electron Emission

When tungsten is irradiated by laser beams, direct photoelectron emission processes can generally (as long as the wavelength is not smaller than about 0.3 μm) not occur, since the laser photon energies are too low to overcome the work function Φ of the material

References for 5.6 on pp. 201/2

($\Phi \approx 4.5$ eV, see Section 4.2.2, pp. 121/3). Nevertheless, considerable electron emission is observed resulting from thermionic and multiple-photon processes.

In most of the early experimental work on laser-induced electron emission from W surfaces, the possibility of multiple-photon processes was not taken into account [1 to 6] or was believed to be unlikely [7, 8], although theoretical calculations already existed that predicted electron emission from such processes to occur [9 to 11]. The experimental results were interpreted in terms of thermionic emission processes in [1 to 8], but it could not be ruled out that other effects occur.

In his study with a Q-switched ruby laser, producing single pulses of about 50 ns duration and 10 to 25 MW/cm^2 peak power density, Ready [12] compared the time dependence of the measured electron current density with the photoelectric current that was calculated from theoretical equations in [10] for a two-photon process. No trace of electron emission from multiple-photon processes was found. Ready concluded that the observed emission was thermal in origin, produced by surface heating and described by the Richardson equation of thermionic emission. The shape and amplitude of the emission current pulses and particularly the delay between the peaks of the laser pulse and the observed emission were in accordance with the expected results for thermionic emission [12]. The time delay is due to the fact that the maximum surface temperature and with it the maximum thermionic electron emission are not reached instantaneously. Power density as well as shape and duration of the laser pulse strongly influence the time for heating up the surface. In [13] the spatial and temporal form of the temperature of laser-heated metals is calculated, and numerical results for W are given. Assuming Gaussian laser pulses of wavelength $\lambda = 1.06$ μm, a maximum temperature change at the W surface of about 30 K is predicted for a 30 ns pulse with maximum irradiance of 1 MW/cm^2, while under the same conditions a 30 ps pulse with maximum irradiance of 1 GW/cm^2 should produce an increase of the surface temperature of several hundred degrees [13].

By comparing the results of an electron emission study using an Nd:YAG laser ($h\nu =$ 1.17 eV) with microsecond pulses with those of a resistively heated surface, it was concluded that the electron emission was purely thermionic [14].

Other authors found in addition to delayed thermionic emission an initial electron emission peak which could not be explained [2, 4] and even was believed to be an experimental artifact [15].

In reviews of Anisimov et al. [16] and Farkas [17], it was pointed out why thermionic and multiple-photon electron emission processes could not be separated in early work. A too long laser pulse duration combined with high intensity masked the multiple-photon processes. As was calculated in [18], there exists a critical duration for the laser pulses (depending on laser intensity) below which thermionic emission plays no longer a role.

Using a 30 ps pulsed Nd:YAG laser with a wavelength of 1.064 μm (corresponding to $h\nu = 1.165$ eV), pure four-photon photoemission from W was observed with laser power densities ranging from about 0.4 to 4 GW/cm^2 [19]. **Fig. 44** (from [19]) shows a plot of the effective current density versus the peak incident irradiance. The theoretical curve was obtained from the four-photon-generalized photoemission theory of Fowler [20] and DuBridge [21]. According to this theory, the electron current density is proportional to the nth power of the incident laser irradiance if a pure multiple-photon electron emission process of nth order occurs (i.e. if an energy of $n \cdot h\nu$ is necessary for an electron to overcome the work function) [19]. The same authors found two-photon photoemission when W was irradiated by 2.33 eV pulses (21.2 ps) from the second-harmonic radiation of their Nd:YAG laser [22]. In a preceding study, using a Q-switched ruby laser ($h\nu = 1.78$ eV) with 13 ns

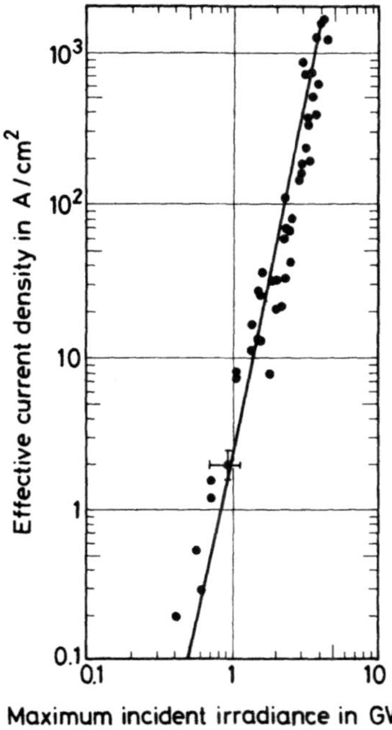

Fig. 44. Electron emission current density from a W surface irradiated by a Nd:YAG laser as a function of the maximum incident irradiance (straight line: theoretical curve according to a four-photon photoemission process) [19].

pulse duration and maximum irradiance in the range of 10 to 100 MW/cm^2, no multiple-photon process could be separated. Instead, a two-photon-assisted thermionic process was proposed to explain the observed data [23]. Under the same experimental conditions as in [19] pure four-photon photoemission was found for power levels below 1 GW/cm^2, but at higher power levels the photocurrent increased much faster than a fourth power law. This effect was attributed to thermally assisted three-photon photoemission enabled by enhanced population of the tail of the Fermi distribution [24].

The polarization and angle-of-incidence dependence of four-photon photoemission from W polycrystals was investigated [25]. The results were interpreted in terms of a combination of a volume and a surface effect. When the samples were irradiated by 5 GW/cm^2 laser pulses ($\lambda = 1.06$ μm) of 30 ps duration, the maximal ratio of surface to (surface + bulk) emission (ca. 0.75) was found at an angle of incidence of about 55° [25].

The electron emission from laser-irradiated W was first interpreted as a result of multiple-photon processes by Arifov and coworkers [26, 27] prior to the studies mentioned above (i.e. [19, 22, 24, 25]), but this interpretation was retracted in a later study [28]. With μs pulses from a ruby laser they had observed spiked 0.1 to 1 μs-wide symmetric emission peaks that were attributed to fourth-order, multiple-photon photoemission processes on account of their dependence on radiation intensity [26, 27]. Additionally, they had found asymmetric emission peaks which were later attributed to the emission of negative ions [29, 30]. To separate the effect of thermionic emission, the heating kinetics of a heated W foil was investigated in another study with 2.5 ms laser pulses of 0.1 MW/cm^2 power density taking into account the change of the reflectivity (R) of the metal surface during laser irradiation. In the same study a linear plot of R vs. T was presented showing a change

References for 5.6 on pp. 201/2

of slope between about 2950 and 3000 °C which was related to the melting of the foil surface, even though this temperature was 400 °C below the melting point of W [31]. The initial interpretation of electron emission peaks as arising from multiple-photon processes was found to be in conflict with some features discussed [28]:

The emission current was found to be proportional to the fourth power of the laser irradiation intensity, although with ruby laser light a three photon process should occur ($3 \cdot h\nu = 5.34$ eV $> \Phi$). Moreover, there was no direct relationship between the electron current and the normal component of the electric vector of the laser irradiation. From the slopes of the current-voltage characteristics [30], an electron temperature of about 21000 K was estimated for a power density of 20 MW/cm^2, whereas the calculated surface temperature did not exceed 2300 K. The electron temperature was found to be proportional to the radiation intensity. Such an emission of "hot" electrons was first reported by Knecht [4] who obtained an electron temperature of about 29500 K when he irradiated a W surface with 40 ns pulses of a Q-switched ruby laser (13.7 MW/cm^2 power density). This shows that the emission of "hot" electrons can be explained neither by multiple-photon emission nor by thermionic emission [28].

In another study, Arifov et al. [32] investigated the electron emission from W in strong electrostatic fields using a 0.1 μs pulsed ruby laser with 8 MW/cm^2 power density. The initial temperature determined with an optical pyrometer ranged from 30 to 1400 °C. The authors presumed that the emission of "hot" electrons is due to a tunneling effect of the excited electrons [32].

A model based on laser irradiation to be the cause for nonequilibrium excitation of the metal electrons is proposed in [33]. The current density of "hot" electrons is shown to be related by a power-law with the intensity of the incident irradiation. It is concluded that the emission of "hot" electrons is a multistage process, whereby the incident radiation pumps the metal electrons through a succession of excited states. Consequently, the energy spectrum of the electrons should have a stepped structure [33]. To substantiate these conclusions, the dependence of the electron emission current from a ruby-laser-irradiated W surface on strong electrostatic fields was investigated [34]. The same authors found that the electron emission is thermionic when a point target is irradiated with quasi-continuous laser radiation [35]; in contrast, when short laser pulses (100 ns) were used as in [34], "hot" electrons were observed. This was shown by the shape of the pulses, the characteristic delay of the emission pulse relative to the laser pulse (about 200 μs), and the current-voltage characteristic [35].

Besides nonequilibrium electrons, the emission of photons in the ultraviolet region was observed during laser irradiation with 75 ns pulses of 7.5 to 370 MW/cm^2. For power densities below 10 MW/cm^2 the photon yield was found to be proportional to the 6th power of the laser intensity [36]. Anomalous thermal radiation from W and Ag produced by 10 and 40 ps pulses of an Nd laser with power densities of 10 to 100 GW/cm^2 was investigated experimentally and theoretically. This emission of photons was explained as nonequilibrium recombination radiation from a narrow layer of about 10 Å above the metal surface that is formed by the emitted ions and the electrons tunneling to them from the metal surface [37].

Ultrashort pulses (75 fs) of 0.62 μm ($h\nu = 2$ eV) radiation from a dye laser were used in [38, 39] to produce electron emission from a W surface. With power densities in the range between about 10 and 100 GW/cm^2, the photoemitted charge was found to increase with about the fourth power of the incident laser pulse intensity, although three-photon processes had been expected to be dominant. According to the authors this enhanced slope is produced by thermally assisted photoemission which implies that a thermal nonequilibrium between electrons and lattice existed in their experiments [38, 39]. In [40] it had been postulated that a nonequilibrium between electrons and phonons can be achieved, if the laser pulses are shorter than or comparable to the electron-phonon energy relaxation time. In this case the smaller heat capacity of the electron gas allows anomalous heating of the electrons to temperatures much higher than that of the lattice [40]. Following these considerations, an electron-phonon energy relaxation time of several hundred femtoseconds and an electron-phonon coupling constant of the order of 5×10^{11} to 1×10^{12} $W \cdot cm^{-3} \cdot K^{-1}$ were estimated from pump-probe measurements on W [39].

The energetical and spatial distributions of thermionic electrons emitted from a W filament were investigated using a cw Ar ion laser ($\lambda = 0.488$ to 0.5145 μm) with comparatively low power densities in the order of 10 kW/cm^2. It was concluded that laser-enhanced thermionic emission yields an electron beam of higher brilliance (i.e., better collimation) and smaller energy distribution than ordinary thermionic emission at the same electron currents [41].

Irradiation of hot W_n clusters with 10 ns pulses from a Q-switched Nd:YAG laser provokes thermionic electron emission. The delayed emission of electrons was recorded as a function of time after the excitation for different cluster sizes varying from $n = 4$ to 40 atoms. A large yield of W_n^+ cluster ions, which were generated as late as microseconds after the laser pulse, was detected for $n > 4$ [42, 43].

References for 5.6:

[1] Honig, R. E.; Woolston, J. R. (Appl. Phys. Lett. **2** [1963] 138/9).
[2] Verber, C. M.; Adelman, A. H. (Appl. Phys. Lett. **2** [1963] 220/2).
[3] Alcock, A. J.; Iannuzzi, M.; Motz, H.; Walsh, D. (J. Electron. Control. **16** [1964] 75/7).
[4] Knecht, W. L. (Appl. Phys. Lett. **6** [1965] 99/100, **8** [1966] 254/6).
[5] Pittaway, L. G.; Smith, J.; Fletcher, E. D.; Nicholls, B. W. (J. Phys. D **1** [1968] 711/9).
[6] Duley, W. W. (Appl. Phys. Lett. **13** [1968] 255/6; Can. J. Phys. **47** [1969] 2419/23).
[7] Giori, F.; MacKenzie, L. A.; McKinney, E. J. (Appl. Phys. Lett. **3** [1963] 25/7).
[8] Lichtman, D.; Ready, J. F. (Phys. Rev. Lett. **10** [1963] 342/5).
[9] Smith, R. L. (Phys. Rev. [2] **128** [1962] 2225/9).
[10] Adawi, I. (Phys. Rev. [2] **134** [1964] A 788/A 798).

[11] Bloch, P. (J. Appl. Phys. **35** [1964] 2052/4).
[12] Ready, J. F. (Phys. Rev. [2] **137** [1965] A 620/A 623).
[13] Bechtel, J. H. (J. Appl. Phys. **46** [1975] 1585/93).
[14] Hiernaut, J.-P.; Magill, J.; Ohse, R. W.; Tetenbaum, M. (High Temp.-High Pressures **17** [1985] 633/8).
[15] Li, P. W.; Isenor, N. R. (Appl. Phys. Lett. **13** [1968] 238).
[16] Anisimov, S. I.; Benderskii, V. A.; Farkas, G. (Usp. Fiz. Nauk **122** [1977] 185/222; Sov. Phys. Usp. [Engl. Transl.] **20** [1977] 467/88).
[17] Farkas, G. (Acta Phys. Austriaca Suppl. **20** [1979] 117/32).
[18] Bunkin, F. V.; Prokhorov, A. M. (Zh. Eksp. Teor. Fiz. **52** [1967] 1610/5; Sov. Phys.-JETP [Engl. Transl.] **25** [1967] 1072/5).
[19] Bechtel, J. H.; Smith, W. L.; Bloembergen, N. (Opt. Commun. **13** [1975] 56/9).
[20] Fowler, R. H. (Phys. Rev. [2] **38** [1931] 45/56).

[21] DuBridge, L. A. (Phys. Rev. [2] **43** [1933] 727/41).
[22] Bechtel, J. H.; Smith, W. L.; Bloembergen, N. (Phys. Rev. [3] B **15** [1977] 4557/63).
[23] Bechtel, J. H.; Franken, P. A. (Phys. Rev. [3] B **11** [1975] 1359/64), Bechtel, J. H. (Diss. Univ. Michigan 1973; Diss. Abstr. Int. **34** [1974] 3962).

[24] Yen, R.; Liu, J.; Bloembergen, N. (Opt. Commun. **35** [1980] 277/82).
[25] Yen, R.; Liu, J.; Dagenais, M.; Bloembergen, N. (Opt. Commun. **31** [1979] 334/9).
[26] Arifov, U. A.; Kazanskii, V. V.; Lugovskoi, V. B.; Kayumova, Z. A. (Izv. Akad. Nauk SSSR Ser. Fiz. **35** [1971] 599/602; Bull. Acad. Sci. USSR Phys. Ser. [Engl. Transl.] **35** [1971] 550/2).
[27] Arifov, U. A.; Kazanskii, V. V.; Lugovskoi, V. B.; Makarenko, V. A. (Izv. Akad. Nauk UzSSR Ser. Fiz.-Mat. Nauk **18** No. 2 [1974] 42/5).
[28] Arifov, U. A.; Lugovskoi, V. B.; Makarenko, V. A. (Fiz. Tverd. Tela [Leningrad] **20** [1978] 1505/10; Sov. Phys.-Solid State [Engl. Transl.] **20** [1978] 867/70).
[29] Arifov, U. A.; Kazanskii, V. V.; Lugovskoi, V. B.; Makarenko, V. A. (Izv. Akad. Nauk UzSSR Ser. Fiz.-Mat. Nauk **19** No. 4 [1975] 49/52; Ref. Zh. Fiz. **1976** No. 1D1187).
[30] Arifov, U. A.; Kazanskii, V. V.; Lugovskoi, V. B.; Makarenko, V. A. (Zh. Tekhn. Fiz. **46** [1976] 417/8; Sov. Phys.-Tech. Phys. [Engl. Transl.] **21** [1976] 239/40).

[31] Arifov, U. A.; Zinov'ev, A. V.; Lugovskoi, V. B.; Makarenko, V. A. (Fiz. Khim. Obrab. Mater. **1977** No. 2, pp. 10/4; C.A. **87** [1977] No. 9955).
[32] Arifov, U. A.; Azizov, S. T.; Lugovskoi, V. B. (Fiz. Tverd. Tela [Leningrad] **20** [1978] 2047/50; Sov. Phys.-Solid State [Engl. Transl.] **20** [1978] 1182/4).
[33] Zinov'ev, A. V.; Lugovskoi, V. B. (Zh. Tekhn. Fiz. **50** [1980] 1635/40; Sov. Phys.-Tech. Phys. [Engl. Transl.] **25** [1980] 953/6).
[34] Azizov, S. T.; Lugovskoi, V. B. (Zh. Tekhn. Fiz. **51** [1981] 2554/9; Sov. Phys.-Tech. Phys. [Engl. Transl.] **26** [1981] 1509/12).
[35] Azizov, S. T.; Lugovskoi, V. B. (Izv. Akad. Nauk UzSSR Ser. Fiz.-Mat. Nauk **26** No. 5 [1982] 63/6; C.A. **98** [1983] No. 64070).
[36] Azizov, S. T.; Ben'kov, A. V.; Lugovskoi, V. B. (Izv. Akad. Nauk UzSSR Ser. Fiz.-Mat. Nauk **25** No. 5 [1981] 55/7; C.A. **96** [1982] No. 94532).
[37] Agranat, M. B.; Anisimov, S. I.; Makshantsev, B. I. (Appl. Phys. B **47** [1988] 209/21, B **55** [1992] 451/61).
[38] Fujimoto, J. G.; Ippen, E. P.; Liu, J. M.; Bloembergen, N. (Springer Ser. Chem. Phys. **38** [1984] 111/3).
[39] Fujimoto, J. G.; Liu, J. M.; Ippen, E. P.; Bloembergen, N. (Phys. Rev. Lett. **53** [1984] 1837/40).
[40] Anisimov, S. I.; Kapeliovich, B. L.; Perel'man, T. L. (Zh. Eksp. Teor. Fiz. **66** [1974] 776/81; Sov. Phys. JETP [Engl. Transl.] **39** [1975] 375/7).

[41] Stamatovic, A.; Stephan, K.; Howorka, F.; Märk, T. D. (Int. J. Mass Spectrom. Ion Processes **85** [1988] R 1/R 5).
[42] Leisner, T.; Athanassenas, K.; Echt, O.; Kandler, O.; Kreisle, D.; Recknagel, E. (Z. Phys. D **20** [1991] 127/9).
[43] Leisner, T.; Athanassenas, K.; Kandler, O.; Kreisle, D.; Recknagel, E.; Echt, O. (Mater. Res. Soc. Symp. Proc. **206** [1991] 259/64; C.A. **115** [1991] No. 268332).

5.7 Electron-Induced Electron Emission

5.7.1 Secondary Electron Emission Yield

Nomenclature. Three groups of electrons are usually distinguished in the spectrum of secondary electrons: (i) the elastically reflected electrons, (ii) inelastically backscattered or "rediffused" electrons, and (iii) so-called true secondary electrons. The boundaries between these groups are somewhat arbitrary and also depend on the effort devoted to their discrimination. For primary electron energies of $E_p \gg 50$ eV, it is common to call secondary electrons with kinetic energies up to $E_s \cong 50$ eV "true" secondaries, their yield coefficient

(=yield per incident electron) being δ. The group (ii) of inelastically backscattered and rediffused electrons involves electrons with energies between 50 eV and E_p (excluding $E_s=E_p$). Their yield coefficient is denoted η. For low primary energies, it is very difficult to discriminate between true secondary and inelastically scattered primary electrons. The yield coefficient of the elastically scattered (reflected) electrons is denoted R. The range of E_s values ($\approx E_p$) accepted for this group is rather ill-defined. In many papers the different groups are not properly distinguished and merely the ratio of secondary to primary electron currents is measured (i.e., the total yield coefficient $\gamma=\delta+\eta+R$). The various authors use the symbols δ, γ, σ, and η quite liberally: the symbols σ and δ are frequently used instead of γ, and η often includes R, i.e., it refers to the sum of all backscattered electrons. To enable a comparison of the different results, the notation of the original papers was changed in the following text in favor of the notation introduced above. The angle of incidence is denoted α ($\alpha=0°$ for normal incidence).

5.7.1.1 Polycrystals

The total yield increases with the primary energy to a maximum value of γ_{max} at $E_p(\gamma_{max})\approx 700$ eV. For higher energies, γ decreases monotonously; see below.

Studies with Low Primary Energies ($E_p<E_p(\gamma_{max})$). The coefficients of total secondary emission, γ, and elastic reflection, R, under normal incidence were measured for primary energies between 1 and 30 eV. A W ribbon was cleaned by heat treatment in vacuum and investigated between 1300 and 1500 K at $p\leq 2\times 10^{-9}$ Torr. The yield of slow electrons ($\gamma-R$) (named δ in the paper) was deduced from the experimental data. It increases with E_p from about zero near 2 eV to 0.35 at 30 eV. R exhibits a maximum around 9 to 10 eV [1]. Similar measurements were performed by [2], where γ and R were studied at about 5×10^{-10} Torr for E_p ranging from ~1.5 to ~30 eV on samples degassed near the melting point at $p<10^{-8}$ Torr. The results of [2] are nearly identical with those reported by [1] and are close to the W(100) results reported by [3] and [4], indicating a (100) texture in the polycrystalline W targets used. R, γ, and ($\gamma-R$) had been earlier determined by [5] for $E_p=0.4$ to 30 eV on samples degassed near 2000 K in $\sim 10^{-8}$ Torr. γ increased from ~0.05 for E_p near zero to 0.4 at $E_p=30$ eV, R showed a maximum around 9 eV, and ($\gamma-R$) increased fairly linearly with E_p [5] in good agreement with [1, 2]. For R vs. E_p between 2 and 10 eV measured at $p<5\times 10^{-9}$ Torr, see [6]. The yield coefficient γ of W heated to 2000 °C is plotted in paper [7] for the E_p range 1 to 10 eV. γ increased from 0.15 at 1 eV to 0.23 at 4 eV and to 0.24 at 10 eV [7]. The reflection coefficient R and the total yield coefficient γ of a rolled W sheet (presumably having a (111) texture) were measured at $\sim 10^{-10}$ Torr using primary energies of 2 to 10 eV. R increased from ~0.07 at 2 eV to ~0.13 at 4 eV; γ increased like R between 2 and 4 eV, but further increased to ~0.29 at 10 eV [8]. For energies ranging up to $E_p=300$ eV, γ increased monotonously to $\gamma(300\text{ eV})=1.31$ [9, 10].

Studies with Primary Energies Including $E_p(\gamma_{max})$. Literature data agree on finding a maximum of $\gamma_{max}=1.2$ to 1.4 in the γ vs. E_p curve of polycrystalline W between 0.6 and 0.8 keV [11 to 21]; see also the review [22]. Slightly smaller values for the position of the maximum were found in [23] (0.5 keV with $\gamma_{max}=1.46$) and in the older paper [24] (0.42 keV with $\gamma_{max}=1.31$). A higher value of $\gamma_{max}=1.57$ ($E_p=0.7$ keV) was reported by [25]. According to the review [26], a maximum occurs at $E_p=0.7$ keV with $\delta_{max}=1.0$ to 1.4 for the yield of "true" secondary electrons ($E_s\leq 50$ eV). The yield of inelastically backscattered and elastically reflected electrons ($E_s\geq 50$ eV) was independent of E_p for $E_p\geq\sim 10$ keV ($\eta+R=0.43$ for $E_p>10$ keV) [26].

References for 5.7.1 on pp. 208/10

Electron beam energies E_p were determined for which electrically floating metal samples do not charge up (i.e., where $\gamma=1$). For polycrystalline, clean W and electrons impacting at an incidence angle of $\alpha=30°$, the condition $\gamma=1$ was fulfilled for $E_p^I=0.22$ keV and $E_p^{II}=4.10$ keV. Under normal incidence, a value of $E_p^{II}=2.9$ keV was obtained. E_p^I does not greatly depend on α ($E_p^I=0.22$ keV for both $\alpha=0°$ and 30°) [27]. $E_p^I=0.23$ keV for $\alpha=0°$ is given in [11].

R vs. E_p was measured below 10^{-9} Torr for $E_p\approx0.1$ to 1 keV. R decreased from ~2.5% at 0.1 keV to ~2% near 0.22 keV, then increased to a broad maximum near 0.8 keV [28]. The yield of backscattered electrons ($\eta+R$) monotonously increased between 0.3 and 3 keV [29].

The dependence of γ and $(\gamma-\delta)$ on the angle of incidence α between 0° and 60° was investigated in [12]. $\gamma(E_p)$ passed a maximum around $E_p=0.8$ keV for all α values studied and increased with α. $(\gamma-\delta)$ increased with energy and seemed to saturate at energies above ~1.5 keV [12].

The angular distribution of inelastically scattered electrons was measured on polished W samples at angles of incidence between $\alpha=0°$ and 85° for $E_p=0.5$ and 3 keV. While the cosine (Lambert's) law was closely followed at normal incidence, there were strong deviations therefrom at near-grazing incidence [30]. The calculated and measured angular distributions of inelastically scattered electrons ($E_p=3$ keV, normal incidence) did not agree very well [31].

No difference was noted (for E_p up to 2 keV) in γ, when the values obtained at room temperature and at 2200 K ($p\leq5\times10^{-9}$ Torr) were compared, although strong thermal emission currents, a factor of 10^4 higher than the secondary emission currents, were observed at the high temperatures [14].

Theoretically calculated $\gamma(E_p)$ curves are discussed in the papers [16, 32, 33]. The secondary yield as a function of E_p and α was treated by a semiempirical sputtering model which allows one to deduce bulk and surface contributions separately from the experimental data [32]. In [33] the secondary electron yield was calculated by a semiempirical theory which suggests the maximum yield to occur at $E_p=0.68$ keV (close to the experimental value of 0.7 keV). For $E_p=0.68$ keV, $\eta+R=0.43$ and $\gamma=1.31$ were calculated [33, p. 626]. These results obviously revise the γ_{max} and E_{max} data given in [34].

The probability P for inelastic backscattering of electrons from W was derived from the inelastic mean free path, giving P=0.2 and 0.17 for $E_p=0.5$ and 1 keV, respectively [35].

Studies with High Primary Energies ($E_p>E_p(\gamma_{max})$). The total yield coefficient γ decreased from ~0.63 at 10 keV to ~(0.35 to 0.4) at 20 keV [36]. Measurements of γ at $E_p=5$ and 28 keV with a microprobe gave $\gamma=0.80$ and 0.51, respectively. For both energies, $\eta=0.45$ was obtained [37]. The η values were confirmed later by the same authors: $\eta=0.473$, 0.433, and 0.434 were found for $E_p=4.5$, 11, and 28 keV, respectively [38]. For a review of the various mechanisms that result in electron emission from solid surfaces in contact with the plasma of fusion reactors (considering the range $E_p=2$ to 40 keV), see [39].

A plot of γ vs. E_p between 0.03 and 0.34 MeV [40, 41] showed that the total yield decreased from $\gamma\approx0.76$ at $E_p=0.04$ MeV to ~0.63 at 0.175 MeV and ~0.62 at 0.34 MeV. The yield in secondaries with $E_s>0.8$ keV increased from about 0.37 at 0.04 MeV to ~0.49 at 0.175 MeV and ~0.50 at 0.34 MeV [40]. The secondary electron yield was measured over the range $E_p=0.02$ to 1.3 MeV in [42, 43]; see also [44]. γ decreased monotonously in this E_p range, but a maximum near $E_p=0.3$ MeV was observed in the yield of high-energy electrons (i.e., those where $E_s > 0.1$ keV [42, 43]). For this yield of high-energy electrons, values of about 0.23, 0.45, and 0.40 were reported for $E_p = 0.02$, 0.3, and 1 MeV, respectively [44]. The maximum energy of backscattered β particles is $\lesssim 1.3$ MeV independent of E_p (for $E_p > 1.3$ MeV), depending linearly on $Z^{1/5}$ [45].

Thick W targets were irradiated by 1, 2, 4, and 10 keV electrons at normal incidence. The intensity of reemitted electrons (including elastically and inelastically scattered electrons and excluding true secondaries) with $E_s \geq 50$ eV is plotted in polar diagrams. The ratio $\chi = a/b$ of the half-axes of the resulting ellipses (characterizing the deviation from a cosine distribution; $\chi = 1$ for a pure cosine distribution) depends on E_p as follows: $\chi \approx 1.4$, 1.4, 1.13, and 1 for $E_p = 1$, 2, 4, and 10 keV, respectively [46, 47].

The backscattering coefficient was calculated by Monte-Carlo methods for 30 keV electrons to be 0.545 ± 0.022 for normal incidence and 0.634 for $\alpha = 45°$ [48]. (For normal incidence the value 0.454 was obtained on thick targets by applying a semiempirical formula given by [49]) [48]. δ vs. E_p was calculated for Maxwellian E_p distributions with $E_p \leq 10$ keV from data published for monoenergetic electron impact [50].

5.7.1.2 Single Crystals

Fine-structure was observed especially in the $R(E_p)$ profiles at low energies (below about 30 eV). Angle-resolved measurements give more insight into the origin of this fine-structure. Such studies and the origin of the fine-structure are discussed in Section 5.7.3, pp. 214/23. In the following the experimental results will be emphasized.

Studies Comparing Different W Surfaces. The coefficients γ (±0.2 to 0.7%), (η + R) (±0.2 to 0.7%), and δ (±0.6 to 1.2%) are plotted vs. E_p in paper [29] for W(100); (η + R) is additionally plotted for W(110), W(111), and for a W polycrystal. (η + R) exceeds 0.5 when $E_p > 1$ keV and 0.4 when $E_p > 0.5$ keV on the single-crystal surfaces. The curves show considerable structure consisting of two kinds of features with widths of several hundreds eV or tens of eV. The finer structures are most significant at low temperatures and smoothed out at higher temperatures [29].

It is remarkable that for each of the three surfaces, W(110), W(211), and W(100), there are E_p ranges where no energy losses occur. This observation shows the existence of a threshold for secondary emission occurring around 5 eV [3, 51]. $R(E_p \approx 0$ eV) was about 0.07 to 0.2 for the three W surfaces studied in [3]. R rose to $R_{max} \approx 0.2$ to 0.4 below 20 eV and decreased quite rapidly to $R < 0.04$ for E_p increasing to 100 eV. γ increased from 0.07 to 0.2 near $E_p = 0$ to a plateau or shallow maximum at $E_p < 20$ eV, then continued a general upward trend to γ values of 0.4 to 0.6 at $E_p = 100$ eV. Many of the fine-structural details of the R profiles recurred in a subdued form in the γ vs. E_p profiles of the corresponding surface [3].

$R(E_p)$ was studied on W(211), W(100), and W(111) below 28 eV in [52]; $R(E_p)$ and $\gamma(E_p)$ for W(110) and W(100) and $\gamma(E_p)$ for W(111) were recorded in paper [53]; intensity profiles for elastically scattered electrons from W(110), W(211), and W(100) having lost less than 0.2 eV or less than 9.5 eV of their axial energy are shown in [54]. The results of these studies are given in the respective sections below.

The azimuthal angle dependence of R was determined and is plotted in polar diagrams for W(110), W(100), and W(111) at $E_p = 0.2$ keV [55].

Studies on W(110) Surfaces. The yield of reflected and true secondary electrons from W(110) surfaces was studied in [1, 3, 29, 51, 53 to 61].

References for 5.7.1 on pp. 208/10

Maxima in γ were found at $E_p=0.7$ and 9.1 eV and minima at $E_p=2.8$ and 13.1 eV. Following the latter minimum, there was a monotonous increase in yield up to 40 eV [56]. γ increased to about 0.6 at 100 eV [3]. The $\gamma(E_p)$ curve paralleled $R(E_p)$ up to ~10 eV, showing a maximum around 3.3 eV, but increased at higher energies to $\gamma=0.5$ for $E_p=30$ eV. $(\delta+\eta)$ exceeded 0.1 even at very low E_p values [57]. The emission coefficient of slow electrons (according to the data given in the paper, this is not equal to $\gamma-R=\eta+\delta$) increased from ~0 near 2 eV to ~0.35 at 28 eV [1]. $(\eta+\delta)$ showed an onset threshold at 3 eV, fine-structure in the range 3 to about 30 eV, and an increase to a value of $(\eta+\delta)=0.47$ at 52 eV, where it then dropped slightly. The R and $(\eta+\delta)$ curves cross in the range 12 to 15 eV. $(\eta+\delta)/R$ increased to ~90 at 60 eV [53].

The coefficient of elastic reflection measured between 1 and 100 eV shows a broad maximum below 20 eV ($R_{max}\approx0.2$) modulated with a fine-structure [3]. In a subsequent paper [51], the energy range below 20 eV was studied in more detail, revealing two maxima located around 3.5 and 13.5 eV. Similar results (maxima around 4 and 14 eV (±1 eV)) were found in [1, 53, 57, 58]. Maxima at the same positions were also observed in curves which include all electrons losing less than 9.5 eV [54]. In [57] an additional R vs. E_p profile for a contaminated W(110) surface was obtained. This profile fairly agrees with the results given in [1] for a "clean" surface.

The angular dependence of the secondary emission is related with the orientation dependence of the emission in Kikuchi patterns as exemplified by a study on W(110) at $E_p=2$ keV (for an intensity vs. α profile covering the range of incidence angles $\alpha=-60°$ to $+60°$, see the paper) [76].

Studies on W(211) Surfaces. The elastic reflection of electrons on W(211) was studied in [1, 3, 28, 51, 52, 54, 62 to 64]. R_{max} is about 0.25 according to [3, 51, 52]. Maxima were observed in R vs. E_p near 4, 9 to 10, and 15 eV in the range below 28 eV [51, 52]. Less distinct maxima were found at 4 and 15 eV in [1]. R only reached to 0.14 at $E_p=5$ to 10 eV according to [62]. At primary energies in the range 100 to 1000 eV, $R(E_p)$ varied between about 0.02 and 0.035 [28].

The "emission coefficient of slow electrons" $(\delta+\eta)$ steadily increased from ~0 near $E_p=2$ eV to ~0.3 at 24 eV [1]. γ increased to about 0.55 at 100 eV [3].

R, γ, and δ were measured as a function of the angle of incidence α. The curves (measured at $E_p=3$ to 20 eV) were asymmetric with respect to $\alpha=0°$ [63].

Studies on W(100) Surfaces. Secondary electron emission from W(100) was studied in [3, 4, 29, 51 to 56, 59, 60, 64 to 71]. The first maximum in γ occurred in the range $E_p=0.9$ to 1.7 eV, the second at 4.7 eV, and minima were at 3.8 and 8.3 eV [56]. Earlier, $\gamma_{max}\approx0.37$ near $E_p=7$ eV, $\gamma_{min}\approx0.25$ near 12 eV, and an increase to a kink near 15 eV with $\gamma\approx0.33$ had been found. The features at 7 and 15 eV appear to be related to maxima in R vs. E_p [4]. γ increased to about 0.6 at 100 eV [3]. γ was studied by determining the energy value E_p^{I} (as a function of α) for which $\gamma=1$. The E_p^{I} vs. α profiles show a rich structure with diffraction-like singularities coinciding with singularities in $\gamma(E_p$ fixed) vs. α profiles. Electron beams polarized normally to the plane of incidence show shifts in E_p^{I} by ΔE; for fixed polarization, $\Delta E=\pm|\Delta E|$ for $\pm\alpha$. At fixed α, but upon reversing the polarization vector, ΔE changed to $-\Delta E$. These results were attributed to spin-orbit interaction [65].

$(\eta+\delta)$ vs. E_p is similar for W(100) and for W(110). In both cases a threshold at 3 eV and a rich fine-structure were found, and R and $(\eta+\delta)$ crossed in the range 12 to 15 eV [53].

R vs. E_p shows maxima around 4 and 8 to 9 eV and a shoulder around 12 eV [51, 52]. Similar curves were obtained by [58] and [54]. In [58], the maxima are located at slightly smaller energies (3 eV, 7 eV, and a weaker peak at 11 eV). A distinct peak between 3.5 and 4.0 eV was observed in [53]. In addition, some smaller peaks were found around 6.5, 10.5, and 16.5 eV [53]. The peak found at 16.5 eV by [53] was also seen by [3]. Different results were obtained in [4]: R vs. E_p increased to $R_{max}=0.16$ at $E_p=6.75$ eV; there was a second, shallow maximum with $R\approx0.1$ near $E_p=15$ eV [4]. R_{max} is about 0.4 according to [3, 51, 52].

R values as high as 90% can be achieved at glancing incidence (milliradian range) for $E_p=1$ to 300 MeV at 0 K<T<600 K according to a numerical study covering various azimuths [66].

The dependence of the secondary emission on the angle of incidence was studied on electropolished and intentionally roughened (etched with HNO_3+HF) W surfaces. When α was increased to 28°, γ varied between ~1.23 and ~1.54 for $E_p=1$ keV and between ~0.9 and ~1.2 for $E_p=4$ keV. The major peaks occurred when the primary beam was aligned with a major zone axis of the crystal, independently of the primary energy [67]. The intensity recorded for energy losses of 0, 12.5, 27.5, and 50 eV depended markedly on the azimuthal angle of incidence for excitation by electrons with $E_p=1$ keV impinging on W(100) at $\alpha\approx45°$. The extrema of the intensity occurred at the same azimuthal angles for the various energy-loss values studied [68]. Electrons at $E_p=10$ keV yielded distinct maxima in $\delta(E_s\leq6$ eV) when impinging along the [011] and [322] azimuths of W(100) with $\alpha=45°$. On rotation around the surface normal, δ revealed a rich fine-structure superposed on the isotropic background [69].

The dependence of the elastic reflection on the azimuthal angle of incidence and the anisotropy $(R_{max}-R_{min})/R_{max}$ (where R_{max} and R_{min} refer to the curves of R as a function of the azimuthal angle of incidence) were plotted vs. E_p (0.5 to 1.5 keV) for the [100], [130], and [110] azimuths of W(100) [70].

With $E_p=0.5$ keV and $\alpha=0°$, several sharp maxima occurred in plots of R vs. polar angle of electron emission, ϑ, along the [010] azimuth. A much broader intensity vs. ϑ distribution was obtained for the $\varphi=34°$ (referred to the [010] direction) azimuth [71].

Studies on W(111) Surfaces. Measurements of the yield of reflected and secondary electrons from W(111) were performed by [8, 29, 52, 53, 55, 57, 61, 62, 72 to 75].

γ below primary energies of 15 eV shows a distinct peak around 3 eV and smaller peaks at about 6 and 9.5 eV [53]. In the early study [72], only a monotonous increase was found for γ being ~0.10 to ~0.28 between 1 and 10 eV on a cleaned (flashed to 2800 K) surface and ~0.10 to ~0.15 on the same surface covered with a vapor-deposited layer of W [72]. The behavior is similar [72] to that of a rolled W ribbon reported by [8].

An R vs. E_p plot below 30 eV shows three maxima at 2.5, 6.5, and 13 eV with R=0.12, 0.15, and 0.16, respectively [57]. The curve of R vs. E_p plotted in [52] shows only weak structures apart from a maximum at about 3 eV [52]. For $\alpha=0°$, R is at a maximum (0.12) near 15 eV and drops when $E_p>20$ eV. Varying α (up to 35°) changed the relative heights of the maxima of the R vs. E_p curve, but did not shift these features. The peaks and shoulders on the R vs. E_p ($\leq$50 eV) curves are attributed to elastic scattering by bulk and surface plasmons (this assignment is made on the basis of $E_p(R_{max})$ being independent of α). Diffraction effects are excluded to be causing the observed features [62].

γ, η, and δ were studied for primary energies between 0.2 and 3 keV in [73] and between 0.1 and 1.5 keV in [61] (note that "η" includes elastically reflected electrons in [61], i.e.,

 References for 5.7.1 on pp. 208/10

it corresponds to $\eta + R$). The structure observed in the curves was explained as arising from diffraction effects and the anisotropy of the conditions for the motion of electrons in the crystal [61, 73].

The dependence of the total electron emission on the angle of incidence α of primary electrons was studied. For $\alpha = 0°$, $\gamma \approx 0.55$ and 1.3 were found at $E_p = 0.12$ and 1 keV, respectively [74]; see also [75].

References for 5.7.1:

[1] Shul'man, A. R.; Ganichev, D. A. (Fiz. Tverd. Tela [Leningrad] **4** [1962] 745/54; Sov. Phys.-Solid State [Engl. Transl.] **4** [1962] 545/52).

[2] Bronshtein, I. M.; Karasik, B. S.; Pronin, V. P.; Khinich, I. I. (Radiotekh. Elektron. [Moscow] **20** [1975] 209/10; Radio Eng. Electron. Phys. [Engl. Transl.] **20** [1975] 147/8).

[3] Khan, I. H.; Hobson, J. P.; Armstrong, R. A. (Phys. Rev. [2] **129** [1963] 1513/23).

[4] Yakubova, Z.; Gorbatyi, N. A. (Izv. Vyssh. Uchebn. Zaved. Fiz. **15** No. 7 [1972] 129/30; Sov. Phys. J. [Engl. Transl.] **15** [1972] 1038/9).

[5] Bronshtein, I. M. (Izv. Akad. Nauk SSSR Ser. Fiz. **22** [1958] 441/6; discuss. p. 447; Bull. Acad. Sci. USSR Phys. Ser. [Engl. Transl.] **22** [1958] 442/7; discuss. pp. 447/8).

[6] Gorodetskii, D. A. (Izv. Akad. Nauk SSSR Ser. Fiz. **20** [1956] 1023/4; Bull. Acad. Sci. USSR Phys. Ser. [Engl. Transl.] **20** [1956] 925/6).

[7] Nemilov, Yu. A. (Dokl. Akad. Nauk SSSR **58** [1947] 1641/4).

[8] Morgulis, N. D.; Gorodetskii, D. A. (Zh. Eksp. Teor. Fiz. **30** [1956] 667/74; Sov. Phys.-JETP [Engl. Transl.] **3** [1956] 535/40).

[9] Treloar, L. R. G. (Proc. Phys. Soc. [London] **48** [1936] 488/96, discuss. p. 497).

[10] Treloar, L. R. G. (Proc. Phys. Soc. [London] **49** [1937] 392/407, discuss. p. 408).

[11] Bronshtein, I. M.; Fraiman, B. S. (Vtorichnaya Elektronnaya Emissiya, Izd. Nauka, Moscow 1969, pp. 340/53).

[12] Bronshtein, I. M.; Segal', R. B. (Fiz. Tverd. Tela [Leningrad] **1** [1959] 1246/9; Sov. Phys.-Solid State [Engl. Transl.] **1** [1959] 1142/5).

[13] Bronshtein, I. M.; Shchuchinskii, Ya. M. (Zap. Leningr. Gorn. Inst. im. G. V. Plekhanova **37** No. 3 [1961] 98/104; C.A. **57** [1962] 207).

[14] Shul'man, A. R.; Morozov, Yu. A.; Petrov, I. N. (Fiz. Tverd. Tela [Leningrad] **5** [1963] 839/42; Sov. Phys.-Solid State [Engl. Transl.] **5** [1963] 614/6).

[15] Morozov, Yu. A.; Shul'man, A. R. (Fiz. Tverd. Tela [Leningrad] **6** [1964] 943/5; Sov. Phys.-Solid State [Engl. Transl.] **6** [1964] 726/7).

[16] Jonker, J. L. H. (Philips Res. Rep. **9** [1954] 391/402).

[17] Petry, R. L. (Phys. Rev. [2] **28** [1926] 362/6).

[18] Krefft, H. E. (Phys. Rev. [2] **31** [1928] 199/214).

[19] Ahearn, A. J. (Phys. Rev. [2] **38** [1931] 1858/70).

[20] Warnecke, R. (J. Phys. Radium [7] **7** [1936] 270/80).

[21] Murray, C. B., Jr. (Physica [Amsterdam] **38** [1968] 549/58).

[22] Kollath, R. (in: Flügge, S.; Handbuch der Physik, Vol. 21, Springer, Berlin-Göttingen-Heidelberg 1956, pp. 232/303).

[23] Suleman, M.; Rashid, Y.; Shah, M. A. (J. Nat. Sci. Math. **22** [1982] 9/15).

[24] Coomes, E. A. (Phys. Rev. [2] **55** [1939] 519/25).

[25] McKay, K. G. (Phys. Rev. [2] **61** [1942] 708/13).

[26] Seiler, H. (J. Appl. Phys. **54** [1983] R 1/R 18).

[27] Bauer, H. E.; Seiler, H. (Scanning Electron Microsc. [Pt. 3] **1984** 1081/8; C.A. **101** [1984] No. 202389).

[28] Rumyantsev, V. V.; Korablev, V. V.; Dubov, V. V.; Morozov, Yu. A. (Izv. Akad. Nauk SSSR Ser. Fiz. **46** [1982] 1336/41; Bull. Acad. Sci. USSR Phys. Ser. [Engl. Transl.] **46** No. 7 [1982] 113/8).
[29] Gomoyunova, M. V.; Aliev, B. Z. (Izv. Akad. Nauk SSSR Ser. Fiz. **35** [1971] 231/6; Bull. Acad. Sci. USSR Phys. Ser. [Engl. Transl.] **35** [1971] 211/6).
[30] Bronshtein, I. M.; Stozharov, V. M.; Pronin, V. P. (Fiz. Tverd. Tela [Leningrad] **14** [1972] 1083/6; Sov. Phys.-Solid State [Engl. Transl.] **14** [1972] 927/30).

[31] Levandovskii, V. G.; Chaika, G. E. (Ukr. Fiz. Zh. [Russ. Ed.] **20** [1975] 1735/7).
[32] Schwarz, S. A. (J. Appl. Phys. **68** [1990] 2382/91).
[33] Ono, S.; Kanaya, K. (J. Phys. D **12** [1979] 619/32).
[34] Kanaya, K.; Kawakatsu, H. (J. Phys. D **5** [1972] 1727/42).
[35] Eckertová, L.; Kleint, Ch. (Surf. Sci. **231** [1990] 168/76).
[36] Bystrov, Yu. A.; Persianov, G. M. (Izv. Leningr. Elektrotekh. Inst. im. V. I. Ul'yanova No. 282 [1981] 46/9; C.A. **97** [1982] No. 228527).
[37] Philibert, J.; Weinryb, E. (C. R. Hebd. Seances Acad. Sci. **255** [1962] 2757/9).
[38] Weinryb, E.; Philibert, J. (C. R. Hebd. Seances Acad. Sci. **258** [1964] 4535/8).
[39] Ertl, K.; Behrisch, R. (NATO ASI Ser. B **131** [1986] 515/38).
[40] Trump, J. G.; van de Graaff, R. J. (Phys. Rev. [2] **75** [1949] 44/5).

[41] Trump, J. G.; van de Graaff, R. J. (J. Appl. Phys. **18** [1947] 327/32).
[42] Miller, B. L. (Rev. Sci. Instrum. **23** [1952] 401/8, 408).
[43] Miller, B. L.; Porter, W. C. (J. Franklin Inst. **260** [1955] 31/9).
[44] Miller, B. L. (J. Chem. Phys. **23** [1955] 599).
[45] Husain, S. A. (Proc. 2nd U.N. Int. Conf. Peaceful Uses At. Energy, Geneva 1958, Vol. 19, pp. 213/4; C.A. **1960** 2974).
[46] Jahrreiss, H.; Oppel, W. (J. Vac. Sci. Technol. **9** [1972] 173/6).
[47] Oppel, W. (Diss. Univ. Köln [Cologne] 1973, 98+4 pp.).
[48] Vicario, E.; Escudie, B. (J. Microsc. Spectrosc. Electron. **5** [1980] 357/70).
[49] Arnal, F.; Verdier, P.; Vincensini, P.-D. (C. R. Seances Acad. Sci. B **268** [1969] 1526/9).
[50] Saussez-Hublet, M. C.; Harbour, P. J. (CLM-R-208 [1980] 7 pp.; C.A. **95** [1981] No. 213801).

[51] Armstrong, R. A. (Surf. Sci. **47** [1975] 666/8).
[52] Armstrong, R. A. (Can. J. Phys. **44** [1966] 1753/64).
[53] Strayer, R. W.; Mackie, W.; Swanson, L. W. (Surf. Sci. **34** [1973] 225/48).
[54] Zollweg, R. J. (Surf. Sci. **2** [1964] 409/17).
[55] Korablev, V. V.; Kudinov, Yu. A.; Sysoev, S. N. (Fiz. Tverd. Tela [Leningrad] **28** [1986] 2648/54; Sov. Phys.-Solid State [Engl. Transl.] **28** [1986] 1485/9).
[56] Silant'ev, V. I.; Shevchenko, N. A.; Melamed, B. Ya. (Ukr. Fiz. Zh. [Russ. Ed.] **24** [1979] 1227/9; C.A. **91** [1979] No. 149943).
[57] Yakubova, Z.; Gorbatyi, N. A. (Izv. Akad. Nauk SSSR Ser. Fiz. **35** [1971] 553/9; Bull. Acad. Sci. USSR Phys. Ser. [Engl. Transl.] **35** [1971] 507/13).
[58] Madey, T. E.; Yates, J. T., Jr. (Nuovo Cimento Suppl. [1] **5** [1967] 483/505, 490/1).
[59] Bronshtein, I. M.; Dyubua, B. Ch.; Khinich, I. I. (Fiz. Elektron. Nauchn. 30th Dokl., Gertsenovskie Chteniya [30th Gertsen Lecture], Leningrad 1977, pp. 3/9; C.A. **90** [1978] No. 32663).
[60] Gomoyunova, M. V.; Aliev, B. Z. (Fiz. Tverd. Tela [Leningrad] **12** [1970] 2472/5; Sov. Phys.-Solid State [Engl. Transl.] **12** [1970] 1981/3).

[61] Arifov, U. A.; Vakhidov, U. V.; Kasymov, A. Kh. (Izv. Akad. Nauk SSSR Ser. Fiz. **35** [1971] 560/1; Bull. Acad. Sci. USSR Phys. Ser. [Engl. Transl.] **35** [1971] 513/5).

[62] Ershova, T. P.; Korablev, V. V.; Morozov, Yu. A. (Fiz. Tverd. Tela [Leningrad] **20** [1978] 1565/7; Sov. Phys.-Solid State [Engl. Transl.] **20** [1978] 905/6).
[63] Morozov, Yu. A.; Ezhova, T. P. (Tr. Leningr. Politekh. Inst. im. M. I. Kalinina No. 328 [1973] 13/8; C.A. **81** [1974] No. 55521).
[64] Bronshtein, I. M.; Karasik, B. S.; Khinich, I. I. (Fiz. Poluprovodn. Elektron. 26th Nauchn. Dokl. Gertsenovskie Chteniya [26th Gertsen Lecture], Leningrad 1973, Vol. 1, pp. 38/41).
[65] Mamaev, Yu. A.; Makarov, B. S.; Mishin, A. N.; Petrov, V. N.; Yakovlev, V. N.; Yashin, Yu. P. (Fiz. Tverd. Tela [Leningrad] **26** [1984] 2181/2; Sov. Phys.-Solid State [Engl. Transl.] **26** [1984] 1322/3).
[66] Vyatkin, E. G.; Dolgikh, V. A.; Vorobiev, S. A. (Radiat. Eff. **100** [1986] 39/50).
[67] Goetz, G. G. (Diss. Univ. Minnesota 1963, 121 pp., pp. 33/5, pp. 63/75; Diss. Abstr. **24** [1964] 4754).
[68] Gomoyunova, M. V.; Zaslavskii, S. L.; Pronin, I. I. (Pis'ma Zh. Tekh. Fiz. **5** [1979] 1009/13; Sov. Tech. Phys. Lett. [Engl. Transl.] **5** [1979] 421/2).
[69] Stevens, H. (Vak. Tech. **21** [1972] 150/4).
[70] Gomoyunova, M. V.; Zaslavskii, S. L.; Pronin, I. I. (Fiz. Tverd. Tela [Leningrad] **24** [1982] 390/5; Sov. Phys.-Solid State [Engl. Transl.] **24** [1982] 221/4).
[71] Gomoyunova, M. V.; Pronin, I. I. (Pis'ma Zh. Tekh. Fiz. **14** [1988] 896/9; Sov. Tech. Phys. Lett. [Engl. Transl.] **14** [1988] 397/8).
[72] Gorodetskii, D. A. (Zh. Eksp. Teor. Fiz. **34** [1958] 7/13; Sov. Phys.-JETP [Engl. Transl.] **7** [1958] 4/9).
[73] Gomoyunova, M. V.; Aliev, B. Z. (Fiz. Tverd. Tela [Leningrad] **11** [1969] 3619/21; Sov. Phys.-Solid State [Engl. Transl.] **11** [1969/70] 3036/7).
[74] Korablev, V. V. (Fiz. Tverd. Tela [Leningrad] **14** [1972] 1789/91; Sov. Phys.-Solid State [Engl. Transl.] **14** [1972] 1539/41).
[75] Korablev, V. V.; Rumyantsev, V. V.; Dubov, V. V. (Fiz. Tverd. Tela [Leningrad] **25** [1983] 3527/31; Sov. Phys.-Solid State [Engl. Transl.] **25** [1983] 2030/2).
[76] Stern, R. M.; Taub, H. (Phys. Rev. Lett. **20** [1968] 1340/3).

5.7.2 Inelastic Secondary Electron Emission Spectra

The energy distribution of secondary electrons from a metal target shows two principal features, namely a large maximum of slow secondary electrons near 2 to 3 eV (often called the "cascade peak") and a sharp maximum at E_p caused by the elastically reflected primary electrons. In addition, fine-structure occurs in the secondary electron emission (SEE) spectra (for some typical spectra from W, see e.g. [1, 2]): On the one hand, there are several sharp maxima at fixed energy differences from the primary energy, generating the so-called energy loss spectra which are discussed in "Tungsten" Suppl. Vol. A 3, 1989, pp. 215/20. On the other hand, a number of additional maxima occur at fixed energy positions (i.e., independent of E_p) between about 10 and 500 eV which are produced by Auger electrons. The positions of these maxima correspond to transition energies between different energy levels of the target material. For a discussion of the Auger electron spectra from W, see "Tungsten" Suppl. Vol. A 2, 1987, pp. 51/3. This chapter deals with electrons in the very-low-energy range around the cascade peak in the SEE spectra. Angle-resolved measurements on single-crystal surfaces revealed some fine-structure in this energy range which is assumed to be related to the one-dimensional density-of-states (DOS) function along the bulk symmetry line normal to the surface investigated; see e.g. [3 to 5]. Thus, the measurement of SEE spectra gives a possibility to study bulk and surface band-structure features above the vacuum level.

5.7.2.1 Polycrystals

The energy distribution of secondary electrons emitted at $\vartheta \approx 30°$ (i.e., under 30° off the surface normal) was measured on W polycrystals under normal-incident ($\alpha = 0°$) electron impact at $E_p = 400$ eV. W polycrystals heated to 300 to 400 °C showed a maximum in their energy distribution at 2.1 eV; samples heated to between 600 and 800 °C showed a maximum at 2.2 eV and a somewhat enhanced emission at higher energies [6]. In an earlier study [7, p. 279], SEE spectra were given for $E_p = 10$ to 600 eV measured at $\vartheta = 0°$. The low-energy region, including the large maximum around 2 eV, was not resolved in these measurements [7].

The low-energy peak is produced by cascade processes: Electrons inside the bulk having received energy from the primary electrons scatter repeatedly before reaching the surface, each electron losing on the average about half its energy for each scattering event. According to theoretical calculations in [8], the position of the cascade peak should be independent of the primary energy as long as this energy is sufficiently high so that primary electrons penetrate the target to a depth which allows the secondary electrons to undergo several scattering events before reaching the surface. To study the behavior of the cascade peak at low energies, SEE spectra were recorded at $\Delta E_p = 1$ eV steps for $E_p = 0$ to 20 eV and at $\Delta E_p = 5$ eV steps between 20 and 100 eV. For $E_p > 20$ eV the low-energy maximum peaks near 2 eV. As E_p decreases the peak position of the low-energy maximum shifts to higher E_s values (for $E_p < 10$ eV, $E_s \approx 3.6$ eV) and its amplitude diminishes. Reflected primary electrons remained but secondary electrons were absent for E_p below 5 eV [9].

5.7.2.2 Single Crystals

Studies Comparing Different W Surfaces. Angle-resolved SEE spectra were measured on W(110), W(100), and W(111) for secondary energies below 23 eV ($p < 5 \times 10^{-10}$ Torr). The spectra consist of the cascade peak (incoherent scattering) superposed on a spectrum due to "coherent" scattering that reflects the band structure of the substrate. The results were compared with calculated data including contributions of the band structure. Theoretical and experimental data agree for the (110) and also for the (100) surface. On W(110) and W(111), the coherent scattering constituted 35% and 5%, respectively, of the total scattering [10]. There was little change in the spectra with E_p varying between 0.05 and 1.5 keV [10] and no significant differences were noted in spectra from W(211), W(110), and W(111) recorded both at room temperature and at ~500 to 1000 K ($p < 4 \times 10^{-10}$ Torr, $E_p = 40$ to 500 eV) [11]. Peculiar features in the range $E_s = 20$ to 30 eV and spectra not in agreement with cascade-type processes were found by [11].

The relation between the measured SEE spectra and the W band structure was extensively discussed by Willis and coworkers [3 to 5, 12 to 18]. Normal-emission ($\vartheta = 0°$) SEE spectra from W(100) and W(111) are compared with one-dimensional DOS calculations in [12]; see also [4, 5]. The kinetic energy distribution extends to the cut-off energy, $h\nu_p \approx 23$ eV, set by the plasma frequency ν_p. The calculated DOS fairly corresponds to the measured W(100) spectrum, but there was an only poor agreement between the calculated DOS and the measured spectrum for W(111). The latter showed two unexpected peaks, a sharp peak at 3.0 eV which shifted to 3.5 eV with increasing ϑ (observed up to 70° off the normal) and a broad peak at 7.5 eV shifting to 9.1 eV for ϑ increasing from 0° to 40° (not observed for $\vartheta > 40°$) [12]. These peaks were explained to be transmission resonances, considering wave-matching conditions at the surface. A peak should occur when the emitting bulk Bloch state has a group velocity equal to the group velocity of the emitted electron [4, 5, 12].

References for 5.7.2 on pp. 213/4

Normal-emission SEE spectra at $E_p = 107$ eV and $\alpha = 45°$ were recorded on W(110), W(100), and W(111) surfaces. These SEE spectra, plotted to $E_s = 46$ eV for W(110), W(100), and W(111), correspond to scans along the Σ, Δ, and Λ lines, respectively [3, 13]. W is characterized by a band gap extending throughout the Brillouin zone and centered near 9 eV above E_F. Except for a small region along the P-H line, this gap exists everywhere in the Brillouin zone. The gap corresponds to the intensity minima in the experimental SEE spectra at 2.5 eV $< E_s <$ 4.5 eV for Δ (corresponding to about 7 to 9 eV above E_F), at 0.8 eV $< E_s <$ 5.4 eV for Σ (corresponding to about 6 to 10.5 eV above E_F), and at 3.2 eV $< E_s <$ 4.4 eV for Λ (corresponding to about 7.5 to 9 eV above E_F) [3, 13] in agreement with the DOS calculated in [19]. The fine-structure in the SEE spectra is attributed to electrons inelastically scattered into excited states throughout the BZ [3, 13]. The studies at $\vartheta = 0°$ [3, 13] were extended in [4] to emission at ϑ as high as 70°. These spectra are compared with band structure calculations in [4] and in more detail in the subsequent paper [5] for W(100) [5, pp. 185/90], W(110) [5, pp. 190/5], and W(111) [5, pp. 195/201]. The gap around $E_F + 9$ eV is much less defined than one should expect on the basis of bulk energy bands. This indicates emission through surface resonances [4].

Studies on W(110) Surfaces. SEE spectra from W(110) were measured for $E_s \leq 35$ eV in a narrow solid angle and over the whole hemisphere above the surface ($E_p = 0.8$ keV, base pressure $p = (2 \text{ to } 5) \times 10^{-11}$ Torr). Spectra within a narrow solid angle along the (100) azimuth for different exit angles ϑ (α was chosen such that $\alpha + \vartheta = 60°$ in each case) and for normal exit ($\vartheta = 0°$) were recorded for E_p ranging from 0.07 to 1.2 keV [20]. Although no quantitative agreement with the calculated data (according to [5]) was found, it is concluded that the principal features in the narrow-solid-angle spectra are governed by the electronic structure of the subsurface region of the crystal [20].

The SEE spectrum ($E_p = 90$ eV, $E_s \leq 30$ eV, $\vartheta = 0° \pm 2.5°$, $p < 5 \times 10^{-10}$ Torr) from a mechanically polished surface, annealed at 1700 °C and flashed to 2000 °C, nearly agrees with the cascade structure calculated for a free-electron gas. The mechanically polished surface is heavily disturbed as indicated by a pronounced structure in the spectrum after an additional electropolish. Also extended heat treatments in 5×10^{-6} Torr O_2 at 1500 °C changed the spectrum significantly [21, 22]. For calculations of the coherent (band structure related) and incoherent (cascade peak) scattering and a comparison with experiment, see [21, 23]. In another paper of the same authors, it was found that the structure in the spectra is suppressed both by Ar^+ ion bombardment and by contamination [24]. Angle-resolved SEE spectra ($0 \leq E_s \leq 22$ eV) on clean W(110) showed a peak near 1 eV, two strong peaks at 7.8 and 10.7 eV, and a broad hump near 18 to 19 eV. These results were compared with spectra obtained on oxygen-covered W(110) surfaces, thus establishing a model (proposed in [10, 25]) for the influence of the electronic structure of the crystal on the shape of the SEE spectra [26].

SEE spectra were recorded for E_p between 4 and 100 eV ($\alpha = 20°$; $p \leq 10^{-9}$ Torr) on W(110) surfaces [27]. For $E_p = 5$ to 10 eV, the spectra showed a sharp elastic peak and a flat maximum due to inelastically scattered electrons and secondary electrons. The maximum occurred at 2.8 eV for $E_p = 80$ to 100 eV and at 4.6 eV for $E_p \approx 19$ eV (where it was most intense). The width $\Delta E_{1/2}$ was 6.6 eV for $E_p = 80$ to 100 eV and 8.5 eV for $E_p \approx 19$ eV. Between $E_p = 80$ and 100 eV, neither the width nor the position of the maximum changed. Maxima at 5.5 and 11 eV below the elastic peak were attributed to one-electron transitions in the valence band from filled to vacant states above the Fermi level (5.5 eV) and to excitations of surface plasmons (11 eV) [27].

Studies on W(100) Surfaces. SEE spectra for $E_s \leq 20$ eV ($E_p = 20$ to 140 eV, $\alpha = 45°$, $\vartheta = 0°$) were studied on W(100) surfaces with a carbon contamination of less than ~0.01 ML. A peak found at 10.5 eV, which was extremely sensitive to contamination (submonolayers of oxygen), was attributed to a surface resonance. This assignment was supported by the fact that calculations disregarding a surface potential barrier did not yield this peak. Maxima and minima in the experimental SEE spectra occurred at 2.8, 5.0, 10.5, and 13.6 eV, and minima at 4.0, 7.2, 11.1, and 17.0 eV [28]. The SEE spectrum measured by [28] was compared with a calculated curve (according to theoretical considerations in [29]) and a reflectivity vs. energy plot of [30]. Excellent agreement was found with respect to the existence and energetic positions of the peaks [28].

Similar experimental results supporting the interpretation given in [28] were obtained by [10, 25]: Maxima occurred in the SEE spectra at about 3, 5, 11, and 13 eV [10, 25] and the maximum at 11 eV became less marked when the surface was contaminated with carbon [10]. Calculated spectra on the basis of the band structure were compared with the low-energy part of the experimental SEE spectra from W(100), recorded along the $\bar{\Sigma}$ azimuth for $\vartheta = 0°$ to 50° and $E_s \leq 22$ eV. (E_p was not given; in the preceding W(110) study [20], the conditions were $E_p = 50$ to 1500 eV and $p \leq (2$ to $5) \times 10^{-11}$ Torr). The main features in the experimental and calculated spectra were well correlated with the exception of the above mentioned split peak near 11 eV for $\vartheta = 0°$ [25].

The dispersion relation of the resonance band found by SEE spectroscopy [15] fits in the two-dimensional band structure predicted by [31]. The spectrum was excited by electrons ($E_p \approx 100$ eV) incident with the azimuth set such that the $\bar{\Gamma}$-$\bar{X}$ ($=\bar{\Delta}$) line could be explored ($\vartheta = 0°$ to 70° in $\Delta\vartheta = 10°$ steps) [15]. Resonance effects are expected for SEE similar to the resonance effects observed in LEED. Surface resonances occurred within a forbidden (E, $\vec{k}$) region of the bulk band structure (energy gap), and wave matching at the crystal/vacuum interface can take place between the plane-wave vacuum states and unbound Bloch states having real $k_{\parallel}$ and complex $k_{\perp}$ wave vector components propagating only on the surface. The band of surface resonances, located in the energy band gap of the unbound electron states of W(100), caused sharp fluctuations in the specular beam profiles and also of the angle-dependent fine-structure observed in SEE [16, 17]. The papers [16 to 18, 32] compare the LEED data of [33 to 35] with the SEE data of [15]. LEED and SEE methods should measure the same dispersion $E(k_{\parallel})$ for the surface resonance on W(100). A comparison between literature data indicates some disagreement between the two methods for the lowest energy values where a comparison is possible [32].

References for 5.7.2:

[1] Harrower, G. A. (Phys. Rev. [2] **102** [1956] 340/7).
[2] Tharp, L. N.; Scheibner, E. J. (J. Appl. Phys. **38** [1967] 3320/30).
[3] Willis, R. F. (Phys. Rev. Lett. **34** [1975] 670/4).
[4] Willis, R. F.; Christensen, N. E. (Phys. Rev. [3] B **18** [1978] 5140/61).
[5] Christensen, N. E.; Willis, R. F. (J. Phys. C **12** [1979] 167/207).
[6] Kollath, R. (Ann. Phys. [Leipzig] [6] **1** [1947] 357/80, 358, 362, 377).
[7] Warnecke, R. (J. Phys. Radium [7] **7** [1936] 270/80).
[8] Wolff, P. A. (Phys. Rev. [2] **95** [1954] 56/66).
[9] Harrower, G. A. (Phys. Rev. [2] **99** [1955] 1650, **104** [1956] 52/6).
[10] Korablev, V. V.; Kudinov, Yu. A.; Sysoev, S. N. (Fiz. Tverd. Tela [Leningrad] **28** [1986] 2648/54; Sov. Phys.-Solid State [Engl. Transl.] **28** [1986] 1485/9).

[11] Kanicheva, I. R.; Alekseev, Yu. V.; Dobromyslov, P. A. (Tr. Leningr. Politekh. Inst. im. M. I. Kalinina No. 397 [1983] 25/7; C.A. **102** [1985] No. 14535).
[12] Christensen, N. E.; Willis, R. F. (Solid State Commun. **25** [1978] 721/4).
[13] Willis, R. F.; Feuerbacher, B. (Surf. Sci. **53** [1975] 144/55).

[14] Willis, R. F. (Faraday Discuss. Chem. Soc. No. 60 [1975] 245/54).
[15] Willis, R. F.; Feuerbacher, B.; Christensen, N. E. (Phys. Rev. Lett. **38** [1977] 1087/9).
[16] Willis, R. F. (Conf. Ser.-Inst. Phys. No. 41 [1978] 233/8).
[17] Willis, R. F.; Christensen, N. E. (Proc. 7th Int. Vac. Congr., Vienna 1977, Vol. 1, pp. 469/72; C.A. **88** [1978] No. 82518).
[18] Willis, R. F. (Phys. Rev. [3] B **17** [1978] 909/11).
[19] Christensen, N. E.; Feuerbacher, B. (Phys. Rev. [3] B **10** [1974] 2349/72).
[20] Korablev, V. V.; Kudinov, Yu. A. (Izv. Akad. Nauk SSSR Ser. Fiz. **49** [1985] 1775/8; Bull. Acad. Sci. USSR Phys. Ser. [Engl. Transl.] **49** No. 9 [1985] 106/9).

[21] Artamonov, O. M.; Smirnov, O. M.; Terekhov, A. N. (Izv. Akad. Nauk SSSR Ser. Fiz. **46** [1982] 1383/8; Bull. Acad. Sci. USSR Phys. Ser. [Engl. Transl.] **46** No. 7 [1982] 158/62).
[22] Artamonov, O. M.; Smirnov, O. M.; Terekhov, A. N. (Fiz. Tverd. Tela [Leningrad] **23** [1981] 1558/60; Sov. Phys.-Solid State [Engl. Transl.] **23** [1981] 913/4).
[23] Artamonov, O. M.; Smirnov, O. M.; Terekhov, A. N. (Deposited Doc. VINITI-1290-82 [1982] 1/30; C.A. **98** [1983] No. 136041).
[24] Artamonov, O. M.; Smirnov, O. M.; Terekhov, A. N. (Fiz. Tverd. Tela [Leningrad] **24** [1982] 2844/6; Sov. Phys.-Solid State [Engl. Transl.] **24** [1982] 1612/3).
[25] Korablev, V. V.; Kudinov, Yu. A.; Sysoev, S. N. (Fiz. Tverd. Tela [Leningrad] **29** [1987] 702/5; Sov. Phys.-Solid State [Engl. Transl.] **29** [1987] 402/4).
[26] Korablev, V. V.; Kudinov, Yu. A.; Ageev, A. I. (Fiz. Tverd. Tela [Leningrad] **30** [1988] 1038/42; Sov. Phys.-Solid State [Engl. Transl.] **30** [1988] 602/4).
[27] Kremkov, M. V. (Poverkhnost **1984** No. 6, pp. 37/42; Phys. Chem. Mech. Surf. [Engl. Transl.] **3** [1985] 1652/63).
[28] Schäfer, J.; Schoppe, R.; Hölzl, J.; Feder, R. (Surf. Sci. **107** [1981] 290/304).
[29] Feder, R.; Pendry, J. B. (Solid State Commun. **26** [1978] 519/21).
[30] Herlt, H. J.; Feder, R.; Meister, G.; Bauer, E. G. (Solid State Commun. **38** [1981] 973/6).

[31] Smith, N. V.; Mattheiss, L. F. (Phys. Rev. Lett. **37** [1976] 1494/7).
[32] McRae, E. G. (Phys. Rev. [3] B **17** [1978] 907/8).
[33] Edwards, D. E., Jr.; Probst, F. M. (J. Chem. Phys. **56** [1972] 3184/5).
[34] Adnot, A.; Carette, J. D. (Phys. Rev. Lett. **38** [1977] 1084/7).
[35] McRae, E. G.; Wheatley, G. H. (Surf. Sci. **29** [1972] 342/54).

5.7.3 Elastic Electron Scattering

5.7.3.1 Introduction

The potential in the neighborhood of the metal/vacuum interface, in general referred to as the "surface potential barrier", is important for the interaction between a charged particle and the surface. For this reason, much effort has been devoted to the elucidation of the potential shape, the knowledge of which is of central importance for understanding many surface processes, including low-energy electron diffraction (LEED), secondary electron emission, thermionic and field emission, photoemission, and inverse photoemission.

At very low electron energies (below about 20 eV), oscillatory fine-structure is found in high-resolution LEED intensity profiles. An analysis of these features was shown to be a powerful method for studying the surface potential barrier; for a review, see e.g. [1 to 3]. A fine-structure had already been observed in early yield studies of elastically scattered electrons (see Section 5.7.1.2, pp. 205/8) and detected for the first time in off-normal specularly reflected beams from a W(100) surface by [4]. Its oscillatory, resonance-like character was clearly resolved for the first time with the aid of high-resolution LEED by Adnot and Carette [5] on W(100) (see **Fig. 45** [5]) and by Baribeau and Carette [6 to 8] on W(110).

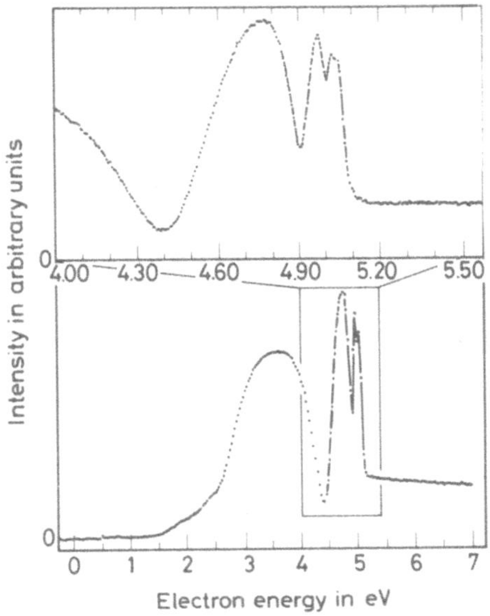

Fig. 45. Intensity of the (00) beam as a function of electron energy for two energy scales from high-energy-resolution LEED measurements on W(001). The electrons were elastically scattered at an incidence angle of 48° along $\bar{\Delta}$ [5].

This fine-structure is associated with the emergence thresholds of nonspecular beams in LEED. At the low-energy side of such a threshold energy, typically a series of narrow peaks of reflected intensity is observed (in the specular beam as well as in nonspecular beams), approximately arranged in a Rydberg series converging on the threshold energy; see e.g. [9, 10]. The origin of the fine-structure lies in the backscattering of nonspecular beams by the surface barrier and their interference with the beam specularly reflected from the surface, as was proposed for the first time by McRae [11]. In earlier studies the structures were interpreted in terms of surface resonances which are related to a forbidden gap in the electronic structure; see e.g. [11 to 14]. Later it was shown that the pure resonance picture is incorrect and that the dominant mechanism is a simple interference between the waves directly reflected and waves which undergo exactly one internal reflection at the surface barrier [9]. For a more detailed discussion, see e.g. [15, pp. 88/91].

Experiments with spin-polarized electrons impinging on a W(100) surface revealed a strong dependence of intensity on spin polarization [16, 17]. These spin-dependent intensity profiles as well as the spin polarization profiles give additional valuable information for surface barrier analysis [16 to 19].

Advanced one-dimensional surface-barrier models, including spin-dependent and relativistic effects, yield profiles agreeing very well with the measured fine-structures for W(110) and W(100); see e.g. [20]. For this reason, the three-dimensional nature of the surface potential barrier has only rarely been taken into account in theoretical studies. Applying an extended multiple-scattering model, the influence of lateral potential variations was demonstrated by [21]. A comparison of measured spin-polarized LEED spectra of W(100) with calculated spectra for a one-dimensional and a three-dimensional barrier clearly favors a three-dimensional surface barrier. But at the same time, a comparable level of agreement

References for 5.7.3 on pp. 223/5

with the experimental data was achieved using a one-dimensional barrier model [22]. Likewise, self-consistent density-functional calculations using the full-potential linearized augmented-plane-wave (FLAPW) film method revealed pronounced three-dimensional features of the surface potential. Nevertheless, a simple one-dimensional three-parameter barrier model provided a good fit to the averaged effective potential in the surface region [2, 3, 23].

5.7.3.2 Fine-Structure in LEED Intensity Profiles

During early experiments on the reflection of slow electrons at normal incidence, it had already been noticed that the major features of the elastic reflection coefficient were associated with the specularly reflected beam. However, the oscillatory character of the fine-structure was not detected due to inadequate experimental conditions used by [24 to 26].

W(110). Many papers on the oscillatory fine-structure in $R(E_p)$ profiles were published by Baribeau et al. [6 to 8, 27 to 32]. R vs. E_p profiles were measured between ~0 and 40 eV for angles of incidence between $\alpha=42°$ and 80° in the azimuths $\varphi=0°$ ($\triangleq[001]$), 90° ($\triangleq[\bar{1}10]$), 54.7° ($\triangleq[\bar{1}11]$), and 35.2°. The characteristic current oscillations are associated with thresholds of Bragg peaks which are given for various α and φ values along with the gradients $d/d\alpha$, $d/d\varphi$, and d/dE_p of the thresholds [27].

Electrons with $E_p \approx 0$ to 12 eV impinged along the [001] azimuth on a clean W(110) surface. The angle of incidence ranged from $\alpha=42°$ to 90°, the energy resolution was $\Delta E \approx 12$ meV, and the angular width was below 0.5°. Under these conditions, more fringes were resolved in profiles with high α values than with lower ones [7]. These fringes were later interpreted as an interference effect by [28]. Note that the beam threshold labeled "$(0\bar{4})$" in [7] actually is the $(0\bar{2})$ beam threshold; see paper [8]. Measured intensity profiles were compared with calculations using the classical image barrier and modifications thereof [8]. The experimental specular beam profiles plotted in [8] were also compared with calculated profiles for $\alpha=60°$ and 75° using parameters that were derived by curve fitting of W(100) data [33].

$R(E_p)$ was also studied along the $[\bar{1}10]$ azimuth (surface rotated through 90° with respect to the previous study [7]). α ranged from 65° to 85° (below 65° the LEED fine-structure was hardly detectable both in the specular beam and in the $(\bar{2}0)$ beam) [28]. High-resolution R vs. E_p plots are displayed for (00) beams at $E_p \approx 13.75$ to 15.6 eV along the $[\bar{1}10]$ azimuth and $\alpha=80°$ in paper [29]. For an attempt to reproduce this profile by model calculations, see [34]. The narrow intensity fluctuations observed for incidence angles $\alpha>65°$ are associated with the emergence of the $(\bar{2}0)$ diffracted beam. This fine-structure of the LEED intensity was explained as being due to interference of the wave directly reflected from the surface and a wave temporarily trapped between substrate and surface potential barrier [28, 30]. For large α values, the experimental results are adequately represented by a four-beam approximation (for this, see e.g. [11], [14, p. 556]), but the fit is much poorer for smaller α values [28, 30].

R vs. E_p profiles ($E_p=0$ to 30 eV) for $\alpha=45°$, 60°, 75° along the $[\bar{1}10]$ azimuth were compared with a surface-barrier model with the mean inner potential of the topmost layer being a parameter. The peak positions are best described by a model that assumes an inner potential of the top layer with an absolute value reduced by about 1 eV compared to the bulk inner potential [31]. The emergence thresholds of $(\bar{1}\bar{1})$ and $(1\bar{1})$ beams given for $E_p \approx 6$ to 17 eV and α between 50° and 80° along the [001] and $[\bar{1}10]$ azimuths were compared in [35] with profiles calculated using the modified image-barrier model of [36].

This comparison indicates that the shape of the modified image barrier is not correct near the topmost atomic layer [35].

The relativistic theory of LEED intensity and polarization profiles developed by [37] on the basis and in extension of the nonrelativistic Korringa-Kohn-Rostocker-Ziman pseudopotentials (for these, see [38]) was augmented by providing a continuous surface potential barrier and a complex δ-function potential at the surface (the complex δ-function potential accounts for inelastic scattering processes such as plasmon creation). This theory achieved fair agreement with respect to the positions of extrema between calculated intensity profiles [39] and experimental data from [40 to 42]. The agreement is rather poor with respect to the relative peak heights for the near normal incidence data [39].

Measured R vs. E_p profiles ($E_p=0$ to 20 eV, $\alpha=0°$) were compared with calculations (according to the relativistic LEED theory [37, 39] mentioned above) for an exponential barrier with the image tail centered at 1.0 Å above the top internuclear plane of W(110). The peak in $R(E_p)$ around $E_p=2$ eV is due to a spin-orbit-induced gap. The small shoulder near 10 eV is associated with the upper edge of the gap at the Γ point. The peak measured at 13.5 eV is accounted for by surface potential barrier effects (in order to shift the peak from the calculated ~14 eV to the measured position, an inner potential of ~15.5 eV would be required). The measured profiles were affected by residual gases even in the low 10^{-11} Torr range; an exposure to ~0.1 L ($\approx 3\times10^{-11}$ Torr·h) CO reduced $R(E_p<4$ eV) by as much as 10% [43].

The following effective, image-force-like potential (from [2]) was used in [3]:

$$V(z) = \{1-\exp(\lambda\cdot[z-z_0])\}/[2(z-z_0)] \qquad \text{for } z<z_0 \qquad \text{and}$$
$$= -U_0/\{1+A\cdot\exp(-B\cdot[z-z_0])\} \qquad \text{for } z\geq z_0 \qquad (z>0\text{: bulk of the metal}).$$

(To simplify the formula, q^2 was set to 1. z_0 denotes the location of the image plane. The parameters A and B are determined by matching V(z) and its derivative at $z=z_0$: $B=U_0/A$ and $A=-1+2\,U_0/\lambda$, U_0 is the inner potential, and λ determines the transition range between the vacuum and bulk potentials. The value of the potential at $z=z_0$ is $-\lambda/2$ (in Ry) [2].) This model barrier was fitted to the data of a density-functional calculation for thin films [3] and to the experimentally observed LEED fine-structure data [44] resulting in the following parameters (in atomic units: Bohr radii a_0 (≈ 0.53 Å) and Rydbergs Ry (≈ 13.6 eV)) [3]:

	z_0	λ	U_0	Ref.
thin-film calculation	−3.28	1.03	1.45	[3]
experimental LEED fine-structure	−3.10	0.90	1.05	[44]

The data base for the analysis in [44] were the experimental results published by [7, 8, 27, 32]. In [44] the inner potential in the outermost atomic layer is assumed to equal the bulk inner potential (in contrast to [32]). The image plane for W(110) is 0.2 to 0.3 a_0 farther away from the outermost atomic layer than the image plane for W(100) [44]. This again is in conflict with [32] who found the image plane for W(110) to lie substantially closer to the first layer than in W(100). The reverse result obtained by [44] is consistent with the self-consistent five-layer density-functional calculations by [23] where the surface barrier in W(110) was found to lie ~0.3 a_0 farther away from the surface than in W(100). This finding was explained with the larger inner potential on the (110) surface [23]. According to more recent thin-film calculations, this difference amounts only to 0.13 a_0 [3].

References for 5.7.3 on pp. 223/5

W(100). The shape of the $R(E_p)$ profile measured in the early study [25] at normal incidence was exactly reproduced in the specular-beam measurements ($\alpha=0°$) of [43]. Three maxima and a shoulder were observed below 20 eV, the maxima being located near 4, 8, and 16 eV and the shoulder around 11 eV [25, 43]. A comparison with calculated profiles indicates that the main peak observed at 8 eV and also the weaker peak at 16 eV are predicted without the need to adjust any parameter if a step-type surface potential barrier is assumed, whereas the peak near 4 eV was not reproduced by theory [45]. In [43] it was shown that the maxima at 8 and 16 eV can be already accounted for in calculations using nonreflecting barriers. But the modeling of the shoulder near 11 eV and the behavior near $E_p=0$ required smoothly reflecting surface barriers. Hence, the shoulder at 11 eV was attributed to a surface-barrier resonance. The peak near 4 eV was ascribed to a spin-orbit-induced band gap nearly 2 eV wide [43]; see also [27]. Previously, the assignment had been controversial; see [46]. In [47] the peak at 4 eV was attributed to multiple scattering. Peaks observed at 7 and 17 eV for R ($0 \leq E_p \leq 27$ eV, $\alpha=0°$) were attributed in [46] to band gaps which are centered near these energies according to [45].

Three sharp peaks in the early specular beam profile ($\alpha=53°$, [10] azimuth) of [4] were attributed to grazing-emergence conditions for the (10), (11), and (20) beams by [11]; see also the review [48]. The sharp intensity fluctuation near 4 eV is identified with a surface-state resonance effect associated with the (10) grazing-emergence condition [11].

R vs. E_p profiles ($E_p \approx 0$ to 40 eV) were measured along the azimuths $\varphi=0°$ ([01] azimuth), 22.5°, and 45° for $42° \leq \alpha \leq 90°$ on surfaces oriented to within about $\pm 1.5°$ [27]. The profiles for $\varphi=0°$ ($\alpha=50°$ and 55°) agree well [27] with the profiles measured at $E_p \approx 0$ to 37 eV ($\alpha=53°$) along the same azimuths by [49, pp. 66, 49, and 99].

A specular beam profile ($\alpha=60°$, $E_p \approx 3.8$ to 11 eV) along the [11] azimuth showed a broad minimum around 6 eV (with a shoulder on the low-energy side), a broad peak near 8 eV, and a sharp minimum near 8.5 eV followed by a sharp peak near 8.8 eV [50]. According to a table in [27], this behavior corresponds to the (11) beam threshold at 8.62 eV.

Intensity and spin-polarization profiles were calculated by a relativistic theory for the E_p range 0 to 30 eV and $\alpha=0°$ [39] (based on [37, 51, 52]). These compare well with the experimental ($E_p<20$ eV, $\alpha=0°$) results by [25]. There is also fair agreement between the relativistic calculations by [39] and the experiments ($\alpha=53°$, $E_p<25$ eV) by [4]. Spin-polarization and intensity profiles were calculated for $\alpha=10°$ along the [10] azimuth with $E_p \leq 2$ Ry (≈ 27.2 eV) [53]. The models tested included (i) electron scattering by muffin tin potentials with weak and strong absorption, (ii) scattering at square surface potential barriers with weak and strong absorption, and (iii) scattering at square surface potential barriers with weak inelastic scattering [53]. Also see the intensity and polarization profiles computed relativistically for the specular and the (10)-reflected beams ($\alpha=0°$, $E_p \leq 4$ Ry ≈ 54.4 eV) suggesting strong polarization effects with unpolarized primary beams for (10) [54].

R($E_p \leq 10$ eV, $\alpha=15°$) was calculated for the [10] azimuth. In addition to interference effects, a resonance mechanism was identified which involves electrons entering above-vacuum energy states of the surface barrier potential [21].

The dispersion of the surface resonances on W(100) observed as narrow fluctuations in R($E_p=2$ to 40 eV, $\alpha \leq 5.5°$) was plotted in [55] (note that the two-dimensional average potential obtained by [55] was challenged by [9, p. 6098]).

R($E_p=2$ to 9 eV) profiles with spin-polarized incident electrons were taken at ten angles of incidence between 15° and 43° along the [01] azimuth of a W(100) surface, held at 500 to 600 °C (i.e., in the (1 × 1) condition) [16]; see also [17]. The electrons were polarized

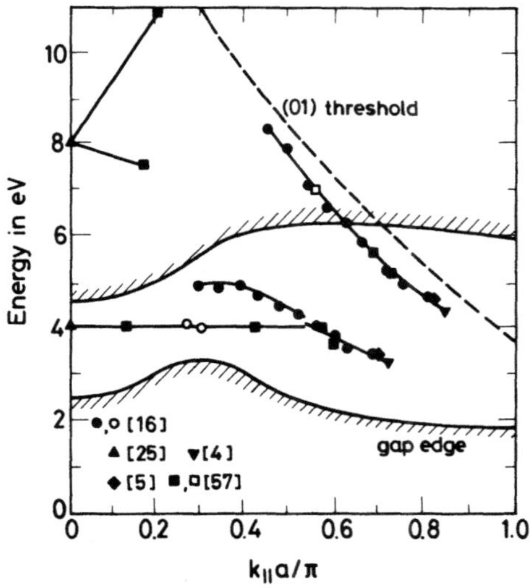

Fig. 46. Positions of the elastic electron reflection (00) intensity peaks on the energy scale and related data for a W(001) surface as a function of the reduced momentum $k_{||} \cdot a/\pi$ (a = 3.16 Å) parallel to the surface along $\bar{\Delta}$. The positions of the band gap edges are as given by [56]. Full lines drawn through points indicate major trends in peak positions. Filled and open symbols indicate positions of peaks and shoulders, respectively [16].

perpendicularly to the plane of incidence ($R^{\uparrow}$ and $R^{\downarrow}$ denoting, respectively, reflectivities with spins up and down with respect to the scattering plane). The structure in the R vs. E_p profiles was attributed to (multiple) scattering of the beam at the sample surface and the surface barrier above it. This scattering is spin-dependent and hence differs for $R^{\uparrow}$ and $R^{\downarrow}$. For the dispersion of the intensity peaks of the (00) beam, see **Fig. 46**, where $k_{||}$ denotes the electron momentum parallel to the W(100) surface [16].

The LEED fine-structure for W(100) was studied for $E_p = 0$ to 35 eV and $\alpha = 45°$ to 70° along the $\varphi = 0° \pm 0.5°$ azimuth [20]. The fine-structure observed for $E_p < 6$ eV can be satisfactorily represented by a one-dimensional surface barrier model (due to [2]). The data for $E_p < 35$ eV are adequately reproduced by this model, provided E_p-dependent values for the mean inner potential are assumed. Strong and E_p-dependent surface damping is indicated by the damping of the threshold features observed for $E_p > 20$ eV. Spin-polarization effects are important, particularly for E_p values above the first threshold. The fine-structure effects are sensitive to surface imperfections and differ for different crystals [20].

The ($\varphi = 0°$, $\alpha = 45°$) profile of [27] excellently agrees with the results published by [5]. At a resolution of $\Delta E = 15$ meV in the range $0 < E_p < 7$ eV, a dip peak feature near 4.4 eV had been found. In addition, two new features had been observed between the threshold for the appearance of the (01) reflection and 125 meV below this threshold with the specularly reflected beam incident at 48° ($\Delta\alpha < 0.8°$); see Fig. 45, p. 215 [5]. The minima in the reflectivity curves are what is important for a comparison with the results of secondary electron emission studies [13]. The preoccupation on reflectivity peaks (rather than minima) that has become common among authors was blamed by [13] being the reason for some confusion in the literature. Commenting on paper [12], it was shown that good agreement exists between the results obtained by secondary electron emission (from [56]) and the LEED data (from [4, 5, 57]) [13]. Rather than in terms of Rydberg series, the reflectivity fluctuations

References for 5.7.3 on pp. 223/5

observed near threshold should be interpreted in terms of a surface resonance in the case of the feature with the lowest energy and in terms of interference effects between (free-electron) surface resonance band states and the first-order diffraction beam in the case of the second and third term of the Rydberg-like series of features [13].

In [9] the fine-structure in the specular beam profiles observed by [5] is explained by multiple beam interference of the electrons reflected between the image potential barrier and the topmost surface layer, making use of potential parameters considered by [43] (particularly a value of $U_0 = 14$ eV (≈ 1.03 Ry) for the surface layer). The much smaller value for U_0 used by [58, 59] (a modified image barrier 0.82 Ry (≈ 11.2 eV) high and ~ 3 Bohr radii off the metal surface) leads to wrong values for the band edges at the forbidden gap [9].

The spin-polarized data [16, 17] and the high-resolution LEED data [5] were analyzed by [18, 19, 60]. In [60] the results were discussed in terms of Rydberg resonances. A barrier with the image plane being 3.3 Bohr radii off the surface and a surface potential amounting to 70% of the bulk inner potential ($U_0 = 1$ Ry) yielded fair agreement [18] between the experimental LEED data [5, 16, 17] and theory [18]. Expanding on paper [18], the normal-incidence LEED measurements of [43] were further analyzed by [19]. While the spin-free calculations yielded inadequate R profiles, the spin-dependent calculations reproduced the data for $\alpha = 0°$ and $E_p = 0$ to 20 eV fairly well (except for a shoulder near 11 eV in [43]). The off-normal ($\alpha = 15°$, $E_p = 0$ to 10 eV) data by Pierce et al. [16, 17] are misrepresented by spin-free calculations. Moreover, the spin-dependent calculations describe these experimental R profiles very poorly only [19].

The models for the modified or saturated image barrier [1, 58] poorly fit the experimental data as long as one assumes that the barrier joins the bulk inner potential at the top-layer edge continuously or with a potential step. It is suspected that the top-layer potential is less attractive than any bulk-layer one. Shifting the top-layer potential upwards significantly improved the fit [61]. A modified image potential proposed by [10] was used in paper [33]. The high mean inner potential value $U_0 = 18.5$ eV and the image potential parameters $z_0 = -0.15$ Å (≈ 0.28 a_0) and $z_1 = -0.95$ Å (≈ 1.80 a_0) were chosen in order to fit the calculated specular beam profile ($E_p = 2$ to 5 eV, $\alpha = 48°$ along the [10] azimuth) to the data of [5] (z_0 and z_1 denote, respectively, the location of the image plane and the starting point of the classical Coulomb potential) [33]. A one-dimensional surface barrier model with two adjustable parameters (the location of the image plane, z_0, and the length scale, λ, for the transition from vacuum to bulk) [2] was fitted to the results of a self-consistent electronic structure calculation [2] of a five-layer W(100) film [2, 3] and to the LEED fine-structure measurements published by [5] and [16]. The same effective image-force-like potential as above for W(110) was used [2, 3]. The following parameters were obtained (for their meaning and for the units used, see p. 217):

	z_0	λ	U_0	Ref.
thin-film calculation	−3.15	0.98	1.35	[3]
experimental LEED fine-structure	−3.20	0.80	1.00	[2]

5.7.3.3 Spin-Polarization Profiles

The topic of spin polarization in low-energy electron diffraction is reviewed in [15, 62 to 66]. The polarization of slow electrons scattered from W surfaces was studied mainly on W(100) surfaces except for an earlier paper [67] where the angular dependence of the spin-polarization of 900 eV electrons elastically scattered at a polycrystalline W foil under 10^{-6} Torr at about 500, 1300, and 1700 °C was investigated. At 500 °C no polarization was observed presumably due to contamination [67]. The other exception is paper [39] where P vs. E_p profiles were calculated for the (01) beam in the $[\bar{1}10]$ azimuth on W(110) in the range 20 to 70 eV for normal incidence. Like with W(100), the polarization peaks occurred at the minima of the corresponding intensity profiles [39].

W(100). A compilation of some typical experimental and theoretical polarization data for the (00) beam from W(100) as a function of energy is shown in **Fig. 47** taken from [15]. The data reveal the equivalence between the polarization generated by scattering

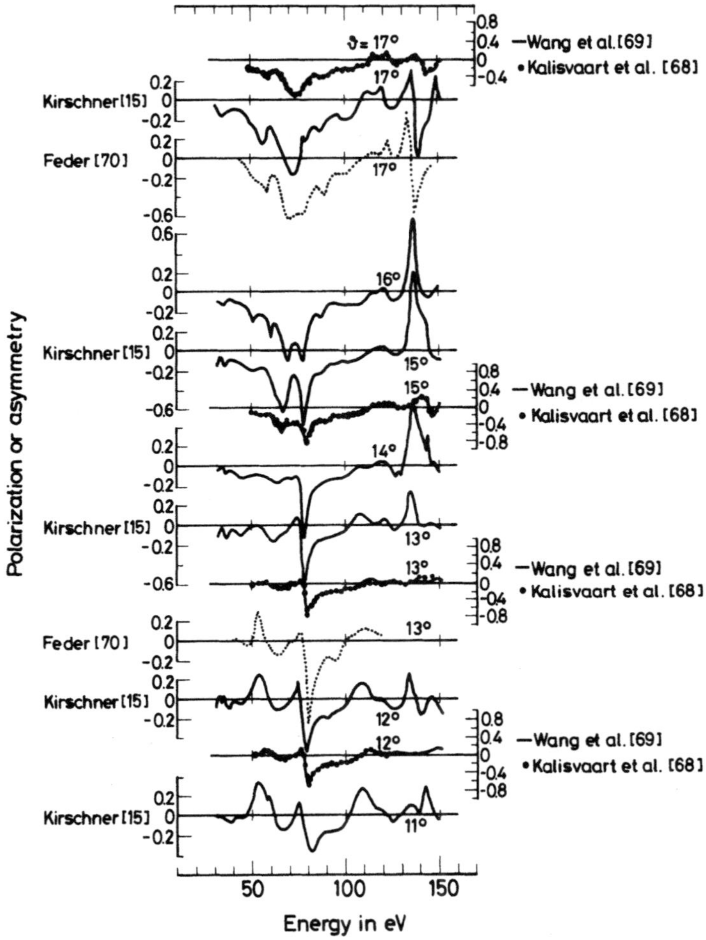

Fig. 47. Comparison of experimental and theoretical polarization data as a function of energy for the (00) beam elastically reflected from W(001) at exit angles between 11° and 17°. The data of Kalisvaart et al. [68] and Kirschner [15] represent the polarization after scattering of an unpolarized beam. The data of Wang et al. [69] represent the intensity asymmetry after scattering for a completely polarized incident beam. The dotted lines (from Feder [70]) show the results of theoretical calculations [15].

References for 5.7.3 on pp. 223/5

of an unpolarized beam [15, 68] and the intensity asymmetry generated by scattering of a completely polarized incident beam [69]. In addition, theoretical results of calculations by Feder [70] assuming this equivalence are given.

Using unpolarized beams at $E_p = 20$ to 150 eV, P(α, φ) was measured on a contaminant-free surface. For scattering planes containing a mirror plane of the crystal, the polarization is always in a direction perpendicular to the scattering plane. Specular beam profiles along [01] were measured at angles of incidence of α = 10° to 18°. For α = 13° such profiles are shown for various azimuths 0° ≤ φ ≤ 45°. Data for α = 0° are also plotted in the paper [68]. (For preliminary P data of (00) beams at oblique incidence and (10) beams at normal incidence, see [71].) The P profiles are more sensitive to changes in α and E_p than are the R profiles. The (00) beam at α = 17° and $E_p \approx 30$ eV changed its polarization from +45% to −35% over an energy range of 2 eV. At α = 13° and $E_p \approx 75$ eV, P changed from +20% to −80% over ΔE = 2 eV. The feature near 75 eV also demonstrates the α sensitivity: P changed from −80% at α = 17° to +55% at α = 19°. A comparison of intensity and polarization profiles for the (00) beam at α = 11° is given in **Fig. 48**. The strong sensitivity to contamination noted for the P profiles was attributed to the associated symmetry breaking [68].

Double-reflection experiments were performed on clean, unreconstructed W(100) surfaces by [72, 73]. Unpolarized electrons impinged at α = 47.5° (±1°) onto a surface kept at 40 to 80 °C during the measurements. The reflected (00) beam in [72] impinged normally (α = 0° ± 0.5°) onto a second W(100) surface. The asymmetry between ($\bar{2}$0) and (20) beams was measured; P vs. φ profiles were recorded and compared with theoretical curves. A maximum degree of polarization of 16% was measured [72]. In the paper [73] these studies were extended to nonspecular beams at normal incidence and to specular beams for α ranging from 11° to 18° in steps of 1°. The results were compared with older experimental work [68, 74] and with theoretical calculations [73]. Quite satisfactory agreement with the older data was found, but the new results exhibited finer structures. Good agreement with the theoretical calculations was achieved, implying a contraction of the topmost interlayer spacing of 7.0 ± 1.5% [73].

A significant temperature dependence of P was found on a clean surface (according to AES) at temperatures ranging from 1000 to 300 °C by [75] (in contrast to [76] whose

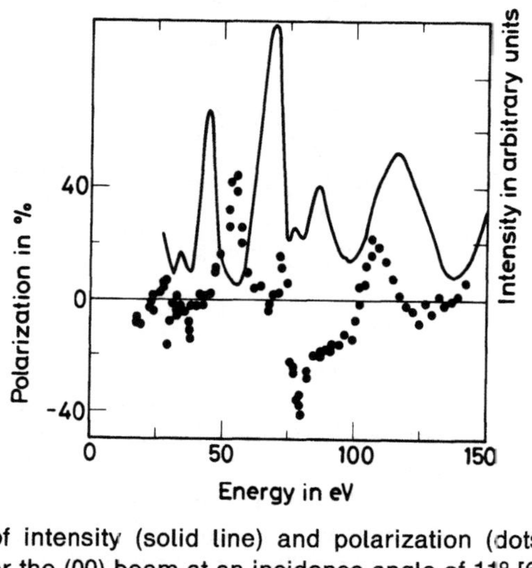

Fig. 48. Comparison of intensity (solid line) and polarization (dots) profiles from W(100) for the (00) beam at an incidence angle of 11° [68].

study was confined to the range 350 to 125 °C). The P vs. E_p profile showed a peak centered near 76 eV. This peak shifted downward by ~1 eV on heating the crystal to 1000 °C. This implies that the degree of polarization has opposite temperature dependences for fixed E_p above and below ~76 eV. The largest change in P was observed at E_p=74 eV with P changing from −45% at 1000 °C to −5% on cooling to 300 °C [75]. In [77] the temperature dependence of P between about 1000 °C and room temperature was measured for different angles of incidence. The results were found to agree with [75], showing a discernible increase and decrease of P with temperature at different energies. A comparison with theoretical results [77] led to the conclusion that the polarization changes arise from modified multiple scattering due to the simultaneous action of thermal lattice expansion and temperature-reduced effective ion-core scattering [77]. For a discussion of the temperature dependence of P, see also [63, pp. 2076/8].

Polarized incident electrons were used for the first time by [78] to study low-energy electron scattering from a W(100) surface. The strength of the spin dependence of the scattering of the electrons was defined by the Sherman function $S(E_p, \alpha)$ with $P_0 \cdot S(E_p, \alpha) = (R[P_0] - R[-P_0])/(R[P_0] + R[-P_0])$. P_0 denotes the initial polarization (parallel or antiparallel to the normal of the scattering plane) and was estimated at 43% for the set-up used. $S(E_p, \alpha)$ was measured for E_p scanned in 1 eV steps between 50 and 150 eV on a W(100) surface heated to between 500 and 650 K (in order to prevent reconstruction) and $\Delta\alpha$ steps of 1° between 10° and 17° [78]. The results were compared with the measurements of [68] where unpolarized incident electrons had been used. Good agreement was found indicating an equivalence of P and S, if the scattering plane is a mirror plane of the crystal [78].

The functions $S(E_p, \alpha)$ and $I_0(E_p, \alpha)$ were determined over a wide α range (I_0 denotes the averaged scattering intensity $(I^\uparrow + I^\downarrow)/2$) for the (01) and $(0\bar{1})$ reflections in the range between about 75 and 130 eV [74]. In addition, the spin dependence of the specular beam was measured under near threshold conditions for the (01) beam emergence (E_p=2 to 8 eV). The observed resonance sequence was explained to result from an interference between the directly reflected beam and the indirect beam temporarily trapped in a quasi-bound state above the vacuum level before being reflected. Spin splitting of the resonance sequence of peaks due to spin-orbit interactions was observed [74]. For a discussion of this spin-dependent splitting within a four-beam-layer multiple scattering approximation, see the subsequent paper [17].

R and P profiles were calculated for the (10) beam with $E_p \leq 2$ Ry (≈27.2 eV) at $\alpha = 10°$ [53] and with $E_p \leq 4$ Ry (≈54.4 eV) at normal incidence [54]. In [79] R and P profiles were calculated by a relativistic LEED theory for the specular beam (E_p=30 to 180 eV) and compared with the experimental data from [71]. Previously, the relativistic LEED formalism developed in [37] had been augmented in [39] by providing a continuous surface potential barrier and a complex δ-function potential which accounts for inelastic scattering processes. (01) beam P profiles were calculated for E_p=15 to 35 eV and normal incidence. P peaked where R was at a minimum. This was also true for the (00) beam P profile ($0 < E_p < 25$ eV, $\alpha = 53°$). Various calculated P profiles (α=40°, 45°, 53°, and 60°) are given for E_p=1 to 6 eV in [39].

References for 5.7.3:

[1] Jennings, P. J.; Jones, R. O. (Appl. Surf. Sci. **22/23** [1985] 35/47).
[2] Jones, R. O.; Jennings, P. J.; Jepsen, O. (Phys. Rev. [3] B **29** [1984] 6474/80).
[3] Jennings, P. J.; Jones, R. O.; Weinert, M. (Phys. Rev. [3] B **37** [1988] 6113/25).
[4] Edwards, D. E., Jr.; Probst, F. M. (J. Chem. Phys. **56** [1972] 3184/5).

[5] Adnot, A.; Carette, J. D. (Phys. Rev. Lett. **38** [1977] 1084/7).
[6] Baribeau, J. M.; Carette, J. D. (Surf. Sci. **111** [1981] L641/L647).
[7] Baribeau, J. M.; Carette, J. D. (Phys. Rev. [3] B **23** [1981] 6201/3).
[8] Baribeau, J. M.; Carette, J. D. (Surf. Sci. **112** [1981] 241/53).
[9] Le Bossé, J. C.; Lopez, J.; Gaubert, C.; Gauthier, Y.; Baudoing, R. (J. Phys. C **15** [1982] 6087/99).
[10] Rundgren, J.; Malmström, G. (J. Phys. C **10** [1977] 4671/87).

[11] McRae, E. G. (Surf. Sci. **25** [1971] 491/512, 506/9).
[12] McRae, E. G. (Phys. Rev. [3] B **17** [1978] 907/8).
[13] Willis, R. F. (Phys. Rev. [3] B **17** [1978] 909/11).
[14] McRae, E. G. (Rev. Mod. Phys. **51** [1979] 541/68).
[15] Kirschner, J. (Springer Tracts Mod. Phys. **106** [1985] 1/158).
[16] McRae, E. G.; Pierce, D. T.; Wang, G. C.; Celotta, R. J. (Phys. Rev. [3] B **24** [1981] 4230/9).
[17] Pierce, D. T.; Celotta, R. J.; Wang, G. C.; McRae, E. G. (Solid State Commun. **39** [1981] 1053/5).
[18] Jennings, P. J.; Jones, R. O. (Solid State Commun. **44** [1982] 17/22).
[19] Jones, R. O.; Jennings, P. J. (Phys. Rev. [3] B **27** [1983] 4702/11).
[20] Baribeau, J. M.; Carette, J. D.; Jennings, P. J.; Jones, R. O. (Phys. Rev. [3] B **32** [1985] 6131/7).

[21] Read, M. N.; Christopoulos, A. S. (Phys. Rev. [3] B **37** [1988] 10407/10).
[22] Tamura, E.; Feder, R. (Solid State Commun. **58** [1986] 729/33).
[23] Jepsen, O.; Jones, R. O. (Phys. Rev. [3] B **34** [1986] 6695/8).
[24] Hobson, J. P.; Khan, I. H. (Phys. Rev. [2] **123** [1961] 1241/2).
[25] Khan, I. H.; Hobson, J. P.; Armstrong, R. A. (Phys. Rev. [2] **129** [1963] 1513/23).
[26] Zollweg, R. J. (Surf. Sci. **2** [1964] 409/17).
[27] Baribeau, J. M.; Roy, D. (Surf. Sci. **166** [1986] 234/48).
[28] Baribeau, J. M.; Carette, J. D. (Phys. Rev. [3] B **25** [1982] 2962/4).
[29] Baribeau, J. M.; Carette, J. D. (J. Vac. Sci. Technol. [2] A **1** [1983] 1174/7).
[30] Baribeau, J. M.; Carette, J. D. (Can. J. Phys. **60** [1982] 1088/95).

[31] Baribeau, J. M.; Le Bossé, J. C.; Lopez, J. (J. Phys. C **18** [1985] L 73/L 78).
[32] Baribeau, J. M.; Lopez, J.; Le Bossé, J. C. (J. Phys. C **18** [1985] 3083/97).
[33] Hall, B. M.; Tong, S. Y.; Mills, D. L. (Phys. Rev. Lett. **50** [1983] 1277/80).
[34] Jennings, P. J.; Jones, R. O. (Surf. Sci. **176** [1986] 691/700).
[35] Le Bossé, J. C.; Lopez, J.; Baribeau, J. M.; Carette, J. D. (Surf. Sci. **137** [1984] 361/72).
[36] Jennings, P. J.; Read, M. N. (Surf. Sci. **41** [1974] 113/24, 117/8).
[37] Feder, R. (Phys. Status Solidi B **49** [1972] 699/710).
[38] Ziman, J. M. (Proc. Phys. Soc. [London] **86** [1965] 337/53).
[39] Feder, R. (Phys. Status Solidi B **62** [1974] 135/46).
[40] Fedorus, A. G. (Fiz. Tverd. Tela [Leningrad] **13** [1971] 3/9; Sov. Phys.-Solid State **13** [1971] 1/6).

[41] Taylor, N. J. (Surf. Sci. **4** [1966] 161/94, 175/7, 185/94).
[42] Farnsworth, H. E.; Bellina, J. (unpublished; private communication to Gafner, G.; Surf. Sci. **19** [1970] 9/20).
[43] Herlt, H. J.; Feder, R.; Meister, G.; Bauer, E. G. (Solid State Commun. **38** [1981] 973/6).
[44] Jennings, P. J.; Jones, R. O. (Phys. Rev. [3] B **34** [1986] 6699/703).
[45] Jennings, P. J.; McRae, E. G. (Surf. Sci. **23** [1970] 363/88, 376/82).
[46] Madey, T. E. (Surf. Sci. **33** [1972] 355/76, 371/2).
[47] Duke, C. B.; Tucker, C. W., Jr. (Phys. Rev. Lett. **23** [1969] 1163/6).

[48] Estrup, P. J.; McRae, E. G. (Surf. Sci. **25** [1971] 1/52, 17/8).
[49] Edwards, D. E., Jr. (Diss. Univ. Illinois, Urbana-Champaign 1970, 277 pp; Diss. Abstr. B **31** [1970/71] 2919).
[50] Woods, J. P.; Erskine, J. L. (J. Vac. Sci. Technol. [2] A **5** [1987] 435/9).

[51] Feder, R. (Phys. Status Solidi B **56** [1973] K 43/K 45).
[52] Feder, R. (Phys. Status Solidi B **46** [1971] K 31/K 34).
[53] Jennings, P. J.; Sim, B. K. (Surf. Sci. **33** [1972] 1/10).
[54] Jennings, P. J. (Surf. Sci. **26** [1971] 509/21).
[55] Stevens, M. A.; Russel, G. J. (Solid State Commun. **34** [1980] 785/8).
[56] Willis, R. F.; Feuerbacher, B.; Christensen, N. E. (Phys. Rev. Lett. **38** [1977] 1087/91).
[57] McRae, E. G.; Wheatley, G. H. (Surf. Sci. **29** [1972] 342/54).
[58] Jennings, P. J. (Surf. Sci. **75** [1978] L 773/L 776).
[59] Jennings, P. J. (Surf. Sci. **88** [1979] L 25/L 28).
[60] Malmström, G.; Rundgren, J. (J. Phys. C **14** [1981] 4937/49, 4944/5).

[61] Lopez, J.; Le Bossé, J. C.; Baribeau, J. M. (J. Phys. C **18** [1985] 2197/207).
[62] Pierce, D. T.; Celotta, R. J. (Adv. Electron. Electron Phys. **56** [1981] 219/89).
[63] Feder, R. (J. Phys. C **14** [1981] 2049/91).
[64] Feder, R. (Polarized Electrons in Surface Physics, World Sci., Singapore 1985).
[65] Kirschner, J. (Springer Ser. Surf. Sci. **5** [1986] 455/82).
[66] Kirschner, J. (Phys. Bl. **42** [1986] 400/6).
[67] Loth, R. (Z. Phys. **203** [1967] 66/70).
[68] Kalisvaart, M.; O'Neill, M. R.; Riddle, T. W.; Dunning, F. B.; Walters, G. K. (Phys. Rev. [3] B **17** [1978] 1570/8).
[69] Wang, G. C.; Celotta, R. J.; Pierce, D. T. (Phys. Rev. [3] B **23** [1981] 1761/70).
[70] Feder, R. (private communication to [15]).

[71] O'Neill, M. R.; Kalisvaart, M.; Dunning, F. B.; Walters, G. K. (Phys. Rev. Lett. **34** [1975] 1167/70).
[72] Kirschner, J.; Feder, R. (Phys. Rev. Lett. **42** [1979] 1008/11).
[73] Feder, R.; Kirschner, J. (Surf. Sci. **103** [1981] 75/102).
[74] Pierce, D. T.; Celotta, R. J.; Wang, G. C. (Vide Couches Minces No. 201 [1980] 1060/3).
[75] Riddle, T. W.; Mahan, A. H.; Dunning, F. B.; Walters, G. K. (J. Vac. Sci. Technol. **15** [1978] 1686/8).
[76] Calvert, R. L.; Russell, G. J.; Haneman, D. (Phys. Rev. Lett. **39** [1977] 1226/9).
[77] Kirschner, J.; Feder, R. (Surf. Sci. **104** [1981] 448/62).
[78] Wang, G. C.; Dunlap, B. I.; Celotta, R. J.; Pierce, D. T. (Phys. Rev. Lett. **42** [1979] 1349/52).
[79] Feder, R. (Phys. Rev. Lett. **36** [1976] 598/600).

5.8 Ion- and Atom-Induced Electron Emission

5.8.1 Introduction

Electron emission from solids bombarded by ions can proceed via the two independent effects of potential electron emission and kinetic electron emission which differ by the mechanism of the energy transfer from the incoming ion to the electrons of the solid.

In potential electron emission, electrons are emitted as a result of an Auger transition (Auger neutralization or resonance neutralization with subsequent Auger deexcitation; for multiply charged ions, autoionization processes are possible in addition) while the ion is outside the solid [1 to 3]. The potential energy of the incident ion must exceed twice the

References for 5.8.1 on pp. 226/7

work function of the solid. The basic principles of potential emission were developed in a semiempirical theory by Hagstrum [1] who also demonstrated its use as a surface-spectroscopical tool (ion-neutralization spectroscopy) [4, 5] and who published comprehensive measurements on the yield and energy distribution of secondary electrons emitted from W by the impact of different noble gases [1, 6, 7]. Additional work on the theory of potential electron emission was done by [8]. Potential emission is the dominating effect for ion velocities below about 10^5 m/s. In more recent work on potential electron emission from W, the interest concentrates on the interaction of multiply charged ions with the surface; see e.g. [9 to 11] and the review [12]. Approaching the surface, the multiply charged ion repeatedly captures electrons resonantly into highly excited states and subsequently relaxes by stepwise deexcitation events, each step lasting typically 10^{-15} s and lowering the ion energy by 15 to 30 eV [3]. For larger energy steps the Auger transitions rapidly become less probable, and for smaller ones the escape probability of electrons from the W surface rapidly drops off [9]. Direct experimental evidence for this behavior is provided by the energy distribution of emitted electrons, which in general seems not to extend beyond about 30 eV; see e.g. [7, 12]. As a consequence, the yield coefficient of secondary electron potential emission, γ_p, is roughly proportional to the total ion potential energy W_q (i.e. the sum of successive ionization energies for the ion with charge q). This proportionality was observed for a large number of different ions impinging on W surfaces [9, 10, 13]; see also the review [12]. It no longer holds if the time available for complete neutralization is too short and the projectile impinging on the surface still is in an ionized state (i.e., if the ion velocity is not sufficiently low or the ion charge is too high) and/or if intra-atomic Auger transitions producing energetic electrons are necessary for completing the neutralization [12]. Potential emission is also observable if metastable atoms are used instead of ions. On the other hand, bombarding the surface with ground-state atoms allows a complete elimination of the effect of potential emission; see e.g. [14, 15]. A comprehensive review on potential emission is given by [16].

Above a certain threshold impact velocity of typically $\sim 10^5$ m/s the second emission process, i.e. kinetic electron emission, can proceed; see e.g. the reviews [17 to 19]. In this case electrons are excited inside the solid by direct transfer of kinetic energy from the impinging particles. Theories on kinetic electron emission are given in [20 to 22]. A linear increase of the yield coefficient of kinetic electron emission, γ_k, for high ion velocities above $\sim 3 \times 10^5$ m/s intersecting the velocity axis at 1.05×10^5 m/s is predicted by the theory of Parilis and Kishinevskii [21]. In the intermediate range between about 1 and 3×10^5 m/s, γ_k increases with the square of the ion velocity, i.e. linear with the ion impact energy [21]. For W good agreement with these theoretical predictions was found experimentally for different kinds of impinging ions; see e.g. [23].

Experimental and theoretical work on ion-induced electron emission performed before 1968 is reviewed in [24] with special emphasis on Russian studies.

The yield coefficients γ_p and γ_k are given in units of electrons per ion (or atom) throughout this chapter. Unless stated otherwise in the following, polycrystalline W was used as target material.

References for 5.8.1:

[1] Hagstrum, H. D. (Phys. Rev. [2] **96** [1954] 336/65).
[2] Hagstrum, H. D.; Becker, G. E. (Phys. Rev. [3] B **8** [1973] 107/21).
[3] Arifov, U. A.; Kishinevskii, L. M.; Mukhamadiev, E. S.; Parilis, E. S. (Zh. Tekh. Fiz. **43** [1973] 181/7; Sov. Phys.-Tech. Phys. [Engl. Transl.] **18** [1973] 118/22).
[4] Hagstrum, H. D. (Phys. Rev. [2] **150** [1966] 495/515).

[5] Hagstrum, H. D.; Becker, G. E. (Phys. Rev. [2] **159** [1967] 572/85).
[6] Hagstrum, H. D. (Phys. Rev. [2] **96** [1954] 325/35).
[7] Hagstrum, H. D. (Phys. Rev. [2] **104** [1956] 317/8).
[8] Sternberg, D. (Diss. Columbia Univ. 1957, 60 pp.; Diss. Abstr. **17** [1957] 2292).
[9] Delaunay, M.; Fehringer, M.; Geller, R.; Hitz, D.; Varga, P.; Winter, H. (Phys. Rev. [3] B **35** [1987] 4232/5).
[10] de Zwart, S. T.; Drentje, A. G.; Boers, A. L.; Morgenstern, R. (Surf. Sci. **217** [1989] 298/316).

[11] Folkerts, L.; Morgenstern, R. (Europhys. Lett. **13** [1990] 377/82).
[12] Varga, P. (Appl. Phys. A **44** [1987] 31/41).
[13] Arifov, U. A.; Mukhamadiev, E. S.; Parilis, E. S.; Pasyuk, A. S. (Zh. Tekh. Fiz. **43** [1973] 375/9; Sov. Phys.-Tech. Phys. [Engl. Transl.] **18** [1973] 240/3).
[14] Berry, H. W. (J. Appl. Phys. **29** [1958] 1219/25).
[15] Dev, B. (Appl. Surf. Sci. **40** [1990] 319/26).
[16] Varga, P.; Winter, H. (Springer Tracts Mod. Phys. **123** [1991] 149/213).
[17] Hasselkamp, D. (Comments At. Mol. Phys. **21** [1988] 241/55).
[18] Devooght, J.; Dehaes, J.-C.; Dubus, A.; Cailler, M. (Springer Tracts Mod. Phys. **122** [1991] 67/128).
[19] Hasselkamp, D. (Springer Tracts Mod. Phys. **123** [1991] 1/95).
[20] Sternglass, E. J. (Phys. Rev. [2] **108** [1957] 1/12).

[21] Parilis, E. S.; Kishinevskii, L. M. (Fiz. Tverd. Tela [Leningrad] **3** [1961] 1219/28; Sov. Phys.-Solid State [Engl. Transl.] **3** [1961] 885/91).
[22] Sigmund, P.; Tougaard, S. (in: Taglauer, E.; Heiland, W.; Inelastic Particle-Surface Collisions, Springer, Berlin 1981, pp. 2/37).
[23] Petrov, N. N.; Dorozhkin, A. A. (Fiz. Tverd. Tela [Leningrad] **3** [1961] 53/60; Sov. Phys.-Solid State [Engl. Transl.] **3** [1961] 38/43).
[24] Arifov, U. A. (Vzaimodeistvie Atomnykh Chastits s Poverkhnost'yu Tverdogo Tela, Nauka, Moscow 1968; Interaction of Atomic Particles with a Solid Surface, Plenum, New York 1969, pp. 1/374).

5.8.2 Electron Emission under Impact of Noble Gases

5.8.2.1 Helium

5.8.2.1.1 He Atoms

The interaction of He 2^3S atoms with polycrystalline W surfaces was studied in [1 to 4]. He 2^1S atoms were used by [5]. Mixtures of these metastable states were used by [2, 6 to 12]. Ground-state He atom beams ($E_p = 2$ keV) with no metastable atoms present were studied by [13].

Emission Yield. The yield coefficient $\gamma(He^*)$ of W(110) was about 1.5% lower than that of W(111) for thermal He* beams impinging on surfaces of the same crystal. The electron yield per ion on the (111) plane was normalized to $\gamma = 0.300$ (see [2]) resulting in $\gamma = 0.296$ for W(110). Evidently, there is little difference between these γ values despite the 1.6 ± 0.3 eV difference in work functions measured by deceleration of slow Xe^+ ions (monitoring the onset and vanishing of the ion current as a function of applied field) [7].

Secondary electron yields of $\gamma = 0.318 \pm 0.030$ and 0.306 ± 0.025 were determined on W polycrystals for thermal He* beams excited by 20.5 and 28 eV electron impact, respectively. Using the known ratio of 2^3S_1 and 2^1S_0 states of He* excited at 28 eV, the yield coefficients $\gamma(2^3S_1) = 0.318 \pm 0.030$ and $\gamma(2^1S_0) = 0.280 \pm 0.060$ were obtained [2]. Good agreement with

References for 5.8.2.1 on pp. 231/2

these data is stated by [6] (the values given in a table of that paper [6, p. 1704] obviously are inadvertent citations from [2]). The much smaller value $\gamma=0.14$ given in [3, 4] is attributed to a calibration error [6]. A recalibration gave $\gamma=0.30$ [6, p. 1707] in excellent agreement with [6] and [2]. The yield $\gamma\approx0.29$ produced by singly charged He^+ ions (indicated by extrapolation to zero velocity of ground-state He^+ ions) [14] well agrees with the γ values given here for He*. This supports the model proposed by [15] predicting that the metastable atoms are initially ionized by tunneling as they approach the surface [6, p. 1705]. With about one monolayer coverage by residual gas, $\gamma<0.01$ was measured for an atom energy of 300 eV increasing fairly linearly to 0.42 at 3.5 keV [10]. Some of the data measured and plotted in [10] are listed in a table in a review of atomic data [16].

A W surface was bombarded with He atom beams of $E_p=1$ to 12 keV produced by charge exchange of neutral gaseous targets with fast ion beams (for the ion production, see [17]). $\gamma=0.41$, 0.66, and 0.88 were measured at $E_p=2$, 4, and 6 keV, respectively [11]. Despite the strong variation of γ with T (factor>3 in the range 0 to 1300 °C studied), the γ values were constant between 600 and 800 °C [12].

The dependence of the secondary electron yield on the angle of incidence α ($\alpha=0°$ to 90°) was studied by [13] using beams with an angular spread of ~0.2°. There were significant departures from the expected $1/\cos(\alpha)$ behavior [13].

Energy Distribution (ED). The ED of electrons emitted by He* from W(110) differed little from that of W(111)-emitted electrons [7]. But there was a strong difference between the He*- and the Ar*-induced distributions. While the upper limits of the electron distributions are about the same for (110) and (111), they differ strongly for the two metastable atom species (13.7±0.3 eV for He* and 6.9±0.6 eV for Ar*) [7]. For an ED measured using He* beams impinging on a polycrystalline, mainly (100)-oriented W ribbon, see [8, p. 1782]. In a later paper [18] it was suggested that the data of [7, 8] result from a contamination of the metastable atom beam with He ions [18]. The secondary electron energy spectrum of electrons emitted from clean W by He 2^3S atoms peaks near 2 eV and vanishes near 13.5 eV [1]. In [9] a similar ED (up to $E_s=12$ eV) of electrons emitted by He* peaking near 2 eV was found. This ED was compared with the much broader distribution expected for Auger neutralization. The authors concluded that incident metastable atoms on W are not resonance-ionized [9]. In contrast an ED measured by [5] is attributed to Auger neutralization.

5.8.2.1.2 He Ions

Singly Charged He Ions

Potential Emission Yield. Fig. 49 (taken from [14]) shows the yields in the range $E_p=0$ to 1000 eV measured at a base pressure of about 10^{-10} Torr for singly charged ground-state ions [14, 19]. Some of the electron yield values measured on He^+-induced electron emission by [14] are listed in a compilation of atomic data [16]. For $E_p=40$, 200, and 1000 eV the values $\gamma_p=0.279$, 0.254, and 0.231, respectively, were calculated by Hagstrum with the aid of his theory of Auger electron emission [15]. These values agree satisfactorily with the respective yields $\gamma_p=0.282$, 0.252, and 0.253 found in the companion experimental study [20].

Hagstrum's results were confirmed by the measurements of Hofer [21] at $E_p=5$ to 500 eV. γ_p decreases monotonously from ~0.276 at $E_p=0$ eV to ~0.233 at 500 eV [21, p. 41]. A value of $\gamma_p=0.26$ (±5%) at 100 eV is given in [22]. $\gamma_p\approx0.25$ can be read from a figure in a study covering the range 300 to 4000 eV [23].

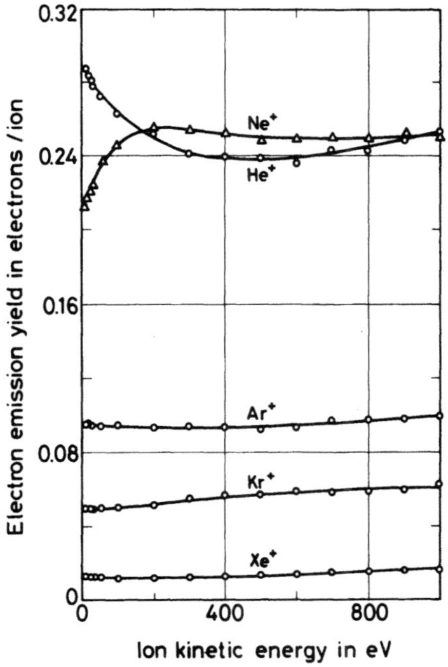

Fig. 49. Total electron yield as a function of ion kinetic energy for singly charged noble gas ions in the ground state [14].

The results of [24, 25] also are in good agreement with Hagstrum's data. $\gamma_p \approx 0.28$, 0.27, and 0.255 were measured at $E_p = 50$, 100, and 200 eV, respectively [24, 25]. For a detailed description of the apparatus used in [24, 25], see [26]. Calculated yields are compared in [27] with the experimental results of [19]. For studies on contaminated surfaces, see [28] and [29].

Assuming that for ground-state He^0 atoms only kinetic emission takes place, the potential emission yield for 2 keV He^+ ions was estimated by subtracting the measured yield produced by He^0 atoms from that of He^+ ions. Values of $\gamma_p \approx 0.4$ and 1.5 resulted for normal and grazing incidence, respectively [13].

The angular distribution of the emitted electrons was studied in the range of potential emission by [29]. The emitted electrons followed a cosine distribution for $E_p = 40$, 400, and 825 eV, with the surface completely gas-covered at 300 K and a monolayer of oxygen present at 1100 K under He^+ impact at $\alpha = 30°$ [29].

Kinetic Emission Yield. Values for the threshold energy of kinetic emission were determined to be $E_{threshold} \approx 0.5$ keV [23] and ~1 keV [30, 31]. In a γ vs. velocity plot, a linear extrapolation from the range of kinetic emission backwards to $\gamma = 0$ gives a threshold velocity of $v_{threshold} \approx 9 \times 10^4$ m/s (i.e. 0.17 keV) [31, 32].

Kinetic electron emission from W was studied as a function of ion velocity by [31]. For $v \geq 2 \times 10^5$ m/s the resulting γ vs. velocity plot is linear, sloping at about 1.5×10^{-6} (electrons/ion)/(m/s) up to $v_{max} \approx 1.6 \times 10^6$ m/s (i.e. 50 keV); see a figure in the paper [31]. A linear increase of γ with ion velocity with the same slope was found in [32]. $(d\gamma/dE_p) \approx 0.105$ (electrons/ion)/keV for E_p ranging from 1 to 4 keV follows from a figure in [23]. In contrast, a plot of γ vs. E_p at higher energies (12 to 32 keV) shows a hyperbola-type curve [13].

References for 5.8.2.1 on pp. 231/2

For a γ vs. velocity curve produced by even higher energetic He^+ ions (80 to 1000 keV), see [33]. For 30 keV He^+ ions incident with $\alpha=0°$, a yield coefficient of $\gamma\approx2$ was measured [34]. γ vs. E_p (about 32 to 127 keV) was studied on a W foil giving yields increasing monotonously from about 1.8 to 3.1 [35]. The yields $\gamma\approx3$, 3.6, 4, and 4.1 for $E_p=100$, 200, 500, and 800 keV, respectively, were taken from a figure in [36]. Some of the electron yield values on He^+-induced electron emission measured in the papers [31, 35] are listed in a table in the atomic data compilation [16].

Pulse height distribution spectra (distribution of the number of electrons emitted per individual, incident ion; individual yields of 1 to 13 electrons/ion being resolved) were recorded for $\alpha=0°$ using $E_p=8$ to 32 keV He^+ ion beams. The number of electrons emitted per incident ion satisfy a Poisson distribution [13].

The angular distribution of electrons emitted from a polycrystalline W surface after bombardment with He^+ ions at $\alpha=60°$ was measured. In the regime of predominant kinetic emission (E_p sufficiently above $E_{threshold}\approx0.5$ keV), the angular distribution of the emitted electrons followed a cosine law with maximum along the surface normal. At lower E_p values, closer to $E_{threshold}$, deviations towards a $\cos^2\vartheta$ distribution were observed, the potential emission peaking more strongly along the surface normal than the kinetic emission. This difference was not observed on gas-covered surfaces which always showed a cosine law behavior [23].

For the γ vs. α dependence, measured at $E_p=2$ keV, see a plot in the paper [13] showing departures from the expected $1/\cos(\alpha)$ behavior [13]; see also the $\gamma(\alpha)\cdot\cos(\alpha)$ plots at $E_p=30$ keV in [34].

Energy Distribution (ED). Fig. 50 (from [14]) shows the ED for electrons emitted from W by 40 eV singly charged ground-state ions [14]. The ED extends to the maximum kinetic energy consistent with an Auger-type excitation mechanism (ionization energy E_i of He^+ minus twice the W work function) [19]. Good agreement with the results of [20] is obtained

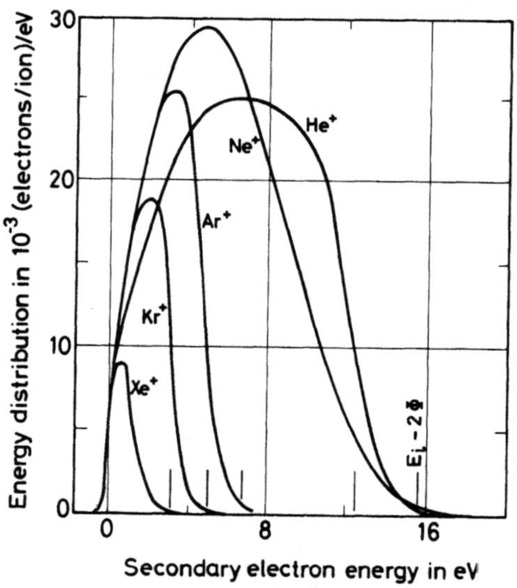

Fig. 50. Energy distribution curves for electrons emitted from W by singly charged noble gas ions in the ground state with an incident kinetic energy of 40 eV. The vertical lines on the abscissa scale indicate the energies $E_i-2\Phi$ (E_i: ionization energy) [14].

for the ED measured with 15 eV ions [21, p. 40], [37]. The ED spectra of electrons emitted by impact of 15 eV He^+ ions were compared in [37] with computed spectra calculated according to the semiempirical theory of Hagstrum [15]. The existence of inelastic scattering and Auger processes involving multiple electrons was assumed to bring the theoretical results in agreement with the experimental data [37]. The ED for 10 eV ions measured by [38] agrees with the results of [20]; see also the ED curves produced by 50, 100, and 200 eV ions in [24, 25]. For ED curves under 20 and 50 eV ion impact, see [22]. The results of [38] are taken as evidence for the ideas advanced by [39] that Auger transitions occur preferentially between levels separated by about 15 to 30 eV [38]. The dependence of the ED on the ion impact energy in the range 5 to 300 eV is discussed in [21, p. 41]. For a calculated secondary electron energy distribution curve extending up to 13 eV with peaks near 7.5 and 9.2 eV (and a shoulder near 10.5 eV), see the paper [40] that refers to 0.05 eV He^+ ions impinging on W(110) [40]. Calculated ED curves for electrons emitted by $E_p \leq 100$ eV impact are given in [27].

The electron emission after collision of 50 to 1000 eV He^+ ions with W(110) under near-grazing incidence was found to be dominated by Auger capture processes involving two W 5d electrons [41].

The energy distribution of electrons emitted from contaminated W foils under He^+ bombardment ($\leq$80 eV) was studied by [28]. Spectra emitted with partially K-covered W(110) surfaces were investigated by [41]. The change of the work function strongly affects the measured ED curves, and at collision energies below 250 eV autodetachment of excited negative ion states ($He^{-*}(1s2s^2)$), formed by resonant two-step capture of two W surface electrons, was observed [41].

Doubly Charged He Ions

The electron emission yield of 0 to 1000 eV He^{2+} ions weakly decreases with ion kinetic energy in contrast to the yields measured for the other doubly charged noble gas ions (showing a monotonous increase with ion energy in all cases) [20]. In the same paper the ED of electrons emitted from W by 200 eV He^{2+} ions is compared to the ED curves produced by Ne, Ar, Kr, and Xe doubly charged ions [20]. An electron energy spectrum generated by 100 eV He^{2+} ions impinging under grazing incidence is discussed in [42]. Auger capture processes, in which two surface electrons are involved, are found to dominate although some formation of He^{**} via resonant capture of two electrons occurs, followed by autoionization [42].

References for 5.8.2.1:

[1] Allison, W.; Dunning, F. B.; Smith, A. C. H. (J. Phys. B **5** [1972] 1175/85).
[2] MacLennan, D. A. (Phys. Rev. [2] **148** [1966] 218/23).
[3] Hasted, J. B.; Mahadevan, P. (Proc. R. Soc. [London] A **249** [1959] 42/50).
[4] Hasted, J. B. (J. Appl. Phys. **30** [1959] 22/4).
[5] Sesselmann, W.; Woratschek, B.; Küppers, J.; Ertl, G.; Haberland, H. (Phys. Rev. [3] B **35** [1987] 8348/59, 8348/50).
[6] Dunning, F. B.; Smith, A. C. H. (J. Phys. B **4** [1971] 1696/710).
[7] MacLennan, D. A.; Delchar, T. A. (J. Chem. Phys. **50** [1969] 1772/8).
[8] Delchar, T. A.; MacLennan, D. A.; Landers, A. M. (J. Chem. Phys. **50** [1969] 1779/87).
[9] Johnson, P. D.; Delchar, T. A. (Surf. Sci. **82** [1979] 237/44).
[10] Berry, H. W. (J. Appl. Phys. **29** [1958] 1219/25).

[11] Devienne, F. M. (J. Phys. [Paris] **28** [1967] 602/10).
[12] Devienne, F. M. (C. R. Séances Acad. Sci. B **264** [1967] 232/4).

[13] Dev, B. (Appl. Surf. Sci. **40** [1990] 319/26).
[14] Hagstrum, H. D. (Phys. Rev. [2] **104** [1956] 317/8).
[15] Hagstrum, H. D. (Phys. Rev. [2] **96** [1954] 336/65).
[16] Barnett, C. F.; Ray, J. A.; Ricci, E.; Wilker, M. I.; McDaniel, E. W.; Thomas, E. W.; Gilbody, H. B. (ORNL-5207 [1977]).
[17] Devienne, M. (C. R. Hebd. Séances Acad. Sci. **254** [1962] 1242/4).
[18] Johnson, P. D.; Delchar, T. A. (Surf. Sci. **77** [1978] 400/8).
[19] Hagstrum, H. D.; D'Amico, C. (J. Appl. Phys. **31** [1960] 715/23).
[20] Hagstrum, H. D. (Phys. Rev. [2] **96** [1954] 325/35).

[21] Hofer, W. (Diss. TU Wien [Vienna] 1983, 103+4 pp.; Phys. Abstr. **88** [1985] No. 109084).
[22] Varga, P.; Winter, H. (Phys. Rev. [3] A **18** [1978] 2453/8).
[23] Klein, H. J. (Z. Phys. **188** [1965] 78/92).
[24] Probst, F. M.; Lüscher, E. (Phys. Rev. [2] **132** [1963] 1037/46).
[25] Probst, F. M. (Diss. Univ. Illinois 1963, 125 pp.; Diss. Abstr. **24** [1964] 3809).
[26] Probst, F. M.; Lüscher, E. (Rev. Sci. Instrum. **34** [1963] 574/9, 578).
[27] Probst, F. M. (Phys. Rev. [2] **129** [1963] 7/11).
[28] Hagstrum, H. D. (J. Appl. Phys. **31** [1960] 897/904).
[29] Abbott, R. C.; Berry, H. W. (J. Appl. Phys. **30** [1959] 871/3).
[30] Arifov, U. A.; Rakhimov, R. R. (Izv. Akad. Nauk UzSSR Ser. Fiz.-Mat. Nauk **1958** No. 6, pp. 49/55; C.A. **1960** 11711).

[31] Arifov, U. A.; Rakhimov, R. R.; Khozinskii, O. V. (Izv. Akad. Nauk SSSR Ser. Fiz. **26** [1962] 1398/402; Bull. Acad. Sci. USSR Phys. Ser. [Engl. Transl.] **26** [1962] 1422/6).
[32] Petrov, N. N.; Dorozhkin, A. A. (Fiz. Tverd. Tela [Leningrad] **3** [1961] 53/60; Soviet Phys.-Solid State [Engl. Transl.] **3** [1961] 38/43).
[33] Hasselkamp, D.; Lang, K. G.; Scharmann, A.; Stiller, N. (Nucl. Instrum. Methods Phys. Res. **180** [1981] 349/56).
[34] Evdokimov, I. N.; Mashkova, E. S.; Molchanov, V. A.; Odintsov, D. D. (Phys. Status Solidi **19** [1967] 407/15).
[35] Large, L. N. (Proc. Phys. Soc. [London] **81** [1963] 1101/3).
[36] Hasselkamp, D.; Hippler, S.; Scharmann, A.; Schmehl, T. (Ann. Phys. [Leipzig] [7] **47** [1990] 555/67).
[37] Hofer, W.; Hetzendorf, G.; Varga, P. (Contrib.-Symp. At. Surf. Phys. **1983** 319/24; C.A. **100** [1984] No. 129291).
[38] Varga, P.; Hofer, W.; Winter, H. (Scanning Electron. Microsc. **1982** No. 3, pp. 967/71).
[39] Arifov, U. A.; Kishinevskii, L. M.; Mukhamadiev, E. S.; Parilis, E. S. (Zh. Tekh. Fiz. **43** [1973] 181/7; Sov. Phys.-Tech. Phys. [Engl. Transl.] **18** [1973] 118/22).
[40] Modinos, A.; Easa, S. I. (Solid State Commun. **63** [1987] 1179/82).

[41] Brenten, H.; Müller, H.; Niehaus, A.; Kempter, V. (Surf. Sci. **278** [1992] 183/92).
[42] Brenten, H.; Müller, H.; Kempter, V. (Surf. Sci. **274** [1992] 309/16).

5.8.2.2 Neon

5.8.2.2.1 Ne Atoms

Thermal velocity beams of metastable Ne* atoms impinging at normal incidence on polycrystalline W were used by [1, 2]. A yield, produced by $Ne(^3P_{0,2})$ and corrected for photoemission, of $\gamma = 0.195 \pm 0.040$ was measured [1] in good agreement with $\gamma = 0.215 \pm 0.020$ found by [2] for Ne* excited by 26 eV electrons. The contribution of each of the two Ne metastable states was not resolved in both determinations [1, 2].

5.8.2.2.2 Ne Ions

Singly Charged Ne Ions

No metastable states of sufficient lifetimes were available with Ne^+ according to [3]. In contrast, an energy distribution from the Auger deexcitation of Ne^{2+} ions showed features taken as indications of a metastable Ne^+ state some 27 eV above the Ne^+ ground state [4].

Potential Emission Yield. Fig. 49 (see p. 229) gives the yields in the range 0 to 1 keV for singly charged noble gas ions in the ground state, measured at a base pressure of about 10^{-10} Torr [5]. A plot of γ_p vs. E_p up to 300 eV in a more recent paper shows very similar results [4], see also [6].

The value $\gamma_p = 0.220$ was calculated for $E_p = 40$ eV by Hagstrum with the aid of his theory of Auger electron emission [7]. This value agrees satisfactorily with a yield of $\gamma_p = 0.232$ found for $E_p = 40$ eV in the companion experimental study [8].

The yield coefficient of potential emission γ_p was found to be nearly independent of E_p on W polycrystals flashed to >2700 K and kept near 1300 K during bombardment with 0.1 to 10 keV Ne^+ ions [9, 10]. A yield coefficient of $\gamma_p \approx 0.225$ was obtained in [10]. The same value was measured by [11] for two different angles of incidence, $\alpha = 60°$ and 0° [11].

Other values for potential emission yields are $\gamma_p(E_p = 15\text{ eV}) = 0.23$ [12, p. 81], $\gamma_p(100\text{ eV}) = 0.245$ (±5%) [13], $\gamma_p(200\text{ eV}) = 0.213$ [14], and $\gamma_p(200\text{ eV}) = 0.2561 \pm 0.0014$ [8].

No effect of nuclear mass was detected when using $^{20}Ne^+$ and $^{22}Ne^+$ ions at $E_p = 200$ eV [8].

Neither γ_p nor γ_k were observed to depend on the sample temperature in the range 300 to 1600 K [15].

Kinetic Emission Yield. The threshold energy of kinetic emission $E_{threshold}$ is about 1 keV for Ne^+; see e.g. [10, 11]. At energies sufficiently above $E_{threshold}$, the measured angular distribution of the emitted electrons exactly follows a cosine law, peaking along the surface normal. Deviations towards a $\cos^2\vartheta$ distribution were found at energies closer to $E_{threshold}$. The same results were observed when He^+ or Ar^+ ions were used [11].

The dependence of γ on the Ne^+ ion velocity is plotted in [16], where the yields for He^+, Ne^+, and Ar^+ impinging at up to ~20 keV normal on the W samples (target at ~1200 °C during the measurements) are compared [16]. W polycrystals were bombarded with 30 keV Ne^+ ions and the yield $\gamma(\alpha)$ and $\gamma(\alpha)\cdot\cos(\alpha)$ were plotted for $\alpha \approx 0°$ to 85°. $\gamma(\alpha = 0°) \approx 2.3$ was found [17]. For emission yields produced by 10 to 140 keV Ne^+ ions and a comparison with the yields of He^+, Ar^+, and Kr^+, see [18]. The electron yield vs. ion velocity is plotted for Ne^+ beams with $E_p = 80$ to 1000 keV in [19]. For 500 keV singly charged Ne ions, a yield coefficient of 7.07 was extrapolated for a clean W surface from measurements on oxygen-covered W [20].

Above the threshold energy the $\gamma(E_p)$ curves increase linearly up to several keV with a slope of 0.075 (electrons/ion)/keV [16, 21]. From a figure in [11] a slope of about 0.097 (electrons/ion)/keV can be extracted [11].

Energy Distribution (ED). The ED of the electrons emitted from W by 40 eV Ne^+ ions is shown in Fig. 50 [5]; see p. 230. For an ED curve under Ne^+ impact with $E_p = 15$ eV, see also [12, p. 40]. Significant differences exist in the shape of the ED curves of electrons emitted under impact of He^+, Ne^+, and Kr^+ ions at various energies (40, 200, and 1000 eV)

References for 5.8.2.2 on p. 235

[8]. A comparison of ED curves for 10 and 40 eV Ne^+ ions revealed an anomalous behavior of Ne^+. For the 40 eV ions the $E_{max}=E_i-2\Phi$ limit (E_i: ionization energy) is clearly violated in the case of Ne [5, 8]. This effect is explained assuming that at higher ion energies, some of the Ne^+ ions are not simply Auger-neutralized but were resonance-neutralized and deexcited in a process which has a higher yield and a larger kinetic energy maximum [7].

An ED curve measured with 10 eV ions by [6] was found to be in good agreement with the results of [8]. ED curves for different ion energies between 10 and 200 eV were measured by [12].

A comparison of ED curves for $Ne^{(1\,to\,3)+}$ ions ($E_p=200$ eV) indicates that the neutralization of the higher charged ions to the ground state occurs predominantly in a series of steps [8].

The ED of electrons emitted under $E_p=500$ keV Ne^+ impact extends from (half-maximum values) about 2 to 12 eV and peaks near 4 eV [20].

Doubly Charged Ne Ions

Emission Yield. γ_p vs. E_p for Ne^{2+} shows a minimum around $E_p=30$ eV in a study covering the range $E_p=5$ to 500 eV [12]; see also [4]. This minimum was taken to be evidence for stepwise neutralization [4, 12].

$\gamma_p=0.68$ ($\pm5\%$) at $E_p=100$ eV is given in [13] in good agreement with [4].

Energy Distribution (ED). The maximum near 2 eV in the ED curves (studied with E_p between 10 and 200 eV) was found to depend sensitively on E_p [12]. Structures observed in the ED curves were taken to result from several combined processes [4, 12]. The ED produced by 10 eV ions in [4, 6] can be well explained assuming the first neutralization step to be either resonance neutralization followed by Auger deexcitation into excited singly charged ion states Ne^{+*} or a direct Auger neutralization into Ne^{+*}. This is followed by Auger deexcitation of Ne^{+*} into the ground state of Ne^+. The electrons emitted in the first deexcitation step acquire at most a kinetic energy of 5 eV and, therefore, the peak in the ED observed around 3 eV is attributed to this step [4]. For ions with 200 eV, the ED curves of $Ne^{(1\,to\,3)+}$ are compared in [8].

Higher Charged Ne Ions. Ne^{3+} to Ne^{9+}

Emission Yield. γ vs. E_p (0 to 1000 eV) for Ne^{3+} ions is shown in [8]. A plot of the yield coefficient vs. the potential energy $W_q=\Sigma_q I_q$ (=sum over the ionization potentials of the q-charged ion) is given in [22] and [23]. The Ne^{q+} (q=4, 5, 6, and 7), Ar^{q+} (q=2, 4, 7, 8, 9, 11, and 12), Kr^{q+} (q=9, 11), and also N^{q+} (q=4, 5, and 6) data fall into one single straight line for an ion velocity of v=40 km/s [22]. A slope of 11.3 (electrons/ion)/keV was read from that plot [22]. For ion velocities v below the threshold value $v_{threshold}\approx100$ km/s straight lines through the origin result for the γ vs. W_q plots. Above the threshold such plots no longer pass through the origin; the nonzero intercept with the ordinate indicates the yield caused by kinetic emission. Thus, it is possible to correct the as-measured γ values and to obtain γ_p vs. velocity plots [23]. The deviations from linearity in γ vs. W_q at the higher velocities seem to be due to the reduction of the interaction time available for Auger-type deexcitation of the ions [22].

Energy Distribution (ED). The ED of secondary electrons for Ne^{3+} ions with $E_p=200$ eV are compared to the results for Ne^+ and Ne^{2+} ions. Maxima occur at secondary electron energies of about 2 and 17 eV [8]. The ED of electrons emitted by the hydrogen-like Ne^{9+}

ions ($E_p = 25$ to 600 eV) was studied on Ar^+ sputter-cleaned W polycrystals. The high-energy structures observed are ascribed to Auger electrons emitted from the projectile [24].

References for 5.8.2.2:

[1] Dunning, F. B.; Smith, A. C. H. (J. Phys. B **4** [1971] 1696/710).
[2] MacLennan, D. A. (Phys. Rev. [2] **148** [1966] 218/23).
[3] Hagstrum, H. D. (Phys. Rev. [2] **104** [1956] 309/16).
[4] Varga, P.; Hofer, W.; Winter, H. (Surf. Sci. **117** [1982] 142/53).
[5] Hagstrum, H. D. (Phys. Rev. [2] **104** [1956] 317/8).
[6] Varga, P.; Hofer, W.; Winter, H. (Scanning Electron. Microsc. **1982** No. 3, pp. 967/71).
[7] Hagstrum, H. D. (Phys. Rev. [2] **96** [1954] 336/65).
[8] Hagstrum, H. D. (Phys. Rev. [2] **96** [1954] 325/35).
[9] Arifov, U. A.; Rakhimov, R. R. (Izv. Akad. Nauk UzSSR Ser. Fiz.-Mat. Nauk **1958** No. 6, pp. 49/55; C.A. **1960** 11711).
[10] Arifov, U. A.; Rakhimov, R. R. (Izv. Akad. Nauk SSSR Ser. Fiz. **24** [1960] 657/63; Bull. Acad. Sci. USSR Phys. Ser. [Engl. Transl.] **24** [1960] 666/72).

[11] Klein, H. J. (Z. Phys. **188** [1965] 78/92).
[12] Hofer, W. (Diss. TU Wien [Vienna] 1983, 103+4 pp.; Phys. Abstr. **88** [1985] No. 109084).
[13] Varga, P.; Winter, H. (Phys. Rev. [3] A **18** [1978] 2453/8).
[14] Hagstrum, H. D. (Phys. Rev. [2] **104** [1956] 672/83).
[15] Arifov, U. A.; Rakhimov, R. R. (Dokl. Akad. Nauk UzSSR **1958** No. 12, pp. 15/8; C.A. **1960** 4163).
[16] Petrov, N. N.; Dorozhkin, A. A. (Fiz. Tverd. Tela [Leningrad] **3** [1961] 53/60; Sov. Phys.-Solid State [Engl. Transl.] **3** [1961] 38/43).
[17] Evdokimov, I. N.; Mashkova, E. S.; Molchanov, V. A.; Odintsov, D. D. (Phys. Status Solidi **19** [1967] 407/15).
[18] Large, L. N. (Proc. Phys. Soc. [London] **81** [1963] 1101/3).
[19] Hasselkamp, D.; Lang, K. G.; Scharmann, A.; Stiller, N. (Nucl. Instrum. Methods Phys. Res. **180** [1981] 349/56).
[20] Hasselkamp, D.; Scharmann, A.; Stiller, N. (Nucl. Instrum. Methods Phys. Res. **168** [1980] 579/83).

[21] Petrov, N. N. (Izv. Akad. Nauk SSSR Ser. Fiz. **26** [1962] 1327/32; Bull. Acad. Sci. USSR Phys. Ser. [Engl. Transl.] **26** [1962] 1350/4).
[22] Delaunay, M.; Fehringer, M.; Geller, R.; Hitz, D.; Varga, P.; Winter, H. (Phys. Rev. [3] B **35** [1987] 4232/5).
[23] Fehringer, M.; Delaunay, M.; Geller, R.; Varga, P.; Winter, H. (Nucl. Instrum. Methods Phys. Res. B **23** [1987] 245/7).
[24] Folkerts, L.; Morgenstern, R. (Europhys. Lett. **13** [1990] 377/82).

5.8.2.3 Argon

5.8.2.3.1 Ar Atoms

Emission Yield. Ar atom impact on W surfaces was investigated in [1 to 7]. W targets, heated at 600 to 800 °C, exhibit a fairly linear γ vs. E_p dependence: $\gamma = 0.17$, 0.33, and 0.50 at $E_p = 2$, 4, and 6 keV, respectively [1]. In a preceding paper the author had measured γ vs. T (0 °C $\leq T \leq$ 1300 °C) for 1 keV Ar atoms impinging on a W ribbon. Despite the strong variation of γ with T (factor >3 in the range studied, obviously due to vacuum and contamination conditions), the γ values were constant between 600 and 800 °C [2]. The production

References for 5.8.2.3 on pp. 239/41

of fast Ar atoms by charge exchange between 3 keV Ar^+ ions and Ar gas had been described in [3].

The secondary emission from W(110) and W(111) upon impact of thermal beams of metastable argon atoms, Ar*, was studied by [4]. The yield on W(111) was found to be about 4% larger than on W(110). This small difference was stated to be astonishing, because for the work functions a difference of 1.6 ± 0.3 eV was measured [4].

Using thermal-energy beams of metastable atoms, γ has been measured for Ar* impinging on a polycrystalline W target heated to 2000 °C. Monitoring the time dependence of γ allowed extrapolation to zero contamination conditions. The yield coefficient measured for $Ar(^3P_{0,2})$ is $\gamma = 0.080 \pm 0.016$ [5]. Only weak secondary electron yields had been obtained with excited Ar atoms by [6]. The CN^- emission observed under N_2^* impact [6] suggests that the target was contaminated at least with carbon.

Energy Distribution (ED). With metastable $Ar(^3P_{0,2})$ thermal atom beams the ED of secondary electrons from clean W extends hardly beyond 4 eV. A contaminated surface produces a much greater yield and a strong peak near 1 eV. This is compared with the behavior of He (2^3S) and Ne* atoms in [7]. Little variation exists between the ED curves of electrons emitted by Ar* from W(110) and from W(111). But there is a strong difference between the He*- and the Ar*-induced distributions. While the upper limits of the ED curves are about the same for (110) and (111), they differ strongly for the two species of metastable atoms (6.9 ± 0.6 eV for Ar* and 13.7 ± 0.3 eV for He*) [4].

5.8.2.3.2 Ar Ions

Singly Charged Ar Ions

Potential Emission Yield. In [8] a theory of Auger electron emission was developed. Yields calculated using this theory were compared in [8] with the experimental yields found in [9]. The ion beams used in [9] contained a small fraction (about 2%) of metastable ions contaminating the beams. The γ_p values corrected for this effect are presented for the range $E_p = 10$ to 1000 eV in Fig. 49 [10]; see p. 229. The appearance of metastable ions was investigated in detail in [11]. The dependence of the electron yield for 40 eV Ar^+ ions on the bombarding electron energy in the ion source was shown [11]. In [12] the corresponding plots for Ar^- and Ar^{2+} impinging at $E_p = 100$ eV on W are given. The data are a good confirmation of the results of [11]. The assumption $\gamma_p(Ar^{+*}) \approx \gamma_p(Ar^{2+})$, made in [11], is justified in view of the results of [13, 14]. The yield coefficients $\gamma_1 = 0.09$, $\gamma_{1,m} = 0.295$, and $\gamma_2 = 0.36$ were obtained for $E_p = 100$ eV (γ_1 and γ_2 denote the coefficients for singly and doubly charged ions, respectively, determined with ion beams virtually free from metastables; $\gamma_{1,m}$ denotes the yield coefficient for metastable, singly charged ions) [14].

According to [15], γ_p does not depend on E_p in the investigated range 5 to 500 eV in good agreement with the results of Hagstrum [10]. $\gamma_p = 0.09$ for ground-state ion beams and $\gamma_p = 0.10$ for beams containing 3.5% metastable ions were found [15, p. 46]. The metastable ion beam fractions were measured for various ion sources releasing singly charged Ar ions [14, 16]. It is assumed that the metastable ions are first (Auger-) deexcited and then neutralized [15]. The electron yields for Ar^+- and Ar^{2+}-ion impact are plotted vs. E_p between 10 and 300 eV in [13].

The following values for the coefficient of the potential emission yield were measured: $\gamma_p(E_p = 10\ eV) = 0.095$ [17], $\gamma_p(15\ eV) = 0.09$ [15, p. 81], $\gamma_p(50\ eV) = 0.098$ [18], $\gamma_p(100\ eV) = 0.090 \pm 0.005$ [12], $\gamma_p(500\ eV) \approx 0.099$ [19], $\gamma_p \approx 0.09$ [20], $\gamma_p \approx 0.10$ [21], and $\gamma_p = 0.135$ [22].

A plot of the yield coefficient γ_p of different metal surfaces vs. their work functions Φ shows that γ_p decreases linearly with Φ (tested with Mg, Ta, Mo, W, Ni, Zr, and Pt) [20].

Neither γ_p nor γ_k depend on the sample temperature [20]. The same result was found in [19] where γ(0.5 keV) and γ(2.4 keV) were measured between about 300 and 2200 K [19].

For γ_p from contaminated W under Ar^+ impact ($\leq$80 eV), see [23].

Kinetic Emission Yield. Above the threshold energy $E_{threshold} \approx 1.5$ keV, kinetic emission commences [20, 21, 24]; see also [22, 25, 26]. γ_k increases linearly with E_p up to several keV as was frequently observed [20 to 22, 24 to 28]. Slopes (in [electrons/ion]/keV) of $d\gamma/dE_p \approx 0.053$ [24], 0.062 [21], 0.06 [27], and 0.0713 (read from a figure in [22]) were obtained.

The experimentally determined threshold velocity is in excellent agreement with a value calculated according to a theory of [29] in [22]. For a comparison of γ vs. velocity plots for He^+, Ne^+, Ar^+, N_1^+, and N_2^+ ions, see [27]. See also the plot of γ vs. ion velocity in a study, where E_p was varied between 80 and 1000 keV [30].

When both Ar^+ and K^+ ions were impinged at $E_p > 1.7$ keV on a W sample heated to 1900 K, γ linearly increased to 6 keV, the highest E_p value studied by [24, 28]. When corrected for the potential emission coefficient of Ar^+ (assumed to be 0.10), the γ vs. E_p curves for Ar^+ and K^+ fall into a single curve [24, 28], see also [21] where the papers [24, 27, 28] are reviewed. In [24, 28] the average data of W and Ta were plotted. This is a reasonable procedure, since W and Ta have very similar yield curves; see [21].

A theory of [29] describes the transfer of the kinetic energy of primary ions incident on a metal surface at 1 to 100 keV in terms of ion-to-metal atom collisions with excitation of bound electrons. The electron yield γ was calculated based on an Auger-type recombination of conduction electrons with (deeper shell) electron holes. The γ vs. velocity relationship of the primaries, calculated by [29], is in excellent agreement with the experimental results by [25] and [27]; see [29].

The dependence of the kinetic emission yield on the angle of incidence was investigated at $E_p = 30$ keV [31], where $\gamma(\alpha)$ and $\gamma(\alpha)\cdot\cos(\alpha)$ were plotted as a function of α between 0° and ~85°. $\gamma \approx 1.7$ was measured for normal incidence ($\alpha = 0°$) [31].

According to the earlier studies [32, 33] using ions with 5 to ~37 keV ($\alpha = 0°$), the kinetic emission yield is larger than 1 for $E_p \geq 20$ keV. γ seemed to saturate around 1.2 for $E_p \geq 30$ keV [32, 33]. The value $\gamma = 1.25$ measured by [34] with $E_p = 5$ keV and $\alpha = 0°$ is obviously obsolete. For an earlier study with 10 to 40 keV Ar^+ ions, see [35]; for a study with about 10 to 15 keV Ar^+ ions impacting at $\alpha \approx 70°$, see [36]. The yields of Ar^+ beams with energies between 10 and 140 keV are given in [37]. Values $\gamma \approx 3.9$, 5.8, 8.2, and 9.9 were measured for Ar^+ ions at $E_p = 100$, 200, 500, and 800 keV, respectively [38].

Energy Distribution (ED). Spectra of electrons emitted from a W(110) surface after bombardment with 50 to 1000 eV Ar^+ ions under near-grazing incidence were presented by [39]. The electron emission is found to be dominated by Auger neutralization processes of the incident ions involving two W 5d electrons [39].

The ED curves produced by Ar^+ ions with $E_p = 10$ eV [40] and 15 eV [15, p. 40] impinging on polycrystalline W ribbons cleaned by flashing at ~2500 K or by noble gas sputtering at $p \approx 10^{-10}$ Torr, agree well with the results of [10], who compared the ED curves of the electrons emitted by singly charged 10 and 40 eV ions. For the ED produced by 40 eV Ar^+ ions, see Fig. 50, p. 230 (from [10]). The ED curves for electrons emitted by 15 eV

References for 5.8.2.3 on pp. 239/41

beams of ground-state ions with and without 3.5% metastable ions were found to be almost identical [15, p. 48], [41].

The ED of electrons from contaminated W foils ($E_p \leq 80$ eV) was studied by [23].

Electron ED curves in the kinetic emission range, i.e. produced by Ar^+ ions with energies of several keV, are discussed in [33, 34, 42 to 45]. For primary energies between about 4 and 7 keV, the energy distribution of the emitted electrons was studied as a function of the Ar^+ incidence angle α on an ion-sputter-cleaned polycrystalline W target. The ED extended beyond $E_s \approx 40$ eV and peaked near $E_s = 1$ eV [43]. ED curves were measured for Ar^+ ions with energies of 35, 40, and 46 keV. The energy resolution for the emitted electrons was $\Delta E_s \approx 1$ eV. The ED curves are fairly well described by $\exp(-E_s/k_BT^*)$. The value for the "characteristic temperature of electron emission" T^* is 33000 K [33, 45]. An ED curve of the electrons produced by Ar^+ ions with an energy of 5 keV was investigated in [34]. 69% of the emitted electrons have energies of $E_s < 5$ eV. The number of electrons with higher energies strongly decreases with energy, and only about 1.5% of the electrons have energies between 20 eV and 30 eV [34].

W(111) single crystal faces were irradiated by 3 keV Ar^+ ions at current densities as high as 0.1 mA/cm^2. The ED of the electrons emitted from the surface opposite to the surface bombarded was measured. The thickness of the crystals ranged from 20 to 100 μm. On prolonged bombardment (~1/2 h), the electron emission strongly decreased (to the detection limit), apparently due to radiation damages produced on the surface. The energy distribution of the secondary electrons emitted from W(111) under Ar^+ bombardment shows several peaks at rather high energies. This effect could not be explained [42].

Multiply Charged Ar Ions. Ar^{2+} to Ar^{12+}

Emission Yield. Potential emission yields of $\gamma_p(Ar^{2+}) = 0.36$ (±5%) and $\gamma_p(Ar^{3+}) = 1.15$ (±5%) were determined using ion beams with $E_p = 100$ eV virtually free of metastables [12]. The yield coefficients for Ar^+ and Ar^{2+} ion impact are plotted vs. E_p (10 to 300 eV) in [13]. $\gamma_p(Ar^{2+})$ was found to decrease strongly with E_p in the range 5 to 500 eV [15, p. 60]. γ_p increases significantly at the threshold voltages for metastable ion production in a plot vs. the ion-source voltage [12]. Curves of γ vs. E_p (0 to 1 keV) for multiply charged ions given in [9] may be flawed by the presence of metastable ions.

Measurements with 200 to 1500 eV Ar^{2+}, Ar^{3+}, Ar^{4+}, and Ar^{6+} ions impinging on a W surface revealed monotonously decreasing emission yields in all cases [46].

A plot of $\gamma(Ar^{9+})$ vs. E_p between ~35 eV and ~250 keV showed the yield decreasing from ~12 near 35 eV to ~8 at 10 keV, then increasing to ~12 at 250 keV, the minimum (near $E_p = 10$ keV) marking the transition from predominantly potential to predominantly kinetic emission [47]. For plots of γ vs. ion kinetic energy (up to about 20 keV) of Ar^{8+} and Ar^{9+} ions, see [48]. At $E_p = 500$ eV, $\gamma(Ar^{9+}) = 10.0 \pm 0.6$ (with a systematic error estimated to be <10%) was found [47]. A somewhat larger value of $\gamma(Ar^{9+}) = 11.3 \pm 0.5$ is given for 500 eV ions by [49]. γ is plotted vs. ion velocity v (≥ 10 km/s) for Ar^{q+} (q up to 12) in [49] and vs. E_p for q up to 9 in [50].

Strictly linear plots resulted for $E_p = 28$ and 70 keV after plotting the "total electron yield" $\eta = \gamma + q$ vs. q^2 [50]; see also the corresponding plot for 20 keV ions with q = 1 to 12 in [48]. The new variable η takes the number q of electrons per ion into account which are needed to neutralize the q-foldly charged ion. A field emission mechanism was proposed to explain this q^2 dependence [48].

At impact velocities up to v = 40 km/s, γ depends linearly on the potential energy W_q of the ions involved. This is shown by a γ vs. W_q plot for v = 40 km/s in [49], where Ne^{q+} (q = 4, 5, 6, and 7), Ar^{q+} (q = 2, 4, 7, 8, 9, 11, and 12), and also N^{q+} (q = 4, 5, and 6) data fall into one single straight line with a slope of ~11.3 (electrons/ion)/keV for v = 40 km/s [49]. A similar result, i.e. a common straight line for different ions of different charge states, is presented in [46] citing data from [9] and adding own results [46]. Note, however, that the γ values indicated in the text of [46] seem to differ from those in a figure of that paper [46]. Another plot of γ vs. W_q of Ar^{q+} ions (up to q = 9) for E_p = 28 and 70 keV in [50] again confirms the linear dependence. At higher velocities, deviations from this linear relationship occur [49, 50], obviously due to a reduction of the interaction time available for Auger-type deexcitation of the ions [49].

Energy Distribution (ED). The ED of electrons emitted from W after low-energy Ar ion impact was extensively discussed by the group around Varga and Hofer [12, 13, 15, 40, 51, 52]. ED curves from 30 eV Ar^+ (with 4.5% metastable ions), Ar^{2+}, and Ar^{3+} are compared in [13, 15]. The structures observed in the curves suggest a stepwise deexcitation of the triply and doubly charged ions. A large peak seen at low energies corresponds to the singly charged Ar^+ ions, while structures at higher E_s values are due to the first deexcitation step of the metastable ions. The first neutralization step for Ar^{2+} is either resonance neutralization followed by Auger deexcitation into excited, singly charged ion states Ar^{+*} or Auger neutralization into Ar^{+*}. This process is followed by Auger deexcitation of Ar^{+*} into the ground state of Ar^+ [13, 15]. These results support the idea of a stepwise neutralization advanced by [53]. For ED curves of electrons emitted by 15 eV Ar^+ and Ar^{2+} ions, see [13, 40, 51].

ED curves are given for Ar^{2+} ion impact at various ion energies ranging from 6 to 200 eV [15, 40, 52]. Increasing E_p results in a considerable loss of structure in the ED curves and the characteristic peaks seem to disappear. This is due to level broadening and level shift which smear out these peaks [40].

Maximum energies of E_s = 16 eV (22 eV) were found for electrons emitted by Ar^{2+} (Ar^{3+}) ions [53].

ED spectra of electrons resulting from Ar^{9+} collisions on polycrystalline W are discussed in [47, 48, 54 to 57]. As in the case of lower charged Ar ions, most of the energy values of the electrons emitted by Ar^{9+} ions are lower than 20 eV [48]. ED curves spanning the E_s range up to several hundred eV reveal Auger electron emission from the projectile particles, whereas Auger electron emission from the W target was not observed [55].

References for 5.8.2.3:

[1] Devienne, F. M. (J. Phys. [Paris] **28** [1967] 602/10).
[2] Devienne, F. M. (C. R. Séances Acad. Sci. B **264** [1967] 232/4).
[3] Devienne, F. M.; Clapier, R.; Roustan, J.-C. (C. R. Séances Acad. Sci. B **254** [1962] 2733/5).
[4] MacLennan, D. A.; Delchar, T. A. (J. Chem. Phys. **50** [1969] 1772/8).
[5] Dunning, F. B.; Smith, A. C. H. (J. Phys. B **4** [1971] 1696/710).
[6] Varney, R. N. (Phys. Rev. [2] **175** [1968] 98/102).
[7] Allison, W.; Dunning, F. B.; Smith, A. C. H. (J. Phys. B **5** [1972] 1175/85).
[8] Hagstrum, H. D. (Phys. Rev. [2] **96** [1954] 336/65).
[9] Hagstrum, H. D. (Phys. Rev. [2] **96** [1954] 325/35).
[10] Hagstrum, H. D. (Phys. Rev. [2] **104** [1956] 317/8).

[11] Hagstrum, H. D. (Phys. Rev. [2] **104** [1956] 309/16).
[12] Varga, P.; Winter, H. (Phys. Rev. [3] A **18** [1978] 2453/8).
[13] Varga, P.; Hofer, W.; Winter, H. (Surf. Sci. **117** [1982] 142/53).
[14] Varga, P.; Hofer, W.; Winter, H. (J. Phys. B **14** [1981] 1341/51).
[15] Hofer, W. (Diss. TU Wien [Vienna] 1983, 103+4 pp.; Phys. Abstr. **88** [1985] No. 109084).
[16] Hofer, W.; Vanek, W.; Varga, P.; Winter, H. (Rev. Sci. Instrum. **54** [1983] 150/7).
[17] Hagstrum, H. D. (Phys. Rev. [2] **104** [1956] 672/83).
[18] Probst, F. M.; Lüscher, E. (Phys. Rev. [2] **132** [1963] 1037/46).
[19] Dorozhkin, A. A.; Petrov, N. N. (Fiz. Tverd. Tela [Leningrad] **7** [1965] 118/22; Sov. Phys.-Solid State [Engl. Transl.] **7** [1965] 88/91).
[20] Arifov, U. A.; Rakhimov, R. R. (Izv. Akad. Nauk SSSR Ser. Fiz. **24** [1960] 657/63; Bull. Acad. Sci. USSR Phys. Ser. [Engl. Transl.] **24** [1960] 666/72).

[21] Petrov, N. N. (Izv. Akad. Nauk SSSR Ser. Fiz. **26** [1962] 1327/32; Bull. Acad. Sci. USSR Phys. Ser. [Engl. Transl.] **26** [1962] 1350/4).
[22] Klein, H. J. (Z. Phys. **188** [1965] 78/92).
[23] Hagstrum, H. D. (J. Appl. Phys. **31** [1960] 897/904).
[24] Petrov, N. N. (Fiz. Tverd. Tela [Leningrad] **2** [1960] 1300/7; Sov. Phys.-Solid State [Engl. Transl.] **2** [1960/61] 1182/8).
[25] Arifov, U. A.; Rakhimov, R. R. (Izv. Akad. Nauk UzSSR Ser. Fiz.-Mat. Nauk **1958** No. 6, pp. 49/55; C.A. **1960** 11711).
[26] Arifov, U. A.; Rakhimov, R. R. (Dokl. Akad. Nauk UzSSR **1958** No. 12, pp. 15/8; C.A. **1960** 4163).
[27] Petrov, N. N.; Dorozhkin, A. A. (Fiz. Tverd. Tela [Leningrad] **3** [1961] 53/60; Sov. Phys.-Solid State [Engl. Transl.] **3** [1961] 38/43).
[28] Petrov, N. N. (Izv. Akad. Nauk SSSR Ser. Fiz. **24** [1960] 673/8; Bull. Acad. Sci. USSR Phys. Ser. [Engl. Transl.] **24** [1960] 682/7).
[29] Parilis, E. S.; Kishinevskii, L. M. (Fiz. Tverd. Tela [Leningrad] **3** [1961] 1219/28; Sov. Phys.-Solid State [Engl. Transl.] **3** [1961] 885/91).
[30] Hasselkamp, D.; Lang, K. G.; Scharmann, A.; Stiller, N. (Nucl. Instrum. Methods Phys. Res. **180** [1981] 349/56).

[31] Evdokimov, I. N.; Mashkova, E. S.; Molchanov, V. A.; Odintsov, D. D. (Phys. Status Solidi **19** [1967] 407/15).
[32] Cousinié, P.; Colombié, N.; Fert, C.; Simon, R. (C. R. Hebd. Séances Acad. Sci. **249** [1959] 387/9).
[33] Simon, R. (Metaux **34** [1959] 204/20).
[34] Slodzian, G. (C. R. Hebd. Séances Acad. Sci. **246** [1958] 3631/4).
[35] Fogel', Ya. M.; Slabopitskii, R. P.; Rastrepin, A. B. (Zh. Tekh. Fiz. **30** [1960] 63/73; Sov. Phys.-Tech. Phys. [Engl. Transl.] **5** [1960] 58/66).
[36] Gaukler, K.-H. (Z. Metallk. **51** [1960] 463/6).
[37] Large, L. N. (Proc. Phys. Soc. [London] **81** [1963] 1101/3).
[38] Hasselkamp, D.; Hippler, S.; Scharmann, A.; Schmehl, T. (Ann. Phys. [7] **47** [1990] 555/67).
[39] Brenten, H.; Müller, H.; Niehaus, A.; Kempter, V. (Surf. Sci. **278** [1992] 183/92).
[40] Varga, P.; Hofer, W.; Winter, H. (Scanning Electron. Microsc. **1982** No. 3, pp. 967/71).

[41] Hofer, W.; Vanek, W.; Varga, P.; Winter, H. (Surf. Sci. **126** [1983] 605/9).
[42] Azizov, E. E.; Babaev, V. G.; Brandt, N. B.; Mashtakov, V. A. (Vestn. Mosk. Univ. Ser. 3 Fiz. Astron. **44** No. 2 [1989] 106/8; Moscow Univ. Phys. Bull. [Engl. Transl.] **44** No. 2 [1989] 123/5).
[43] Burrekoven, H. (Diss. Univ. Köln [Cologne, FRG] 1956, 34+XIV pp.).

[44] Fan, Ch.; Yu, Zh.; Chen, X. (J. Vac. Sci. Technol. [2] A **5** [1987] 1206/8).
[45] Pradal, F.; Simon, R. (C. R. Hebd. Séances Acad. Sci. **247** [1958] 438/41).
[46] Arifov, U. A.; Mukhamadiev, E. S.; Parilis, E. S.; Pasyuk, A. S. (Zh. Tekh. Fiz. **43** [1973] 375/9; Sov. Phys.-Tech. Phys. [Engl. Transl.] **18** [1973] 240/3).
[47] de Zwart, S. T.; Drentje, A. G.; Boers, A. L.; Morgenstern, R. (Surf. Sci. **217** [1989] 298/316).
[48] Delaunay, M.; Geller, R.; Debernardi, J.; Ludwig, P.; Sortais, P. (Surf. Sci. **197** [1988] L 273/L 280).
[49] Delaunay, M.; Fehringer, M.; Geller, R.; Hitz, D.; Varga, P.; Winter, H. (Phys. Rev. [3] B **35** [1987] 4232/5).
[50] Delaunay, M.; Benazeth, C.; Benazeth, N.; Bourg, F.; Geller, R.; Ludwig, P.; Mayoral, C.; Melin, G. (J. Phys. Colloq. [Paris] **50** [1989] C1-277/C1-284).

[51] Hofer, W.; Varga, P. (Nucl. Instrum. Methods Phys. Res. B **2** [1984] 391/5).
[52] Varga, P. (Appl. Phys. A **44** [1987] 31/41).
[53] Arifov, U. A.; Kishinevskii, L. M.; Mukhamadiev, E. S.; Parilis, E. S. (Zh. Tekh. Fiz. **43** [1973] 181/7; Sov. Phys.-Tech. Phys. [Engl. Transl.] **18** [1973] 118/22).
[54] de Zwart, S. T.; Fried, T.; Jellen, U.; Boers, A. L.; Drentje, A. G. (J. Phys. B **18** [1985] L 623/L 628).
[55] Delaunay, M.; Fehringer, M.; Geller, R.; Varga, P.; Winter, H. (Europhys. Lett. **4** [1987] 377/82).
[56] de Zwart, S. T. (Nucl. Instrum. Methods Phys. Res. B **23** [1987] 239/41).
[57] Folkerts, L.; Morgenstern, R. (J. Phys. Colloq. [Paris] **50** [1989] C1-541/C1-547).

5.8.2.4 Krypton

Results on Kr atoms impinging on W are not reported in the literature. All of the following describes results obtained with singly and multiply charged Kr ions.

Singly Charged Kr Ions

Potential Emission Yield. Fig. 49 (see p. 229) gives the electron yields produced by impact of singly charged 0 to 1 keV noble gas ions in the ground state [1]. The values for $\gamma(E_p = 0$ to 1 keV) given in the previous paper [2] were about 10% too large due to metastable ions present in the beams; see the note added in proof in [2, p. 335]. Plotting the electron yield of 40 eV Kr^+ ions as a function of the bombarding electron energy in the ion source, a threshold for the appearance of metastable Kr^+ ions at about 30 eV was detected [3]. Similar measurements with a Kr^+ impact energy of 100 eV confirmed this result [4]. Based on the theory of potential emission of [5], the emission yield of metastable Kr^+ ions was determined to be $\gamma_{1,m} = 0.250$ for an ion impact energy of 100 eV [6]. The metastable ion beam fractions for various ion sources were determined by measuring the secondary electron spectra in the region of potential emission [7].

The theoretically calculated emission yield $\gamma_p(E_p = 40$ eV$) = 0.027$ [5] strongly deviates from the corrected experimental yield $\gamma_p(40$ eV$) = 0.043$ found in the companion paper [2]. Further corrected values from [2] are $\gamma_p(200$ eV$) = 0.046$ and $\gamma_p(1$ keV$) = 0.053$. Other measured potential emission yields are $\gamma_p(15$ eV$) = 0.049$ [8, p. 81] and $\gamma_p(100$ eV$) = 0.049$ [4, 6]. For a theoretical calculation of emission yields, see also [9].

Plots of γ vs. E_p (0.1 to 10 keV) were obtained on samples kept at 1300 K during the measurements. Between 0.1 and 1 keV, γ does not depend on E_p, showing the range of potential emission [10, 11]. A value of $\gamma_p \approx 0.05$ can be read from a figure in [10].

References for 5.8.2.4 on p. 243

The emission yield of contaminated W foils under Kr^+ ion impact at $E_p \leq 80$ eV was studied by [12].

Kinetic Emission Yield. In the range of kinetic emission, a linear increase of γ from about 0.05 at $E_p = 1$ keV to about 0.22 at 10 keV was found in [10, 11]. Measurements with incident ion energies varying from about 16 to 64 keV revealed a linear increase of γ from 0.4 to 1.7 [13]. With 25 keV Kr^+ ions and incidence angles α varying from 0° to ~85°, the electron yields $\gamma(\alpha)$ were measured. $\gamma(\alpha)$ and $\gamma(\alpha) \cdot \cos(\alpha)$ were plotted showing deviations from the expected $1/\cos(\alpha)$ law at $\alpha > 42°$. For normal incidence ($\alpha = 0°$), $\gamma \approx 0.7$ was measured [14].

Energy Distribution (ED). Fig. 50 (see p. 230) gives the ED for 40 eV singly charged ions in the ground state [1]. It replaces the data given previously in [2] which were obtained with ion beams containing ~2% metastable Kr^+ ions. Significant differences exist in the shape of the ED curves for impact energies of 40 eV, 200 eV, and 1 keV. An extension of the distribution to higher electron energies is observed at higher ion energies. This was explained to result from the broader distribution of processes occurring closer to the metal as is the case where the ion approaches more rapidly [2].

Good agreement with the results of [2] was found for the ED of 15 eV Kr^+ ions [8] and 10 eV Kr^+ ions [15]. An ED curve shown in [16] was obtained with a 15 eV Kr^+ ion beam containing 2.4% metastables, which produced some structure in the ED above 5 eV [16]. The ED curves for 15 eV ground-state Kr^+ and $Kr^+ + 2.8\%\ Kr^{+*}$ beams were discussed in [8].

The ED of secondary electrons emitted from contaminated W foils under Kr^+ impact ($E_p \leq 80$ eV) was studied by [12].

Doubly Charged Kr Ions

Emission Yield. The emission yield as a function of E_p (0 to 1 keV) for Kr^{2+} ions impinging on a W ribbon is plotted in [2]. There is a monotonous decrease of the yield from about 0.33 near 0 eV to 0.25 at 1 keV [2]. An emission yield of $\gamma = 0.31$ was measured for 100 eV Kr^{2+} ions [4, 6]. A plot of $\gamma(E_p = 100$ eV) vs. ion-source voltage reveals a significant increase at the threshold voltages for metastable ion production [4].

Energy Distribution (ED). ED curves of electrons emitted during 15 eV Kr^{2+} bombardment are shown in [8, 16]. The ED from Kr^{2+} extends to a maximum energy of about 16 eV for the emitted secondary electrons [8, 16], which is somewhat more than the maximum energy of 13 eV reported by [17]. The structures seen in the curves suggest that the distributions result from several combined processes [8, 16]. With increasing E_p the structure in the curves disappears [8]. For a comparison of the Kr^{2+} ED curve with those of Kr^+, Kr^{3+}, and Kr^{4+} ions impinging with $E_p = 200$ eV, see [2].

Higher Charged Kr Ions. Kr^{3+} to Kr^{11+}

Emission Yield. The emission yields produced with Kr^{3+} and Kr^{4+} ions impinging on W with energies between 0 and 1 keV are plotted in [2]. γ decreases monotonously from about 0.90 and 2.00 at zero energy to about 0.75 and 1.68 at 1 keV for Kr^{3+} and Kr^{4+} ions, respectively [2]. $\gamma(E_p = 100$ eV$) = 0.90$ was determined for Kr^{3+} ions by [4].

The linear relationship between γ and W_q discussed above for multiply charged Ar ions (see p. 238) is also valid for Kr^{q+} ions with q=4 to 7 [18] and q=9 and 11 [19]. Deviations from this behavior at higher ion velocities are also discussed in [19].

Energy Distribution (ED). The ED curves produced by 200 eV $Kr^{(1 \text{ to } 4)+}$ ions are compared in [2]. For Kr^{3+} and Kr^{4+} the experimental ED curves extended to maximum energies of the emitted secondary electrons of about 18 and 24 eV, respectively [2], whereas the theoretically calculated maximum energies are 16.4 and 22 eV, respectively [17].

References for 5.8.2.4:

[1] Hagstrum, H. D. (Phys. Rev. [2] **104** [1956] 317/8).
[2] Hagstrum, H. D. (Phys. Rev. [2] **96** [1954] 325/35).
[3] Hagstrum, H. D. (Phys. Rev. [2] **104** [1956] 309/16).
[4] Varga, P.; Winter, H. (Phys. Rev. [3] A **18** [1978] 2453/8).
[5] Hagstrum, H. D. (Phys. Rev. [2] **96** [1954] 336/65).
[6] Varga, P.; Hofer, W.; Winter, H. (J. Phys. B **14** [1981] 1341/51).
[7] Hofer, W.; Vanek, W.; Varga, P.; Winter, H. (Rev. Sci. Instrum. **54** [1983] 150/7).
[8] Hofer, W. (Diss. TU Wien [Vienna] 1983, 103+4 pp.; Phys. Abstr. **88** [1985] No. 109084).
[9] Sternberg, D. (Diss. Columbia Univ. 1957, 60 pp.; Diss. Abstr. **17** [1957] 2292).
[10] Arifov, U. A.; Rakhimov, R. R. (Izv. Akad. Nauk SSSR Ser. Fiz. **24** [1960] 657/63; Bull. Acad. Sci. USSR Phys. Ser. [Engl. Transl.] **24** [1960] 666/72).

[11] Arifov, U. A.; Rakhimov, R. (Izv. Akad. Nauk UzSSR Ser. Fiz.-Mat. Nauk **1958** No. 6, pp. 49/55; C.A. **1960** 11711).
[12] Hagstrum, H. D. (J. Appl. Phys. **31** [1960] 897/904).
[13] Large, L. N. (Proc. Phys. Soc. [London] **81** [1963] 1101/3).
[14] Evdokimov, I. N.; Mashkova, E. S.; Molchanov, V. A.; Odintsov, D. D. (Phys. Status Solidi **19** [1967] 407/15).
[15] Varga, P.; Hofer, W.; Winter, H. (Scanning Electron. Microsc. **1982** No. 3, pp. 967/71).
[16] Varga, P.; Hofer, W.; Winter, H. (Surf. Sci. **117** [1982] 142/53).
[17] Arifov, U. A.; Kishinevskii, L. M.; Mukhamadiev, E. S.; Parilis, E. S. (Zh. Tekh. Fiz. **43** [1973] 181/7; Sov. Phys.-Tech. Phys. [Engl. Transl.] **18** [1973] 118/22).
[18] Arifov, U. A.; Mukhamadiev, E. S.; Parilis, E. S.; Pasyuk, A. S. (Zh. Tekh. Fiz. **43** [1973] 375/9; Sov. Phys.-Tech. Phys. [Engl. Transl.] **18** [1973] 240/3).
[19] Delaunay, M.; Fehringer, M.; Geller, R.; Hitz, D.; Varga, P.; Winter, H. (Phys. Rev. [3] B **35** [1987] 4232/5).

5.8.2.5 Xenon

5.8.2.5.1 Xe Atoms

Thermal beams ($E_p = 25$ to 200 eV) of metastable Xe* atoms showed only weak secondary electron emission from a W ribbon which was said to be clean (even though it yielded CN^- ions on impact of N_2 metastables) [1].

5.8.2.5.2 Xe Ions

Singly Charged Xe Ions

Emission Yield. The yields for Xe^+ ions in the ground state in the range 0 to 1 keV are given in Fig. 49 (see p. 229) [2]. In an earlier paper of the same author [3], results for beams containing about 2% of metastable ions Xe^{+*} were presented, showing yields about 30% higher than for ground-state ions; see the note added in proof [3, p. 335]. For $E_p = 40$ eV, $\gamma_p = 0.017$ was found for the mixed beam ($Xe^+ + 2\%\ Xe^{+*}$) [3], whereas $\gamma_p = 0.013$ was obtained for a pure ground-state ion beam [2, 4]. The same value is given for ground-

References for 5.8.2.5 on p. 245

state Xe^+ ions with an energy of 10 eV in [5, p. 673] and 15 eV in [6, p. 81]. The value $\gamma_p(E_p = 50\ eV) = 0.017$ given in [7] presumably was taken from [3]. For a 100 eV Xe^+ ion beam virtually free from metastables $\gamma_p = 0.013$ was obtained [8, 9]. From Hagstrum's theory of Auger electron emission [10] $\gamma_p(E_p = 40\ eV) = 0.012$ followed in good agreement with the experimental values. Another theoretical study of emission yields is given by [11].

The appearance potential for the production of metastable Xe^{+*} ions was investigated by [4, 9]. In both studies a strong increase in the electron emission yield was observed above a bombarding electron energy in the ion source of about 24 eV [4, 9]. The measurement of metastable ion beam fractions was discussed for various ion sources by [12]. The yield coefficient for metastable, singly charged ions at $E_p = 100$ eV is $\gamma_{1,m} = 0.153$ [8].

For the onset of kinetic electron emission, a threshold energy of about 3 keV was measured [13].

Energy Distribution (ED). ED curves for 15 eV ground-state Xe^+ and (Xe^+ + 2.2% Xe^{+*}) beams were measured by [6]. Good agreement with the results of [3] was obtained by [6] and [14] where 10 eV Xe^+ ions were used. For a comparison of the ED curves of Xe^+ with those of other singly charged noble gas ions, see Fig. 50, p. 230 [2]. The ED curves of electrons emitted from W by 200 eV $Xe^{(1\ to\ 5)+}$ ions were compared in the paper [3].

ED curves for different E_p between 10 and 300 eV are given for Xe^+ (with 2% Xe^{+*}) ions in [15]. The ED of the electrons allows one to distinguish pure ground-state ion beams from beams containing metastables [9, 15]; see also the results obtained for 15 eV ions in [6, p. 51] and a comparison between the ED curves of Xe^{2+} and (Xe^+ + 2.2% Xe^{+*}) beams in [6, p. 65]. Structures observed at higher energies are due to the first deexcitation step of the metastable ions [15].

The electron emission after collision of 50 to 1000 eV Xe^+ ions with W(110) under near-grazing incidence was found to be dominated by Auger capture processes involving two W 5d electrons [16].

ED curves of electrons produced by Xe^+ ions impinging at 2 to 8 keV on a W surface are presented in [13]. With increasing primary energy, the ED broadens and the maximum (centered at about 2.4 eV for $E_p = 2$ keV) shifts to higher energies (about 3.7 eV for $E_p = 8$ keV) [13].

Doubly Charged Xe Ions

Emission Yield. The yield coefficient of doubly charged ions was determined for 10 to 500 eV Xe^{2+} ion beams, virtually free from metastables. A value of $\gamma = 0.21$ was found for 100 eV Xe^{2+} ions [8, 9]. For a plot of γ vs. E_p (0 to 1 keV), see [3].

Energy Distribution (ED). An ED of electrons emitted by 200 eV Xe^{2+} ions is presented in [3]. A maximum energy for the emitted electrons of 10 eV was observed [3]. This value is in good agreement with the 10 eV limit calculated by [17]. A measured ED curve for electrons emitted by 15 eV Xe^{2+} ions is shown in the paper [18]. The ED is dominated by Auger deexcitation of Xe^{+*} ions which have an excitation energy of about 12 eV. The high-energy tail of the ED is explained by 20% resonance neutralization followed by Auger deexcitation to the ground state of the singly charged Xe ion ($E \approx 16$ eV) and by 10% Auger neutralization ($E \approx 21$ eV) [18]; see also [15]. A decrease in the ED structure with E_p increasing from 5 to 500 eV is shown for Xe^{2+} in [6, p. 65] supporting the view that the structures seen in the measured ED curves result from several combined processes [6].

Higher Charged Xe Ions. Xe^{3+} to Xe^{8+}

Emission Yield. The emission yields as a function of E_p for 0 to 1 keV Xe^{3+}, Xe^{4+}, and Xe^{5+} ions are given in [3]. γ monotonously decreases in all three cases [3]. The linear relationship between γ and W_q discussed above for multiply charged Ar ions (see p. 238) is also valid for Xe^{q+} (q = 3 to 8) ions [19].

Energy Distribution (ED). The ED curves of electrons emitted from W by 200 eV Xe^{3+}, Xe^{4+}, and Xe^{5+} ions are plotted in [3]. Maximum energies of the emitted secondary electrons of 15, 18, and 24 eV are cited in [17] from [3] and compared with the calculated values 14.3, 17.2, and 17 eV for Xe^{3+}, Xe^{4+}, and Xe^{5+} ions, respectively [17].

References for 5.8.2.5:

[1] Varney, R. N. (Phys. Rev. [2] **175** [1968] 98/102).
[2] Hagstrum, H. D. (Phys. Rev. [2] **104** [1956] 317/8).
[3] Hagstrum, H. D. (Phys. Rev. [2] **96** [1954] 325/35).
[4] Hagstrum, H. D. (Phys. Rev. [2] **104** [1956] 309/16).
[5] Hagstrum, H. D. (Phys. Rev. [2] **104** [1956] 672/83).
[6] Hofer, W. (Diss. TU Wien [Vienna] 1983, 103+4 pp.; Phys. Abstr. **88** [1985] No. 109084).
[7] Probst, F. M.; Lüscher, E. (Phys. Rev. [2] **132** [1963] 1037/46).
[8] Varga, P.; Hofer, W.; Winter, H. (J. Phys. B **14** [1981] 1341/51).
[9] Varga, P.; Winter, H. (Phys. Rev. [3] A **18** [1978] 2453/8).
[10] Hagstrum, H. D. (Phys. Rev. [2] **96** [1954] 336/65).

[11] Sternberg, D. (Diss. Columbia Univ. 1957, 60 pp.; Diss. Abstr. **17** [1957] 2292).
[12] Hofer, W.; Vanek, W.; Varga, P.; Winter, H. (Rev. Sci. Instrum. **54** [1983] 150/7).
[13] Bonanno, A.; Amoddeo, A.; Oliva, A. (Nucl. Instrum. Methods Phys. Res. B **46** [1990] 456/9).
[14] Varga, P.; Hofer, W.; Winter, H. (Scanning Electron. Microsc. **1982** No. 3, pp. 967/71).
[15] Varga, P.; Hofer, W.; Winter, H. (Surf. Sci. **117** [1982] 142/53).
[16] Brenten, H.; Müller, H.; Niehaus, A.; Kempter, V. (Surf. Sci. **278** [1992] 183/92).
[17] Arifov, U. A.; Kishinevskii, L. M.; Mukhamadiev, E. S.; Parilis, E. S. (Zh. Tekh. Fiz. **43** [1973] 181/7; Sov. Phys.-Tech. Phys. [Engl. Transl.] **18** [1973] 118/22).
[18] Hofer, W.; Varga, P. (Nucl. Instrum. Methods Phys. Res. B **2** [1984] 391/5).
[19] Arifov, U. A.; Mukhamadiev, E. S.; Parilis, E. S.; Pasyuk, A. S. (Zh. Tekh. Fiz. **43** [1973] 375/9; Sov. Phys.-Tech. Phys. [Engl. Transl.] **18** [1973] 240/3).

5.8.3 Electron Emission under Impact of Other Nonmetals

5.8.3.1 Hydrogen

Neutral Hydrogen Atoms and Molecules

Plots of γ vs. E_p (about 1 to 12 keV) are given for H, H_2, and H_3 beams impinging on W targets heated to between 600 and 800 °C [1]. The data conform well to the scaling law $\gamma(H_3) = 2\,\gamma(H_2) - \gamma(H)$, observed for positive ions by [2]. $\gamma(H_2) = 0.71$, 1.16, and 1.48 were determined for $E_p = 2$, 4, and 6 keV, respectively [1]. Despite the strong variation of γ with T, the γ values are constant between 600 and 800 °C for 1 keV H_2 beams [3]. With 25 to 200 eV beams of excited H_2^*, weak secondary electron currents and strong Lyman α emission were observed on a surface that was presumably contaminated with carbon (since the impact of N_2^* resulted in CN^- emission) [4]. From calculations using the secondary electron yield of $C_6H_6^+$, $\gamma(H) = 0.54$ was determined for v = 400 km/s [5].

References for 5.8.3.1 on pp. 247/8

Positive Hydrogen Ions

With H^+, H_2^+, and H_3^+ a linear dependence of γ on the ion velocity was observed for velocities exceeding the threshold value of $\sim$200 km/s. The curves of γ vs. velocity sloped least for H^+ and strongest for H_3^+ [6].

With respect to the yield, only the number of particles impinging on the surface was found to be relevant regardless of their bonding state. $\gamma(H_2^+) \approx 2\ \gamma(H^+)$ and $\gamma(H_3^+) \approx 3\ \gamma(H^+)$ were measured [7]. The yields found with 100, 200, 500, and 800 keV H^+, H_2^+, and H_3^+ ions conform to the scaling law $\gamma(H_3^+) = 2\ \gamma(H_2^+) - \gamma(H^+)$ (given in [2]) [8].

Electron emission spectra produced by slow H^+ and H_2^+ ions (E_p below 100 eV) impinging under grazing incidence on W(100) are shown in [9]. Auger capture processes involving two W surface electrons were found to dominate the ion-surface interaction [9].

Protons. Measurements on the secondary electron emission yield from W bombarded with H^+ ions were performed by [7 to 19]. A theoretical calculation gave a value of $\gamma = 0.035$ for slow protons impinging on a W surface [20, p. 39].

The E_p dependence of the yields from a W(100) surface [10, 11] was compared with the values obtained for a polycrystalline surface [12] for ion energies ranging from 50 to 225 keV. For the W(100) surface as well as for the polycrystalline surface, the yield peaks at $E_p = 125$ keV (γ(W(100)) = 1.51 and γ(polycr. W) = 1.65). For all ion energies the yield on polycrystalline W is about 10% higher than on the (100) surface, which is attributed to the higher work function of W(100) [10, 11]. In [13] ($E_p = 32$ to 127 keV) a maximum in the yield curve ($\gamma \approx 1.6$) was also observed; it is located at about 100 keV [13].

From a figure in paper [8, p. 557], $\gamma \approx 1.69$, 1.57, 1.24, and 0.99 for $E_p = 100$, 200, 500, and 800 keV, respectively, were extracted. A value of $\gamma \approx 1.67$ for $E_p = 100$ keV was measured by [14]; see also the review [15]. Plots of γ vs. ion velocity are given in [13] and [16, p. 353]. A linear relationship between the electronic stopping power $(dE/dx)_e$ [in eV/Å] and the yield γ [in electrons/ion] of the form $\gamma = 0.07 \cdot (dE/dx)_e$ was found for H^+ ions with E_p ranging from 80 to 1000 keV [16]. This was refined to $\gamma/(dE/dx)_e = 0.074 \pm 0.004$ in [8].

The energy distribution of the secondary electrons emitted by 4 to 7 keV protons and the dependence on the incidence angle α was studied by [17]. The energy distribution curve extends beyond $\sim$40 eV and peaks near 1 eV [17].

H_2^+ Molecular Ions. The interaction of H_2^+ ions with W was studied in [7 to 9, 13, 16, 21 to 25]. $\gamma(H_2^+) \approx 0.026$, 0.03, and 0.04 were obtained at $E_p = 50$, 100, and 200 eV, respectively [21, 22]. These results on polycrystalline W are in good agreement with the W(100) data of [23] where, however, a stronger focusing of the emitted electrons along the surface normal was found [23, 24]. When a W target, heated to just below the onset of thermionic emission, was bombarded with H_2^+ ions, γ increased from about 0.3 near 3 keV to about 1 for 15 keV ion impact energy. At room temperature slightly higher γ values were obtained [7]. The γ vs. T dependence of H_2^+ ions impinging at normal incidence onto W(111) is plotted in [25] between $\sim$1700 and $\sim$2100 K and for $E_p = 4$ to 16 keV. γ slightly increases with T. At 1700 K γ becomes 0.3, 0.54, 0.75, and 0.9 for $E_p = 4$, 8, 12, and 16 keV, respectively [25].

The yield of H_2^+ ions is about twice the yield of H^+ ions having the same velocity. For a comparison of the yields of H^+ and H_2^+ ions, measured at $E_p \leq 140$ keV, see [13]. Similarly, yields are compared by a γ vs. ion velocity plot in the energy range from 80 to 1000 keV [16]. $\gamma \approx 3.19$ was measured for 200 keV H_2^+ ions [14]. Some of the yield data measured by [7, 13, 22] are listed in the data compilation [26].

Energy distributions of secondary electrons induced by slow (E_p below about 200 eV) H_2^+ ions were investigated by [9, 22 to 24]. The shape of the distribution curves of electrons emitted from W(110) [9], W(100) [23, 24], and polycrystalline W [22] show no significant differences. The spectra are dominated by Auger capture processes involving two 5d W surface electrons [9].

H_3^+ Molecular Ions. $\gamma \approx 4.56$ was obtained for 300 keV H_3^+ ions [14]. γ was measured for 1 to 30 keV H_3^+ ions and compared with the yields measured for H^+ and H_2^+ ions [7].

Negative Hydrogen Ions

The theory by [27] on kinetic electron emission is extended to the case of negative ions in [28]. The calculated γ vs. velocity plots for H^- ions give a threshold velocity of $v_{threshold} \approx 100$ km/s and a slope of $d\gamma/dv \approx 2.7 \times 10^{-7}$ (electrons/ion)/(m/s) [28].

References for 5.8.3.1:

[1] Devienne, F. M. (J. Phys. [Paris] **28** [1967] 602/10).
[2] Hasselkamp, D.; Hippler, S.; Scharmann, A.; Schartner, K.-H. (Z. Phys. D **6** [1987] 269/74).
[3] Devienne, F. M. (C. R. Séances Acad. Sci. B **264** [1967] 232/4).
[4] Varney, R. N. (Phys. Rev. [2] **175** [1968] 98/102).
[5] Beuhler, R. J.; Friedman, L. (J. Appl. Phys. **48** [1977] 3928/36, 3929, 3933/4).
[6] Petrov, N. N. (Izv. Akad. Nauk SSSR Ser. Fiz. **26** [1962] 1327/32; Bull. Acad. Sci. USSR Phys. Ser. [Engl. Transl.] **26** [1962] 1350/4).
[7] Dorozhkin, A. A.; Petrov, N. N. (Zh. Tekh. Fiz. **33** [1963] 350/5; Sov. Phys.-Tech. Phys. [Engl. Transl.] **8** [1963] 257/61).
[8] Hasselkamp, D.; Hippler, S.; Scharmann, A.; Schmehl, T. (Ann. Phys. [Leipzig] [7] **47** [1990] 555/67).
[9] Müller, H.; Hausmann, R.; Brenten, H.; Kempter, V. (Surf. Sci. **284** [1993] 129/38).
[10] Ewing, R. I. (SC-DC-1559 [1966] 7 pp.; C.A. **68** [1968] No. 34187).

[11] Ewing, R. I. (Phys. Rev. [2] **166** [1968] 324/5).
[12] Ewing, R. I. (Phys. Rev. [2] **139** [1965] A 1840/A 1843).
[13] Large, L. N. (Proc. Phys. Soc. [London] **81** [1963] 1101/3).
[14] Hippler, S.; Hasselkamp, D.; Scharmann, A. (Nucl. Instrum. Methods Phys. Res. B **34** [1988] 518/20).
[15] Hasselkamp, D. (Comments At. Mol. Phys. **21** [1988] 241/55, 248).
[16] Hasselkamp, D.; Lang, K. G.; Scharmann, A.; Stiller, N. (Nucl. Instrum. Methods Phys. Res. **180** [1981] 349/56).
[17] Burrekoven, H. (Diss. Univ. Köln [Cologne, FRG] 1956, 34 + XIV pp.).
[18] Gaukler, K.-H. (Z. Metallk. **51** [1960] 463/6).
[19] Fogel', Ya. M.; Slabopitskii, R. P.; Rastrepin, A. B. (Zh. Tekh. Fiz. **30** [1960] 63/73; Sov. Phys.-Tech. Phys. [Engl. Transl.] **5** [1960] 58/66).
[20] Sternberg, D. (Diss. Columbia Univ. 1957, 60 pp.; Diss. Abstr. **17** [1957] 2292).

[21] Probst, F. M. (Diss. Univ. Illinois 1963, 125 pp.; Diss. Abstr. **24** [1964] 3809).
[22] Probst, F. M.; Lüscher, E. (Phys. Rev. [2] **132** [1963] 1037/46).
[23] Prince, R. H. (AD-675206 [1968] 120 pp.; C.A. **70** [1968] No. 81182).
[24] Prince, R. H.; French, J. B. (AD-682373 [1968] 130 pp.; C.A. **71** [1968] No. 16846).
[25] Dorozhkin, A. A.; Petrov, N. N. (Fiz. Tverd. Tela [Leningrad] **16** [1974] 800/3; Sov. Phys.-Solid State [Engl. Transl.] **16** [1974] 517/9).

[26] Barnett, C. F.; Ray, J. A.; Ricci, E.; Wilker, M. I.; McDaniel, E. W.; Thomas, E. W.; Gilbody, H. B. (ORNL-5207 [1977]).
[27] Parilis, E. S.; Kishinevskii, L. M. (Fiz. Tverd. Tela [Leningrad] **3** [1961] 1219/28; Sov. Phys.-Solid State [Engl. Transl.] **3** [1961] 885/91).
[28] Ghosh, S. N.; Majumdar, B. K. (Indian J. Phys. A **54** [1980] 246/57, 254).

5.8.3.2 Oxygen

Positive Oxygen Ions

The yields of O^+ and O_2^+ ions are compared in [1]. γ increases from about 2.5 to 5.0 for O^+ and from about 3.0 to 6.8 for O_2^+ when E_p increases from 32 to 127 keV. $\gamma(O_2^+) \approx 2\gamma(O^+)$ results for the same ion velocity [1]. Some yield values for O^+ and O_2^+ ion impact taken from the curves in [1] are listed in [2]. $\gamma(O_2^+) \approx 0.011$, 0.013, and 0.015 was determined for 50, 100, and 200 eV, respectively [3]. This was slightly revised to $\gamma(O_2^+) \approx 0.011$, 0.015, and 0.017 for $E_p = 50$, 100, and 200 eV in [4], where also energy distributions of the secondary electrons emitted by O_2^+ ions are plotted [4]. For ion beams consisting of (O^+ plus O_2^{2+}) mixtures, $\gamma \approx 0.08$, 0.09, and 0.12 was obtained for 50, 100, and 200 eV, respectively [3].

Energy distribution curves of secondary electrons emitted from W by impact of 100 eV O^{q+} ions (q = 3 to 7) are shown in [5]. The position of the maximum increases from $E_s \approx 25$ eV for O^{3+} to about 60 eV for O^{7+} [5].

Negative Oxygen Ions

By extending the theory of kinetic electron emission [6] to the case of negative ions, γ vs. ion velocity plots were calculated for O^- ions, indicating a threshold velocity of $v_{threshold} \approx 100$ km/s for electron emission. The slope of the calculated γ vs. v plot is about 1.8×10^{-6} (electrons/ion)/(m/s) [7].

References for 5.8.3.2:

[1] Large, L. N. (Proc. Phys. Soc. [London] **81** [1963] 1101/3).
[2] Barnett, C. F.; Ray, J. A.; Ricci, E.; Wilker, M. I.; McDaniel, E. W.; Thomas, E. W.; Gilbody, H. B. (ORNL-5207 [1977]).
[3] Probst, F. M. (Diss. Univ. Illinois 1963, 125 pp.; Diss. Abstr. **24** [1964] 3809).
[4] Probst, F. M.; Lüscher, E. (Phys. Rev. [2] **132** [1963] 1037/46).
[5] Folkerts, L.; Morgenstern, R. (Europhys. Lett. **13** [1990] 377/82).
[6] Parilis, E. S.; Kishinevskii, L. M. (Fiz. Tverd. Tela [Leningrad] **3** [1961] 1219/28; Sov. Phys.-Solid State [Engl. Transl.] **3** [1961] 885/91).
[7] Ghosh, S. N.; Majumdar, B. K. (Indian J. Phys. A **54** [1980] 246/57, 254).

5.8.3.3 Nitrogen

Neutral Nitrogen Atoms and Molecules

For plots of γ vs. E_p (about 1 to 9 keV) of N-, N_2-, and N_3-induced electron emission, see [1]. For neutral N_2 molecules impinging on W polycrystals heated to between 600 and 800 °C, $\gamma = 0.45$, 0.89, and 1.34 were obtained at $E_p = 2$, 4, and 6 keV, respectively [1]. The temperature dependence of $\gamma(E_p = 1$ keV) in the range 0 to 1300 °C was studied on a polycrystalline W surface. There was a strong variation of γ with T, but $\gamma(N_2)$ was constant between 600 and 800 °C [2]. A W ribbon was exposed to slow beams of excited N_2^* molecules. The surface of the ribbon used presumably was contaminated with carbon (since CN^- ions were emitted under N_2^* impact). No electron emission was observed in that study [3].

Nitrogen Ions

N^+, N_2^+, and N_2^{2+} Ions. The interaction of N^+ and N_2^+ ions with W surfaces was investigated in [4 to 12]. Pertinent work performed at the Leningrad Polytechnical Institute was reviewed by [13].

$\gamma(N^+) \approx 0.035$, 0.038, 0.028, and 0.032 was measured on W(100) for $E_p = 15$, 20, 25, and 30 eV, respectively [4, 5]. In regards to N_2^+ impact on W(100), good agreement with the data from a study on polycrystalline W [6] was stated by [4, 5]. However, the values measured by [4] indicate a stronger focusing of the emitted electrons along the surface normal than had been found by [6]; see [4]. For (N^+ and N_2^{2+}) mixtures that could not be separated, $\gamma \approx 0.0095$ and 0.011 were obtained for ion impact energies of 50 to 100 eV and of 200 eV, respectively [7]. $\gamma(N_2^+) \approx 0.025$, 0.023, and 0.022 were measured by the same author for $E_p = 50$, 100, and 200 eV, respectively [6, 7].

The plots of γ vs. E_p (1 to 10 keV) for N^+ and N_2^+ nearly coincide according to [8]. In contrast, the yield induced by N_2^+ ions in the range 32 to 140 keV is significantly larger than that for N^+ ions. $\gamma(N^+)$ increases from about 2.3 to 4.8 and $\gamma(N_2^+)$ from about 2.8 to 7.0 for $E_p = 32$ to 140 keV. For the same ion velocity, γ is approximately twice as large for N_2^+ ions than for N^+ ions [9]. $\gamma(N^+) \approx 2.9$ was found for an impact energy of 30 keV and normal incidence [10]. $\gamma(N_2^+) = 2$ was obtained for an ion impact energy of 20 keV [8, 13]. For 5 keV ions "from the air", $\gamma = 2$ was measured [11].

$d\gamma/dE_p \approx 0.135$ (electrons/ion)/keV was found for N^+ and N_2^+ in the range of kinetic emission studied up to 20 keV [8, 13].

The energy distribution, obtained with 15 eV N_2^+ ions immediately after heating the sample to ~2300 K, is considerably sharper [4] than the distribution obtained by [6] using 50 eV ions [6]. The difference may be due to the difference in E_p or due to the different samples used (W(100) vs. polycrystal) [4, 5].

$\gamma(\alpha)$ and $\gamma(\alpha) \cdot \cos(\alpha)$ are plotted for N^+ ions. There are significant deviations from the $1/\cos(\alpha)$ law for $\alpha > 49°$ [10].

Higher Charged N Ions. N^{4+} to N^{6+}. The yield induced by $N^{(4 to 6)+}$ ions was discussed by [14 to 16]. The energy distribution produced by these ions is dealt with in [14, 17, 18]. No physically relevant information was obtained for W surfaces. The electron emission processes are dominated by ion-specific neutralization processes. For a review on such neutralization processes, see e.g. [14].

References for 5.8.3.3:

[1] Devienne, F. M. (J. Phys. [Paris] **28** [1967] 602/10).
[2] Devienne, F. M. (C. R. Séances Acad. Sci. B **264** [1967] 232/4).
[3] Varney, R. N. (Phys. Rev. [2] **175** [1968] 98/102).
[4] Prince, R. H. (AD-675206 [1968] 120 pp.; C.A. **70** [1968] No. 81182).
[5] Prince, R. H.; French, J. B. (AD-682373 [1968] 130 pp.; C.A. **71** [1968] No. 16846).
[6] Probst, F. M.; Lüscher, E. (Phys. Rev. [2] **132** [1963] 1037/46).
[7] Probst, F. M. (Diss. Univ. Illinois 1963, 125 pp.; Diss. Abstr. **24** [1964] 3809).
[8] Petrov, N. N.; Dorozhkin, A. A. (Fiz. Tverd. Tela [Leningrad] **3** [1961] 53/60; Sov. Phys.-Solid State [Engl. Transl.] **3** [1961] 38/43).
[9] Large, L. N. (Proc. Phys. Soc. [London] **81** [1963] 1101/3).

[10] Evdokimov, I. N.; Mashkova, E. S.; Molchanov, V. A.; Odintsov, D. D. (Phys. Status Solidi **19** [1967] 407/15).

[11] Slodzian, G. (C. R. Hebd. Séances Acad. Sci. **246** [1958] 3631/4).
[12] Gaukler, K.-H. (Z. Metallk. **51** [1960] 463/6).
[13] Petrov, N. N. (Izv. Akad. Nauk SSSR Ser. Fiz. **26** [1962] 1327/32; Bull. Acad. Sci. USSR Phys. Ser. [Engl. Transl.] **26** [1962] 1350/4).
[14] Varga, P. (Appl. Phys. A **44** [1987] 31/41).
[15] Fehringer, M.; Delaunay, M.; Geller, R.; Varga, P.; Winter, H. (Nucl. Instrum. Methods Phys. Res. B **23** [1987] 245/7).
[16] Delaunay, M.; Fehringer, M.; Geller, R.; Hitz, D.; Varga, P.; Winter, H. (Phys. Rev. [3] B **35** [1987] 4232/5).
[17] Delaunay, M.; Fehringer, M.; Geller, R.; Varga, P.; Winter, H. (Europhys. Lett. **4** [1987] 377/82).
[18] Folkerts, L.; Morgenstern, R. (Europhys. Lett. **13** [1990] 377/82).

5.8.3.4 Halogens

Fluorine. In an early paper [1] F^+ ions were impinged at $E_p=0.3$ to 1.25 keV under near normal incidence on samples that were considered not to be clean (although degassed). $\gamma \approx 0.315$ was obtained for an ion impact energy of 0.78 keV [1]. A γ value of 2.95 assumed for atomic fluorine impinging at 400 km/s was consistent with the yields measured with $C_{10}F_{17}^+$ ions [2].

Chlorine. Negative ions have been used by [1] and [3] and considered in a theoretical study [4]; positive ions were used by [1]. In a study at $E_p=0.4$ to 6.5 keV using $^{35}Cl^-$ ions, a negligible electron yield γ was found below the threshold energy $E_p=0.8$ keV. γ linearly increased from about 0 at 0.8 keV to 0.22 at 6.5 keV ($d\gamma/dE_{kin} \approx 0.0386$ [electrons/ion]/keV) [3]. Fair agreement with the $K^+ \rightarrow W$ data of [5] was obtained [3]. Plots of γ vs. ion velocity were calculated by [4], extending the theory on kinetic emission developed by [6]. These plots indicate a threshold velocity for electron emission of $v_{threshold} \approx 100$ km/s and a slope of $d\gamma/dv \approx 3.8 \times 10^{-6}$ (electrons/ion)/(m/s) for Cl^- ions [4]. At $E_p=0.78$ keV, $\gamma(Cl^-)=0.79$ and $\gamma(Cl^+)=0.49$ were measured (note the qualifying remark with F^+) [1].

Iodine. I^+ and I^- ion beams with energies between 0.3 and 1.25 keV were directed on a polycrystalline W surface. For $E_p=0.78$ keV, $\gamma(I^+)=0.15$ and $\gamma(I^-)=0.26$ were obtained (note the qualifying remark with F^+) [1].

References for 5.8.3.4:

[1] Zandberg, E. Ya. (Zh. Tekh. Fiz. **25** [1955] 1386/98; C.A. **1956** 1449).
[2] Beuhler, R. J.; Friedman, L. (J. Appl. Phys. **48** [1977] 3928/36, 3929, 3933/4).
[3] Arifov, U. A.; Kashimov, N. M. (Radiotekh. Elektron. [Moscow] **8** [1963] 316/21; Radio Eng. Electron. Phys. [Engl. Transl.] **8** [1963] 274/9).
[4] Ghosh, S. N.; Majumdar, B. K. (Indian J. Phys. A **54** [1980] 246/57, 254).
[5] Petrov, N. N. (Izv. Akad. Nauk SSSR Ser. Fiz. **24** [1960] 673/8; Bull. Acad. Sci. USSR Phys. Ser. [Engl. Transl.] **24** [1960] 682/7).
[6] Parilis, E. S.; Kishinevskii, L. M. (Fiz. Tverd. Tela [Leningrad] **3** [1961] 1219/28; Sov. Phys.-Solid State [Engl. Transl.] **3** [1961] 885/91).

5.8.3.5 Carbon

$\gamma=2.51$ for atomic C was consistent with the secondary electron yield of complex ions like $C_6H_6^+$ striking a W surface at a velocity of 400 km/s [1]. Energy distributions of the electrons emitted under impact of the hydrogenlike ion C^{5+} were recorded for $E_p=150$ eV [2] and 6 keV [3].

References for 5.8.3.5:

[1] Beuhler, R. J.; Friedman, L. (J. Appl. Phys. **48** [1977] 3928/36, 3929, 3933/4).
[2] Folkerts, L.; Morgenstern, R. (Europhys. Lett. **13** [1990] 377/82).
[3] de Zwart, S. T.; Drentje, A. G.; Boers, A. L.; Morgenstern, R. (Surf. Sci. **217** [1989] 298/316).

5.8.3.6 Nonmetal Compounds

The electron emission yield of a W target under impact of $C_6H_6^+$ and $C_{10}F_{17}^+$ ions was measured for ion velocities $v\approx300$ to 450 and 90 to 180 km/s, respectively. The fairly linear γ vs. v plots indicate a threshold velocity for kinetic emission around 50 km/s for both ions. $\gamma(C_6H_6^+)$ increases from ~12.5 to ~20.5 and $\gamma(C_{10}F_{17}^+)$ increases from ~11 to ~32.5 in the velocity ranges studied [1].

Reference for 5.8.3.6:

[1] Beuhler, R. J.; Friedman, L. (J. Appl. Phys. **48** [1977] 3928/36).

5.8.4 Electron Emission under Impact of Metals

5.8.4.1 Lithium

The yield under impact of neutral Li atoms increased linearly from about 2.2 at $E_p=20$ keV to about 4.2 at 60 keV. It was about 10 to 20% larger than that of Li^+ ions (increasing from about 2 at 20 keV to 3.5 at 60 keV). Much larger yields ($\gamma=4$ to 6) were obtained on an oxide-covered surface [1].

No electron emission was observed for 1 keV Li^+ ions incident at $\alpha=5°$ on W(110) [2]. Measurements on a W sample heated to 1300 K indicated a threshold energy for the onset of kinetic electron emission of $E_{threshold}\approx1.75$ keV when the 2 to 6 keV data were extrapolated to small energy values. Between 2 and 6 keV, γ increased linearly at a rate of $d\gamma/dE_p\approx0.144$ (electrons/ion)/keV. At 293 K, however, the obviously contaminated sample showed no threshold and had a much higher yield than the hot surface [3]. γ was measured for Li^+ impact energies between 150 and 1500 eV. Electron emission probabilities increase from about 10^{-6} to 6.2×10^{-2}. No threshold was observed within this energy range [4].

The energy distribution of electrons emitted by 1 keV Li^+ ions (incident at 70°) is plotted in [4]. It peaks near 1.5 eV with ~0.009 (electrons/ion)/eV, the FWHM being about 2 eV. The high-energy tail of the distribution extends to near 11 eV [4].

5.8.4.2 Sodium

The yields under 0.3 to 1.25 keV impact of Na^- and Na^+ ions were measured by [5] on surfaces that were considered not to be clean. The values $\gamma=0.41$ and 0.25 were obtained for 0.78 keV Na^- and Na^+ ions, respectively [5]. A plot of γ vs. E_p for 0.05 to 10 keV Na^+ ions showed a threshold for secondary electron emission near 1 keV for a W target kept at about 1300 K [6].

References for 5.8.4 on p. 254

5.8.4.3 Potassium

K^+ ions at $E_p=1$ to 6 keV were impinged under normal incidence on a W foil. Care was taken to avoid effects of implanted (entrapped) K. By extrapolating back from energies $E_p \leq 1.8$ keV, a threshold for secondary electron emission of $E_{threshold}=0.82$ keV was deduced, but back-extrapolating from the linear part of the yield curve between $E_p=3$ and 6 keV placed the intercept with the E_p axis at 1.8 keV [7]. Earlier studies [6, 8 to 12] yielded threshold values consistent with this higher value: plots of γ vs. E_p for 0.05 to 10 keV indicated $E_{threshold} \approx 1.5$ keV [6]; by linear back extrapolation of the γ vs. E_p curve, $E_{threshold}=1.8$ keV was obtained for a target temperature of 1450 K in a study covering the range from $E_p=1$ to 8 keV [8]; $E_{threshold} \approx 1.6 \pm 0.1$ keV [9], 1.7 keV [10], and 1.5 keV [11, 12] resulted from studies performed with ion impact energies of up to 6 keV at $T \approx$ 1500 K by Petrov [9 to 12]; (the value of ~2.5 keV found at 1200 K by [13] is obviously obsolete). At 1800 K and $E_p=0.84$ keV, only thermionic emission from the filament was obtained rather than secondary electron emission [14]. The discrepancy between $E_{threshold}=0.82$ keV [7] and the about twice as large earlier results just cited might be due to potassium entrapped below the topmost surface layer; see the discussion in [7].

No potential emission was found when alkali ions impinged on W. For both K^+ and Ar^+ ions impinging at $E_p>1.7$ keV, Petrov [12] measured a linear increase in γ with E_p up to 6 keV (the highest E_p value studied) with $d\gamma/dE_p \approx 0.058$ (electrons/ion)/keV for a W sample heated to 1900 K [12] ($d\gamma/dE_p \approx 0.057$ [9], 0.053 [15], and 0.06 [10] (electrons/ion)/keV were given in other papers of the same author). When corrected for the potential emission coefficient of Ar^+ (assumed to be 10%), the γ vs. E_p curves for Ar^+ and K^+ fall into a single curve according to Petrov [12, 15]; see also his review [11]. A slope of $d\gamma/dE_p \approx 0.052$ (electrons/ion)/keV resulted for E_p between 3 and 6 keV [7]. $d\gamma/dE_p \approx 0.063$ (electrons/ion)/keV was given in the earlier paper [13] which formed the experimental basis for the discussion of kinetic emission by [16]. The yields measured at 1450 K using 1 to 8 keV K^+ ions impinging under $\alpha=45°$ were somewhat larger than the results obtained for normal incidence [8].

5.8.4.4 Rubidium

A plot of γ vs. E_p for 0.05 to 10 keV Rb^+ ions impinging on a W target kept at ~1300 K indicates a threshold for secondary electron emission of $E_{threshold} \approx 2$ keV [6]. This agrees with the earlier observation where no secondary electrons were emitted under impact of 0.3 keV Rb^+ ions [17].

5.8.4.5 Caesium

A threshold energy $E_{threshold}=2.5$ keV for kinetic emission under Cs^+ ion bombardment was found by backward extrapolation in a study covering the range between $E_p=3$ and 21 keV [18]. This agrees well with the value $E_{threshold}=2.7$ keV found in a study performed with the target at 1450 K and ion impact energies from 1 to 8 keV [8]. In a very early paper [19] yields of $\gamma=0.012$ at $E_p=0.5$ keV and ~0.06 at 1 keV were found. No threshold was observed in [19] and in studies at $E_p=0.15$ to 1.5 keV [4, 20].

Between 3 and 21 keV, γ increased strictly linearly with E_p from $\gamma=0.02$ at 3 keV to $\gamma=0.74$ at 21 keV with a slope of $d\gamma/dE_p=0.04$ (electrons/ion)/keV [18]. $d\gamma/dE_p=0.045$ (electrons/ion)/keV follows from data measured at $\alpha=45°$ [8].

The energy distribution from Cs^+-induced electron emission is similar to the energy distribution produced by Li^+ impact, but the high-energy tail of the Cs^+-induced spectrum is somewhat shorter than in the Li^+ case [4].

5.8.4.6 Calcium

The secondary electron emission under Ca^+ ion impact for $E_p \lesssim 20$ keV and normal incidence was studied by [21]. In the range of kinetic emission, a slope of $d\gamma/dE_p \approx 0.073$ (electrons/ion)/keV was obtained [21]. $d\gamma/dE_p = 0.072$ (electrons/ion)/keV is given in the review [11].

5.8.4.7 Zinc

The secondary electron emission from a W sample showed hardly any dependence on E_p in the range studied. $\gamma(Zn^+) \approx 0.016$ [22] (0.016 to 0.019 [23]) was measured between 0.4 and 1.6 keV.

5.8.4.8 Cadmium

Using 0.3 to 1.0 keV Cd^+ ions, the secondary electron emission from W was investigated. As in the case of Zn^+, hardly any E_p dependence was noted. $\gamma(Cd^+) \approx 0.012$ was measured between 0.3 and 1 keV [22, 23].

5.8.4.9 Mercury

Metastable Hg* atoms were thermally diffused to a W surface heated to above 1800 K in the early study [24]. It was concluded that no more than one or two metastable atoms out of a hundred caused the emission of an electron [24].

Hg^+ ions impinging on polycrystalline surfaces were used by [25 to 27]. In addition to the secondary yield by Hg^+ ions, the yields under Hg^{2+} and Hg^{3+} ion impact were measured in [25]. The impact energy ranges studied were 0.5 to 2.8 keV [26], 1 to 10 keV [25], and 700 to 2500 keV [27].

With the sample at 2100 K, $\gamma(Hg^+) \approx 0.026$ was obtained at 0.5 to 2.8 keV. At $E_p = 1$ keV, γ increased from about 0.011 at 300 K gradually to about 0.015 near 1500 K, then the γ vs. T curves sloped upward more strongly. This T dependence of γ was believed not to be due to contamination, since flashing the sample immediately before the data acquisition did not change the results. In addition, the Ar^+ data obtained in this study agreed with published γ values for clean W. While $\gamma(Hg^+)$ was tripled on heating from 300 to ~2000 K, $\gamma(Ar^+)$ remained constant [26]. W samples, heated at 1100 K, were studied by [25]. For $E_p \leq 3$ keV, γ was constant (~0.036). At higher E_p values up to 10 keV, γ increased at a rate of ~0.007 (electrons/ion)/keV [25]. Between 700 and 2350 keV, γ increased linearly with E_p [27]. $\gamma(Hg^+) = \gamma(Hg^{2+}) - 18\% = \gamma(Hg^{3+}) - 35\%$ was obtained for an Ni sample; W was said to behave similarly [25].

5.8.4.10 Bismuth

The secondary electron yield of polycrystalline W samples under Bi^+ and Bi^- ion impact under near normal incidence was measured at $E_p = 300$ to 1250 eV, but the degassed samples

References for 5.8.4 on p. 254

were considered to be not clean. Yield values obtained for $E_p = 780$ eV were $\gamma = 0.06$ and 0.18 for Bi^+ and Bi^-, respectively [5].

References for 5.8.4:

[1] Bethge, K.; Lexa, P. (Br. J. Appl. Phys. **17** [1966] 181/6).
[2] Brenten, H.; Müller, H.; Knorr, K. H.; Kruse, D.; Schall, H.; Kempter, V. (Surf. Sci. **243** [1991] 309/16).
[3] Eremeev, M. A.; Shestukhina, V. V. (Zh. Tekh. Fiz. **22** [1952] 1268/70).
[4] Waters, P. M. (Phys. Rev. [2] **111** [1958] 1053/61).
[5] Zandberg, E. Ya. (Zh. Tekh. Fiz. **25** [1955] 1386/98; C.A. **1956** 1449).
[6] Arifov, U. A.; Rakhimov, R. (Izv. Akad. Nauk UzSSR Ser. Fiz.-Mat. Nauk **1958** No. 6, pp. 49/55; C.A. **1960** 11711).
[7] Breunig, J. L. (Phys. Rev. [3] B **6** [1972] 687/94).
[8] Makarov, V. V.; Petrov, N. N. (Izv. Akad. Nauk SSSR Ser. Fiz. **28** [1964] 1416/22; Bull. Acad. Sci. USSR Phys. Ser. [Engl. Transl.] **28** [1964] 1324/9).
[9] Petrov, N. N. (Nauchno-Tekh. Inform. Byull. Leningr. Politekh. Inst. **1960** No. 3, pp. 63/71).
[10] Petrov, N. N. (Fiz. Tverd. Tela [Leningrad] **2** [1960] 949/58; Sov. Phys.-Solid State [Engl. Transl.] **2** [1960] 865/71).

[11] Petrov, N. N. (Izv. Akad. Nauk SSSR Ser. Fiz. **26** [1962] 1327/32; Bull. Acad. Sci. USSR Phys. Ser. [Engl. Transl.] **26** [1962] 1350/4).
[12] Petrov, N. N. (Izv. Akad. Nauk SSSR Ser. Fiz. **24** [1960] 673/8; Bull. Acad. Sci. USSR Phys. Ser. [Engl. Transl.] **24** [1960] 682/7).
[13] Eremeev, M. A.; Shestukhina, V. V. (Zh. Tekh. Fiz. **22** [1952] 1262/7).
[14] Arifov, U. A.; Ayukhanov, A. Kh.; Starodubtsev, S. V.; Khadzhimukhamedov, Kh. Kh. (Dokl. Akad. Nauk SSSR **124** [1959] 60/2; Sov. Phys.-Dokl. [Engl. Transl.] **4** [1959] 86/8).
[15] Petrov, N. N. (Fiz. Tverd. Tela [Leningrad] **2** [1960] 1300/7; Sov. Phys.-Solid State [Engl. Transl.] **2** [1960/61] 1182/8).
[16] Izmailov, S. V. (Zh. Tekh. Fiz. **28** [1958] 2209/16; Sov. Phys.-Tech. Phys. [Engl. Transl.] **3** [1958] 2031/7).
[17] Arifov, U. A.; Ayukhanov, A. Kh. (Izv. Akad. Nauk SSSR Ser. Fiz. **20** [1956] 560/1; Bull. Acad. Sci. USSR Phys. Ser. [Engl. Transl.] **35** [1971] 513/5).
[18] Bosch, S. H.; Kuskevics, G. (Phys. Rev. [2] **134** [1964] A 1356/A 1358).
[19] Paetow, H.; Walcher, W. (Z. Phys. **110** [1938] 69/83).
[20] Waters, P. M. (Phys. Rev. [2] **109** [1958] 1466/7).

[21] Petrov, N. N.; Dorozhkin, A. A. (Fiz. Tverd. Tela [Leningrad] **3** [1961] 53/60; Sov. Phys.-Solid State [Engl. Transl.] **3** [1961] 38/43).
[22] Petrov, N. N. (Nauchno-Tekh. Inform. Byull. Leningr. Politekh. Inst. **1960** No. 3, pp. 46/52).
[23] Petrov, N. N. (Fiz. Tverd. Tela [Leningrad] **2** [1960] 940/8; Sov. Phys.-Solid State [Engl. Transl.] **2** [1960] 857/64).
[24] Sonkin, S. (Phys. Rev. [2] **43** [1933] 788/803).
[25] Dorozhkin, A. A. (Tr. Leningr. Politekh. Inst. im. M. I. Kalinina No. 277 1966 [1967] 89/92; C.A. **69** [1968] No. 62694).
[26] Dorozhkin, A. A.; Petrov, N. N. (Fiz. Tverd. Tela [Leningrad] **7** [1965] 118/22; Sov. Phys.-Solid State [Engl. Transl.] **7** [1965] 88/91).
[27] Linford, L. H. (Phys. Rev. [2] **47** [1935] 279/82).

6 Field Evaporation

6.1 Survey

Field desorption was first observed with Ba adsorbed on W and described in the paper [1]. A direct proof for the "evaporation" phenomenon in field evaporation was obtained when a second tip was held over a tip subjected to a high field. The second tip showed craters which could have been produced only by accelerated ions and not by thermal atoms. The reduction of the first tip by the applied field, confirmed by electron and light microscopy, was due to evaporation into the space above the tip (there was clearly no migration off the apex along the tip shank) [2].

The term field desorption is used to describe the desorption of any atom by the action of an applied field, both for foreign adatoms and for substrate atoms. The latter process usually is termed field evaporation. To give a feeling for the fields required, the (0 K) values 10 and 8 V/Å for W(111) and W(110), respectively, are quoted from an early paper [3]. A more detailed treatment will be given on pp. 262/4. For reviews of field desorption and field evaporation, see [4] and [5], respectively. The use of field ion microscopy (FIM) in studies of field evaporation is reviewed in [6]. FIM is reviewed in [7, 8]. Atom-probe FIM (see [9]) and field-desorption microscopy are reviewed in [4]. Field desorption will be discussed in the following only to the extent necessary to clarify statements about, or findings with, field evaporation. Field adsorption, which is reviewed in [10, pp. 145/9], will be treated on pp. 266/8 addressing mainly its effect in promoting field evaporation (older field evaporation work usually was done in the presence of image gases) and an attempt will not be made to provide a comprehensive coverage of the topic. The observation that the residual gas at base pressures any higher than 2×10^{-10} Torr strongly affected the rate data [11], however, illustrates its importance. Hydrogen from the residual gas seems to be field-adsorbed at low temperatures and decreases the activation energy involved from, e.g., 0.26 to 0.06 eV [11]. For a discussion of activation energies and their dependence on temperature and field, see pp. 261/2. Field evaporation combines thermal activation, atomic or ionic tunneling, and electronic transitions. The observed temperature dependence of the change of the relative abundance of ionic species (ionic charges) is much too strong for a purely thermally activated process [12] (to which the considerations by [13], [14, pp. 100/5] had been confined). Tunneling and thermal activation are equally important near 72 K (read from a figure in the paper [12]). The value of ~705 K [15, p. 818] is obviously obsolete. More recent values for the cross-over temperature are near 50 K; see pp. 258/60.

Early FEM studies showed the field evaporation rate ϱ to be (1.7 to 4) $\times 10^{14}$ atoms/(cm$^2\cdot$s) at room temperature using a field of 5 V/Å. At 900 K a field evaporation rate of $\varrho=50\times10^{14}$ atoms/(cm$^2\cdot$s) was found with a field of 4 V/Å. He-FIM studies (unaware of the promoting effect of image gas) indicated an extreme field dependence of ϱ [13, p. 620]. (Note that the field dependence of ϱ is much less with evaporation in vacuum than when using an image gas [16]). Under conditions where the desorption of an atomic plane lasted 10 s under 4.98 V/Å, it took 1 s at 5.05 V/Å and 0.2 s at 5.1 V/Å [13, p. 620]. For more recent rate measurements studying the influence of various parameters, see pp. 260/1.

A major application of field evaporation is the production of perfect surfaces by evaporation at temperatures low enough to avoid migration of lattice imperfections to the surface [3]. (For an application of field evaporation to investigate a screw dislocation as it extends into the bulk of a W crystal, see [17].) Field-evaporated tips have a rounded-off shape, the radius of curvature varying in the same sense as the work function [3]. For a comparison of thermal endform and field-evaporated endform tips, see, e.g., the FIM (ion) and FEM (electron) micrographs given in the paper [18]. The endform of field-evaporated tips is dealt with on pp. 270/2.

References for 6 on pp. 272/5

6.2 Charge States and Charge State Distributions

In field evaporation of tungsten, all of the charge states W^{1+} through W^{6+} were observed. The probability for a W surface atom to be desorbed in a particular charge state depends on the temperature, field, and crystallographic characteristics of the evaporating site; see below. Under usual field evaporation conditions, the most frequently observed tungsten ion species is W^{3+}; see, e.g. [19] (and with less abundance W^{4+}, W^{2+}, and W^{5+} ions; see, e.g. [20]). In the following papers, field desorption of individual W ion species was established experimentally: W^{1+}: [19, 21], W^{2+}: [19 to 28] (in [22] W^{2+} was obtained as a dissociation product of WH^{3+}), W^{3+}: [11, 16, 19 to 36], W^{4+}: [11, 16, 19 to 25, 29 to 31], W^{5+}: [20, 23, 24] (in [24] a WC substrate was used), and W^{6+}: [23].

The mean charge on the W ions field-evaporated around W(531) at base pressures below 5×10^{-10} Torr was investigated. The following ion charge distributions were found [24]:

T in K	dc evap. field in V/Å	field during pulse in V/Å	W^{2+} in %	W^{3+} in %	W^{4+} in %	mean charge
295	5.1 ± 0.2	5.4 ± 0.2	–	99.2 ± 0.2	0.7 ± 0.2	3.01 ± 0.01
93	5.7 ± 0.2 [a]	6.0 ± 0.2	0.5 ± 0.1	89.5 ± 0.6	10.0 ± 0.6	3.10 ± 0.02

[a] 78 K value from [37]; apparently used to calibrate the data.

Field evaporation was studied using a pulsed-laser atom-probe (PLAP) FIM at $p<10^{-9}$ Torr by Kellogg [21, 26, 29, 38]. Keeping the evaporation rate constant on heating, there is a monotonic shift in the charge state of the field-evaporated W ions from 4+ through 3+ to 2+ and eventually to the 1+ state. The $W^{4+}/(W^{3+}+W^{4+})$ ratio decreases with increasing temperature, from 12% at 60 K to <2% at 400 K. While the evaporation rate at 60 K increases by a factor of 160 when the field is increased by a mere 0.14 V/Å, the associated change in the $W^{4+}/(W^{3+}+W^{4+})$ ratio is only from 9.6% to 10.6% [29]. W evaporates as W^{3+} (~90%) and W^{4+} (~10%) at a field of 5.5 V/Å [21]. In fair agreement with the post-ionization theory advanced by [39], this distribution shifts to lower charge states as the field decreases to ~3 V/Å where the dominant species is W^{2+}; see **Fig. 51** (taken from

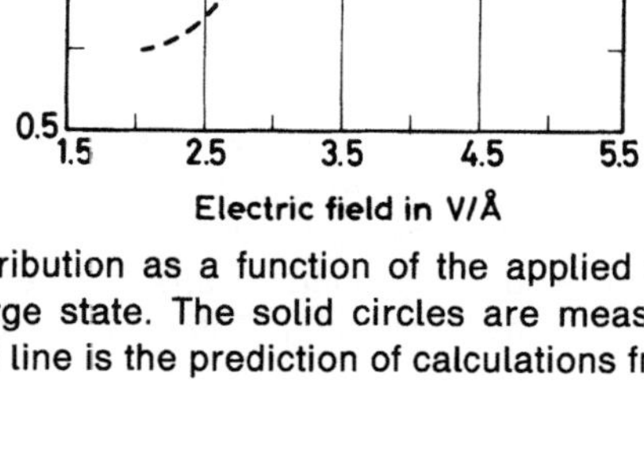

Fig. 51. Charge state distribution as a function of the applied field [26]. f_n is the fraction of ions in the n-fold charge state. The solid circles are measured values from [26] and the dashed line is the prediction of calculations from [39, 40].

[26]). On further reduction of the field also W^{1+} (and W_2^{3+} [21]) appears, but the average value of the ion charges remains fairly constant near two [21, 26, 29]. The field dependence of the charge state distribution of W ions from a tip biased at 2 to 15 kV (dc) was investigated in the PLAP study [26]. With increasing voltage, the W^{3+} signal increased at the expense of the W^{2+} signal; at 2.9 kV only W^{2+} was observed, at 3.2 kV the W^{3+} and W^{2+} signals were nearly equal, while at 3.6 kV there was a strong W^{3+} and a weak W^{2+} signal. The dashed line in Fig. 51 [26] is the prediction of calculations by [39, 40]. The theoretical curve is based on the assumption that all ions initially evaporate in a singlefold charge state. The disagreement between the prediction and the PLAP time-of-flight spectral results may be due to field enhancement at local protrusions formed at high temperatures (note that low fields correspond to high temperatures or high laser power) [26]. At lower F values (high T values), no W^{4+} was detected and the activation energy Q corresponded to desorption of W^{3+} alone, while at higher fields (at 5.9 V/Å) ~10% of the ions were W^{4+} ions and separate determinations of Q were possible for W^{3+} and W^{4+} [11].

For W^{6+} an abundance ratio of $W^{6+}/(W^{3+}+W^{4+}) \approx 1.3 \times 10^{-5}$ was observed in a series of measurements along the ⟨111⟩ zone (presumably at 85 K) [23].

W^{5+} ions occurred only under conditions where W^{4+} was abundant, i.e., at evaporation rates of about 10^9 layers/s. Sampling various planes along the ⟨111⟩ zone, 36 W^{5+} ions (any isotope) were found along with 28 W^{4+} and 196 W^{3+} ions of isotope 180. In view of the 0.142% abundance of the isotope 180, this result corresponds to a $W^{5+}/(W^{3+}+W^{4+})$ ratio of 0.00023 (while the W^{4+}/W^{3+} ratio was 0.143 in the probed region). Along the ⟨100⟩ zone the $W^{5+}/(W^{3+}+W^{4+})$ ratio decreases to 0.00013 although W^{4+} almost doubles with respect to W^{3+}. Among the 30000 W^{3+} and 2000 W^{4+} ions, no W^{5+} ions were found around the high yield W^{4+} regions (661) and (441). Most of the data were recorded at 85 K, but there seems to be no significantly higher W^{5+} yield at 28 K. W^{5+} ions were observed not only at the extremely high evaporation rates usually used (~1 ML/ns), but also at much lower rates (lower fields) characterized by a yield of 136 W^{2+}, 6300 W^{3+}, 137 W^{4+}, and 2 W^{5+} ions (≈0.03% W^{5+} abundance) [23].

W^{4+} ions evaporate nearly exclusively from the (110) plane center and on the ⟨111⟩ zone lines, while the W^{3+} ions evaporate rather uniformly from the surface (in presence of He; also with Ne at 78 K) [16]. With atom-probe FIM under 5×10^{-6} Torr of He at 28 to 85 K, the bright FIM regions around (111), (211), (221), (433), and (411) yield W^{3+} and an equal number of He^+ and some W^{4+} ions. The abundance of W^{4+} is a few % on the (211) plane. The W^{4+} abundance increases with the evaporation rate. At the high evaporation rate of one (110) layer per pulse it ranged from 30% to 40% in the bright regions along the ⟨100⟩ zone (on planes like (210), (310), and (520)) and also on (522) and (322) planes and was a few % along ⟨111⟩ and on (211) [25]. Nitrogen adsorption changed the W^{3+}-to-W^{4+} ratio to about unity while the distributions typically found (at 80 to 300 K, base pressure: 2×10^{-8} Torr, presumably in the presence of ~1 mTorr He) were like the following: out of 281 evaporated atoms, 2 were W^{4+}, 275 were W^{3+}, 3 were W^{2+}, and 1 was W^+. Below 300 K, the charge state ratio did not depend on the temperature [19] (this result was confirmed for the abundances of W^{2+}, W^{3+}, and W^{4+} ions in the later study [26]).

Atom-probe FIM was used to study the field evaporation of W ions from W(110) planes between ~20 K and ~250 K. The accumulated number of field-evaporated W ions plotted vs. the number of voltage pulses applied shows plateaus that are identified with smooth planes formed by layer-by-layer field evaporation. The ions evaporated are usually W^{3+}. In the last pulse preceding a plateau, W^{4+} ions also occur in addition to one or more W^{3+} ions. W^{4+} ions were never observed otherwise; also no charge state other than 3+

 References for 6 on pp. 272/5

or 4+ was noticed. It appears that W^{4+} ions form at high local fields that are present just before the last protrusions (single adatoms) on a plane are field-desorbed. This assumption is also consistent with the observed temperature dependence of the W^{4+} formation probabilities which seems to depend on the local field that is smaller at high temperatures than at low ones [30].

Predictions are made for the relative abundances of W^{n+} ions (n=2 to 5) based on the assumption that the W atoms are field-evaporated in a given charge state (1+, 2+, or 3+) and then post-ionized in applied fields of 2 to 12 V/Å [41].

W atoms bound to W surfaces are essentially neutral except for polarization effects in high applied fields. On W(111), W atoms at kink sites may, in fact, have a charge higher than 0.55 elementary charge at the evaporation field of 5.7 V/Å [42].

6.3 Post-Ionization

It is virtually certain that the field evaporation of W occurs by a Gomer-type process (see [43, 44]) with simultaneous ionization and escape in a surface charge-exchange process (thermally activated escape into a 1+ or possibly 2+ state, followed by one or more post-ionization events) [38, 39, 45, 46]. For a discussion of post-ionization, see [10, pp. 136/7]. The initial stage of field evaporation consists of charge-draining out of the first ionization level and results in a singly charged ion. Then post-ionization to higher charge states occurs with a field-dependent probability [47]. The "location of ionization" is assumed at 0.21±0.02 times the neutral-atom bond distance above the surface [48]. Evaluating the data of [49] in terms of the intersection model (see [32, 43, 50]), a critical distance (where atomic and ionic potential curves intersect) of $x_c \approx 1$ Å fits the experimental data for W(110) [51]. For W^{2+} and W^{3+}, $x_c = 1.94$ and 1.71 Å, respectively, were calculated [52]; see also the potential energy curves calculated by [53]. With F=6 V/Å, a W^{2+} ion escapes without activation and turns into W^{3+} some 1.5 Å above the image plane [53].

The measured charge state distribution [26, 29] (see Fig. 51, p. 256) agrees well with the following field dependence of the post-ionization probability P calculated by [39]:

P in %	$W^{1+} \rightarrow W^{2+}$ in V/Å	$W^{2+} \rightarrow W^{3+}$ in V/Å	$W^{3+} \rightarrow W^{4+}$ in V/Å
0.1	1.9	2.6	4.6
10	2.4	3.1	5.8
99.9	3.1	4.0	7.2

The data given in **Fig. 52** taken from the paper [26] confirm earlier results given in [21]. The improved data in [26] represent constant field measurements (F=3.85 V/Å; circles) and constant temperature measurements (T not given; squares) with F increasing from 3.36 to 4.21 V/Å. The W^{3+}/W^{2+} ratio does not change with increasing temperature (at constant field). This indicates that the activation energy for field evaporation is the same for both W^{3+} and W^{2+} ions, suggesting that the same process produces the parent species for both types of ions with post-ionization occurring away from the tip surface [26]. These results support the post-ionization theory advanced by [39].

6.4 Tunneling vs. Thermal Activation

The straight line behavior of $1/F_{ev}$ vs. $T^{1/2}$ above ~50 K, where F_{ev} is the onset-of-evaporation field (usually simply termed "evaporation field"), indicates that 1.) field evapo-

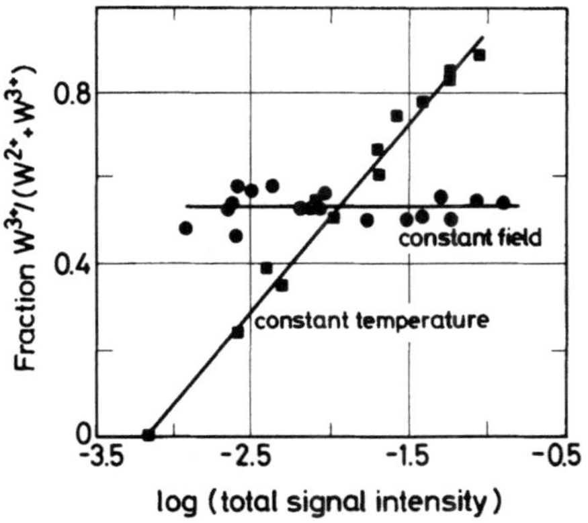

Fig. 52. Change in fraction of field-evaporated W^{3+} ions as a function of the log of the total signal intensity (proportional to the log of the evaporation rate) at constant field strength and constant temperature [26].

ration is thermally activated above ~50 K, 2.) a Gomer-type escape mechanism operates, 3.) a curve-intersection formalism is adequate, and 4.) the bottom of the surface atom bonding well is satisfactorily described by a parabola (harmonic potential) [46].

The evaporation fields F_{ev} recorded between ~20 and ~130 K by [49, p. 450] were replotted as $\sqrt{T}$ vs. $1/F_{ev}$ by [54]. They follow a straight line above ~60 K. A $\sqrt{T}$ vs. $1/F_{ev}$ replot of the data of [29] also yields reasonably straight lines between 50 and 200 K, but there seems to be a change in behavior above 250 K [54]. The data recorded by [55] between 20 and 300 K suggested that above ~180 K the field-evaporation mechanism changes [54]. (This change in behavior might be associated with the ~1 eV barriers for surface diffusion, neglected in the harmonic oscillator model used by [54].)

Below 50 ± 10 K the data points from [49], replotted in the form $1/F_{ev}$ vs. $\sqrt{T}$, show a deviation from linearity which may be associated with ion-core tunneling (as was suggested by [49] and for a long time was suspected as dominating the field evaporation at low temperatures) [46]; see also the model calculations by [56].

Denoting with F^z_{ev} the zero-Q evaporation field, the relationship $\sqrt{T}=\sqrt{\Theta}\times[(F^z_{ev}/F_{ev})-1]$, based on the Gomer-type charge-exchange mechanism, is established (and supported by the reevaluation of the measured data of [49]). The parameters $\Theta = 940 \pm 120$ K and $F^z_{ev} = 6.91 \pm 0.1$ V/Å were recalculated from the data of [49] by [54]. Restandardization gave $F^z_{ev} = 7.42 \pm 0.11$ (statistical error) ± 1 (calibration error) V/Å [54] (the value $F_{ev} = 5.36$ V/Å evaluated from the data of [49] for T = 78 K differs from the usual value of 5.7 V/Å, giving rise to a restandardization of the values of the field strength [54]). The cross-over temperature 50 ± 10 K for the change from ion tunneling to thermally activated ion jumps over the classical barrier is compared in the paper with various estimates made earlier in the literature that mostly ranged between 40 and ~70 K [54, p. 201].

Potential energy (E) curves were calculated as explicit functions of the electrical field (F), charge transfer (q), and distance (x) from the image plane of the metal. The results are given as plots of E vs. x for q = 0, 1, 2, 3, 4 and F = 0, 2, 4, 6 V/Å. For F = 2 V/Å the minimum potential energy curve corresponds to q = 1, and field evaporation with a very large activation energy would occur by a process similar to the image-hump model [53].

The charge-exchange model (developed for field desorption of Ba^+, Ba^{2+}, etc. by [43]) was improved by [57] and theoretical arguments are given against the Müller-Schottky

References for 6 on pp. 272/5

mechanism of field evaporation in [58]. The expected charge states and evaporation fields derived in [59] using the concepts developed in [14] thus are obsolete. Nonetheless, the evaporation fields of W found at 78 to 350 K by FIM are said to be consistent with the image-force model [60]. The repetition of the preliminary study [50] at much improved vacuum conditions and with better electronics indicated that the concept of image potential fails at distances closer to the surface than ~1 Å (the analysis was performed for W^{3+} ions) [32]. The image-force model is considered inadequate and the charge-transfer model (by [44]) more plausible [32, pp. 4206/8].

6.5 Evaporation Rates

The evaporation rate at 60 K increased by a factor of 160 when the field was increased by 0.14 V/Å [29]. For the very steep increase of the field-evaporation rate with field, studied for the range 4.8 to <6.6 V/Å (at fixed temperatures of 50, 200, 300, and 418 K), see a figure in the paper [11]. There was an increase of the evaporation rate by seven orders of magnitude on raising the field by only 10% (determined by removal of single adatoms that were deposited at $\sim 10^{-10}$ Torr on field-evaporated equilibrium tips carefully prepared by removing ~0.01 layers/s). Note that the field decreases by ~10% from the edge to the center of a (110) plane (this figure is in good agreement with [61] and the significant disagreement with [62] might be due to smaller W(110) surfaces used by the latter researchers) [32]. In an FIM study of field evaporation of W(110) at 21 to 180 K under pulsed field conditions (0.2 to 20 μs pulses), the topmost layer was desorbed at 78 K to such an extent that a cluster of 15 to 10 Å radius remained. There was collective desorption of this remainder for tip radii of curvature r>200 Å (e.g., 300 Å). On the other hand, with r below 100 Å, e.g., 90 Å, there was rapid though not jumplike removal of the cluster [63]. (Despite the bad vacuum conditions (10^{-7} to 10^{-6} Torr), clean conditions were assumed to prevail since double pulses were applied, the first for sample cleaning and the second for analysis [63].) There was a linear relationship between the number of field pulses required to desorb the topmost W(110) layer and the number of W atoms of which this layer consists. This was shown with 0.8 μs pulses (0.4 μs rise time, 0.1 μs flat top; typically 8.5 kV constant dc voltage and 1.6 kV pulsed voltage) [64]. The r dependence of the evaporation field observed for r<150 Å was attributed to the influence of the surface free energy in the early paper [65]. While the evaporation field (at 77 K) exceeded the He best imaging field (4.4 V/Å) by 27% for r=1000 Å, these fields were equal at r=40 Å [65]. (For a critique, see [66] as cited by [15, p. 808].) The evaporation rate (in layers/s) depended exponentially on the number of (110) layers evaporated (associated with the relationship between the radius of curvature and the number of planes evaporated). The position of the straight line in the log(rate) vs. number plot depended on both the voltage applied to the tip and its temperature [50]. For removal rates of 10^{-2} to 10^{7} W(110) layers/s using both continuous and pulsed field evaporation at base pressures of $\sim 10^{-8}$ Torr and temperatures of 77.3 and 90.1 K, see [67]. An Arrhenius-type behavior of the evaporation rate was observed by [68].

The time dependence of the (110) ring diameter d (of the image of the annular edge observed on FIM patterns) was studied at 78 K under field-evaporation conditions. The stepped d vs. t curves indicate that during the evaporation of an atomic layer there exist different more or less stable configurations. For a comparison with computer simulations of the field-evaporation process, see the paper [69]. For a computer simulation of field desorption of a ⟨110⟩-oriented W tip under 5.7 V/Å, see the paper [70].

Pulsed field evaporation (presumably at 78 K) on W(211) starts at 5.68 V/Å (in agreement with the 5.7 V/Å found by [20]) according to FIM. The following W(211) values for the evaporation rate refer, at least for the lower fields, to the removal of edge atoms [71]:

rate in atoms/μs	0.4 ± 0.2	3.0 ± 2.0	10.9 ± 3.9	13.3 ± 8.8	33.6 ± 10.3
field in V/Å	5.68	5.73	5.90	6.05	6.19

The propensity of various W atoms on a W(111) surface for field evaporation was studied. The number of third-nearest neighbors is said to be decisive (studied with 5 to 9 atoms in the third-nearest neighborhood [72] and obviously supersedes the earlier conclusion in [73] that the critical bonds are those to first- and fourth-nearest neighbors [73]). The effects of the various bonds are not additive and simple bond energies between atoms on unrelaxed sites do not apply [72] (this confirms similar conclusions drawn by [62] and [61]). Kink-site atoms field-evaporate with a higher probability than do ledge atoms. The evaporation sequence is reversed, however, in the case of divacancies (introduced by irradiation with 180 to 230 keV Xe^+ ions near 80 K). The combined effect of a reduced binding and an increased field gives rise to the faulty evaporation sequence observed at (111) [74].

The dependence of the field evaporation rate on the field strength was measured for W(110) at 78 K in $\sim 1.5 \times 10^{-3}$ Torr He (base pressure: $< 10^{-10}$ Torr). At 5.5 V/Å the evaporation rate was 0.01 layers/s and the charge state of the field-evaporated atoms was +3 [75]. These results excellently fit the calculated field-dependent evaporation rate for W(110) at 78 K of [76] using a theory based on Gomer's charge-exchange model [44]. Using Morse and Born-Mayer potentials as well as values for the n-th ionization energy, the work function, and the polarizability (all taken from the literature), the only adjustable parameter left is the penetration depth of the field. With this single adjustable parameter, the calculated temperature dependence (0 to 500 K) of the evaporation field also agrees well with the experimental data given in [29] and [49] (note that the discrepancies between the experimental data due to different definitions of the evaporation rate are greater than are the discrepancies between calculated and measured curves). For a discussion of the somewhat large field penetration depth calculated, see the paper [76].

6.6 Activation Energies and Frequency Pre-exponential Factors

The activation barrier for field evaporation decreases monotonically with increasing field. The pre-exponential factor increases up to F = 4.5 V/Å due to the increasing ionization probability, but then decreases due to changes in the surface potential. The results of a perturbation-theoretical calculation for W(111) are compared in [77] with the experimental results by [11] which, though referring to the whole tip, are mainly due to evaporation from W(110) steps [11]. The following table gives values for the activation energy Q and frequency pre-exponential factors k_0 as a function of field F, measured by the total $(W^{4+} + W^{3+})$ signal between ca. 100 and 500 K at $p < 2 \times 10^{-10}$ Torr [11], along with the corresponding calculated values from [77]:

F in V/Å	Q_{exp} in eV	Q_{calc} in eV	k_{0-exp} in 10^{11} s^{-1}	k_{0-calc} in 10^{11} s^{-1}
4.70	0.90 ± 0.04	1.17	300000	2760
4.93	0.60 ± 0.01	0.85	100	2600
5.10	0.52 ± 0.01	0.58	80	2210
5.30	0.35 ± 0.003	0.41	7	2250
5.47	0.31 ± 0.008	0.31	7	2060
5.72	0.20 ± 0.004	0.20	3	1660
5.90	0.12 ± 0.012	0.12	4	546

References for 6 on pp. 272/5

At 5.92 V/Å, Q = 0.11 ± 0.005 and 0.12 ± 0.003 eV was found from Arrhenius plots for W^{4+} and W^{3+}, respectively. This agreement (within the experimental uncertainty) supports the concept of post-ionization [11]. For field evaporation of W from W(110), the activation energy Q is estimated at 0.15 ± 0.06 eV for a field of 5.5 V/Å [48]. In (100) layers, Q seems to vary from site to site [68].

The activation energy for the evaporation of W^{n+} ions, Q_n, and the approximately linear temperature dependence of the evaporation field are discussed in terms of the charge-exchange model in [78]. For W^{3+}, $Q_3 = 0.16$ eV at 5.7 V/Å and removal rates of 0.01 layers/s were found at 77.3 K [67, p. 24]. For similar conditions, the pre-exponential factor in the Arrhenius expression amounts to 10^7 s^{-1} [67, p. 25]. The much higher Q values given in [59] and [79] are obviously obsolete.

There was poor agreement between the Q vs. F relationships calculated for parabolic and Morse bonding potentials [80] when compared with the experimental results of [11, 29, 49].

For kink-site atoms the diffusion to terrace sites prior to desorption is assumed to play a major role in field evaporation [81]. It appears that atoms from highly coordinated sites (e.g., from ledge sites) move by field-induced migration to sites with low coordination (e.g., to terrace sites) from which the atoms evaporate. The nearly temperature-independent evaporation field found at the lowest temperatures might indicate conditions of field evaporation from high-coordination sites directly into the vacuum without intermediate field-induced surface migration [78]. At evaporation rates of 10^8 layers/s at liquid hydrogen temperatures, there does not seem to be sufficient time for a kink site atom to be thermally activated over the potential barrier to a nearby metastable site [50].

6.7 Evaporation Fields

For the evaporation field F_{ev} at low temperatures, usually a value around 5.7 V/Å was observed, see, e.g. [20, 37]. The value 7.86 V/Å given for the region around W(110) in [82, p. 2811] (cited as 5.79 V/Å in [52, p. 242]) is obviously obsolete. F_{ev} for W(211) is 94% that of W(111) according to FIM at $\sim 10^{-10}$ Torr [83]. In the presence of the imaging gas He, F_{ev}(78 K) is 5.5, 6.2, and 5.7 V/Å for (110), (111), and (211), respectively. At 21 K the respective fields (in He) are enhanced to 5.7, 6.3, and 5.9 V/Å [20]. In presence of H_2, D_2, or Kr, the values at 78 K are 5.5, 6.1, and 5.7 (±0.1) V/Å for (110), (111), and (211), respectively [84]. The true (UHV) F_{ev} values are somewhat higher than given by [84], since the imaging gases H_2, D_2, and Kr present for calibration purposes (the fields are determined from the free-space field ionization energy deficits) reduce the F_{ev} by a few percent according to [11]. Also, field-adsorbed imaging gases reduce the evaporation field below the value in vacuum [50, 67]. At 63 K, field evaporation of a W tip starts (in 10^{-3} Torr He) at ~5.7 V/Å [18]. For the promoting effect of image gases on field evaporation, see Chapter 6.9, pp. 266/8.

The temperature dependence of F_{ev} was investigated for W(110) evaporating at a rate of 0.1 layers/s. The results were discussed in terms of the charge-exchange model. The constant rate of 0.1 layers/s was controlled by the applied voltage in the presence of 7.5×10^{-5} Torr image gas (He). The evaporation field F_{ev} decreased from ~5.7 V/Å near 20 K to ~5 V/Å near 130 K [78]. See also the temperature dependence of the onset-of-evaporation field treated in Chapter 6.4, p. 258/60. From studies between ca. 17 and 150 K, above ~50 K the evaporation field decreases with temperature at a rate of about −0.0061 (V/Å)/K [81]. The field required to achieve a field-evaporation rate of 0.5 W(110) layers/s linearly decreases from 50 K to 200 K by 13.5% and continues to decrease linearly up to 400 K (the upper limit of the temperature range studied) [85, p. 1186]. However, a linear decrease of F_{ev} with the square of the temperature was found between ca. 150 and 500 K in an earlier paper [55]. Measuring $F_{ev}(T)/F_{ev}(77\ K)$ in the range from ~20 to ~300 K, a significant decrease between ~20 K and ~100 K was observed, followed by a weak decrease up to room temperature [86, 87]. At ~20 K, an evaporation field of about 6 V/Å was found [86]. $F_{ev}(T)/F_{ev}(77\ K)$ was measured between 77 and ~950 K in the (310)-to-(121) region. By back-extrapolation to 0 K, $F_{ev}(T)$ at 750 and 1060 K is ~62% and ~47% of $F_{ev}(0\ K)$ [88]. With (110) the desorption field at 145 K is ~3% lower than with the tip at 20 K, while on (211) the reduction in the desorption field is 11% [62].

Field evaporation was studied using a pulsed-laser atom-probe FIM at $p<10^{-9}$ Torr between ca. 60 and 500 K. The nitrogen laser used ($\lambda=337$ nm) had a pulse width of 10 ns. A calibration scale for the emitter temperature between ca. 60 and 500 K was established by the evaporation field decreasing with increasing temperature. Near 60 K the applied voltage required to produce pulsed field evaporation at rates of 10^9 ions/s is only 0.1 kV (i.e., a few % of the total voltage applied) higher than the dc voltage producing 100 ions/s. At higher temperatures, maintaining that difference in evaporation rates requires larger differences in the voltages involved. There is a nearly linear decrease of the dc evaporation field from ~5.7 V/Å at 60 K to ~3.3 V/Å at 500 K. Between 60 and 300 K the reduction of this field is ~21% [29] (the value of ~16% found by [55] presumably is due to the presence of ~10^{-5} Torr He in that experiment). With pulsed field evaporation the reduction in evaporation field with temperature is more gradual than with dc voltage. Pulsed-laser field evaporation at a constant evaporation rate (i.e., under conditions where both temperature and field change simultaneously) indicates that the changes in the charge state ratio (studied with W^{2+} and W^{3+}) are brought about by the changing field strength rather than by the temperature change [29]; see also Fig. 52, p. 259.

Pulsed evaporation using 5 ns field pulses requires a field higher by 6% than does dc field evaporation to achieve comparable mean removal rates [51, p. 66]. The decrease of F_{ev} with increasing temperature is much stronger with dc conditions (e.g., 0.2 layers/s) than at the high rates occurring under pulsed field conditions (e.g., 30 ions per ~0.07 µs); see **Fig. 53** [11]. For the dc- and pulsed-field evaporation fields at 93 and 295 K, see also the table on p. 256 [24].

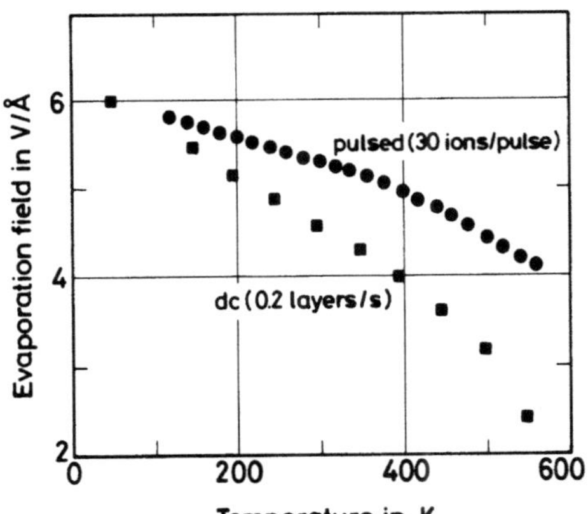

Fig. 53. Temperature dependence of the evaporation field for pulsed- and dc-field evaporation [11].

References for 6 on pp. 272/5

W adatoms field-desorb from different surfaces of W tips (field-evaporated at 21 K) with increasing field in the order (110), (211), (111), (321), and (100). These measurements were done in the presence of typically 1.5 mTorr He [61] (for a critique of such binding energy determinations, see [33]). For T=0 K the low value of $F_{ev}=4.8$ V/Å was given for W^{2+}. The following low F_{ev} values were given for a tip field-evaporated at 78 K: 3.45, 4.52, 5.01, 4.79, and 4.98 V/Å for (110), (211), (100), (111), and (321), respectively [61].

W adatoms were deposited onto a field-evaporated endform tip and field evaporation of these adatoms was followed by FIM at ~0.5 mTorr He. At 20 K, the observed desorption field is 6.1 V/Å. The W adatoms clear off at increasing tip voltage first from (110), then from (211), (111), and (411) almost simultaneously, then (310), and finally (321) [62, p. 1471]. On the (321) surface some adatoms are removed even from the center of the plane (where the field is lowest) at voltages some 20% below those that clear the plane. Except for (110), similar though less distinct behavior was also found with the other planes (where ~1 out of 10 adatoms desorbed at voltages ~7% smaller than the clear-off field) [62].

Single W adatoms were deposited onto a clean tip field-evaporated at 78 K. The onset of field evaporation was observed by heating for 1 min under He best image voltage (He BIV) conditions at preselected temperatures between ~150 and ~350 K. The temperature dependence of F_{ev} for the (211) step site atoms is given by $\ln(F_{ev} \cdot nm/V) = 4.05 - 0.00103 \cdot T$. The temperatures for the onset of field evaporation are as follows for stepsite atoms under He BIV conditions [89]:

(hkl)	(211)	(321)	(431)	(111)	(110)
T in K	240	240	240	230	268

For W tips with average tip radii between 200 and 900 Å, F_{ev}(79 K) ranged from 5.6 to 5.8 V/Å (as determined by scaling to the field emission). The values of F_{ev}(21 K) were 4 to 6% higher than F_{ev}(79 K) [90]; see also [91].

The (evaporation) field felt by the edge atoms of (111) layers is 1.28 times that felt by edge atoms of (110) faces. The ratio of the (ionization) fields at 2 to 4 Å above (111) and (110) faces is 1.30. These electrolytic trough and analog computer results are consistent with best image voltages determined by FIM [92].

At 79 K field evaporation of thermally annealed tips occurs at 5.7±0.1 V/Å for (411), (321), and (111). A much higher evaporation field is required to remove atoms from the field-evaporated endform tip. With (411) and (110) at 79 K, evaporation fields of 7.3±0.1 and 6.8±0.2 V/Å resulted, respectively [66].

With the measured evaporation fields of 5.6 and 7.45 V/Å for W(110) and W(111), respectively, polarization contributions to the binding energy of atoms on these planes of 0.57 and 2.71 eV are calculated. FIM suggests that the polarization binding may stabilize atoms in low-coordination sites (3-fold vs. 4-fold coordination). The "polarization binding" is responsible for the decoration of the ⟨100⟩ zone found with endforms of W tips field-evaporated at 21 K. This decoration is *not* a general property of the lattice type, rather it is typical for the particular metal (due to electronic properties). For instance, Ta behaves totally different from W [93].

6.8 Energy Distributions and Energy Deficits

The pulsed-laser ($\lambda = 337$ nm; 0.3 ns; 0.05 mJ per pulse) time-of-flight (ToF) atom-probe FIM with a flight path of 7.78 m is described in [94]. ToF spectra are given for W^{3+} ions (mass numbers 182, 183, 184, and 186). The critical energy deficits for these ions give an average value of $\Delta E_c = 45.51 \pm 0.47$ eV for kink-site atoms field-desorbed from W(110) [94]. These critical energy deficits $\Delta E_c^{3+}/3$ had been measured earlier as 14.6, 16.3, 15.0, and 17.1 eV, respectively (average: 15.8 ± 1.0 eV) [34]. (Other values given by the same authors are $3 \times 15.7 \pm 1.0$ eV [95] and 45.36 ± 0.51 eV per W^{3+} from W(110) below 100 K using 0.05 J pulses of 0.3 ns [96].) This compares well with the value 3×15.1 eV estimated for the heat of evaporation for an atom plus the sum of energies of ionization for the first, second, and third ionization stages minus the work function released in the absorption of three electrons. Thus, there is no evidence for a photoexcitation effect in pulsed-laser stimulated field evaporation [34]. The critical energy deficit equals the binding energy of the kink-site atom corrected for the ionization energy required to produce W^{3+} ions and considering the energy transfer by the detached electrons releasing the work function (times the charge number of the ions). The value 9.03 ± 0.47 eV is given for the kink-site atom binding energy (to be compared with 8.80 ± 0.51 eV cited from [96]). The polarization term increasing the binding energy and the reduction in binding energy due to reduced binding electron density seem to cancel one another [94].

Pulsed-laser stimulated field-desorbed metal ions show a high energy tail in the energy distribution. The excess energies are well over 300 eV indicating that the ions are ejected from the surface at initial kinetic energies of a few hundred eV [10, p. 144].

Resonant tunneling peaks (Jason peaks) are seen in the energy distributions (ED) of W^{3+} and WHe^{3+} ions, pulsed-laser ($\lambda = 337$ nm) field-desorbed at ~0.96 and ~0.93 V/Å, respectively (and also of Ne^+ ions field-desorbed at ~4.5 V/Å from W(110) lattice steps). Under conditions chosen such that only triply charged ions are present, W^{3+} and WHe^{3+} are desorbed with an FWHM of ~14.4 and ~13.8 eV at $F \approx 0.96$ and ~0.93 V/Å, respectively (using 0.3 ns pulses of $\lambda = 337$ nm). No low energy tail is seen in the ED of these triply charged ions. Comparison with the ED of ions having a lower charge indicates that the critical energy deficits are not associated with surface plasmon generation (which would result in much greater deficits than observed) [35].

The energy analysis (using a Möllenstedt energy analyzer) of field-desorbed He^+, Ne^+, Ar^+, and H_2^+ ions confirmed the existence of energy loss peaks [97] (previously found by [98] using an electrostatic retarder, see also [99]). A theory is given for the Jason peaks in [100]. It is based on an improved version of Jason's approach [98]. The electron scattering by the crystal surface is described in [100] by a complex reflection coefficient evaluating the LEED data of [101]. The energy vs. field relationship found by [98] was qualitatively confirmed. The periodic structure of the spectra is best developed in those spectra originating from (110) showing up to seven peaks. The spectra from (111), (211), (411), and higher index planes, on the other hand, show only two or three peaks clearly [97].

The field-ion appearance energies (= energies of ions that just managed to escape off the surface at vanishing kinetic energy and experiencing no anomalous change in energy during the escape process) are computed (by the image-potential approximation) for ions arriving at the detector as W^+, W^{2+}, W^{3+}, or W^{4+} (bound at the surface as W^0, W^{1+}, W^{2+}, or W^{3+}). Good agreement was obtained with the value of 51 ± 9 eV (see [102]) recalculated from the energy loss data by [103] refering to (presumably) W^{3+}. No conclusion, however, was possible as to the charge state of the W atoms to be desorbed from the surface, since the calculated standard field-ion appearance energies (for detected W^{3+}) are 58, 57, and 61 eV for W^0, W^+, and W^{2+} bond states [104].

The energy distributions of He^+ ions field-desorbed from W(100) at 21 K in the low 10^{-10} Torr range show a broad feature in the 1 to 2 eV energy deficit region. This feature is explained in terms of electron states above E_F. Similarly, the distribution of Ne^+ ions field-desorbed at 78 K from W(110) shows a broad feature around the ~1 eV deficit [105].

 References for 6 on pp. 272/5

The energy distribution of field-desorbed He, Ne, Ar, and hydrogen ions was measured for tips at 21 to 78 K. With the rare gases, singly peaked distributions resulted with half widths of 0.5 to 0.7 eV. Two-peak distributions resulted with hydrogen, the two peaks being attributed to H^+ and H_2^+ [106]. Energy distributions of He^+ ions field-desorbed from (110), (100), and (111) surfaces were measured with the tip at 21 K. On (110) the position of the peak in the distribution varies by ~0.7 eV depending on the part on the surface probed (possibly due to protruding W atoms). From (100) a broad distribution (width ~4 eV) emerges with a broad energy deficit peak near 2 eV. This result of [107] is in conflict with [105]. The energy distributions of the emitted ions might reflect in some way the band structure of the substrate [107].

Pulsed-laser field desorption from W cryotips heated by 0.3 ns pulses of 337 nm radiation was studied at base pressures of $\sim 10^{-8}$ Torr. The temperature rise due to the laser pulses was typically 200 to 300 K. Energy distributions of N_2^+ ions detached from W(110) lattice steps and terraces, of Ne^+ detached from W(110) lattice steps, and of Ar^+ desorption are given in the paper [35].

6.9 Field Adsorption and Promotion of Field Evaporation by Image and Residual Gases

Field Adsorption. In the review [4] the importance of the field-induced adsorption on the FIM imaging mechanism is described. Minute admixtures of Ne to the imaging gas He strongly affect the imaging process, obviously due to the larger polarizability of Ne [4]. The substrate atoms at protruding sites (kinks, adatoms on terraces, and the like) are decorated by field-adsorbed imaging gas atoms [108]. Partial pressures as low as 10^{-8} Torr Ne reduce the intensity of $^{4}He^*$ while $^{20}Ne^*$ and $^{22}Ne^*$ become strong even before Ne^+ formation in FIM [4].

Field adsorption of He occurs above 2.0 and 3.3 V/Å at 21 and 78 K, respectively. At 78 K, Ne field adsorption is restricted to the field range of 2.3 to 5.0 V/Å [33]. No Ne^+ or He^+ signal was detected with (111)-oriented W tips at 300 K while desorption images formed at 80 K [36].

The field adsorption of He, Ne, and Ar on W(211) surfaces was studied by probe-hole FIM at base pressures above 10^{-9} Torr, image gas pressures of 5×10^{-6} Torr, and temperatures between 30 and ~300 K controlled to within ±0.5 K. For adsorption energies and further details, see [109]. For the kinetics of field adsorption, see [110].

The calculated temperature dependence is plotted in [111] for the probability that at any instant of time an inert gas atom is adsorbed on the apex of a substrate surface atom and compared with the experimental results for He on W(411) at 4.9 and 5.6 V/Å and for Ne at 5.0 V/Å on W(110) and W(322). In addition, the calculated field dependence of the He field adsorption at 20 and 80 K on kink-site atoms on the W(110) plane at 2×10^{-5} and 2×10^{-3} Torr is plotted in the paper [111].

Promotion of Field Evaporation by Image and Residual Gases. The promotion of field evaporation by physisorbed He and Ne is explained by the potential-barrier reduction due to induced dipoles of the adsorbed gases [52]. The results from a theoretical treatment of field adsorption of He on W(111) and of field evaporation of W (as well as its promotion by adsorbed hydrogen and adsorbed He) are in line with the results published by [11] for W(110) [112]. For the promotion of field evaporation by adsorbed image gas, especially by Ne, see [113]. Depending on the arrangement of neighbors, different evaporation probabilities were observed by FIM for the atoms forming a (111), (332), (433), or (411) surface and evaporating therefrom in the presence of He and Ne at 20 and 78 K [114].

As much as 80% of the field-desorbed heavy ions from the dim regions at and around W(110) are WHe^{3+} ions rather than pure W ions according to atom-probe FIM at 5×10^{-6} Torr of He or Ne (usually at 85 K, sometimes at 28 K) [25]. With He, the WHe^{3+} ions desorb rather uniformly from the surface. The He^+ images show some structure and are similar to the W^{4+} images [16]. The bright FIM regions around (111), (211), (221), (433), and (411) yield W^{3+} and an equal number of He^+ and some W^{4+} ions. The temperature dependence of the abundance ratio $WHe^{3+}/(W^{3+}+WHe^{3+})$ peaks at 80 K for the dim regions. Field-evaporated ions from the dim regions are nearly all triply charged [25]; see also [31].

Time-of-flight (ToF) spectra of W^{3+}, WHe^{3+}, and WHe_2^{3+} are given in the paper [35]. At 31 K and 5.0 to 5.1 V/Å in 6×10^{-6} Torr He, the triply charged ions in the ToF spectra consist of ~21% W^{3+}, ~78% WHe^{3+}, and ~1% WHe_2^{3+} ions. By pulsed-laser field evaporation of W in 10^{-6} Torr He, more than 90% of the W may be desorbed as helide ions. Helide ions are the more abundant the lower the temperature and the higher the field. (The few candidates for tri- and tetrahelide ions could not be unambiguously separated from WC^{3+} and WO^{3+} ions.) The nearly identical energy distributions observed with W^{3+} and WHe^{3+} ions indicate that both ions are formed right at the tip surface and only a few W^{3+} ions form by field dissociation above the surface [35]. In the presence of 10^{-6} Torr He, the ion $(WHe)^{3+}$ was abundant, but none of the four- or fivefold charged ions were WHe compounds [23].

In pulsed field desorption, He^+ (and Ne^+) appears only under conditions where metal ions are also observed. The compound ions WHe^{3+} and WHe_2^{3+} appear in addition to W^{3+}. With small additions of H_2 (e.g., $<10^{-7}$ Torr H_2 added to 10^{-4} Torr He), the ions W^{3+}, WHe^{3+}, and WHe_2H^{3+} were observed at 21 K [33]. The field ionization of He in an FIM occurs at 4.5 V/Å, but only 3 V/Å are required in the presence of hydrogen. The promotion of He ionization might be due to rearrangements of the surface charge from protruding W atoms to adjacent, adsorbed hydrogen [115]. In a mixture of 80% He and 20% Ne, Ne^+ dominates in pulsed field desorption for fields $F<4.5$ V/Å while at higher field values He^+ becomes more abundant. He^+ is the only noble gas ion observed above 5 V/Å. The atom-probe FIM study showed that the noble gas ions originated from the edges of W(110) or W(211) surfaces [33]. With mixtures of 10^{-4} Torr He plus 10^{-8} Torr Ne, the field desorption starts at ca. 70%, 77%, 85%, and 74% of the field evaporation voltage on W(111), W(211), W(110), and W(310), respectively. With W(110) the Ne desorption at 78 K starts near 4.7 V/Å [116]. Ne desorbs preferentially from the high field areas of the tip [116] and no molecular ions were found [33].

W tips in 0.1 to 0.5 mTorr Ne at ~80 K yield ~90% W^{3+} and 10% W^{2+} ions. Mixing the Ne with 0.01 to 3 mTorr H_2 results in a total of ~8% of triply and doubly charged tungsten neide molecular ions in addition to the metal ions if the dc voltage is lowered to allow for the approach of the gases through the ionization zone to the surface [27]; see also [28, pp. 64/5, 82/3]. For field desorption characteristics observed with Ne, see the paper [25]. At 78 K, nearly all of the W^{4+} ions evaporate from the (110) plane center and on the ⟨111⟩ zone lines in the presence of Ne gas while the W^{3+} (or WNe^{3+}) and Ne^+ ions desorb rather uniformly from the surface [16].

Ne^+ ions observed in a study on pulsed field desorption originate from the edges of W(110) or W(211) surfaces [33]. Predominantly doubly charged argon ions codesorb with W^{4+} and W^{3+} (apparently at 10^{-4} Torr Ar). With hydrogen (added to Ar ?) WH^{3+} occurs and dissociates into W^{2+} and H^+ [22].

The evaporation field of W(110) linearly decreased with the increasing logarithm of the Ne pressure between ca. 10^{-8} and 10^{-5} Torr (at higher Ne pressures there was scatter

References for 6 on pp. 272/5

in the data) [15]. The reduction of the evaporation field below the value in vacuum by the presence of an image gas [82] was explained by the impacts of imaging gas atoms on a surface of metal atoms, the binding energy of which had been reduced by the adsorbate [33]. This suggests that early surface binding-energy determinations need a careful reconsideration. (In many cases single metal atoms had been deposited on the FIM tip and were subsequently field-evaporated in the presence of adsorbed image gases. Notably the papers [6], [61], and [62] seem to report binding-energy values affected by the presence of image gas [33].) These conclusions of [33] are not invalidated by the different, more recent explanation for the promoting effect in terms of dipole-induced barrier reduction due to the physisorbed gases (see, e.g. [52]).

At base pressures any higher than 2×10^{-10} Torr, the residual gas strongly affected the rate data [11]. In image-gas-promoted field evaporation, the evaporation rates usually change by a factor of ~30 for an ~1% change in field. The field dependence of the evaporation rate is much lower with evaporation in vacuum than under an image gas. At temperatures well below 20 K the nanosecond-pulsed evaporation (and desorption) rates of W (and He) become extremely sensitive functions of the applied field, by which a change of 0.25% may alter the evaporation rate from less than 0.1 planes/s to several planes/s. Under these conditions the evaporation sometimes oscillates between low-field and high-field regions on the specimen and is no longer as uniform as it is at higher temperatures [16]. For 78 K, 4×10^{-5} Torr He admitted to a base pressure below $\sim 2 \times 10^{-11}$ Torr, and a field of ~5.7 V/Å, the evaporation rate was ~0.1 layers/s at the center of the largest W(110) plane obtainable [117]. The effect of image gases (2 to 3 mTorr He at 62 or 88 K; 2 mTorr He plus 0.2 mTorr Ne at 62 K) on the desorption rate had been studied earlier by [118].

Imaging Ne^+ ions (FIM) have a very narrow energy distribution (half-width: ~0.7 eV) corresponding to an ionization zone only 1/7 Å wide some 4 Å above the surface. Added He or Ne reduce the vacuum evaporation field; the reduction is independent of the imaging gas pressure (between 0.1 and 2 mTorr). At 21 K this reduction amounts to ~1.3% with He and to ~10.7% with Ne. At 78 and 300 K the respective reductions are 6% and 4% with Ne and <0.5% for both temperatures with He [82].

The final tip shape ("endform") is nearly independent of the evaporation rate under vacuum conditions, but it is altered under pulsed evaporation by the introduction of a gas such as He or Ne [16]. Field-evaporated endforms produced at base pressures below 10^{-10} Torr in the presence of $\sim 10^{-3}$ Torr imaging gas He differ in shape depending on evaporation temperatures of 78 K or 215 K. At 78 K the axial ratio of the ellipse formed by the oval plane edge is 0.87 for the first ring and 0.90 for the second and third rings. At 215 K the axial ratio is 0.92 for the first ring and 0.95 (i.e., nearly circular) for the second and third rings. Above 140 K the field evaporation differs from that at 78 K in that it becomes very anisotropic leaving oriented clusters as residuals on the topmost tip after incomplete field evaporation [119]. The field-evaporated endform in Ne differs from that in He. With Ne the endform shows greatly varying local radii of curvature. The endform cap in Ne does not show the many high index net planes exposed by the endform in He. Rather, in Ne it consists mostly of ring-shaped (011) plane edges [82].

6.10 Field Desorption (Evaporation) Micrographs

Continuous-evaporation images were taken at evaporation rates less than one plane/s sustained for tens of seconds and at rates up to some hundreds of planes/s sustained for ~1 s. There is a rich variety of bright and dark lines in the images. The bright zone lines observed above 80 K differ from those commonly observed in FIM where they are the result of preferential ionization of image gas over atoms in particular surface sites (such a structure is unexpected with field evaporation and instead a uniform shell of material would be expected to evaporate). The evaporation images differ from FIM images in that they contain intricate crystallographic-related networks of dark lines. In addition, the centers of prominent poles frequently are darker than are their environments. Also, these poles are frequently surrounded by sets of concentric dark rings that differ from the rings seen in FIM images. Pulsed-evaporation images were taken in UHV at rates of 10^7 to 10^8 planes/s. The dark plane centers are more pronounced than with continuous evaporation images. The bright $\langle\bar{1}11\rangle$ zone lines through (110) seen in the continuous evaporation images are absent with pulsed evaporation. The details are much better resolved at 20 K than at 78 K with both continuous and pulsed-field evaporation. The ring structure is much sharper at 20 K and distinctly polygonal. The dark and bright lines in the field evaporation images are explained by diffusion towards the pole centers along preferred crystallographic directions. The kink-site atoms on the $\langle\bar{1}11\rangle$ zone lines have only one near neighbor in the top plane and cannot easily diffuse towards the plane center. This explains (at least in part) their preferred desorption as W^{4+} ions rather than as W^{3+} ions [16]. A bright $\langle100\rangle$ zone line decoration occurs (in FIM) due to atoms in metastable sites with field evaporation at slow rates ($\sim$0.01 layers/s) for both liquid nitrogen and hydrogen temperatures. At evaporation rates of 10^8 layers/s, the decoration is less distinct. At these high rates, there does not seem to be sufficient time for a kink-site atom to be thermally activated over the potential barrier to a nearby metastable site. Short-pulsed field evaporation at liquid nitrogen temperatures does not remove the zone line decoration [50]. Field desorption micrographs (apparently He ion images) taken at 4.4 and 5.9 V/Å are shown in the paper [71]. When the field is increased above the evaporation field (5.68 V/Å), bright and dark regions are interchanged, obviously due to migration of the surface atoms [71].

A sequence of time-gated field evaporation images of W(110) is shown with the tip bias incremented by 5 V between subsequent pictures (mean bias: $\sim$11.4 kV). The pictures made up of W^{3+} and W^{4+} ion image spots show the shrinkage and eventual disappearance of a (110) layer as the tip bias increases ("net plane collapse") [120, p. 242]. Atom-probe He desorption images and gated W^{3+} and W^{4+} field (ion) evaporation micrographs (FIEMs) of a W emitter tip are shown. The FIEMs are rather uniform showing no crystallographic structure while the He desorption image appears much like an FIMicrograph (although actually recorded at 0.6 μTorr) [121]. FIEMs are also compared with FIMs in [122]. For desorption images taken between cryogenic and ambient temperatures, see [123]. The W^{3+} field evaporation images taken from W(111) at 4×10^{-10} Torr are compared with Ne^+ and He^+ ion images produced while field evaporating the tip in 7.2×10^{-10} Torr Ne or 4.3×10^{-10} Torr He, respectively. In any case, the low-ion-yield regions are the same. For the (110)-type planes, WHe^{3+} is assumed to be abundant [120, p. 253]. The dark and bright strands forming regular patterns in the FIEMs are attributed to defocusing and focusing ion-optical conditions on the tip, the shape of which deviates from a true hemisphere. The lack of these strands in FIM is explained by the FIM image of the tip being formed somewhat above the tip surface and thus originating from a surface that is more closely hemispherical [122]. Similarly, parallax effects had been used to explain shifts between the image gas formed FIM image and the specimen (W) atom-probe signal [124]. The FIEMs show with low contrast the concentric ring patterns well known from FIM images. While hundreds of atomic layers are removed during the observation of evaporation images, the tip remains smooth and round as proven by FIM images taken immediately after the recording of the evaporation images. This shows that the brighter parts of the evaporation images are not due to enhanced evaporation. It is speculated that the contrast seen in the evaporation images is due to

References for 6 on pp. 272/5

atoms bonded by forces oblique to the surface normal (this would result in a deflection of the evaporating ion) [125].

Multilayer evaporation images that show a prominent ring structure and a complex network of intersecting dark and bright lines indicate a strong variation in evaporation probability over the surface at 18 K and 10^{-9} Torr. The dark zone-decorating lines appear more distinctly the higher is the number of evaporated planes. Field evaporation images taken with W^{3+} and W^{4+} ions in the presence of 10^{-5} Torr He show strong to complete randomization of the image suggesting that much of the structure seen in the multilayer evaporation images is due to field-induced surface migration prior to evaporation. Focusing and defocusing alone are not sufficient to bring about the observed structures [120, p. 246].

The field desorption micrographs (FDM) taken at 78 K from a ⟨110⟩-oriented W tip are by and large independent of vacuum conditions (2 to 500×10^{-10} Torr). Only minor variations exist between FDM taken by evaporation pulses of 15 ns, 0.08 ms, 0.5 ms, or by rapid continuous evaporation. The patterns from ⟨110⟩-oriented tips show ring structure, zone decoration, and radial bright lines extending to the (211)-type surfaces [126]. A later study [16], however, yielded different results for continuous and pulsed-field evaporation near 80 K, both in vacuum and in low-pressure rare gas environments [16].

Anomalous features are seen in the FIM images from W tips evaporated with pulsed fields at 20 or 78 K. These features are bright bands extending from (110) to the surrounding (211)-type planes. They are not observed upon field evaporation at 195 K, although features produced at 78 K persist on warming to 320 K. These anomalous features are observed only with pulsed field evaporation and the pulse duration must not exceed a value somewhere between 0.05 and 1 ms. The formation of the bright bands depends on the number of atom planes removed but not on the removal rate (planes/s). The bright bands observed are consistent with decoration of tip regions by "field-stabilized" atoms. The removal of W atoms at slow evaporation rates is promoted by adsorbed image gas atoms. In pulsed-field evaporation most atoms are desorbed without promotion, the supply with imaging gas being too poor (the results are consistent with He surface diffusion being activated by ~0.15 eV) [127].

6.11 Field-Evaporated Endform

The change of the field-evaporated endform in the presence of image gases is discussed on p. 268.

The endforms produced at 580 K (F=3.0 V/Å) and at 300 K (F=4.5 V/Å) are different as shown by the average work functions of 3.73 and 4.28 eV, respectively, derived from F-N plots. Annealing of the tips at 1200 K restores the usual F-N slopes indicating that the difference is not associated with large differences in the tip radius [128]. The combined FEM/FIM study [129] corroborates the earlier findings by [128] and, due to strict UHV conditions obeyed, rules out contamination as a possible reason for the decrease in apparent average work function of the tip [129].

The high work function regions are more flatly developed in the endform of tips produced at very high evaporation rates (studied between 0.01 and 10^7 to 10^8 layers/s under ~0.0015 Torr He and near 80 K) [50]. The distribution of local radii of curvature, r, along the ⟨111⟩, ⟨110⟩, and ⟨100⟩ zone lines is given for the field-evaporated endform of W tips in [90]; see also [91]. The distribution of r along the ⟨110⟩ zone line is plotted in another paper [86] for field evaporation at 78, 158, 179, and 193 K. Field evaporation at 300 and 400 K results in changes of the tip shape (due to changes in r) [86].

Field-evaporated endforms produced at different temperatures below 273 K do not differ in appearance from one another, though the local radii of curvature, r, vary somewhat between 77 and 193 K. r is smallest near (100) and next smallest around (111) with field evaporation at 470 K [92], or below 600 K (i.e., below the assumed onset of surface diffusion) [130]. The small r value near (111) observed below 620 K is no longer seen at 670 K [92]. Above 370 K, r is smallest near (100) and below 370 K, r is smallest near (111) being ~1/6 of the value near (110) [131]. The following sequence of cohesive energies was derived from the above evidence and is neither consistent with work function data nor with atomic density data: (100) (largest), (111), (110), and (211) (weakest) [130]; see also [92]. The binding energies of the adatoms on top of the net planes decrease in the sequence $E_{411} > E_{111} > E_{211} > E_{100} > E_{110}$ as calculated using Morse potentials. This sequence agrees with the experimental result except for the (100) planes, whose abnormally small local radius of curvature is difficult to understand [55].

With field evaporation at 470 K, disorder is noted in the "high field" regions (100), (610), (611), (111), and (433). With field evaporation at 620 K, the ⟨100⟩ zone line disappears in the FIM patterns and the disorder extends over the whole tip image [86]. An anomalous field enhancement seems to exist at (100) surfaces with W tips field-evaporated at 600 K. It might be due to disorder and/or reduction in the local radius of curvature [129]. Under field evaporation at 670 K, clusters arise at any site of the tip while steps, two or three atoms high, disappear [92]. Field evaporation above 90 K results in perfectly circular FIM images of (211) surfaces while these are distorted after field evaporation at lower temperatures [86]. Slow field evaporation at 60 K of tips prepared by field evaporation below 600 K reveals much disorder around (211), but none around (110) surface poles [130].

A nearly perfect, apparent (1 × 1) surface consisting of 77 atoms was prepared removing the serrated edges of a (100) plane by field evaporation at ~15 K. The tip voltage then was lowered from 9 kV to 6 kV and the tip temperature raised to 430 K. At 430 K the voltage was raised for 30 s to 6.5 kV. After cooling to ~15 K the tip was imaged with 9 kV. The treatment had produced vacancies in the plane and repetitions of this treatment eventually resulted in a p(2 × 2) configuration. Raising the voltage from 6.5 to 7.0 kV at high temperature caused further preferential field evaporation eventually resulting in a perfect c(2 × 2)-R 45°($\sqrt{2} \times \sqrt{2}$) domain containing 19 atoms. Except for the p(2 × 2) and c(2 × 2) structures occasionally not being single domains, the described sequence was very reproducible [132]. (Often, atoms inside a plane but close to the plane edge evaporate before the kink-site atoms. This creates an artifactual c(2 × 2) atomic structure near the plane edge [133].) The p(2 × 2) structure formation was accompanied by field-assisted surface diffusion of atoms that were subsequently field-evaporated. The preferential loss of atoms from alternate interior sites and alternate edge sites of the (100) planes differs from the usual shrinkage of the (hkl) planes. It might be due to periodic vertical displacements and/or periodic variations in the binding energy of the surface atoms (reconstructed surface) [132]. Certain interior atoms evaporate before the outer edge atoms. This indicates that some interior atoms have a lower zero-field binding energy and/or are slightly elevated above the surrounding surface plane [68].

There was a significant difference in the appearance of the (100) planes and any other plane. Their edges nearly always became ragged due to the selective field evaporation of certain edge atoms (at 70 to 100% of the best imaging field with He/10% Ne mixtures). (100) planes comprising initially 70 to 100 W atoms were selected for study. Between 15 and 580 K, a few atoms were field-evaporated, then the specimen was cooled to 15 K and imaged, heated, field-evaporated, cooled, imaged, and so forth until the plane was completely removed. The results differentiated the temperature ranges 15 to 100 K, 100 to 400 K, and 400 to 580 K: At 15 to 100 K, only in the latest stage when only 4 × 4 atoms had survived, preferential removal of alternate interior atoms was observed leading to

References for 6 on pp. 272/5

a c(2×2) configuration for the last few atoms. At 100 to 400 K, especially near 400 K, occasionally interior vacancies formed near the plane edges. At 400 to 580 K the initial (1×1) configuration changed to a p(2×2) structure by preferential field evaporation. The production of a vacancy during the p(2×2) formation is accompanied by the appearance of a W atom adsorbed nearby the vacancy atop the (100) plane. This observation is attributed to field-assisted surface diffusion of an atom from the site of the subsequently observed vacancy. Similar experiments were performed on (111), (110), (211), and (310) planes. No anomalies similar to those on (100) were found on these planes (i.e., the atoms were always field-evaporated at the circumference of the planes), except for (310) which consists of rows of (100) plane ledges. At these ledges there occurred preferential field evaporation of alternate atoms in a row. The imaging gas had either no or negligible effect on the observed behavior [68].

References for 6:

[1] Müller, E. W. (Naturwissenschaften **29** [1941] 533/4).
[2] Drechsler, M.; Vanselow, R. (Verh. Dtsch. Phys. Ges. **8** [1957] 181).
[3] Müller, E. W. (Acta Metall. **6** [1958] 620/30).
[4] Müller, E. W. (Chem. Phys. Solid Surf. **1977** 1/25).
[5] Tsong, T. T. (Progr. Surf. Sci. **10** [1980] 165/248).
[6] Müller, E. W. (Z. Elektrochem. **61** [1957] 43/7).
[7] Müller, E. W. (Science [Washington, D. C.] **149** [1965] 591/601).
[8] Müller, E. W. (Abh. Dtsch. Akad. Wiss. Berlin Kl. Math. Phys. Tech. [1967] No. 1, pp. 107/19).
[9] Müller, E. W.; Panitz, J. A.; McLane, S. B. (Rev. Sci. Instrum. **39** [1968] 83/6).
[10] Tsong, T. T. (Surf. Sci. Rep. **8** [1988] 127/209).

[11] Kellogg, G. L. (Phys. Rev. [3] B **29** [1984] 4304/12).
[12] Tsong, T. T. (Surf. Sci. **10** [1968] 102/17).
[13] Müller, E. W. (Phys. Rev. [2] **102** [1956] 618/24).
[14] Müller, E. W. (Adv. Electron. Electron Phys. **13** [1960] 83/179).
[15] Brandon, D. G. (Philos. Mag. [8] **14** [1966] 803/20).
[16] Waugh, A. R.; Boyes, E. D.; Southon, M. J. (Surf. Sci. **61** [1976] 109/42).
[17] Carroll, R. W.; Schubert, C. C. (J. Appl. Phys. **39** [1968] 2339/42).
[18] Sugata, E.; Izeki, K. (J. Electron Microsc. **14** [1965] 189/95).
[19] Brenner, S. S.; McKinney, J. T. (Appl. Phys. Lett. **13** [1968] 29/32).
[20] Tsong, T. T. (Surf. Sci. **70** [1978] 211/33).

[21] Kellogg, G. L. (Phys. Rev. [3] B **24** [1981] 1848/51).
[22] Müller, E. W.; Krishnaswamy, S. V.; McLane, S. B. (Surf. Sci. **23** [1970] 112/29, 125/6).
[23] Müller, E. W.; Krishnaswamy, S. V. (Phys. Rev. Lett. **37** [1976] 1011/4).
[24] Andrén, H.-O.; Henjered, A.; Kingham, D. R. (Surf. Sci. **138** [1984] 227/36).
[25] Krishnaswamy, S. V.; McLane, S. B.; Müller, E. W. (J. Vac. Sci. Technol. **13** [1976] 665/9).
[26] Kellogg, G. L. (Surf. Sci. **120** [1982] 319/33).
[27] Kapur, S.; Müller, E. W. (Surf. Sci. **62** [1977] 610/20).
[28] Kapur, S. (Diss. Pennsylvania State Univ. 1976, 127+8 pp.; Diss. Abstr. B **37** [1977] 5713).
[29] Kellogg, G. L. (J. Appl. Phys. **52** [1981] 5320/8).
[30] Murakami, K.; Adachi, T.; Kuroda, T.; Nakamura, S. (Surf. Sci. **140** [1984] L 253/L 258).

[31] Müller, E. W.; Krishnaswamy, S. V.; McLane, S. B. (Phys. Rev. Lett. **31** [1973] 1282/4).

[32] Tsong, T. T. (J. Chem. Phys. **54** [1971] 4205/16).
[33] Müller, E. W.; McLane, S. B.; Panitz, J. A. (Surf. Sci. **17** [1969] 430/8).
[34] Tsong, T. T. (Phys. Rev. [3] B **30** [1984] 4946/61, 4952/4).
[35] Tsong, T. T.; Kinkus, T. J. (Phys. Rev. [3] B **29** [1984] 529/42).
[36] Panitz, J. A. (J. Vac. Sci. Technol. **12** [1974] 210/2).
[37] Sakurai, T.; Müller, E. W. (Phys. Rev. Lett. **30** [1973] 532/5).
[38] Kellogg, G. L. (Bull. Am. Phys. Soc. [2] **26** [1981] 426/7).
[39] Haydock, R.; Kingham, D. R. (Phys. Rev. Lett. **44** [1980] 1520/3).
[40] Kingham, D. R. (Surf. Sci. **116** [1982] 273/301, 291/7).

[41] Haydock, R.; Kingham, D. R. (Surf. Sci. **104** [1981] L 194/L 198).
[42] Forbes, R. G. (J. Phys. D **15** [1982] 1301/22, 1314/6).
[43] Gomer, R. (J. Chem. Phys. **31** [1959] 341/5).
[44] Gomer, R.; Swanson, L. W. (J. Chem. Phys. **38** [1963] 1613/29).
[45] Forbes, R. G. (J. Phys. D **15** [1982] L 99/L 104).
[46] Forbes, R. G. (J. Phys. D **15** [1982] L 149/L 152).
[47] Kingham, D. R. (Vacuum **32** [1982] 471/6).
[48] Forbes, R. G.; Biswas, R. K.; Chibane, K. (Surf. Sci. **114** [1982] 498/514).
[49] Wada, M.; Konishi, M.; Nishikawa, O. (Surf. Sci. **100** [1980] 439/52).
[50] Tsong, T. T.; Müller, E. W. (Phys. Status Solidi A **1** [1970] 513/33).

[51] Konishi, M.; Wada, M.; Nishikawa, O. (Surf. Sci. **107** [1981] 63/74).
[52] Nishikawa, O. (Surf. Sci. **131** [1983] 239/44).
[53] Kingham, D. R. (J. Phys. Colloq. [Paris] **47** [1986] C 2-11/C 2-16).
[54] Chibane, K.; Forbes, R. G. (Surf. Sci. **122** [1982] 191/215).
[55] Nakamura, S.; Kuroda, T. (Surf. Sci. **17** [1969] 346/58).
[56] Kingham, D. R. (J. Phys. D **15** [1982] 2537/44).
[57] Forbes, R. G. (Surf. Sci. **70** [1978] 239/54).
[58] Biswas, R. K.; Forbes, R. G. (J. Phys. D **15** [1982] 1323/38).
[59] Brandon, D. G. (Surf. Sci. **3** [1964/65] 1/18).
[60] Lenkow, W.; Parnicka, R. (Mater. Semin. Nauk. Wydz. Mat.-Przyr. Wyzsza Szk. Pedagog. Czestochowie **3** [1979] 27/9 from C. A. **95** [1981] No. 156747).

[61] Plummer, E. W.; Rhodin, T. N. (J. Chem. Phys. **49** [1968] 3479/96).
[62] Ehrlich, G.; Kirk, C. F. (J. Chem. Phys. **48** [1968] 1465/80).
[63] Mikhailovskii, I. M.; Dranova, Zh. I.; Ksenofontov, V. A.; Kul'ko, V. B. (Zh. Eksp. Teor. Fiz. **76** [1979] 1309/15; Sov. Phys.-JETP [Engl. Transl.] **49** [1979] 664/8).
[64] Suvorov, A. L.; Lazarev, N. E.; Bobkov, A. F.; Zaitsev, S. V. (Poverkhnost **1985** No. 6, pp. 69/76; Phys. Chem. Mech. Surf. [Engl. Transl.] **4** [1986] 1741/55).
[65] Müller, E. W.; Young, R. D. (J. Appl. Phys. **32** [1961] 2425/8).
[66] van Oostrom, A. G. J. (Diss. Univ. Amsterdam 1965; Philips Res. Rep. Suppl. **1966** No. 1, pp. 102+6 pp.).
[67] Tsong, T. T.; Müller, E. W. (AD-697280 [1969] 46 pp.; C.A. **72** [1970] No. 94506).
[68] Tung, R. T.; Graham, W. R.; Melmed, A. J. (Surf. Sci. **115** [1982] 576/98).
[69] Suvorov, A. L.; Razinkova, T. L.; Kukavadze, G. M.; Bobkov, A. F.; Kuznetsov, B. Ya.; Kuznetsov, V. A. (Zh. Eksp. Teor. Fiz. **68** [1975] 1460/70; Sov. Phys.-JETP [Engl. Transl.] **41** [1975/76] 729/34).
[70] Suvorov, A. L.; Razinkova, T. L. (Surf. Sci. **65** [1977] 354/60).

[71] Bobkov, A. F.; Suvorov, A. L. (Zh. Eksp. Teor. Fiz. **79** [1980] 1376/84; Sov. Phys.-JETP [Engl. Transl.] **52** [1980] 695/9).
[72] Moore, A. J. W.; Spink, J. A. (Surf. Sci. **17** [1969] 262/6).
[73] Moore, A. J. W.; Brandon, D. G. (Philos. Mag. [8] **18** [1968] 679/89).

[74] Stiller, K.; Andrén, H.-O. (Surf. Sci. **114** [1982] L 57/L 61).
[75] Tsong, T. T. (J. Phys. F **8** [1978] 1349/52).
[76] Sakai, A.; Sakurai, T. (J. Phys. Colloq. [Paris] **47** [1986] C 2-17/C 2-21).
[77] Wang, L. C.; Ehrlich, G. (Surf. Sci. **237** [1990] 337/46).
[78] Wada, M. (J. Phys. Colloq. [Paris] **45** [1984] C 9-89/C 9-94).
[79] Brandon, D. G. (Br. J. Appl. Phys. **16** [1965] 683/8).
[80] Chibane, K.; Forbes, R. G. (J. Phys. Colloq. [Paris] **45** [1984] C 9-99/C 9-104).

[81] Wada, M. (Surf. Sci. **145** [1984] 451/65).
[82] Nishikawa, O.; Müller, E. W. (J. Appl. Phys. **35** [1964] 2806/12).
[83] Menand, A.; Gallot, J. (Rev. Phys. Appl. **9** [1974] 323/9).
[84] Sakurai, T.; Müller, E. W. (J. Appl. Phys. **48** [1977] 2618/25, 2620/2).
[85] Kellogg, G. L.; Tsong, T. T. (J. Appl. Phys. **51** [1980] 1184/93, 1186, 1190/2).
[86] Kuroda, T.; Yagi, H.; Nakamura, S. (Mem. Inst. Sci. Ind. Res. Osaka Univ. **26** [1969] 101/14; C.A. **71** [1969] No. 74504).
[87] Kuroda, T.; Nakamura, S.; Tamaki, S. (Shinku **12** [1969] 210/23; C.A. **71** [1969] No. 129997).
[88] Mordyuk, V. S.; Ivanov, Yu. I.; Mal'tsev, A. N. (Fiz. Met. Metalloved. **66** [1988] 546/50; Phys. Met. Metallogr. [Engl. Transl.] **66** No. 3 [1988] 119/23).
[89] Sadakane, Y.; Abe, T.; Hirano, K. (Proc. 7th Int. Vac. Congr., Vienna 1977, Vol. 3, pp. 2641/4; C.A. **88** [1978] No. 114005).
[90] Nakamura, S. (Mem. Inst. Sci. Ind. Res. Osaka Univ. **24** [1967] 71/81).

[91] Nakamura, S. (J. Electron Microsc. **15** [1966] 279/85).
[92] Tamaki, S.; Kuroda, T. (Jpn. J. Appl. Phys. **7** [1968] 1202/9).
[93] Müller, E. W. (Surf. Sci. **2** [1964] 484/94).
[94] Tsong, T. T. (Surf. Sci. **231** [1990] 81/9).
[95] Tsong, T. T.; McLane, S. B.; Liou, Y. [Liu, J.] (J. Phys. Colloq. [Paris] **45** [1984] C 9-71/C 9-75).
[96] Liou, Y. [Liu, J.]; Tsong, T. T. (J. Phys. Colloq. [Paris] **49** [1988] C 6-105/C 6-110).
[97] Müller, E. W.; Krishnaswamy, S. V. (Surf. Sci. **36** [1973] 29/47).
[98] Jason, A. J. (Phys. Rev. [2] **156** [1967] 266/85).
[99] Jason, A. J.; Burns, R. P.; Parr, A. C.; Inghram, M. G. (J. Chem. Phys. **44** [1966] 4351/2).
[100] Appelbaum, J. A.; McRae, E. G. (Surf. Sci. **47** [1975] 445/68).

[101] Armstrong, R. A. (Surf. Sci. **47** [1975] 666/8).
[102] Forbes, R. G. (J. Phys. D **9** [1976] L 191/L 193).
[103] Waugh, A. R.; Southon, M. J. (J. Phys. D **9** [1976] 1017/23, plates).
[104] Forbes, R. G. (J. Phys. D **13** [1980] 1357/63).
[105] Utsumi, T.; Smith, N. V. (Phys. Rev. Lett. **33** [1974] 1294/7).
[106] Tsong, T. T.; Müller, E. W. (J. Chem. Phys. **41** [1964] 3279/84).
[107] Hanson, G. R.; Inghram, M. G. (Surf. Sci. **55** [1976] 29/39).
[108] Tsong, T. T.; Müller, E. W. (Phys. Rev. Lett. **25** [1970] 911/3).
[109] Kinkus, T. J.; Tsong, T. T. (J. Vac. Sci. Technol. [2] A **3** [1985] 1521/4).
[110] Tsong, T. T. (Surf. Sci. **140** [1984] 377/92).

[111] Tsong, T. T.; Müller, E. W. (J. Chem. Phys. **55** [1971] 2884/9).
[112] Kreuzer, H. J.; Nath, K. (Surf. Sci. **183** [1987] 591/608).
[113] Moore, A. J. W.; Spink, J. A. (Surf. Sci. **44** [1974] 11/20).
[114] Moore, A. J. W.; Spink, J. A. (Surf. Sci. **44** [1974] 198/212).
[115] Müller, E. W. (Surf. Sci. **7** [1967] 462/73).
[116] Rendulic, K. W. (Surf. Sci. **28** [1971] 285/98, 296).

[117] Wang, S. C.; Tsong, T. T. (Phys. Rev. [3] B **26** [1982] 6470/5).
[118] Brandon, D. G. (Br. J. Appl. Phys. **14** [1963] 474/84).
[119] Nishigaki, S.; Nakamura, S. (Jpn. J. Appl. Phys. **14** [1975] 769/77).
[120] Panitz, J. A. (Prog. Surf. Sci. **8** [1978] 219/62, 241/62).

[121] Panitz, J. A. (J. Vac. Sci. Technol. **11** [1974] 206/10).
[122] Krishnaswamy, S. V.; Martinka, M.; Müller, E. W. (Surf. Sci. **64** [1977] 23/42).
[123] Walko, R. J.; Müller, E. W. (Phys. Status Solidi A **9** [1972] K 9/K 10; separate plate in the A **9** No. 1 issue).
[124] Brenner, S. S.; McKinney, J. T. (Surf. Sci. **23** [1970] 88/111).
[125] Moore, A. J. W. (Philos. Mag. [8] A **43** [1981] 803/14).
[126] Waugh, A. R.; Boyes, E. D.; Southon, M. J. (Nature **253** [1975] 342/3).
[127] Hren, J. J.; Moore, A. J. W.; Spink, J. A. (Surf. Sci. **29** [1972] 331/41).
[128] Schmidt, L.; Gomer, R. (J. Chem. Phys. **42** [1965] 3573/98, 3582/3).
[129] Feldman, U.; Gomer, R. (J. Appl. Phys. **37** [1966] 2380/90, 2387/90).
[130] Tamaki, S.; Kuroda, T. (Jpn. J. Appl. Phys. **7** [1968] 180/1).

[131] Tamaki, S.; Nakamura, S. (Jpn. J. Appl. Phys. **6** [1967] 1249).
[132] Melmed, A. J.; Tung, R. T.; Graham, W. R.; Smith, G. D. W. (Phys. Rev. Lett. **43** [1979] 1521/4).
[133] Tsong, T. T.; Sweeney, J. (Solid State Commun. **30** [1979] 767/70, 767).

Physical Constants and Conversion Factors

Avogadro constant N_A (or L) = 6.02214×10^{23} mol^{-1}
Faraday constant F = 9.64853×10^{4} C/mol
molar gas constant R = 8.31451 $J \cdot mol^{-1} \cdot K^{-1}$
molar volume (ideal gas) $V_m = 2.24141 \times 10^{1}$ L/mol
(273.15 K, 101325 Pa)

Planck constant h = 6.62608×10^{-34} $J \cdot s$
elementary charge e = 1.60218×10^{-19} C
electron mass $m_e = 9.10939 \times 10^{-31}$ kg
proton mass $m_p = 1.67262 \times 10^{-27}$ kg

1 kg = 2.205 pounds
1 m = 3.937×10^{1} inches = 3.281 feet
1 m^3 = 2.642×10^{2} gallons (U.S.)
1 m^3 = 2.200×10^{2} gallons (Imperial)

Force	N	dyn	kp
1 N	1	10^{5}	1.019716×10^{-1}
1 dyn	10^{-5}	1	1.019716×10^{-6}
1 kp	9.80665	9.80665×10^{5}	1

Pressure	Pa	bar	kp/m^2	at	atm	Torr	lb/in^2
1 Pa = 1 N/m^2	1	10^{-5}	1.019716×10^{-1}	1.019716×10^{-5}	9.86923×10^{-6}	7.50062×10^{-3}	1.450378×10^{-4}
1 bar = 10^6 dyn/cm^2	10^{5}	1	1.019716×10^{4}	1.019716	9.86923×10^{-1}	7.50062×10^{2}	1.450378×10^{1}
1 kp/m^2 = 1 mm H_2O	9.80665	9.80665×10^{-5}	1	10^{-4}	9.67841×10^{-5}	7.35559×10^{-2}	1.422335×10^{-3}
1 at (technical)	9.80665×10^{4}	9.80665×10^{-1}	10^{4}	1	9.67841×10^{-1}	7.35559×10^{2}	1.422335×10^{1}
1 atm = 760 Torr	1.01325×10^{5}	1.01325	1.033227×10^{4}	1.033227	1	7.60×10^{2}	1.469595×10^{1}
1 Torr = 1 mm Hg	1.333224×10^{2}	1.333224×10^{-3}	1.359510×10^{1}	1.359510×10^{-3}	1.315789×10^{-3}	1	1.933678×10^{-2}
1 lb/in^2 = 1 psi	6.89476×10^{3}	6.89476×10^{-2}	7.03069×10^{2}	7.03069×10^{-2}	6.80460×10^{-2}	5.17149×10^{1}	1

Work, Energy, Heat	J	kW·h	kcal	Btu	eV
1 J = 1 W·s = 1 N·m = 10^7 erg	1	2.778×10^{-7}	2.39006×10^{-4}	9.4781×10^{-4}	6.242×10^{18}
1 kW·h	3.6×10^6	1	8.604×10^2	3.41214×10^3	2.247×10^{25}
1 kcal	4.1840×10^3	1.1622×10^{-3}	1.	3.96566	2.6117×10^{22}
1 Btu (British thermal unit)	1.05506×10^3	2.93071×10^{-4}	2.5164×10^{-1}	1	6.5858×10^{21}
1 eV	1.602×10^{-19}	4.450×10^{-26}	3.8289×10^{-23}	1.51840×10^{-22}	1

1 $cm^{-1} \triangleq 1.239842 \times 10^{-4}$ eV
1 Hz $\triangleq 4.135669 \times 10^{-15}$ eV
2 Rydberg (Ry) = 1 hartree = 27.2114 eV
1 eV $\triangleq$ 23.0578 kcal/mol

Power	kW	hp	kp·m·s^{-1}	kcal/s
1 kW = 10^3 J/s	1	1.35962	1.01972×10^2	2.39006×10^{-1}
1 hp (horsepower, metric)	7.3550×10^{-1}	1	7.5×10^1	1.7579×10^{-1}
1 kp·m·s^{-1}	9.80665×10^{-3}	1.333×10^{-2}	1	2.34384×10^{-3}
1 kcal/s	4.1840	5.6886	4.26650×10^2	1

References:

Mills, I. (Ed.), International Union of Pure and Applied Chemistry, Quantities, Units and Symbols in Physical Chemistry, Blackwell Scientific Publications, Oxford 1988.
The International System of Units (SI), National Bureau of Standards Spec. Publ. 330 [1972].
Landolt-Börnstein, 6th Ed., Vol. II, Pt. 1, 1971, pp. 1/14.
ISO Standards Handbook 2, Units of Measurement, 2nd Ed., Geneva 1982.
Cohen, E. R., Taylor, B. N., Codata Bulletin No. 63, Pergamon, Oxford 1986.

Key to the Gmelin System of Elements and Compounds

System Number	Symbol	Element
1		Noble Gases
2	H	Hydrogen
3	O	Oxygen
4	N	Nitrogen
5	F	Fluorine
6	**Cl**	**Chlorine**
7	Br	Bromine
8	I	Iodine
8a	At	Astatine
9	S	Sulfur
10	Se	Selenium
11	Te	Tellurium
12	Po	Polonium
13	B	Boron
14	C	Carbon
15	Si	Silicon
16	P	Phosphorus
17	As	Arsenic
18	Sb	Antimony
19	Bi	Bismuth
20	Li	Lithium
21	Na	Sodium
22	K	Potassium
23	NH_4	Ammonium
24	Rb	Rubidium
25	Cs	Caesium
25a	Fr	Francium
26	Be	Beryllium
27	Mg	Magnesium
28	Ca	Calcium
29	Sr	Strontium
30	Ba	Barium
31	Ra	Radium
32	**Zn**	**Zinc**
33	Cd	Cadmium
34	Hg	Mercury
35	Al	Aluminium
36	Ga	Gallium

HCl $ZnCl_2$

System Number	Symbol	Element
37	In	Indium
38	Tl	Thallium
39	Sc, Y La–Lu	Rare Earth Elements
40	Ac	Actinium
41	Ti	Titanium
42	Zr	Zirconium
43	Hf	Hafnium
44	Th	Thorium
45	Ge	Germanium
46	Sn	Tin
47	Pb	Lead
48	V	Vanadium
49	Nb	Niobium
50	Ta	Tantalum
51	Pa	Protactinium
52	**Cr**	**Chromium**
53	Mo	Molybdenum
54	W	Tungsten
55	U	Uranium
56	Mn	Manganese
57	Ni	Nickel
58	Co	Cobalt
59	Fe	Iron
60	Cu	Copper
61	Ag	Silver
62	Au	Gold
63	Ru	Ruthenium
64	Rh	Rhodium
65	Pd	Palladium
66	Os	Osmium
67	Ir	Iridium
68	Pt	Platinum
69	Tc	Technetium[1]
70	Re	Rhenium
71	Np, Pu ...	Transuranium Elements

$CrCl_2$ $ZnCrO_4$

Material presented under each Gmelin System Number includes all information concerning the element(s) listed for that number plus the compounds with elements of lower System Number.

For example, zinc (System Number 32) as well as all zinc compounds with elements numbered from 1 to 31 are classified under number 32.

[1] A Gmelin volume titled "Masurium" was published with this System Number in 1941.

A Periodic Table of the Elements with the Gmelin System Numbers is given on the Inside Front Cover

GPSR Compliance

The European Union's (EU) General Product Safety Regulation (GPSR) is a set of rules that requires consumer products to be safe and our obligations to ensure this.

If you have any concerns about our products, you can contact us on ProductSafety@springernature.com

In case Publisher is established outside the EU, the EU authorized representative is:

Springer Nature Customer Service Center GmbH
Europaplatz 3
69115 Heidelberg, Germany

Batch number: 10620854

Printed by Printforce, the Netherlands